VOLUME ONE HUNDRED AND SIXTY FIVE

PROGRESS IN MOLECULAR BIOLOGY AND TRANSLATIONAL SCIENCE

Brain Imaging

VOLUME ONE HUNDRED AND SIXTY FIVE

PROGRESS IN MOLECULAR BIOLOGY AND TRANSLATIONAL SCIENCE

Brain Imaging

Edited by

JAMES T. BECKER, PhD
Professor of Psychiatry, Neurology and Psychology
University of Pittsburgh, Pittsburgh, PA, United States

Associate Center Director and
Director of the Neuroimaging Core
Alzheimer's Disease Research Center
University of Pittsburgh

ANN D. COHEN
Department of Psychiatry
University of Pittsburgh
Pittsburgh, PA, United States

ELSEVIER

ACADEMIC PRESS
An imprint of Elsevier

Academic Press is an imprint of Elsevier
50 Hampshire Street, 5th Floor, Cambridge, MA 02139, United States
525 B Street, Suite 1650, San Diego, CA 92101, United States
The Boulevard, Langford Lane, Kidlington, Oxford OX5 1GB, United Kingdom
125 London Wall, London, EC2Y 5AS, United Kingdom

First edition 2019

ISBN: 978-0-12-818361-8
ISSN: 1877-1173

For information on all Academic Press publications
visit our website at https://www.elsevier.com/books-and-journals

Publisher: Zoe Kruze
Acquisition Editor: Ashlie M. Jackman
Editorial Project Manager: Joanna Collett
Production Project Manager: James Selvam
Cover Designer: Matthew Limbert

Typeset by SPi Global, India

Contents

Contributors

Sarah B. Berman
University of Pittsburgh, Department of Neurology; Pittsburgh Institute for Neurodegenerative Diseases; University of Pittsburgh, Clinical and Translational Science Institute, Pittsburgh, PA, United States

Tobey J. Betthauser
Wisconsin Alzheimer's Disease Research Center, University of Wisconsin School of Medicine and Public Health, Madison, WI, United States

Ricardo Bruña
Laboratory of Cognitive and Computational Neuroscience (UCM-UPM), Centre for Biomedical Technology (CTB), Technical University of Madrid (UPM); Department of Experimental Psychology, Complutense University of Madrid (UCM), Madrid; Biomedical Research Networking Center in Bioengineering, Biomaterials and Nanomedicine (CIBER-BBN), Zaragoza, Spain

Owen Carmichael
Pennington Biomedical Research Center, Baton Rouge, LA, United States

Jaisalmer de Frutos-Lucas
Laboratory of Cognitive and Computational Neuroscience (UCM-UPM), Centre for Biomedical Technology (CTB), Technical University of Madrid (UPM); Biological and Health Psychology Department, Universidad Autonoma de Madrid, Madrid, Spain; School of Medical and Health Sciences, Edith Cowan University, Joondalup, WA, Australia

Peter J. Gianaros
Center for the Neural Basis of Cognition; Department of Psychology, University of Pittsburgh, Pittsburgh, PA, United States

Julie Gonneaud
Centre for Studies on the Prevention of Alzheimer's Disease, Douglas Mental Health University Institute; McGill University, Montréal, Canada

Sarah Gregory
Huntington's Disease Centre, UCL Queen Square Institute of Neurology, University College London, London, United Kingdom

Bieneke Janssen
Department of Radiology, Perelman School of Medicine, University of Pennsylvania, Philadelphia, PA, United States

Eileanoir B. Johnson
Huntington's Disease Centre, UCL Queen Square Institute of Neurology, University College London, London, United Kingdom

Dana R. Jorgensen
University of Pittsburgh Medical Center, Pittsburgh, PA, United States

Hyejin Kim
Department of Health and Community Systems, University of Pittsburgh School of Nursing; Alzheimer Disease Research Center, University of Pittsburgh School of Medicine, Pittsburgh, PA, United States

Brandon J. Lew
Department of Neurological Sciences, University of Nebraska Medical Center (UNMC); Center for Magnetoencephalography, UNMC, Omaha, NE, United States

Jennifer H. Lingler
Department of Health and Community Systems, University of Pittsburgh School of Nursing; Alzheimer Disease Research Center, University of Pittsburgh School of Medicine, Pittsburgh, PA, United States

David López-Sanz
Laboratory of Cognitive and Computational Neuroscience (UCM-UPM), Centre for Biomedical Technology (CTB), Technical University of Madrid (UPM); Department of Experimental Psychology, Complutense University of Madrid (UCM), Madrid, Spain

Robert H. Mach
Department of Radiology, Perelman School of Medicine, University of Pennsylvania, Philadelphia, PA, United States

Fernando Maestú
Laboratory of Cognitive and Computational Neuroscience (UCM-UPM), Centre for Biomedical Technology (CTB), Technical University of Madrid (UPM); Department of Experimental Psychology, Complutense University of Madrid (UCM), Madrid; Biomedical Research Networking Center in Bioengineering, Biomaterials and Nanomedicine (CIBER-BBN), Zaragoza, Spain

Melissa McSweeney
Centre for Studies on the Prevention of Alzheimer's Disease, Douglas Mental Health University Institute; McGill University, Montréal, Canada

Adam P. Mecca
Department of Psychiatry and Alzheimer's Disease Research Unit, Yale University School of Medicine, New Haven, CT, United States

Joseph Mettenburg
Department of Radiology, University of Pittsburgh, Pittsburgh, PA, United States

Pierre-François Meyer
Centre for Studies on the Prevention of Alzheimer's Disease, Douglas Mental Health University Institute; McGill University, Montréal, Canada

Cameron Miller-Patterson
University of Pittsburgh, Department of Neurology, Pittsburgh, PA, United States

Robert Newton Jr.
Pennington Biomedical Research Center, Baton Rouge, LA, United States

Erin O'Connor
University of Maryland School of Medicine, Baltimore, MD, United States

Michael T. Rezich
Department of Neurological Sciences, University of Nebraska Medical Center (UNMC); Center for Magnetoencephalography, UNMC, Omaha, NE, United States

Caterina Rosano
Department of Epidemiology, Graduate School of Public Health; Center for the Neural Basis of Cognition, University of Pittsburgh, Pittsburgh, PA, United States

C. Elizabeth Shaaban
Department of Epidemiology, Graduate School of Public Health; Center for the Neural Basis of Cognition, University of Pittsburgh, Pittsburgh, PA, United States

Rachel K. Spooner
Department of Neurological Sciences, University of Nebraska Medical Center (UNMC); Center for Magnetoencephalography, UNMC, Omaha, NE, United States

Sylvia Villeneuve
Centre for Studies on the Prevention of Alzheimer's Disease, Douglas Mental Health University Institute; McGill University, Montréal, Canada

Jennifer L. Whitwell
Department of Radiology, Mayo Clinic, Rochester, MN, United States

Alex I. Wiesman
Department of Neurological Sciences, University of Nebraska Medical Center (UNMC); Center for Magnetoencephalography, UNMC, Omaha, NE, United States

Tony W. Wilson
Department of Neurological Sciences, University of Nebraska Medical Center (UNMC); Center for Magnetoencephalography, UNMC, Omaha, NE, United States

Thomas Zeffiro
University of Maryland School of Medicine, Baltimore, MD, United States

Preface

Understanding the changes in brain structure and function that are associated with specific clinical and pathological conditions is critical to the study of brain aging and neurodegeneration. Over the past two decades, brain imaging technology has developed rapidly allowing for the visualization of brain structure, function, and specific brain pathologies in vivo. Brain imaging technologies are also critical for disentangling the clinical effects of brain neurodegeneration and associated comorbid conditions. As imaging technologies have developed, this has allowed for improvement in early detection, diagnosis, clinical staging, and participant selection for clinical trials in a variety of neurodegenerative diseases.

This issue focuses on the newest developments in imaging technologies in Alzheimer's disease (Chapter 1), including structural (1a) and functional (1b) neuroimaging. Additionally, Chapter 1 will also provide insights into PET imaging of amyloid (1c), tau (1d), and synaptic dysfunction (1c), and the development of PET imaging agents to detect the pathology of Alzheimer's disease has been transformative to the field of Alzheimer's research over the last decade and has substantially changed the way the disease is classified and diagnosed. Additional chapters focus on others dementias (Chapter 3) and neurodegenerative diseases (Chapter 2 and 6), as well as conditions known to affect brain aging, such as cerebrovascular disease (Chapter 4) and HIV (Chapter 5). Finally, two chapters focus on emerging issues in the field of brain imaging, development of PET tracers for neuroinflammation (Chapter 7), and disclosure of biomarker results (Chapter 8).

The Editors wish to thank all those who made this book possible. Special thanks to Dr. David Teplow, the serial editor of *Progress in Molecular Biology and Translational Science* for his advice and support in the development of this edition. Thanks also to Joanna M. Collett and Ashlie M. Jackman from Elsevier for their important administrative and editorial work. Finally, we wish to thank all the authors for their hard work and timely contributions, and this edition would not be possible without their contributions. We hope that this edition has provided reader with a snapshot into the current state of brain neuroimaging in aging and neurodegeneration.

Ann D. Cohen, PhD
James T. Becker, PhD
University of Pittsburgh, School of Medicine
Department of Psychiatry, Pittsburgh, PA, USA

Chapter 1 AD

1a

Brain MRI findings related to Alzheimer's disease in older African American adults

Owen Carmichael*, Robert Newton Jr.
Pennington Biomedical Research Center, Baton Rouge, LA, United States
*Corresponding author: e-mail address: owen.carmichael@pbrc.edu

Contents

Abstract

Although a substantial body of research has identified brain MRI measures as important markers of Alzheimer's disease (AD) risk, progression, and treatment response, most of that research has been performed in non-Hispanic white American populations, leading

Progress in Molecular Biology and Translational Science, Volume 165
ISSN 1877-1173
https://doi.org/10.1016/bs.pmbts.2019.04.002

to questions about the utility of the brain MRI measures among individuals of other races or ethnicities. African American individuals in particular are under-represented in AD research, and may exhibit differences in prevalence of AD risk factors, prevalence of AD, incidence of AD, the clinical course of cognitive decline, and AD neuropathology, each of which could influence the utility of brain MRI markers. Unfortunately, while current evidence suggests that African Americans exhibit poorer brain health late in life based on brain MRI measurements, several other aspects of brain MRI markers in this population are unclear, including trajectories of brain MRI markers leading up to old age, relationships between traditional brain health risk factors and brain MRI findings, and the status of brain MRI markers as correlates of cognitive impairment. This unclear state of affairs highlights the urgency of future research in which large numbers of older African American adults contribute longitudinal brain MRI measurements concurrent with clinical, cognitive, and molecular biomarker measurements, ideally in the context of AD preventive or therapeutic trials.

1. Introduction

The goal of this chapter is to provide an overview of the current state of knowledge about brain MRI findings in Alzheimer's disease (AD) among African American individuals, and how those findings differ from corresponding findings among non-Hispanic white Americans. There are two reasons to focus on this topic. First is that the large and steadily increasing public health burden of Alzheimer's disease in the United States is well known,[1] and brain MRI has played a prominent role in the effort to identify valid in vivo markers of AD risk, progression, and treatment response.[2,3] Second is that although substantial progress has been made in establishing brain MRI based markers of AD by collecting brain MRI scans from large numbers of adults, the vast majority of those scans were collected from non-Hispanic white American individuals. Because various aspects of AD risk, progression, and treatment response could potentially differ by race, it is possible that the optimal brain MRI markers that capture various aspects of the pathway to AD could differ by race. Our goal is to present the current state of evidence supporting this possibility, by reviewing background information about AD risk, clinical course, and neuropathology in African Americans, as well as MRI markers of aging, AD risk factors, and the clinical expression of AD in African Americans. We will present information from studies of African American adults as well as cohort studies that directly compare African American and non-Hispanic white American individuals. The chapter concludes with a summary of major knowledge gaps in this area and suggestions for future work.

Throughout the chapter, we will be focused on published studies of MRI markers in older African American adults. The MRI markers reported in studies to date have primarily been traditional and structural: predominantly summaries of gray matter and white matter volume from T1-weighted sequences, as well as summaries of white matter hyperintensity from T2-weighted imaging. White matter connectivity characteristics from diffusion MRI sequences also appear in this literature, and to a lesser extent functional activation and connectivity measures from blood oxygenation level dependent (BOLD) functional MRI sequences.

2. Differences between African Americans and non-Hispanic white Americans in Alzheimer's disease

This section provides background information about differences in AD risk, progression, and biology between African Americans and non-Hispanic white Americans, each of which could potentially be reflected in race differences in brain MRI markers of the disease. Specifically, we review the evidence for race differences in AD incidence, prevalence, risk factors, neuropathology, and molecular markers.

2.1 Race differences in AD incidence

The incidence of AD has been shown to be higher in African Americans compared to non-Hispanic white Americans. The Washington Heights-Inwood Columbia Aging Project[4] showed an incident rate of 4.2% in African Americans and 1.9% in non-Hispanic white Americans. The annual incidence rate of AD in African Americans was 2.5% in the Indianapolis-Ibadan study.[5] US census standardized estimates of AD incidence was 4% in African Americans and 2% in non-Hispanic white Americans according to the Chicago Health and Aging Project.[6] Not all studies echo the higher incidence of AD in African Americans.[7] It has been shown that African Americans are at 1.6,[8] 1.9,[9] and 2.4[4] times the risk of incident dementia compared to non-Hispanic white Americans. Some data suggest that the incidence of AD is declining over time in both African Americans and non-Hispanic white Americans.[6,10,11] One conjectured reason for the decline is the increase in the level of educational attainment among African American adults over time; greater educational attainment is associated with lower AD risk. Despite gradual reductions in incidence, the number of African American individuals with AD continues to rise given the continued growth of the aging population. Overall, the preponderance of evidence suggests that the incidence of AD is higher in African Americans compared to non-Hispanic white Americans.

2.2 Race differences in AD prevalence

A systematic review has suggested that the prevalence of AD is approximately twice as high in African Americans as it is in non-Hispanic white Americans.[12] However, while several studies including African American and non-Hispanic white American participants have shown this pattern, the prevalence rates themselves have been variable, from 20–27%[6,9] to 10–20%[11] to <10%.[13,14] Longitudinal studies suggest that the prevalence of AD among African Americans appears to be stable over time.[6,13] The relatively small number of studies suggesting similar prevalence rates across ethnic groups[15,16] tended to have smaller sample sizes and shorter follow-up periods than those reporting higher prevalence among African Americans. Taken together, current literature suggests that both the incidence and prevalence of AD is higher in African American compared to non-Hispanic white Americans. Race differences in incidence and prevalence might possibly reflect differences in how the disease unfolds biologically, and therefore could possibly influence the utility of brain MRI measures as markers of risk, progression, and treatment effects.

2.3 Race differences in AD risk factors

A multi-factorial array of risk factors for AD have been identified, and refinement of knowledge of these risk factors remains one of the primary AD research endeavors to date. Race differences in a set of the most well-known and widely replicated risk factors have been explored, with mixed results as described below.

2.3.1 Age

Age is the single greatest risk factor for AD.[6] Approximately 3% of adults age 65–74, 17% between 75 and 84 old, and 32% of individuals ≥85 are diagnosed with the disease.[17] One study has suggested that the effect of age on AD risk was similar across ethnic groups[15] and another study has suggested that greater age increases risk specifically in an African American cohort.[5] Although the number of studies is small, this evidence suggests that the effects of age on AD risk may be similar between African Americans and non-Hispanic white Americans.

2.3.2 Genetics

The Apolipoprotein E (APOE) gene is the most replicated genetic risk factor for sporadic AD.[18] Among African Americans specifically, it has been shown that being ε4 homozygous[16,19,20] or heterozygous[20] substantially increases the risk of AD over non ε4 carriers. Furthermore, African Americans who are homozygotes compared to heterozygotes may have as much as four times the risk of developing AD.[16]

Data consistently show that African Americans have a greater prevalence of the ε4 allele compared to non-Hispanic white Americans.[9,16,19] In addition, while African American ε4 homozygotes had a six-fold increased risk of developing AD compared to corresponding non ε4 carriers (HR = 6.3, 95% CI = [2.76,4.45]), the increase was four-fold in non-Hispanic white Americans,[16] suggesting a stronger effect of the ε4 in African Americans. In contrast to these findings, one study showed no ethnic differences in APOE ε4 carriers or non-carriers.[9] Recently, several other genes have been implicated in AD risk in African Americans, including ABCA7,[21] TOMM40 '523,[16] TREM2,[22] Thr92AlaD2,[23] DRD2,[24] and ADAP9,[25] but replication is needed. These findings suggest that while APOE ε4 prevalence is greater among African Americans, it is not yet clear if APOE ε4 carriage is a more potent risk factor for AD in African Americans compared to non-Hispanic white Americans. However, the search for genetic contributions to AD risk continues to advance over time.

2.3.3 Education

At a population level, educational attainment among African Americans is significantly lower than that of non-Hispanic white Americans, by approximately 2 years of schooling on average.[9] Higher levels of education are associated with reduced risk of developing AD among African-Americans and non-Hispanic white Americans.[9,26] Lower education (<12 years) is associated with poorer cognitive function among cognitively healthy African Americans and non-Hispanic white Americans.[27] However, each year of education above 12 years is positively associated with cognitive function in African Americans, but not with non-Hispanic white Americans.[27] Education also interacts with environmental factors to influence cognitive status. In a recent study, it was shown that education is associated with reduced risk of developing AD in those who grew up in rural populations, but there was no effect of education for those in an urban population.[28] Rural residence was shown to be associated with increased odds for developing AD in African Americans and this effect was nearly four times stronger in those with lower education compared to those with higher education.[29] Not all studies show ethnic differences in the association between education and cognitive function.[30] Quality of education may also influence cognitive performance in normal[31, 32] and AD patients.[33] However, when education is controlled for ethnicity differences in AD risk are substantially reduced or non-existent.[9, 34] These findings strongly suggest that both the quantity and quality of education have a significant impact on ethnic differences in AD risk.

2.3.4 Chronic disease

African Americans have higher rates of diabetes, cardiovascular disease, and obesity compared to non-Hispanic white Americans.[35–37] Therefore, it is believed that these factors play a particularly detrimental role in dementia risk in African Americans. Most data have shown that diabetes diagnosis and elevated blood glucose increase risk for dementia,[38] poor cognitive function,[39] and cognitive decline[40] similarly across ethnic groups. The data are equivocal with respect to cardiovascular disease. Some studies have shown that cardiovascular risk factors, such as hypertension, increase risk for dementia similarly between race and ethnic groups.[41,42] However, other studies suggest that stroke[43] and elevated blood pressure[41,44] have different associations with cognitive function in different race and ethnic groups. In some multi-ethnic cohort studies assessing the relationship between body weight and AD or cognitive function, race and ethnic differences were non-significant or such differences were not explored.[45,46] Yet, data exists suggesting that body weight is a stronger predictor of dementia among non-Hispanic white versus African Americans.[47]

2.3.5 Lifestyle

African Americans have poor dietary quality and low levels of physical activity.[48,49] Several studies show an inverse association between diet quality and cognitive function[50–52] and cognitive impairment,[53] with a lack of race and ethnic differences in these associations. However, at least one study has shown that such relationships are only significant among African Americans.[54] Similarly disparate findings have also been shown for various food groups and macronutrients.[55,56] The findings for physical activity mirror those of most lifestyle factors demonstrating either ethnic differences[57] or similarities[58] in its association with dementia and cognitive function. The most compelling data are derived from a study of objectively measured physical activity that showed a more robust positive association between moderate to vigorous physical activity and cognitive function in non-Hispanic white Americans compared to African Americans.[59]

2.4 Race differences in AD clinical course

Lower cognitive test scores are associated with increased risk of developing AD in African Americans from threefold 13–17 years prior to ninefold 1 year prior.[60] It has been robustly shown that African Americans typically score lower on tests of cognitive function compared to non-Hispanic white Americans,[30,33,61] although these differences are largely accounted for by differences in the quantity and quality of education. A slower rate of cognitive decline has also been demonstrated in African Americans compared to non-Hispanic white Americans after the diagnosis of AD.[12,61] The slower rate of decline in African Americans has been shown in terms of both global cognitive functioning and some individual domains.[62] However, some longitudinal studies have shown no race differences in cognitive decline in some cognitive domains[30] or across normal, MCI, and AD patients.[63]

2.5 Race differences in neuropathology

Post mortem neuropathological studies have the potential to play an important role in clarifying why older African American adults appear to display poorer brain characteristics compared to older non-Hispanic white Americans. But even the results of these studies are mixed, with some suggesting that older African Americans and non-Hispanic white Americans carry a similar burden of common neuropathological features,[64–67] and others suggesting that certain neuropathological features are more prevalent in one race compared to the other.[68–70] Two of these latter studies have suggested that African Americans may be more likely to exhibit cerebrovascular pathology at autopsy relative to non-Hispanic white Americans[68,69]; one of them further suggests that Alzheimer's and Lewy body pathologies are also more prevalent in African Americans, while TDP-43 and FTLD tau is less prevalent, relative to non-Hispanic white Americans.[69] A third study has suggested that the prevalence of mixed pathology—the co-occurrence of multiple pathological findings including Alzheimer, cerebrovascular, Lewy body, and other pathologies—is greater among African Americans relative to non-Hispanic white Americans.[70] Findings from these three studies suggest that a greater burden of underlying neuropathology among African Americans may contribute to poorer in vivo brain structural and functional characteristics, but given mixed findings to date, clearly a larger number of studies is needed to further clarify race differences in neuropathology.

2.6 Race differences in in vivo molecular markers of neurodegenerative diseases

In vivo studies of molecular markers of neurodegenerative disease have the potential to provide complementary information about the biological substrates for race differences in late life brain MRI measures. However, studies that have carefully assessed race differences in CSF based or PET imaging based markers of neurodegenerative disease processes have been rare, as have studies assessing race differences in associations between these markers and either risk factors or cognitive outcomes. Two in vivo studies have suggested that older African Americans had a greater risk of elevated levels of either cerebral amyloid or tau, compared to corresponding non-Hispanic white Americans.[71,72] An additional study suggested that greater amyloid deposition is associated with especially fast cognitive decline over the preceding decade among African Americans compared to non-Hispanic white Americans.[73] Finally, one study concluded that while midlife vascular risk factors elevated the risk of late-life amyloid positivity in a large biracial cohort.[74] Given the small number of publications on this topic, clearly much additional study is needed to clarify the distribution of in vivo markers of neurodegenerative processes in older African American adults.

3. Differences between African Americans and non-Hispanic white Americans in brain MRI measures

The preceding section provided evidence that the prevalence of AD and several of its most important risk factors may be greater among older African Americans. There is also mixed evidence that the potency of established AD risk factors may differ in African Americans, and that there are race differences in the biological and clinical course of AD development. The current section describes evidence that these possible race differences in AD manifest themselves as race differences in AD-related brain MRI markers. Specifically, we survey evidence of race differences in developmental trajectories of brain MRI characteristics, as well as race differences in associations between established AD risk factors and brain MRI characteristics. Next, we similarly survey race differences in brain MRI markers and relationships between risk factors and brain MRI markers late in life. Finally, we summarize data on associations between brain MRI markers and cognitive functioning in older African Americans.

3.1 Race differences in brain MRI characteristics in childhood and early adulthood

Any assessment of differences between African Americans and non-Hispanic white Americans in late-life brain characteristics begins with the question of whether individuals of these two races enter their later years with pre-existing brain differences that are driven by differences in either genetic factors or environmental exposures. Developmental studies of race differences in brain developmental trajectories in childhood are scarce. There are scattered reports from relatively small-scale studies of race differences in brain structure and function in young adulthood and middle age, including larger individual brain regions,[75] smaller total brain volume,[76] and smaller cranial vault size among healthy young to middle-aged African American adults.[77] The latter two differences are concerning, given a substantial literature linking smaller brain size and smaller cranial vault size to adverse outcomes in old age including Alzheimer's disease.[78–80] To our knowledge, only one study to date has addressed the relevant question of whether such race differences are more likely to represent race differences in genetic characteristics as opposed to environmental exposures.[77] In this study of African American and non-Hispanic white American twins, cranial vault size was significantly larger among the non-Hispanic white American' twins, in agreement with earlier reports of race differences in head size.[81] In addition, cranial vault size was more heritable among non-Hispanic white Americans than among African Americans; common environmental factors had a greater influence on cranial vault size among African Americans. Together, this small set of studies provides preliminary suggestions that African American adults may enter old age with pre-existing, pre-clinical differences in brain characteristics that are associated with adverse outcomes, and that these differences are more strongly driven by environmental exposures rather than genetic factors.

3.2 Race differences in risk factor—Brain MRI relationships in childhood and young adulthood

Race differences in relationships between risk factors and brain outcomes, identified at earlier points in the lifespan could further influence late-life race differences in brain characteristics and their relationships to risk factors. The number of such analyses is exceptionally small and disparate, with only two studies suggesting that young or middle aged African American adults may bear a disproportionately greater burden of brain effects of certain systemic risk factors. Specifically, there has been one report that smoking may be associated with disproportionately greater brain hyperactivation to smoking-related visual cues in African American adults compared to non-Hispanic white Americans,[82] and an additional report that African Americans diagnosed with HIV may bear a greater burden of diffuse white matter disease than corresponding non-Hispanic white Americans diagnosed with HIV.[83] In addition, two other disparate studies in African American populations have suggested that relationships between risk factors and brain characteristics are similar to those reported earlier in non-Hispanic white American populations. Specifically, there has been one report that lower socioeconomic status (SES) may be associated with less cortical and deep gray matter volumes among African American babies[84]; and cocaine addiction may be associated with less regional gray matter volumes among African American adults.[85] Although these studies provide an initial hint that African Americans may bear disproportionately grave brain consequences of certain common risk factors early in adulthood, thus setting the stage for further negative consequences of risk factors emerging in middle age and old age, it should be emphasized that race differences in relationships between risk factors and brain outcomes are very much underexplored among young and middle-aged adults.

3.3 Race differences in brain MRI characteristics late in life

Two small, very early studies of race differences in late-life brain characteristics suggested that African Americans and non-Hispanic white Americans with dementia show similar patterns of brain deficits compared to their cognitively intact peers, including brain atrophy and hippocampus atrophy[86,87]; and one similarly early and small study suggested that older African Americans and non-Hispanic white Americans enrolled in an antidepressant medication trial showed no significant baseline differences in white matter hyperintensity or ventricular size.[88] In addition, one early large-scale epidemiological study identified no differences in semi-quantitative neuroradiological ratings of brain MRI scans between cognitively intact African American and non-Hispanic white American older adults.[89] However, since then, the preponderance of large-scale epidemiological studies of aging with significant African American and non-Hispanic white American representation have reported significantly poorer MRI-based indices of brain health in older African American adults compared to non-Hispanic white American peers. The MRI indices in these studies have included ventricular expansion,[90]

MRI infarcts,[91,92] white matter hyperintensities,[93–97] and gray matter integrity.[96] It should be noted that not every result from these large-scale studies suggested poorer brain outcomes in African Americans; for example, greater cerebral blood flow[94] and total brain volumes[95] among African Americans relative to non-Hispanic white Americans has been reported, as well as a lack of race differences in gray matter atrophy.[96] But the preponderance of results nonetheless suggest poorer brain characteristics among older African Americans compared to non-Hispanic white American peers.

3.4 Race differences in Risk Factor—Brain MRI relationships late in life

One natural explanation for the observation of poorer brain outcomes among older African Americans is that African Americans carry a greater burden of certain risk factors for adverse late-life brain outcomes, including cardiometabolic risk factors, low socioeconomic status, lack of access to medical care, and others, as described above. But an additional possibility is that such risk factors may have particularly potent negative effects on the African American brain. Data on this question has been both sparse and mixed. On one hand, two studies in largely or entirely African American cohorts identified associations between risk factors and brain outcomes that were similar to those previously reported among non-Hispanic white Americans. Specifically, among older African Americans diagnosed with type 2 diabetes, poorer kidney function was associated with greater white matter hyperintensity volume and less gray matter volume, as has been reported earlier in non-Hispanic white American populations.[98] In addition, in a largely African American population, a greater burden of vascular risk factors was associated with greater fMRI hyper-activation to a Flanker task, again similar to earlier reports.[99] Furthermore, in a large bi-racial cohort, associations between multiple cardiovascular risk factors and MRI measures were not substantially modified by race.[91, 100] However, there are at least two reports that older African Americans bear more severe consequences of risk factor carriage. In one bi-racial cohort, an epigenetics-based measure of greater age acceleration was associated with greater white matter hyperintensity burden among African American but not non-Hispanic white American members of the cohort.[101] In another study, greater vascular disease burden was associated with especially smaller brain volumes and higher white matter hyperintensity volumes among the older African Americans of the cohort.[95] Although these reports are concerning, it should be emphasized that the amount of data available on race differences in risk factor effects is quite low.

3.5 Race differences in Brain MRI—Cognition relationships late in life

Associations between poorer brain characteristics and poorer cognitive functioning in older adults are well documented in predominantly non-Hispanic white American cohorts.[102–107] If such brain-cognition associations are similar among older African Americans, we would expect a higher burden of poor brain characteristics to culminate in an especially high burden of cognitive decline in this population. However, current data on race differences in brain-cognition associations has been mixed. Two studies have suggested that these associations may be substantially different: the first suggested that hippocampus volume, one of the cardinal correlates of memory function across a range from normal cognition to frank dementia[108–110] in largely non-Hispanic white American studies, is more weakly associated with memory in African Americans compared to non-Hispanic white Americans[111]; the second suggested that hippocampus volume is not significantly associated with subsequent change in memory and other cognitive domains in a racially diverse cohort.[112] Both studies emphasized the importance of white matter hyperintensities as correlates of cognitive function: in the first study, they were significantly associated with several domains of cognitive functioning among African Americans only; in the second study, white matter hyperintensities predicted subsequent cognitive change in the racially diverse cohort. In a third study, differences in brain measures between individuals who were cognitively normal, mildly impaired, and clinically demented were largely similar across African American and non-Hispanic white American groups.[113] However, relative to mildly impaired peers, African Americans with dementia had greater white matter hyperintensity burden than corresponding non-Hispanic white Americans. In addition, African Americans had relatively more intact hippocampus volume at the mildly impaired stage, compared to non-Hispanic white Americans. In contrast to these findings of race differences in brain-cognition relationships, a lack of such differences has been reported in two large bi-racial cohorts.[114,115] Thus, currently the question of race differences in cognitive outcomes associated with certain brain MRI findings remains unsettled.

4. Future work

The preceding sections clearly demonstrate a need for additional studies that clarify essentially all aspects of brain MRI markers of AD in older African American adults. In addition, the following critical questions about brain MRI markers are especially under-explored to date.

4.1 Race differences in genetics-brain MRI associations

Studies of race differences in associations between genetic markers and brain characteristics have been rare, with one small study suggesting that TTR genotype may be associated with similar brain outcomes in non-Hispanic white Americans and African Americans,[116] and another study finding that IL-6 genotype associated with a greater burden of white matter hyperintensity and infarction in non-Hispanic white Americans but not African Americans.[117]

4.2 Race differences in MRI markers of drug treatment response

To our knowledge, no studies to date have assessed whether pre-existing poorer brain structural or functional characteristics predispose older African American adults to poorer responses to Alzheimer's disease pharmacological treatments; or whether African Americans are more likely to experience poorer pharmacological treatment responses as measured by in vivo brain structural or functional markers. To our knowledge, only one study to date has identified poorer brain structural outcomes to drug treatment among African Americans relative to non-Hispanic white Americans, but this study was in multiple sclerosis.[118] The lack of literature on this topic likely reflects the overall low participation of older African American adults in Alzheimer's disease clinical trials.[119,120]

4.3 Race differences in brain responses to lifestyle interventions

Physical activity promotion has been the most commonly utilized form of lifestyle intervention that has been applied to Alzheimer's disease prevention and treatment. However, most studies have either not included a substantial number of African Americans or did not provide race specific data.[121–131] One large multicomponent lifestyle program (i.e., diet, physical activity) has been conducted, but it occurred in Finland, and no racial data was reported.[132] One physical activity study has been completed in older African American adults, with promising results.[133] A second trial appears to be ongoing.[134] However, these were conducted in exclusively African American cohorts so ethnic comparisons cannot be made. Therefore, there is a need for lifestyle interventions that include representative numbers of African American and non-Hispanic white Americans in order to determine the efficacy of these interventions across ethnic groups. Within these trials, race differences in brain responses to the intervention, including brain MRI markers, would be of especially high interest.

References

1. Brookmeyer R, Johnson E, Ziegler-Graham K, Arrighi HM. Forecasting the global burden of Alzheimer's disease. *Alzheimers Dement.* 2007;3(3):186–191.
2. Kantarci K, Jack CR. Neuroimaging in Alzheimer disease: an evidence-based review. *Neurol Clin.* 2003;13(2):197–209.
3. Radanovic M, Pereira FRS, Stella F, et al. White matter abnormalities associated with Alzheimer's disease and mild cognitive impairment: a critical review of MRI studies. *Expert Rev Neurother.* 2013;13(5):483–493.
4. Tang MX, Cross P, Andrews H, et al. Incidence of AD in African-Americans, Caribbean Hispanics, and Caucasians in northern Manhattan. *Neurology.* 2001;56(1): 49–56.
5. Ogunniyi A, Hall KS, Gureje O, et al. Risk factors for incident Alzheimer's disease in African Americans and Yoruba. *Metab Brain Dis.* 2006;21(2–3):235–240.
6. Rajan KB, Weuve J, Barnes LL, Wilson RS, Evans DA. Prevalence and incidence of clinically diagnosed Alzheimer's disease dementia from 1994 to 2012 in a population study. *Alzheimers Dement.* 2019;15(1):1–7.
7. Fillenbaum GG, Heyman A, Huber MS, et al. The prevalence and 3-year incidence of dementia in older black and white community residents. *J Clin Epidemiol.* 1998;51(7):587–595.
8. Steenland K, Goldstein FC, Levey A, Wharton W. A meta-analysis of Alzheimer's disease incidence and prevalence comparing African-Americans and Caucasians. *J Alzheimers Dis.* 2016;50(1):71–76.
9. Weuve J, Barnes LL, Mendes de Leon CF, et al. Cognitive aging in black and white Americans: cognition, cognitive decline, and incidence of Alzheimer disease dementia. *Epidemiology.* 2018;29(1):151–159.
10. Gao S, Ogunniyi A, Hall KS, et al. Dementia incidence declined in African-Americans but not in Yoruba. *Alzheimers Dement.* 2016;12(3):244–251.
11. Noble JM, Schupf N, Manly JJ, Andrews H, Tang MX, Mayeux R. Secular trends in the incidence of dementia in a multi-ethnic community. *J Alzheimers Dis.* 2017;60(3):1065–1075.
12. Mehta KM, Yeo GW. Systematic review of dementia prevalence and incidence in United States race/ethnic populations. *Alzheimers Dement.* 2017;13(1):72–83.
13. Hall KS, Gao S, Baiyewu O, et al. Prevalence rates for dementia and Alzheimer's disease in African Americans: 1992 versus 2001. *Alzheimers Dement.* 2009;5(3):227–233.
14. Hendrie HC, Osuntokun BO, Hall KS, et al. Prevalence of Alzheimer's disease and dementia in two communities: Nigerian Africans and African Americans. *Am J Psychiatry.* 1995;152(10):1485–1492.
15. Katz MJ, Lipton RB, Hall CB, et al. Age-specific and sex-specific prevalence and incidence of mild cognitive impairment, dementia, and Alzheimer dementia in blacks and whites: a report from the Einstein aging study. *Alzheimer Dis Assoc Disord.* 2012;26(4):335–343.
16. Yu L, Lutz MW, Wilson RS, et al. APOE epsilon 4-TOMM40 '523 haplotypes and the risk of Alzheimer's disease in older Caucasian and African Americans. *PLoS One.* 2017;12(7):e0180356.
17. Hebert LE, Weuve J, Scherr PA, Evans DA. Alzheimer disease in the United States (2010–2050) estimated using the 2010 census. *Neurology.* 2013;80(19):1778–1783.
18. Reitz C, Mayeux R. Genetics of Alzheimer's disease in Caribbean Hispanic and African American populations. *Biol Psychiatry.* 2014;75(7):534–541.
19. Kwon OD, Khaleeq A, Chan W, Pavlik VN, Doody RS. Apolipoprotein E polymorphism and age at onset of Alzheimer's disease in a quadriethnic sample. *Dement Geriatr Cogn Disord.* 2010;30(6):486–491.

20. Hendrie HC, Murrell J, Baiyewu O, et al. APOE epsilon4 and the risk for Alzheimer disease and cognitive decline in African Americans and Yoruba. *Int Psychogeriatr.* 2014;26(6):977–985.
21. Reitz C, Jun G, Naj A, et al. Variants in the ATP-binding cassette transporter (ABCA7), apolipoprotein E 4,and the risk of late-onset Alzheimer disease in African Americans. *JAMA.* 2013;309(14):1483–1492.
22. Jin SC, Carrasquillo MM, Benitez BA, et al. TREM2 is associated with increased risk for Alzheimer's disease in African Americans. *Mol Neurodegener.* 2015;10:19.
23. McAninch EA, Rajan KB, Evans DA, et al. A common DIO2 polymorphism and Alzheimer disease dementia in African and European Americans. *J Clin Endocrinol Metab.* 2018;103(5):1818–1826.
24. Blum K, Badgaiyan RD, Dunston GM, et al. The DRD2 Taq1A A1 Allele may magnify the risk of Alzheimer's in aging African-Americans. *Mol Neurobiol.* 2018;55(7):5526–5536.
25. Logue MW, Schu M, Vardarajan BN, et al. Two rare AKAP9 variants are associated with Alzheimer's disease in African Americans. *Alzheimers Dement.* 2014;10(6): 609–618.e611.
26. Liu H, Gao S, Hall KS, et al. Optimal blood pressure for cognitive function: findings from an elderly African-American cohort study. *J Am Geriatr Soc.* 2013;61(6):875–881.
27. Barnes LL, Wilson RS, Hebert LE, Scherr PA, Evans DA, Mendes de Leon CF. Racial differences in the association of education with physical and cognitive function in older blacks and whites. *J Gerontol B Psychol Sci Soc Sci.* 2011;66(3):354–363.
28. Hendrie HC, Smith-Gamble V, Lane KA, Purnell C, Clark DO, Gao S. The association of early life factors and declining incidence rates of dementia in an elderly population of African Americans. *J Gerontol B Psychol Sci Soc Sci.* 2018;73(suppl 1): S82–S89.
29. Hall KS, Gao S, Unverzagt FW, Hendrie HC. Low education and childhood rural residence: risk for Alzheimer's disease in African Americans. *Neurology.* 2000;54(1): 95–99.
30. Wilson RS, Capuano AW, Sytsma J, Bennett DA, Barnes LL. Cognitive aging in older black and white persons. *Psychol Aging.* 2015;30(2):279–285.
31. Manly JJ, Schupf N, Tang MX, Stern Y. Cognitive decline and literacy among ethnically diverse elders. *J Geriatr Psychiatry Neurol.* 2005;18(4):213–217.
32. Carvalho A, Rea IM, Parimon T, Cusack BJ. Physical activity and cognitive function in individuals over 60 years of age: a systematic review. *Clin Interv Aging.* 2014;9:661.
33. Chin AL, Negash S, Xie S, Arnold SE, Hamilton R. Quality, and not just quantity, of education accounts for differences in psychometric performance between african americans and white non-hispanics with Alzheimer's disease. *J Int Neuropsychol Soc.* 2012;18(2):277–285.
34. Masel MC, Raji M, Peek MK. Education and physical activity mediate the relationship between ethnicity and cognitive function in late middle-aged adults. *Ethn Health.* 2010;15(3):283–302.
35. Ogden CL, Carroll MD, Flegal KM. Prevalence of obesity in the United States. *JAMA.* 2014;312(2):189–190.
36. Prevention CfDCa. *National Diabetes Statistics Report, 2017.* Centers for Disease Control and Prevention; 2017.
37. Dorans KS, Mills KT, Liu Y, He J. Trends in prevalence and control of hypertension according to the 2017 American College of Cardiology/American Heart Association (ACC/AHA) guideline. *J Am Heart Assoc.* 2018;7(11):e008888.
38. Noble JM, Manly JJ, Schupf N, Tang MX, Luchsinger JA. Type 2 diabetes and ethnic disparities in cognitive impairment. *Ethn Dis.* 2012;22(1):38–44.

39. Luchsinger JA, Reitz C, Patel B, Tang MX, Manly JJ, Mayeux R. Relation of diabetes to mild cognitive impairment. *Arch Neurol.* 2007;64(4):570–575.
40. Rajan KB, Arvanitakis Z, Lynch EB, et al. Cognitive decline following incident and preexisting diabetes mellitus in a population sample. *Neurology.* 2016;87(16): 1681–1687.
41. Gottesman RF, Schneider AL, Albert M, et al. Midlife hypertension and 20-year cognitive change: the atherosclerosis risk in communities neurocognitive study. *JAMA Neurol.* 2014;71(10):1218–1227.
42. Gottesman RF, Albert MS, Alonso A, et al. Associations between midlife vascular risk factors and 25-year incident dementia in the atherosclerosis risk in communities (ARIC) cohort. *JAMA Neurol.* 2017;74(10):1246–1254.
43. Clark DG, Boan AD, Sims-Robinson C, et al. Differential impact of index stroke on dementia risk in African-Americans compared to whites. *J Stroke Cerebrovasc Dis.* 2018;27(10):2725–2730.
44. Lamar M, Rubin LH, Ajilore O, et al. What metabolic syndrome contributes to brain outcomes in African American & caucasian cohorts. *Curr Alzheimer Res.* 2015;12(7):640–647.
45. Thacker EL, Gillett SR, Wadley VG, et al. The American Heart Association life's simple 7 and incident cognitive impairment: the reasons for geographic and racial differences in stroke (REGARDS) study. *J Am Heart Assoc.* 2014;3(3):e000635.
46. Arvanitakis Z, Capuano AW, Bennett DA, Barnes LL. Body mass index and decline in cognitive function in older black and white persons. *J Gerontol A Biol Sci Med Sci.* 2018;73(2):198–203.
47. Hu G, Horswell R, Wang Y, et al. Body mass index and the risk of dementia among Louisiana low income diabetic patients. *PLoS One.* 2012;7(9):e44537.
48. James DC. Factors influencing food choices, dietary intake, and nutrition-related attitudes among African Americans: application of a culturally sensitive model. *Ethn Health.* 2004;9(4):349–367.
49. Williams WM, Yore MM, Whitt-Glover MC. Estimating physical activity trends among blacks in the United States through examination of four national surveys. *AIMS Public Health.* 2018;5(2):144–157.
50. Morris MC, Evans DA, Tangney CC, Bienias JL, Wilson RS. Associations of vegetable and fruit consumption with age-related cognitive change. *Neurology.* 2006;67(8): 1370–1376.
51. Tsivgoulis G, Judd S, Letter AJ, et al. Adherence to a Mediterranean diet and risk of incident cognitive impairment. *Neurology.* 2013;80(18):1684–1692.
52. Petruski-Ivleva N, Kucharska-Newton A, Palta P, et al. Milk intake at midlife and cognitive decline over 20 years. the atherosclerosis risk in communities (ARIC) study. *Nutrients.* 2017;9(10):1134.
53. Tangney CC, Kwasny MJ, Li H, Wilson RS, Evans DA, Morris MC. Adherence to a Mediterranean-type dietary pattern and cognitive decline in a community population. *Am J Clin Nutr.* 2011;93(3):601–607.
54. Koyama A, Houston DK, Simonsick EM, et al. Association between the Mediterranean diet and cognitive decline in a biracial population. *J Gerontol A Biol Sci Med Sci.* 2015;70(3):354–359.
55. Wright CB, Lee HS, Paik MC, Stabler SP, Allen RH, Sacco RL. Total homocysteine and cognition in a tri-ethnic cohort: the Northern Manhattan Study. *Neurology.* 2004;63(2):254–260.
56. Sundermann EE, Katz MJ, Lipton RB, Lichtenstein AH, Derby CA. A brief dietary assessment predicts executive dysfunction in an elderly cohort: results from the einstein aging study. *J Am Geriatr Soc.* 2016;64(11):e131–e136.

57. Rajan KB, Barnes LL, Skarupski KA, Mendes de Leon CF, Wilson RS, Evans DA. Physical and cognitive activities as deterrents of cognitive decline in a biracial population sample. *Am J Geriatr Psychiatry*. 2015;23(12):1225–1233.
58. Brewster PW, Melrose RJ, Marquine MJ, et al. Life experience and demographic influences on cognitive function in older adults. *Neuropsychology*. 2014;28(6):846–858.
59. Zhu W, Wadley VG, Howard VJ, Hutto B, Blair SN, Hooker SP. Objectively measured physical activity and cognitive function in older adults. *Med Sci Sports Exerc*. 2017;49(1):47–53.
60. Rajan KB, Wilson RS, Weuve J, Barnes LL, Evans DA. Cognitive impairment 18 years before clinical diagnosis of Alzheimer disease dementia. *Neurology*. 2015;85(10):898–904.
61. Rajan KB, Wilson RS, Barnes LL, Aggarwal NT, Weuve J, Evans DA. A cognitive turning point in development of clinical alzheimer's disease dementia and mild cognitive impairment: a biracial population study. *J Gerontol A Biol Sci Med Sci*. 2017;72(3):424–430.
62. Wilson RS, Rajan KB, Barnes LL, et al. Terminal decline of episodic memory and perceptual speed in a biracial population. *Neuropsychol Dev Cogn B Aging Neuropsychol Cogn*. 2018;25(3):378–389.
63. Wilson RS, Aggarwal NT, Barnes LL, Mendes de Leon CF, Hebert LE, Evans DA. Cognitive decline in incident Alzheimer disease in a community population. *Neurology*. 2010;74(12):951–955.
64. Mehta K, Yaffe K, Pérez-Stable EJ, et al. Race/ethnic differences in AD survival in US Alzheimer's disease centers. *Neurology*. 2008;70(14):1163–1170.
65. Wilkins CH, Grant EA, Schmitt SE, McKeel DW, Morris JC. The neuropathology of Alzheimer disease in African American and white individuals. *Arch Neurol*. 2006;63(1):87–90.
66. Sandberg G, Stewart W, Smialek J, Troncoso JC. The prevalence of the neuropathological lesions of Alzheimer's disease is independent of race and gender. *Neurobiol Aging*. 2001;22(2):169–175.
67. Riudavets MA, Rubio A, Cox C, Rudow G, Fowler D, Troncoso JC. The prevalence of Alzheimer neuropathologic lesions is similar in blacks and whites. *J Neuropathol Exp Neurol*. 2006;65(12):1143–1148.
68. Miles TP, Froehlich TE, Bogardus Jr ST, Inouye SK. Dementia and race: are there differences between African Americans and Caucasians? *J Am Geriatr Soc*. 2001;49(4):477–484.
69. Graff-Radford NR, Besser LM, Crook JE, Kukull WA, Dickson DW. Neuropathologic differences by race from the National Alzheimer's Coordinating Center. *Alzheimers Dement*. 2016;12(6):669–677.
70. Barnes LL, Leurgans S, Aggarwal NT, et al. Mixed pathology is more likely in black than white decedents with Alzheimer dementia. *Neurology*. 2015;85(6):528–534.
71. Howell JC, Watts KD, Parker MW, et al. Race modifies the relationship between cognition and Alzheimer's disease cerebrospinal fluid biomarkers. *Alzheimers Res Ther*. 2017;9(1):88.
72. Gottesman RF, Schneider AL, Zhou Y, et al. The ARIC-PET amyloid imaging study: brain amyloid differences by age, race, sex, and APOE. *Neurology*. 2016;87(5):473–480.
73. Gu Y, Razlighi QR, Zahodne LB, et al. Brain amyloid deposition and longitudinal cognitive decline in nondemented older subjects: results from a multi-ethnic population. *PLoS One*. 2015;10(7):e0123743.
74. Gottesman RF, Schneider AL, Zhou Y, et al. Association between midlife vascular risk factors and estimated brain amyloid deposition. *JAMA*. 2017;317(14):1443–1450.

75. Isamah N, Faison W, Payne ME, et al. Variability in frontotemporal brain structure: the importance of recruitment of African Americans in neuroscience research. *PLoS One*. 2010;5(10):e13642.
76. Rushton JP, Ankney CD. Brain size and cognitive ability: correlations with age, sex, social class, and race. *Psychon Bull Rev*. 1996;3(1):21–36.
77. Rushton JP, Osborne RT. Genetic and environmental contributions to cranial capacity in black and white adolescents. *Dermatol Int*. 1995;20(1):1–13.
78. Schofield P, Logroscino G, Andrews HF, Albert S, Stern Y. An association between head circumference and Alzheimer's disease in a population-based study of aging and dementia. *Neurology*. 1997;49(1):30–37.
79. Silbert LC, Quinn J, Moore M, et al. Changes in premorbid brain volume predict Alzheimer's disease pathology. *Neurology*. 2003;61(4):487–492.
80. Farias ST, Mungas D, Reed B, et al. Maximal brain size remains an important predictor of cognition in old age, independent of current brain pathology. *Neurobiol Aging*. 2012;33(8):1758–1768.
81. Rushton JP. Sex and race differences in cranial capacity from International Labour Office data. *Dermatol Int*. 1994;19(3):281–294.
82. Okuyemi KS, Powell JN, Savage CR, et al. Clinical and imaging study: enhanced cue-elicited brain activation in African American compared with caucasian smokers: an fMRI study. *Addict Biol*. 2006;11(1):97–106.
83. Haddow LJ, Dudau C, Chandrashekar H, et al. Cross-sectional study of unexplained white matter lesions in HIV positive individuals undergoing brain magnetic resonance imaging. *AIDS Patient Care STDs*. 2014;28(7):341–349.
84. Betancourt LM, Avants B, Farah MJ, et al. Effect of socioeconomic status (SES) disparity on neural development in female African-American infants at age 1 month. *Dev Sci*. 2016;19(6):947–956.
85. Weller RE, Stoeckel LE, Milby JB, et al. Smaller regional gray matter volume in homeless African American cocaine-dependent men: a preliminary report. *Open Neuroimaging J*. 2011;5:57.
86. Charletta D, Gorelick PB, Dollear T, Freels S, Harris Y. CT and MRI findings among African-Americans with Alzheimer's disease, vascular dementia, and stroke without dementia. *Neurology*. 1995;45(8):1456–1461.
87. Sencakova D, Graff-Radford NR, Willis FB, et al. Hippocampal atrophy correlates with clinical features of Alzheimer disease in African Americans. *Arch Neurol*. 2001;58(10):1593–1597.
88. Lesser IM, Smith MW, Wohl M, Mena R, Mehringer CM, Lin K-M. Brain imaging, antidepressants, and ethnicity: preliminary observations. *Psychopharmacol Bull*. 1996;32(2):235–242.
89. Longstreth Jr W, Arnold AM, Manolio TA, et al. Clinical correlates of ventricular and sulcal size on cranial magnetic resonance imaging of 3,301 elderly people. *Neuroepidemiology*. 2000;19(1):30–42.
90. Carmichael OT, Kuller LH, Lopez OL, et al. Acceleration of cerebral ventricular expansion in the cardiovascular health study. *Neurobiol Aging*. 2007;28(9):1316–1321.
91. Knopman DS, Penman AD, Catellier DJ, et al. Vascular risk factors and longitudinal changes on brain MRI: the ARIC study. *Neurology*. 2011;76(22):1879–1885.
92. Prabhakaran S, Wright CB, Yoshita M, et al. Prevalence and determinants of subclinical brain infarction: the Northern Manhattan study. *Neurology*. 2008;70(6):425–430.
93. Liao D, Cooper L, Cai J, et al. The prevalence and severity of white matter lesions, their relationship with age, ethnicity, gender, and cardiovascular disease risk factors: the ARIC Study. *Neuroepidemiology*. 1997;16(3):149–162.
94. Launer LJ, Lewis CE, Schreiner PJ, et al. Vascular factors and multiple measures of early brain health: CARDIA brain MRI study. *PLoS One*. 2015;10(3):e0122138.

95. Brickman AM, Schupf N, Manly JJ, et al. Brain morphology in older African Americans, caribbean hispanics, and whites from northern Manhattan. *Arch Neurol.* 2008;65(8):1053–1061.
96. Liu G, Allen B, Lopez O, et al. Racial differences in gray matter integrity by diffusion tensor in black and white octogenarians. *Curr Alzheimer Res.* 2015;12(7):648–654.
97. Gardener H, Scarmeas N, Gu Y, et al. Mediterranean diet and white matter hyperintensity volume in the Northern Manhattan study. *Arch Neurol.* 2012;69(2): 251–256.
98. Sink KM, Divers J, Whitlow CT, et al. Cerebral structural changes in diabetic kidney disease: African American–diabetes heart study MIND. *Diabetes Care.* 2015;38(2): 206–212.
99. Chuang Y-F, Eldreth D, Erickson KI, et al. Cardiovascular risks and brain function: a functional magnetic resonance imaging study of executive function in older adults. *Neurobiol Aging.* 2014;35(6):1396–1403.
100. Power MC, Deal JA, Sharrett AR, et al. Smoking and white matter hyperintensity progression: the ARIC-MRI Study. *Neurology.* 2015;84(8):841–848.
101. Raina A, Zhao X, Grove ML, et al. Cerebral white matter hyperintensities on MRI and acceleration of epigenetic aging: the atherosclerosis risk in communities study. *Clin Epigenetics.* 2017;9(1):21.
102. Gunning-Dixon FM, Brickman AM, Cheng JC, Alexopoulos GS. Aging of cerebral white matter: a review of MRI findings. *Int J Geriatr Psychiatry.* 2009;24(2):109–117.
103. Van Petten C. Relationship between hippocampal volume and memory ability in healthy individuals across the lifespan: review and meta-analysis. *Neuropsychologia.* 2004;42(10):1394–1413.
104. Debette S, Markus HS. The clinical importance of white matter hyperintensities on brain magnetic resonance imaging: systematic review and meta-analysis. *BMJ.* 2010;341:c3666.
105. Vermeer SE, Longstreth Jr WT, Koudstaal PJ. Silent brain infarcts: a systematic review. *Lancet Neurol.* 2007;6(7):611–619.
106. Fjell AM, Walhovd KB. Structural brain changes in aging: courses, causes and cognitive consequences. *Rev Neurosci.* 2010;21(3):187–222.
107. Raz N, Rodrigue KM. Differential aging of the brain: patterns, cognitive correlates and modifiers. *Neurosci Biobehav Rev.* 2006;30(6):730–748.
108. Persson J, Nyberg L, Lind J, et al. Structure–function correlates of cognitive decline in aging. *Cereb Cortex.* 2005;16(7):907–915.
109. Shi F, Liu B, Zhou Y, Yu C, Jiang T. Hippocampal volume and asymmetry in mild cognitive impairment and Alzheimer's disease: meta-analyses of MRI studies. *Hippocampus.* 2009;19(11):1055–1064.
110. Geuze E, Vermetten E, Bremner JD. MR-based in vivo hippocampal volumetrics: 2. Findings in neuropsychiatric disorders. *Mol Psychiatry.* 2005;10(2):160.
111. Zahodne LB, Manly JJ, Narkhede A, et al. Structural MRI predictors of late-life cognition differ across African Americans, hispanics, and whites. *Curr Alzheimer Res.* 2015;12(7):632–639.
112. Carmichael O, Mungas D, Beckett L, et al. MRI predictors of cognitive change in a diverse and carefully characterized elderly population. *Neurobiol Aging.* 2012;33(1): 83–95.
113. DeCarli C, Reed BR, Jagust W, Martinez O, Ortega M, Mungas D. Brain behavior relationships among African Americans, whites, and Hispanics. *Alzheimer Dis Assoc Disord.* 2008;22(4):382–391.
114. Aggarwal NT, Wilson RS, Bienias JL, et al. The association of magnetic resonance imaging measures with cognitive function in a biracial population sample. *Arch Neurol.* 2010;67(4):475–482.

115. Mosley Jr TH, Knopman DS, Catellier DJ, et al. Cerebral MRI findings and cognitive functioning: the atherosclerosis risk in communities study. *Neurology*. 2005;64(12): 2056–2062.
116. Cuenco KT, Friedland R, Baldwin CT, et al. Association of TTR polymorphisms with hippocampal atrophy in Alzheimer disease families. *Neurobiol Aging*. 2011;32(2): 249–256.
117. Fornage M, Chiang YA, O'Meara ES, et al. Biomarkers of inflammation and MRI-defined small vessel disease of the brain: the cardiovascular Health Study. *Stroke*. 2008;39(7):1952–1959.
118. Cree BA, Al-Sabbagh A, Bennett R, Goodin D. Response to interferon beta-1a treatment in African American multiple sclerosis patients. *Arch Neurol*. 2005;62(11): 1681–1683.
119. Harris Y, Gorelick PB, Samuels P, Bempong I. Why African Americans may not be participating in clinical trials. *J Natl Med Assoc*. 1996;88(10):630.
120. Olin JT, Dagerman KS, Fox LS, Bowers B, Schneider LS. Increasing ethnic minority participation in Alzheimer disease research. *Alzheimer Dis Assoc Disord*. 2002;16: S82–S85.
121. Callisaya M, Daly R, Sharman J, et al. Feasibility of a multi-modal exercise program on cognition in older adults with type 2 diabetes–a pilot randomised controlled trial. *BMC Geriatr*. 2017;17(1):237.
122. Hariprasad V, Koparde V, Sivakumar P, et al. Randomized clinical trial of yoga-based intervention in residents from elderly homes: effects on cognitive function. *Indian J Psychiatry*. 2013;55(Suppl 3):S357.
123. Lautenschlager NT, Cox KL, Flicker L, et al. Effect of physical activity on cognitive function in older adults at risk for Alzheimer disease: a randomized trial. *JAMA*. 2008;300(9):1027–1037.
124. Cassilhas RC, Viana VA, Grassmann V, et al. The impact of resistance exercise on the cognitive function of the elderly. *Med Sci Sports Exerc*. 2007;39(8):1401–1407.
125. Rathore A, Lom B. The effects of chronic and acute physical activity on working memory performance in healthy participants: a systematic review with meta-analysis of randomized controlled trials. *Syst Rev*. 2017;6(1):124.
126. Vidoni ED, Johnson DK, Morris JK, et al. Dose-response of aerobic exercise on cognition: a community-based, pilot randomized controlled trial. *PLoS One*. 2015;10(7):e0131647.
127. Napoli N, Shah K, Waters DL, Sinacore DR, Qualls C, Villareal DT. Effect of weight loss, exercise, or both on cognition and quality of life in obese older adults. *Am J Clin Nutr*. 2014;100(1):189–198.
128. Muscari A, Giannoni C, Pierpaoli L, et al. Chronic endurance exercise training prevents aging-related cognitive decline in healthy older adults: a randomized controlled trial. *Int J Geriatr Psychiatry*. 2010;25(10):1055–1064.
129. Williams P, Lord SR. Effects of group exercise on cognitive functioning and mood in older women. *Aust N Z J Public Health*. 1997;21(1):45–52.
130. Smiley-Oyen AL, Lowry KA, Francois SJ, Kohut ML, Ekkekakis P. Exercise, fitness, and neurocognitive function in older adults: the "selective improvement" and "cardiovascular fitness" hypotheses. *Ann Behav Med*. 2008;36(3):280–291.
131. Erickson KI, Voss MW, Prakash RS, et al. Exercise training increases size of hippocampus and improves memory. *Proc Natl Acad Sci U S A*. 2011;108(7):3017–3022.
132. Ngandu T, Lehtisalo J, Solomon A, et al. A 2 year multidomain intervention of diet, exercise, cognitive training, and vascular risk monitoring versus control to prevent cognitive decline in at-risk elderly people (FINGER): a randomised controlled trial. *Lancet*. 2015;385(9984):2255–2263.

133. Allard JS, Ntekim O, Johnson SP, et al. APOEε4 impacts up-regulation of brain-derived neurotrophic factor after a six-month stretch and aerobic exercise intervention in mild cognitively impaired elderly African Americans: a pilot study. *Exp Gerontol.* 2017;87:129–136.
134. Lukach AJ, Jedrziewski MK, Grove GA, et al. Rhythm experience and Africana culture trial (REACT!): a culturally salient intervention to promote neurocognitive health, mood, and well-being in older African Americans. *Contemp Clin Trials.* 2016;48:41–45.

1b

Magnetoencephalography applied to the study of Alzheimer's disease

David López-Sanz[a,b], Ricardo Bruña[a,b,c], Jaisalmer de Frutos-Lucas[a,d,e], Fernando Maestú[a,b,c,*]
[a]Laboratory of Cognitive and Computational Neuroscience (UCM-UPM), Centre for Biomedical Technology (CTB), Technical University of Madrid (UPM), Madrid, Spain
[b]Department of Experimental Psychology, Complutense University of Madrid (UCM), Madrid, Spain
[c]Biomedical Research Networking Center in Bioengineering, Biomaterials and Nanomedicine (CIBER-BBN), Zaragoza, Spain
[d]Biological and Health Psychology Department, Universidad Autonoma de Madrid, Madrid, Spain
[e]School of Medical and Health Sciences, Edith Cowan University, Joondalup, WA, Australia
*Corresponding author: e-mail address: fernando.maestu@ctb.upm.es

Contents

Progress in Molecular Biology and Translational Science, Volume 165
ISSN 1877-1173
https://doi.org/10.1016/bs.pmbts.2019.04.007

1. What is MEG and EEG

1.1 Physiological phenomena

Magnetoencephalography (MEG) and electroencephalography (EEG) are electrophysiological techniques aimed to study brain activity. Both techniques are based on the measurement, from different points of view, of the electrochemical activity originated by the action potential generated by neurons in the brain.

The action potential is the mechanism used by neurons for the transmission of information. A neuron at rest presents a negative potential of about 65 mV with respect to the surrounding medium. This difference of potential is maintained by means of active channels in the cellular membrane. On the other hand, the chemical stimulation occurring in the synapses modifies the permeability of the membrane. Excitatory stimulation of the membrane increases its permeability, reducing the difference of potential and creating a shocking wave that propagates through the neuron, from the dendrites to the soma, and from the soma to the axon. This shocking wave is similar to an electrical current propagating inside the neuron and is usually referred to as primary (or intra-cellular) current.[1] The active channels restore, after a short time, the equilibrium, generating a second shocking wave that propagates outside the neuron, usually termed secondary (or volume) current. Both phenomena (the primary and secondary currents) are similar in nature and, due to the principle of conservation of charge, have the same intensity. The primary current is confined inside the neuron and cannot be directly measured in vivo. The secondary current propagates through the whole of the conducting volume, and can, in theory, be measured at any point.

1.2 Measurement of EEG and MEG

In the case of the electrophysiology of the central nervous system, the secondary current, generated inside the cranium, can be observed at the level of the scalp. This measurement is named electroencephalography. However, in order to reach the scalp, the secondary current must propagate through all the different structures forming the head, following a convoluted path determined by the electric conductivity of the different tissues that it comprises. As a consequence, to accurately estimate the origin of the EEG signal, as measured in the scalp, the complete conductivity profile of the head must be known.

According to Ampère's law of electromagnetism, any electrical current generates a proportional magnetic field, which propagates perpendicularly to the current. This is also true for electrophysiological signals. Even when confined inside of the neuron, the primary current generates a magnetic field that can be measured outside of the head. This measurement is named magnetoencephalography. Secondary currents also generate associated magnetic fields but, as these currents distribute over a larger volume, the generated magnetic field is, in general, weaker. In addition, this magnetic field is not affected by the different tissues of the head, propagating always in a straight line. Thus, the estimation of the origin of a magnetic field measured outside of the scalp is relatively easy.

In order to obtain EEG or MEG data, the signal of interest (secondary current or magnetic field, respectively) must propagate far from the neuron. This only occurs in open-field neurons, especially in long neurons as pyramidal neurons.[2] In addition, the current generated by a single neuron is too weak to be observed outside of the head. According to estimations, at least 10,000 neurons must activate synchronously so that electrophysiological activity can be correctly measured. The larger group of neurons which fulfills these requirements in the human brain is the population of pyramidal neurons in the cortex. The dendrites of these cells are distributed in a parallel grid (allowing for direct spatial summation of their signal) and large patches of neurons activate simultaneously (which allows temporal summation). From this standpoint, models indicate that these neurons are the main (but not only) contributors to EEG and MEG signals.

1.3 Differences between EEG and MEG

EEG and MEG signals are originated by the same physiological phenomena. In ideal conditions, assuming similar noise levels and a favorable geometry, the information provided by both techniques should be the same. However, they have slightly different properties in real practice.

The first difference, which has already been mentioned, is that while MEG is a measure of (mainly) the primary currents, EEG is a measure of only secondary currents. As secondary currents follow a convoluted path to reach the electrodes, the identification of the original of the activity (usually termed source reconstruction) in EEG is far more complicated than in MEG. This usually leads to higher spatial resolution in MEG, when compared to EEG, under similar circumstances.

Other important difference between EEG and MEG is that, while MEG measures the magnetic field at a single point, EEG cannot directly measure the electrical current in the scalp. On the contrary, EEG estimates this current from the difference of potential between two points in the scalp as. According to Ohm's law, if there exists a difference of potential between two connected points, there must be a current flowing between them, and vice-versa. EEG is, thus, a reference-based measure, while MEG is reference free. If the underlying distribution of primary currents is estimated, this difference is not important. However, if the activity is analyzed at the sensor level the reference electrode will effectively modify the measurement and must be taken into account.

The third difference between EEG and MEG is related to the different sensitivity of both techniques. EEG measures the secondary currents in the scalp and can detect the echo related to any open field neuron. On the other hand, MEG measures the magnetic field generated by both primary and secondary currents. While, in general, the secondary magnetic field is weaker than the primary magnetic field, in situations with high symmetry the secondary current can orientate in such a way that their magnetic field cancels the one generated by the primary current. These very specific configurations are termed silent sources.[3] One classical example of a silent source is a radial primary current in an (approximately) spherical conductor. As the head can be roughly approximated as a sphere, MEG is usually said to be insensitive to radial sources, while EEG is not.

2. Metrics for signal analysis in M/EEG

As indicated in the previous section, M/EEG signal is, originally, a time series of electrical activity in the brain. This time series must be processed to extract any information about the underlying neurological processes. For the study of dementia due to Alzheimer's disease (AD), in this chapter, we will summarize the main findings using four different analysis methods: the spectral analysis, sometimes referred to as power analysis; the complexity of the time series itself; the functional connectivity, estimated as the synchronization between time courses; and the network analysis, derived from the previous one.

2.1 Spectral analysis

Spectral analysis is based on Fourier decomposition of a signal. According to Fourier analysis, any "natural" signal (i.e., real, continuous, derivable and without discontinuities) can be rewritten as an (infinite) sum of out-of-phase cosines.[4] This decomposition is unique and invertible, and can be performed using the Fourier transform:

$$X(f) = \int_{-\infty}^{\infty} x(t) \cdot e^{-j2\pi ft} dt$$

$$x(t) = \int_{-\infty}^{\infty} X(f) \cdot e^{j2\pi ft} df$$

In Fourier analysis, $x(t)$ represents the original signal, and $X(f)$ its Fourier decomposition, also termed Fourier spectrum. From this decomposition, it is possible to deal with each frequency component separately, as $X(f_i)$ is independent of $X(f_j)$ for any $i \neq j$. One common compact representation of the spectrum is by the power associated to each spectral component:

$$P(f) = \langle E\{X(f)\}^2 \rangle \approx |X(f)|^2$$

This representation is termed power spectral density (PSD) of $x(t)$, as it summarizes the power related to each spectral component of the signal.

The M/EEG signal has, since its origins, been described in terms of oscillations,[5,6] so the representation of these signals using Fourier spectrum is a natural step. The whole spectrum is classically divided into frequency bands: delta, for frequencies below 4 Hz; theta, for frequencies between 4 and 8 Hz; alpha, for frequencies between 8 and 12 Hz; beta, for frequencies between 12 and 30 Hz; and gamma, for frequencies above 30 Hz. These band limits are not strictly defined and can vary in different studies. The PSD of the neurophysiological signal can be analyzed directly (power analysis, using $P(f)$) or normalized (relative power analysis, using $P_n(f)$). In the second case, power is normalized in such a way that the total spectral power (for the considered broad band) is 1, or 100%, and the power for each frequency (or band) is modified accordingly:

$$P_n(f) = \frac{P(f)}{\int_{-\infty}^{\infty} P(f) df}$$

This normalization ignores spurious differences in amplitude (as, for example, differences in the distance to the MEG sensor array) but could also remove real differences in the amplitude of the electrophysiological signal.

Fourier analysis requires that the represented signal is stationary, rendering it, a priori, invalid for the study of the ever-changing neurophysiological activity. A common approach to deal with this limitation is to apply Fourier analysis over segments of the data, in what is called time-frequency analysis. The neurophysiological signal is divided into contiguous (maybe overlapping) temporal segments, and the spectrum of each segment is then calculated. This operation assigns a different PSD to each segment, allowing for the application of spectral analysis in non-stationary data, as is the case of task-related brain activity. This method is termed short-time Fourier transformation (STFT). A popular variation over STFT is wavelet analysis[7] in which the segmented signal is decomposed into non-sinusoidal mother functions other than cosines.

Both STFT and wavelet analysis provide similar information, allowing to study the variations in spectral content of non-stationary signals over time. These variations are, usually in the context of a cognitive task, frequency-wise corrected with respect to the brain pre-stimulus (or, in some cases, resting-state) activity, and the results are reported in terms of relative activations, or ratios.

2.2 Complexity

Complexity is a concept extracted from information theory describing how unpredictable, or complex, a stochastic process is. Complexity is a metric closely related to entropy,[8] but with a different mathematical definition: a stochastic process with high level of randomness would be associated with high entropy and would likely show a high level of complexity; a process highly predictable would be associated with a low entropy, and, with it, low complexity. For the special case of M/EEG signal, complexity can be estimated directly from the time course of the electrophysiological activity (both in sensor- or source-space) or from its power spectral density.

In the first case, the time series is transformed to a series of symbols, and the complexity is estimated as the ability to predict a new symbol from the past ones. Popular examples of these metrics are Lempel-Ziv complexity (LCZ[9]) and sample entropy (SampEn[10]).

In the second case, the normalized power spectral density $P_n(f)$ is treated as a probability density function (PDF), where each frequency bin has a given probability. From this PDF one can then estimate the level of complexity, usually using Shannon's spectral entropy (SSE)[8]:

$$\mathrm{SSE} = \int_{-\infty}^{\infty} \mathrm{P}_n(f) \cdot \log(\mathrm{P}_n(f))df$$

There are several variations over SSE, like Tsallis (TSE)[11] and Rényi (RSE)[12] spectral entropies. In all the cases, the spectral entropy increases with the unpredictability of the signal and is thus an indicator of its complexity.

2.3 Functional connectivity

Functional connectivity (FC) is a concept popularized by Friston.[13] The concept is a generalization of the connectome theory, where different parts of the brain interact to carry out neurological functions. The brain has been proved to show anatomical connectivity at several levels, having physical connections between different brain regions. FC defines the connectivity in a more computational way, were areas share some information without the need of a physical link between them. The limitation of this definition is that it is not possible to directly measure FC, as we cannot explain how this information was transferred. However, it can be estimated through synchronization, this is, the simultaneous activation of different brain regions not explainable by chance.

FC is, thus, based on a statistical estimation. In order to estimate FC, the time series obtained from M/EEG is checked for synchronizations, generally using three models: the linear model, the phase synchronization (PS) model and the generalized synchronization (GS) model.

The linear model is based on direct correlations between time series. As the electrophysiological signal from M/EEG is very noisy, the direct calculation of correlations is not very informative. The effect of this noise is sometimes minimized correlating the power of the signal M/EEG, instead of its time series. This instantaneous power is, in most cases, obtained from the amplitude envelope of the signal (amplitude envelope correlation, AEC[14]) but can also be extracted from the temporal variation of the time-frequency response for a given frequency (or frequency band). When the instantaneous power of two M/EEG time series show a high correlation, the underlying neurophysiological processes carried out by the different brain regions are assumed to be functionally coupled.

The PS model is based on a different idea, proposed by Rossenblum and colleagues.[15] These methods assess if the phases of two time series, and not their amplitudes, are synchronized. Accordingly, they require the extraction of the phase of the signal, using Fourier or Hilbert approaches, and then the consistency of the phase difference between two time series is evaluated: if the phase difference is stable, the time series are phase-locked, and thus synchronized; if the phase difference is random, the time series are not synchronized. The most popular metrics for calculating PS are coherency,[16] phase locking value (PLV[17]) and phase lag index (PLI[18]).

Last, the GS model uses a broader idea for synchronization, which cannot be calculated mathematically. GS refers to a synchronization of events, that is, if a certain event in one time series is followed by a related event in the second time series. This relation can be of any kind. One example of GS is the relation between the succession of the seasons and the life cycle of plants: in spring (event in the first system) the plants develop flowers (event in the second system). There is a relation between both events, but this relation is not apparent at first sight; however, the study of the repeating patterns of both events makes this relation evident. GS can also include linear relations or PS relations and can be interpreted as a higher-level model for synchronization. Popular metrics to calculate GS include mutual information (MI[19]) and synchronization likelihood (SL[20]).

2.4 Network analysis

Network analysis is a concept derived from communication networks science. In this field, different entities, termed as nodes, are connected through a series of links, termed edges. The characteristics of a given network determine if all the nodes are connected with all the remaining nodes, or if only a subset of possible connections are stablished. In a similar way, some edges can have larger bandwidth than others, creating routes with different levels of efficiency. In this complex scenario, network analysis arises as a way to identify optimal routes connecting pairs of nodes, and to identify critical nodes in the network, usually termed hubs.

From the definition of FC, the brain itself can be studied as a communications network, where different entities (sensors, brain regions or individual sources of activity) are connected through a set of edges (the functional connections themselves). The level of FC between two nodes can be interpreted as the bandwidth of the edge, and edges with a synchronization below the statistical threshold can be interpreted as not connected.

One very simple, yet useful, parameter used in network analysis is the degree, this is, the number of edges that connect each node with the rest of the network. In weighted networks, where the edges are not binary, the equivalent metric is the strength of a node, and it is calculated as the average value of all the edges connecting the node with the rest of the network. In both cases, it is possible to estimate the average for all the nodes of the network, getting the average degree or average strength for the whole network.

Another useful parameter is the path length, represented as the minimum number of steps needed to travel between two nodes of the network. In the case of weighted networks, the path length is the minimum sum of the distances (usually the inverse of the bandwidth) of each traveled edge. As the path length is defined between each pair of nodes, it is represented as a matrix, with zeros in the diagonal (the distance from one node to itself is always 0). This information can be summarized into the characteristic path length of the network, that is, the average of all the path lengths between pairs of nodes of the network.

Modularity is a different parameter receiving increasing attention in network studies. It quantifies the number of "independent" and non-overlapping sub-networks that comprise the original network. It is typically calculated using optimization algorithms rather than computed exactly. In the context of neuroscience, modular subdivision corresponds to some degree to modules with a specialized function.

One of the main parameters in the field is the clustering of a network or a certain node. It expresses the fraction of triangles around a given node and reflects the proportion of node's neighbors that are also neighbors of each other. This parameter, in combination with the characteristic path length of a given network, defines the concept of small-world network. This network structure is characterized by a high clustering (nodes neighbors are highly connected between them) while preserving a relatively short path length, ensuring optimal communication between distant nodes.

3. Alzheimer's disease and MEG: Principal findings from resting state activity

Dementia, and AD in particular, involves deficits in a variety of cognitive domains, specially memory and executive functioning. However, the study of idling awake restful activity has become a research topic of remarkable interest in MEG AD-related literature. In fact, most of the papers searching for a specific biomarker among the electrophysiological properties of AD brain activity make use of this paradigm. The study of resting state brain function with neuroimaging techniques implies the characterization of the different patterns of activation and interactions between regions in the absence of an explicit external task. Brain fluctuations in this context typically reflect the baseline activity when there is no goal-directed behavior. However, regardless of whether human brain is carrying out a particular task or not, brain activity exhibits certain spatio-temporal dynamics.[21] These dynamical properties in resting state are not a mere reflection of a relaxed and inactive brain. On the contrary, they are influenced by previous inputs and also modulate future responses.[22] Furthermore, silent problem-solving strategies may be taking place during the resting state, for instance, in the form of insight processes reflected in the spectral properties of brain activity. Another significant source of support for the understanding of resting state as a marker of cognitive performance comes from studies reporting associations between intellectual and cognitive level with certain parameters of brain network configuration at rest, such as small-world or clustering coefficients,[23,24] or with brain spontaneous connectivity between certain regions.[25]

An extensive line of evidence has demonstrated that brain activity during resting state is organized and consistent in the healthy brain. Consequently, certain regions tend to show similar trajectories of co-activation, thus activating and deactivating synchronously. These sets of regions are referred to as resting state networks (RSNs) and they are known to be consistent across subjects, and also over time for the same subject.[26] The first study in this regard found two sets of brain regions: one of them showing task-related activations and the other exhibiting task-related deactivations.[27] These two initial anti-correlated networks led to the discovery of multiple subnetworks from these broad original sets of regions, which relate to different cognitive and sensorimotor functions, like the motor network,[28] visual or language networks[29] and attentional networks among others.[30] The task-negative network, nowadays known as default mode network (DMN), has attracted a lot of attention in resting state research and its alterations have become one of the main research interests in the field of AD. The DMN includes areas such as cingulate cortices, inferior parietal and medial prefrontal regions. This group of regions shows higher levels of synchrony during rest compared to task performance, and they are widely known as preferential areas for AD pathology accumulation. DMN activity is typically related with stimulus-independent internal thoughts.[31]

Furthermore, it has also been linked to internal processing, linking already stored personal experiences with planning and external world monitorization.[32]

Given that resting state structure and dynamics can unveil information about cognitive function in normal healthy states, it is a reasonable assumption to consider that it might also be informative in pathological conditions of the brain. In fact, resting state abnormalities have been largely reported as potential biomarkers of AD and these alterations have been associated with cognitive decline progression throughout the course of the illness. Multiple studies have reported changes in a plethora of metrics that can describe the features of resting state brain activity in AD, from complexity and non-linear brain dynamics, to power spectral properties, functional connectivity characteristics or topological features of brain networks. The standardization and homogenization of concrete thresholds or cut-offs for what normal values should be considered in each of the metrics is far from completed. There are still major discrepancies in acquisition, pre-processing and analysis procedures that make this task difficult to accomplish at present. However, there has been a huge growth in the number of works describing abnormalities in the profiles of AD electrophysiological activity, as measured by MEG that will be narratively summarized in the following sections.

3.1 Non-linear analysis of MEG brain activity complexity and entropy in Alzheimer's disease

This group of metrics understands brain activity as a set of nonlinear couplings between different neuronal populations. In general, this type of analyses estimate the predictability of brain signals to infer, in the case of complexity algorithms, the number of frequency components underlying the observed activity,[33] under the assumption that AD pathology would compromise this intricate pattern of interacting synapses and axonal conductive properties. Accordingly, its normal dynamics at a macroscopic level should be disrupted, and electrophysiological techniques, such as MEG, would represent a useful tool to detect those alterations.

One of the first works studying changes in correlation dimension in AD patients reported a frequency-dependent mixed pattern of changes in AD. This pattern consisted of a significant decrease in this metric in low frequency bands and a significant increase in faster brain rhythms.[34] A subsequent work using this same metric failed to replicate the increase in correlation dimension and only reported generalized decreases of this parameter in this group of patients.[35] This finding was interpreted as a loss of complexity in the brain activity of AD patients detectable by MEG. In a similar vein, several works by different groups demonstrated a consistent reduction in automutual correlation decrease rate when studying MEG brain signals, thus showing a general increase in signal predictability associated to dementia due to AD. Given the fact that the decline rate of automutual information is somewhat related to the entropy of that signal (i.e., the predictability of the brain activity in this particular case) the authors suggested that brain activity in AD could show lower entropy values. Accordingly, posterior work demonstrated that brain activity is indeed abnormally predictable compared to age- and sex-matched controls.[36] Interestingly, the increase in signal regularity measured in AD brain activity has been consistently replicated in multiple studies using a variety of metrics such as approximate entropy, sample entropy or multiscale entropy,[37,38] and different measures of spectral entropy.[39,40] Gomez and colleagues reported an increase in long-range self-similarity in non-stationary data by means of detrended fluctuation analysis and backward detrending moving average.[41] Both measures presented a significant increase in dementia, supporting previous work showing increases in signal self-predictability in patients affected by the condition. This increase in brain activity regularity measured by MEG in AD patients has been traditionally explained in the literature by a decrease in the dynamic complexity of different parts of the brain. However, its implications for AD pathology and symptoms are not yet fully understood. Authors have pointed out that some of the physiological mechanisms underlying signal regularity alterations in AD could be neuronal death, loss of dynamical brain responsivity to stimuli, alterations in neurotransmitter regulation or a deficit in local networks connection due to nerve cell death. However, this increase in signal regularity has been consistently interpreted as representative of the cognitive dysfunction in AD.

Some studies have put forward that AD pathology produces not only an increase in predictability and signal regularity as above-mentioned, but also a significant decrease in signal complexity. Interestingly, this finding has been replicated using different metrics such as Higuchi's fractal dimension,[42] or Lempel-Ziv complexity.[43] Remarkably, both studies demonstrated that complexity measures were able to correctly distinguish AD from controls with more than 85% accuracy.

Studies characterizing the non-linear and complexity changes in the preclinical stages of the disease are very scarce compared with the relatively large number of works conducted with patients in the dementia stage. One of the first studies including a group of patients with mild cognitive impairment (MCI) and a group of AD showed a significant decrease in Lempel-Ziv complexity of MEG brain signals in the latter.[44] Although not significantly, MCI patients showed a trend toward lower complexity values, thus remaining at an intermediate state between healthy elders and AD. This tendency toward simpler brain activity implies an increase in signal predictability and was interpreted as a loss of independent oscillators underlying brain activity already evident in the early stages of the disease. Similarly, a subsequent study reported a pronounced drop in spectral entropy in a group of AD patients. Again, MCI patients showed minor decreases, particularly over right lateral sensors. This decrease was associated with a reduction in brain signal information content.[45] Interestingly, this study suggested that MCI patients were able to compensate some of the AD-related damage, as reflected by an increase in a disequilibrium measure and a complexity metric based on Lopez Ruiz-Mancini-Calbet index (LMC), hypothesized to precede the well-known decrease in complexity observed in AD. However, these discrepancies could likely be caused by methodological inconsistencies between studies. In this sense, recent works studying MCI patients seem to support again classical results in AD (i.e., a decrease in complexity) by using different metrics such as Jensen's divergence and wavelet turbulence.[46,47]

Most of the literature in the field converges in the same finding, brain activity in AD becomes more regular, predictable and less complex. However, few studies have addressed the pattern of changes in the preclinical stages, thus our knowledge about the evolution of non-linear brain interactions as measured by MEG remains inconclusive yet. All the studies conducted to date within this approach have been carried out in the sensor space, which is, nowadays, one of the major limitations of the field that hinders the capability to make concrete regional and spatial interpretation of the findings reported here.

3.2 Power spectral analysis of MEG brain activity in Alzheimer's disease

The quantification of power spectral density is thought to reflect the status of local synchronization mechanisms in the brain. Hence, it represents a convenient tool for the study of brain pathology, and in particular for AD, as spectral power calculation and interpretation are relatively straightforward, thus simplifying and reducing the amount of methodological approaches, fostering homogenization between studies.

Literature regarding power spectral properties of MEG brain signals in AD patients is highly convergent and has consistently demonstrated a slowing of the background brain activity. This activity slowdown has been observed using different metrics such as a reduction in the frequency of the main power spectrum peak, that is, the point of the individual spectrum with the highest amplitude, or alpha peak.[34,48,49] Also, other works have demonstrated this brain activity slowdown as a reduction of the median frequency of the spectrum, implying that the frequency at which the spectrum is divided into two sections with the same amount of total power tends to be lower in AD patients compared to controls.[50] Different studies have showed that both measures: the frequency of the peak and the median frequency of the spectrum are able to significantly classify AD patients and controls at the individual level.[51,52] Remarkably, both studies agreed that the median frequency was the best performing measure among those included in each respective work, as it described best the tendency toward lower frequencies taking place in the condition. It has been pointed out that the cholinergic system, known to modulate spontaneous cortical activity in the theta and alpha bands, might be at least partly responsible for this shift to the left in the power spectrum, given that cholinergic deficit involves a loss of the neurotransmitter aceltylcholine. This decrease in aceltylcholine would produce local disconnection of several groups of neurons, reflected as an increase in the relative contribution of lower frequencies to the MEG spectrum. Nevertheless, AD is a heterogeneous condition, henceforth, this particular alteration might not be enough to fully explain this observation. High frequency power reductions might as well be related to neuronal death, as MEG power is closely connected with the number of active synapses in a specific frequency range,[53] and also with delays and impairment in axonal conductance, hindering synchronization in higher frequency rhythms specifically. Furthermore, simulation studies have demonstrated that synaptic damage over cortico-thalamo-cortical circuits results in the observed slowdown of the alpha peak,[54] and thus MEG findings should probably be reflective of such synaptic damage, even in the preclinical stages as will be discussed shortly.

Multiple studies have consistently highlighted that this shift to the left of the main power spectrum frequency is driven by an increase of both the contribution of the lower frequencies in the brain, delta (2–4 Hz) band and theta (4–8 Hz) band,[55,56] accompanied by a decrease in faster brain rhythms, being beta (13–30 Hz) and gamma (above 30 Hz) bands the most commonly reported.[39,57] Alpha band activity has also been reported to decrease in AD patients, and remarkably, those AD patients showing lower alpha power demonstrated a worse cognitive status in the MMSE.[58] The slowdown is not homogeneously distributed across the brain in AD patients, as parietal, occipital and temporal regions, particularly over the left hemisphere, are the ones mainly reported in most of the aforementioned studies. Interestingly, this slowing of the background MEG activity has been observed to correlate with a worse functional and cognitive status and, more importantly, with gray matter loss over hippocampal regions, a well-known marker of AD pathology.[59,60] All these changes show the pathological nature of brain activity slowing. Furthermore, power spectrum slowdown has been proven useful in the detection of individuals with AD, with areas under the curve typically over 0.8.[58,61] However, since multiple pathological brain conditions are also associated to this phenomenon, it cannot be considered pathognomonic of the condition, and hence cannot represent a valid biomarker for the disease in isolation.

MEG is a very promising tool in the context of early AD detection. Several studies have already investigated the oscillatory power disruptions occurring in early stages of the disease, before the onset of dementia. Results with MCI patients suggest a progressive decline in synaptic functioning as highlighted by López et al.[62] In this study MCI patients exhibited a similar pattern to that described in AD, with relative power increases in low frequency oscillations and decreases in faster brain rhythms. Interestingly, alterations were more pronounced in the multidomain MCI subgroup compared to the single domain, emphasizing the relevance of these synaptic malfunction on cognitive status. Of note, this pattern of changes has shown great accuracy distinguishing between the different stages of the disease, from healthy older adults, to prodromic stages and demented patients.[63,64] The slowing observed in AD patients can already be observed in the predementia stages.[65] Moreover, this alteration seems to increase throughout the disease continuum.[66] A recent study was able to characterize spectral alterations in healthy older adults with subjective cognitive decline consisting of a widespread decrease in alpha power, significantly correlated with worse cognitive performance within the sample.[67] Interestingly, theta band alterations have also been reported in totally asymptomatic healthy older adults carriers of the APOE ε4 allele, one of the major risk factors for AD known to date, who showed a significant relative power increase in relation to worse cognitive performance,[68] consistent with observations in the AD continuum. All the above-mentioned findings highlight the relevance of spectral analysis in the context of dementia due to AD. Of particular interest are the findings concerning the preclinical stages of the disease, in which MEG could be a very useful tool for early detection as reported by multiple studies.

3.3 Functional connectivity analysis of MEG brain activity in Alzheimer's disease

Successful cognitive processing is heavily dependent on distributed brain activity in which multiple regions are synchronously working toward the same goal, either performing a similar task in a coordinated fashion or regulating different mental processes occurring across the cortex.[69] Synchronized activity is captured through different statistical methods already described, and collectively known as FC metrics. One of the hypothesized mechanisms underlying cognitive decline in AD is the FC breakdown, which would progressively arise throughout its continuum. In fact, AD pathology impairs neuronal synapses through different mechanisms, thus limiting the brain's normal temporal coordination. Consequently, AD has been conceptualized as a *disconnection syndrome*.[70]

Very few studies have reported widespread FC increases in AD patients using MEG. While a minority of them found FC increases affecting broadband or high frequency bands,[71,72] most of the studies reporting hyperconnectivity in AD reported it for slow brain rhythms such as delta and theta bands.[72,73] However, the main body of literature addressing FC changes in demented patients support the notion of AD as a disconnection syndrome. Berendse et al.[55] reported a decrease in coherence affecting all the frequency bands in one of the first studies conducted with this population making use of this approach, while Alonso et al.[72] reported a decrease in FC in faster brain rhythms such as beta and gamma using coherence, particularly affecting parietal, posterior and temporal sensors. Disconnection in faster brain rhythms has been reported in multiple studies, and it has also been observed in the alpha band.[74–76] A study conducted by Francioti and colleagues[77] was able to distinguish both AD patients and patients affected by dementia with Lewy bodies based on FC decreases in the alpha band. In fact, lower FC between temporal interhemispheric regions has been shown to correlate with a worse general cognitive performance in the MMSE,[71] which highlights the relevance of FC deterioration as an underlying feature of cognitive decline in dementia.

While the exact interpretation and meaning of FC changes is not yet completely understood, it is crucial to bear in mind that sensor space FC analysis, like the above-mentioned, might be severely compromised due to the cross-talk occurring between neighboring sensors, i.e., the magnetic activity from a single source may arise at different sensors virtually at the same time. As a result, FC findings in the sensor space should be cautiously considered and interpreted, as differences in signal power in specific frequency bands could contribute to the observed statistical dependency between different sensors.

More recent studies conducted after brain level activity reconstruction have confirmed FC alterations previously reported in AD patients. Koelewijn et al.[78] compared FC patterns in AD patients to that exhibited by healthy older adults and a group of young controls. Results showed that while FC increased with normal aging, particularly over left parieto-temporal cortices, AD patients demonstrated a significant decrease in these same regions. Accordingly, the authors interpreted that the connectivity changes observed in AD are qualitatively different from a merely exacerbated or accelerated aging process. Making use of directed phase transfer entropy, an effective connectivity metric able to detect directionality in brain interactions, Engels et al.[79] confirmed previous findings in healthy subjects by reporting that information tends to flow from posterior to anterior brain regions in faster brain rhythms while it mainly flows in the opposite direction in the theta band. Remarkably, AD patients' information flow from posterior to anterior brain regions was heavily affected in the beta band, particularly in those links emerging from the precuneus, a well-known region for its implication in AD development.

The DMN is an extensively described resting state network, whose regions are known to be highly connected in wakeful resting state, while reducing their coupling whenever the subject engages in a cognitive task as previously mentioned.[80] The DMN mainly comprises the synchronous activity of the medial prefrontal cortices, bilateral inferior parietal lobules, medial temporal structures, such as the hippocampus, and particularly the precuneus/posterior cingulate cortex, and its alteration has been consistently associated to AD progression given the high rates of AD-related pathology accumulation in these regions.[81] Surprisingly, the DMN has been very scarcely studied in the context of AD using MEG, and it has been mainly described using functional magnetic resonance imaging. However, a recent study using a novel analytic approach based on a Hidden Markov Model was able to describe a highly spatiotemporal precise pattern of transitions between the different *discrete states* of the brain, i.e., the temporal alternation between the different resting state networks. In agreement with other imaging modalities, the study showed that AD patients' activity in the DMN was altered. However, given the extraordinary temporal resolution of MEG, this study reported for the first time the specific timing of these alterations, proving that the states associated with the DMN in AD patients occurred less often, and were both shorter and less intense in average compared to healthy controls.

Research on the evolution of FC during the preclinical stages of AD has demonstrated that FC alterations are already present in MCI patients, as observed by a decrease in mean FC in the beta band in the sensor space,[82] confirmed using granger causality[83] and subsequently localized to temporal, supramarginal and precentral regions.[84] Alpha band disconnection was also observed in the latter, mainly over temporal and parietal regions, critically related to AD pathogenesis as already mentioned. Alpha band disconnection has been proposed as a good candidate to track progression throughout the AD continuum, as it correlated with disease severity in a sample of MCI and AD participants.[85] DMN connectivity is also affected before dementia onset, as decreases have been reported in both SCD and MCI elders.[86] This study demonstrated that SCD and MCI patterns of alterations were highly similar, and resembling of that previously described in AD with other imaging modalities.[87] A combined pattern of hyper-synchronization over anterior brain regions and desynchronization in posterior areas was observed in both groups. This result is consistent with the cascading network failure hypothesis, proposing that hyper-synchronization should not be understood as a compensatory attempt to improve task or behavioral performance, as traditionally interpreted in the literature, but rather as a pathological sign of network hyper-excitability, probably associated to Aβ deposition in the brain. Under this framework, the prevailing disconnection observed in AD patients would be preceded by a phase of pathological hyperactivity of the networks that would eventually produce a breakdown of the system.[88] This is consistent with studies showing alpha band hyper-connectivity in MCI patients was predictive of future conversion to dementia.[89]

FC may play a crucial role in the understanding of AD pathology, as the neuropathology underlying disease progression seems to be tightly associated with neuronal disconnection. However, although several progresses have been made in the field, contradictory results between studies currently hinder clinical application of these metrics. Plenty of different FC metrics are available, each of them measuring and reflecting slightly different aspects of brain synchronization, which together with heterogeneity of signal processing and analytical approaches may account for most of the variability observed in this complex field of study. A better understanding of the implications and nuances of each metric is needed to gain insight into what apparent contradictory results among measures may actually mean to the field of AD neurophysiology.

3.4 Network topological analysis of MEG brain activity in Alzheimer's disease

A substantial body of literature supports the notion that healthy information processing depends on interactions between distant brain regions, as already mentioned. Network analyses, applied to neuroscience, offer a tool to simplify and reduce the vast information contained in the connectome to its most basic, yet informative, organizational properties associated with specific aspects of human brain structure and function.[90]

The first approach to AD brain organization within this framework using MEG revealed a more random network structure in this group of patients, characterized by a decrease in both path length and clustering coefficient in the low alpha band.[76] The authors observed that the pattern of alterations observed in AD were better explained by a *selective attack* than by a *random attack*. Henceforth, according to their interpretation, specific nodes of the network are more likely to be affected by AD neuropathology than others. Graph theory has revealed that certain nodes of a given network are particularly crucial for the distribution and integration of information across its different parts. This kind of nodes are known as hubs, and brain networks are not an exception to this type of organization. Regions such as the precuneus, inferior parietal lobe and some temporal areas are critical hubs for brain network topology.[90] A recent study combined the information contained in the five classical frequency bands in a novel way to obtain a multi-layer network representation of brain connectivity in healthy older adults and AD patients.[91] Yu and colleagues described that posterior hub regions, including the above-mentioned, are selectively vulnerable to AD pathology, decreasing their nodal centrality (i.e., their relative contribution in network structure) in this group of patients. These results suggest that the disease process in AD may specifically affect association fibers connecting brain areas that are highly connected to the rest of the brain, that is: higher order association areas. This study further supported the notion of a *targeted attack* to particular brain regions critical to brain network maintenance. Interestingly, these changes were positively correlated with cognitive deterioration, proving the relevance of this phenomena in the process of cognitive decline.[76] The association between decreased centrality of medial temporal regions and worse cognitive status was also corroborated in a different study.[92] This work also showed that AD patients exhibit a loss of network robustness as demonstrated by a decrease in algebraic connectivity and spectral radius affecting several bands. The spatial pattern of alterations has been specifically associated to the type of dementia, being temporal damage specific to AD, according to a recent study.[93] Lastly, modular partition, i.e., the subdivision of brain activity into functionally segregated brain clusters, is also affected in AD patients.[94] Graph theoretical analysis of AD brain networks revealed weakened connections within, but particularly between different brain modules. This finding highlights the impairment of the specialized functional architecture of brain networks in AD.

Although very scarce, some studies have also reported a tendency toward more random network organization in patients with MCI as shown by a significant reduction in small-worldness, clustering and transitivity coefficient and a more segregated hierarchical brain organization, which supports the presence of inter-module connectivity alterations in the early stages of the disease.[95] Remarkably, this same study observed that older adults with SCD exhibited intermediate values between MCI and healthy controls for most of the parameters. Clustering reductions were also found in a pharmacological model of AD, with healthy subjects that were under scopolamine effects.[96] However, some inconsistent results have also been reported as no significant differences in either clustering or path length were reported in a recent study with MCI patients.[97]

These findings highlight that brain network structure, at its most fundamental level, is seriously impaired in dementia due to AD, and that some of these alterations can already be observed in the preclinical stages of dementia. Furthermore, MEG is a well-suited technique to measure such subtle alterations, as shown by the literature reviewed. Some inconsistent results hinder the interpretability of the results in this specific field, probably due to the high variability in the methodological approaches employed by different authors.

3.5 Potential utility of MEG as a biomarker for Alzheimer's disease

The number of studies available studying the electrophysiological changes in AD as captured by MEG is ever-increasing. Even though MEG is a relatively novel technique, some considerable advancements have already been made in this relatively young field of research characterizing AD at the group level, as already discussed. However, despite the efforts made to date, the question of whether MEG can be used as an independent biomarker for the diagnostic and prognosis of AD remains to be convincingly answered yet.

In order to obtain a valid biomarker for any disease it should be sensitive to its pathology. Several studies have demonstrated that MEG is able to distinguish AD patients from healthy controls, not only at the group level, but also classifying at the individual level with high accuracy. Different studies made use of a variety of electrophysiological alterations to this aim, such as the lower complexity scores observed in AD patients at rest or the pattern brain activation measured during the performance of a Sternberg memory task.[98] Resting state power spectral measures, and even MEG functional connectivity informed with diffusion tensor imaging, have shown great utility in distinguishing AD patients and MCIs, respectively, from healthy older adults.[99] However, most of the studies aimed at distinguishing and diagnosing the different stages in the disease continuum suffer from the same methodological problem, the absence of replication samples in which testing the hypothesis or the specific findings of the study. In this regard, FC measurements obtained from MEG activity have proven to be reliable across different laboratories, as shown in a multicentre study.[99a] Maestú and colleagues analyzed sensor-level synchronization patterns of several participants (MCI and healthy matched controls) coming from different MEG systems distributed across the world. With this dataset they trained a classifier that was subsequently tested in a completely independent sub-sample of the study. Crucially, as the first blind randomized study done with MEG, it was able to demonstrate reliability across MEG sites and high accuracy in the classification of unseen data, an important step-forward in the field. In this same line, Escudero et al.[51] also used resting state activity in a group of participants subdivided into training and validation samples obtaining classifications with areas under the curve of around 0.8 using different metrics such as median frequency, spectral entropy, sample entropy or Lempel-Ziv complexity.

MEG measurements are also sensitive to disease progression as repeatedly demonstrated by different studies. In a study conducted by Fernández et al.[100] a 4.5-fold risk increase for AD development was observed in those MCI patients with higher delta power at rest, which has been consistently regarded as a classical feature of AD as above-mentioned. This emphasizes not only its validity in the detection, but also its prognostic value. Moreover, increased occipital theta power has been also related to an increased risk of progression to AD in a different sample of MCI participants.[101] Similarly, MCI patients showing at baseline recording higher levels of activation during the performance of a Sternberg memory task, in both the ventral and the dorsal pathways, were more likely than their counterparts to develop AD in a 2-year follow-up study. This pattern of hyper-activation as a clinical sign of progression has been also reported in FC studies. A study conducted by Bajo et al.[102] using a modified version of the Sternberg's task demonstrated that higher values of synchronization likelihood between parieto-occipital sensors in the alpha and beta bands during task performance were associated with conversion over a 2-year follow-up in a sample of MCI participants. Interestingly, similar results were also observed in a different sample of MCI patients during resting state.[89] In this particular study, it was reported that progression from MCI to AD could be reliably predicted within a 2 years period using five links, connecting anterior cingulate cortex with posterior brain regions, more specifically bilateral temporo-occipital areas. Even though dementia stages of AD have been widely characterized as a disconnection syndrome, multiple preclinical studies highlight the relevance of hyper-excitability in the early stages of the disease as a potential marker for disease progression. Recent work has proposed a new interpretation to these findings, showing that soluble Aβ oligomers and early plaques may lead to a disruption in the excitatory/inhibitory balance of the networks, which could be the initial event triggering the cascade of alterations leading to AD.[103] Accordingly, hypo-synchronization could appear after the initial observed increase in FC, and therefore both types of changes could be interpreted as a result of neurodegeneration, which seems to be supported by its relationship with hippocampal atrophy.[86]

The aforementioned hypothesis found further support by some of the scarce studies carried out using MEG in combination with measurements of classically AD-related proteins (i.e., Aβ and tau). A recent study reported that healthy older adults, at risk of AD because of their high Aβ brain-deposit levels, mainly demonstrated an increase in FC between the precuneus with bilateral inferior parietal cortices, critical areas of the DMN in the theta and beta bands.[104] Interestingly, studies conducted employing samples at a more advanced stage in the disease continuum reported consistent results with this view. In this vein, a study of MCI FC patterns in relation to Aβ levels reported a mixed pattern of increased and decreased FC in the Aβ positive group,[105] while a study with AD patients reported a marked impairment of the posterior hubs, characterized by a disconnection of these brain regions, significantly correlated with Aβ levels in CSF.[91] All these findings suggest a non-linear trajectory of the FC changes across the continuum of the disease, with initial hyper-synchronization followed by a pronounced desynchronization between brain regions after dementia onset.

Alterations in the power spectrum in the AD continuum have been repeatedly associated to other classical markers of dementia due to AD. Increases in slow frequency power have been associated to medial temporal gray matter atrophy in areas such as hippocampus and entorhinal cortices.[60,106] And also with decreases in posterior cingulate cortex metabolism,[106] a classical sign of AD pathology. Lastly MEG alterations have been studied in relation to white matter disruption. FC disconnection in the posterior DMN as measured by MEG in a sample of MCI was associated to structural deterioration of the underlying white matter tracts.[107] Furthermore, fractional anisotropy decreases correlated with clustering alterations in a sample of MCI patients, which highlights the relationship of MEG findings with white matter alterations observed in AD.

In conclusion, the most stable findings in MEG literature such as the increase in slow frequency power, FC alterations along the continuum and network disruption, have been associated to classical markers of the disease, including cortical atrophy, protein accumulation and hypometabolism. These studies verified that electrophysiological alterations characterized by MEG are relevant to the condition, as they are associated with current standards for its diagnosis. However, although interesting and critical for the establishment of MEG signals as a candidate biomarker in AD, extensive efforts should be made in this particular area of the field. MEG seems to be sensitive to detect AD pathology reflected at the synaptic level as consistently settled by multiple studies. However, the exact relationship between the underlying pathophysiology leading to AD development, and the observed alterations remain elusive. Non-linear trajectories for this relationship across the different stages of the disease have been hypothesized, and the need for a deeper understanding of such relationships is urgent in order to establish MEG activity as a potential candidate marker of the disease.

4. Alzheimer's disease and MEG: Principal findings from task performance activity

The study of magnetic brain activity during resting state has been demonstrated to be far more stable than could be initially thought. This state is now known to be characterized by the activation of "default" resting state networks that can be systematically assessed and compared within and between subjects. However, employing concrete cognitive tasks, aimed at activating certain brain regions known to be involved in AD pathophysiology, could intuitively be expected as a more sensitive way to demonstrate abnormalities in such condition.[108] During recent years, the use of resting state for the identification of a sensitive biomarker for AD pathology is raising in popularity, mainly due to its great ecological validity and ease of use. However, multiple studies have addressed the changes occurring in AD patients, and in its prodromal stages, while performing different types of task. In the present section, a brief summary of the findings and most commonly employed paradigms in the field will be presented.

One of the simplest, yet highly informative, tasks employed in classical neurological assessments is the study of brain rhythms reactivity to eyes opening/closing. Different authors have applied this procedure to groups of AD patients. The more commonly reported outcome using this paradigm is a decrease in the reactivity to eyes opening. Consequently, the expected alpha desynchronization, largely associated to eyes opening, has been observed to be reduced in AD patients.[55] Moreover, this absence of reactivity has been proposed as potentially sensitive in distinguishing AD patients from controls, but more interestingly, in distinguishing severely demented patients from early dementia stages.[77] However, although supporting the impaired reactivity to eyes opening in AD patients, Kurimoto et al.[109] were unable to find differences in the MCI stage. Some contradictory results have been also reported, since a recent study found no differences in the event-related synchronization of the alpha rhythm after eyes closing.[110] However, most of the studies in the field seems to support an impairment of the mechanisms underlying alpha reactivation after eyes closing.

Irrelevant sensory inhibition has been also typically studied in the context of dementia. AD patients displayed larger amplitudes of M50 response compared to both a group of healthy young and older adults. Furthermore, while healthy aging as well as AD produced a slowing and a decrease in amplitude of the mismatch negativity response the slowing was more pronounced in the AD group.[111] It is thought that increased M50 responses are related to decreased inhibition of redundant inputs in mild AD. Interestingly, a recent study argued that auditory gating strongly depends on a specific region localized in the prefrontal cortex that might exert a modulatory effect on the anticipated generators in bilateral auditory cortices.[112] Remarkably, this study demonstrated that the medial prefrontal cortex modulation upon temporal regions was completely lacking in the group of AD patients while preserved in the age-matched controls. Authors interpreted that the deterioration of the cholinergic system associated to AD could underlie these deficits in medial prefrontal gating, thus giving rise to the already mentioned inhibitory impairment of redundant sensory information characteristic of the disease.

Different task settings making use of a variety of cognitive approaches have reported increases in activation in MCI population. MCI patients showed power increases, particularly over left hemisphere brain regions in a shallow encoding task,[113] or increased FC in delta, theta, beta and gamma bands in an internally directed cognitive task consisting in counting backward López et al.[114] Similarly, a different paradigm assessing working memory with a modified version of the well-known Sternberg's task showed power increases in MCI in the ventral pathway which the authors interpreted as a compensatory sign.[115]

Network structure seems to be also affected in the prodromal stages of the disease since the core hub regions in controls exhibited greater changes in MCI patients during a working memory task.[116] Moreover, network complexity was reduced in this same sample in the theta and alpha bands being the latter frequency band able to distinguish MCI and healthy controls with high accuracy.[117]

Several studies making use of different cognitive paradigms have consistently replicated that AD patients exhibit impaired synaptic activity during the performance of mental processes such as memory, sensory processing or executive functioning. This approach might shed light into the specific alterations driving the observed cognitive deterioration in the different domains known to be affected in this population.

5. Neurophysiological underpinnings of the relationship between lifestyle and Alzheimer's disease

In spite of the great effort made by numerous research groups around the world in the past few decades, pharmacological approaches barely offer any hope to patients and caregivers, and no cure has been found for AD. As a study from 2014 reveals, from the substances that were in clinical trials between 2002 and 2012, 99.6% failed to prove their efficacy.[118]

A potential reason for this lack of progress, as it has been previously explained, is closely related to the fact that the brains of Alzheimer's patients start to deteriorate even decades before the very first clinical symptoms arise.[119,120] This means that by the time the disease is diagnosed, the extension of the damage is so large that there is little substrate for these drugs to act. Attention has therefore shifted toward AD prevention and this is where modifiable lifestyle factors could come into play.

As people get older, specially individuals at higher risk for AD, it becomes more likely that their brains present AD pathology to some extent.

However, some people will never manifest any symptoms. In other words, they will die from other causes before the accumulated pathology overcomes the threshold that triggers cognitive impairment. Education, diet or physical activity are some of the lifestyle variables that have been suggested to have an impact on both: how well our brain resists pathology and how much pathological burden it can deal with before such damage begins to show. Hence lifestyle interventions could provide patients and people at risk with greater quality of life for longer periods. At best, such changes could even push AD onset beyond the individual's life expectancy.

MEG, due to its precision and non-invasiveness features, is an excellent tool to monitor the effect of the proposed lifestyle changes. Its high spatial and temporal resolution has already been effectively applied to the study of the neurophysiological hallmarks of AD. Being completely harmless, it is safe to repeat the procedure as many times as required to track the potential effect of an intervention in the long-term, especially now that such programs target younger and healthier populations.

One of the most widely studied lifestyle factors is physical activity. Physical exercise has been proposed to delay the onset of AD and improve cognitive functioning in both healthy older adults and patients. Cross-sectional and longitudinal studies employing varying neuroimaging techniques reveal that physical activity levels are associated with hippocampal volumes and functional connectivity within the default mode network, the frontal executive network and the fronto-parietal network.[121,122] A few studies have been conducted employing MEG and EEG to describe the effect that exercise has on the brain. For example, in an EEG study Choi et al.[123] describe how the alpha wave is reduced in dementia patients after an exercise rehabilitation program, together with an improvement in cognition and daily activities performance. A recent study found that a whole-body vibration intervention resulted in greater EEG activation in women with senile dementia.[124] Interestingly, in a sample of MCI seniors,[125] realized that precuneus/posterior cingulate cortex activity in the delta, theta and beta bands was increased after an intervention program, but only when this program involved combined physical and cognitive training. Once again, such increase correlated with cognitive functioning. In healthy older adults, MEG data showed that higher spontaneous physical activity was associated with higher individual alpha peak frequencies, but such association only remained significant among those who were not at increased genetic risk for AD.[126] On the contrary, physically active older adults at increased genetic risk for AD showed greater right temporal lobe activation than their sedentary counterparts in matching probe trials during a working memory task measured via MEG.[127] In that same study, physical activity was not associated with temporal activation among non-carriers of that genetic risk factor. Another study including healthy seniors found that incidental physical activity (the result of unstructured daily activities) was related to slower activity in frontal, central and temporal regions in the alpha, beta and theta bands.[128] At the network level, a combined physical and cognitive intervention in MCI patients produced a network reorganization in the beta band, mainly in the dorsal but also in the ventral streams, measured through EEG.[129] Additionally, physical fitness (in terms of cardiovascular fitness) has been associated with increased inter-modular connectivity in the beta band and reduced within-modules connectivity in an MEG study.[130]

Other lifestyle variables have been less widely considered from a neurophysiological approach. One such factor is diet and nutrition (an extensive review on this topic can be found in Ref. 131 More specifically, a high consumption of monounsaturated and polyunsaturated fat acids (mainly omega-3), together with a low consumption of saturated fat acids, has been generally associated with delayed cognitive decline in relationship with age. Similarly, antioxidants have drawn the attention of scientist around the world as a protective factor against aging and dementia. However, diet patterns and not specific nutrients are nowadays the focus of interventional studies within this field. In this sense, Mediterranean diet is linked to slower cognitive impairment and greater gray matter volumes among older adults. The utility of MEG for the assessment of nutritional interventions in AD was already proposed almost a decade ago.[132,133] Nevertheless, only a few studies have been carried out with this aim. Suvenaid is a medical food that has been designed to provide a series of nutrients that might help to counteract the deleterious effects of synapses loss in AD. The efficacy of this supplement was proven in mild AD patients through an EEG study in which they found that after a 24-week intervention, the active group presented better memory functioning, reduced peak frequency slowing and differential mean PLI trajectories in the delta band, compared to the control group.[134] In that same study, the active group was seemingly more resilient to network disruption in the beta band during the 24-week intervention.[135] Still, these results could not be replicated employing MEG, probably due to the smaller sample size.[136] In addition, higher adherence to Mediterranean diet has been reported to correlate with increased functional connectivity between the anterior and posterior default mode network in healthy older adults, as evidenced by a preliminary MEG study.[136a]

Besides physical activity and diet, several proxies of cognitive reserve have been analyzed as potential protective factors for the development of AD.

Cognitive reserve has been defined as the inter-individual differences in the ability of the brain to preserve an optimal functioning in spite of an undergoing neuropathological process.[137] Educational attainment, occupation, bilingualism or socioeconomic status are some of the variables that are thought to delay the onset of AD. This means that although they do not counteract the neuropathological process at the cellular and molecular level, they can modify the trajectory of its clinical and cognitive outcomes. For example, an MEG study revealed that ex-literate older women (having learnt to read after the age of 50 years) presented less brain functional asymmetry recorded with MEG than those with 10 years of regular schooling during the execution of a language recognition test, which was thought to reflect the greater effort that the former group encountered when performing the task, although they scored similarly at the behavioral level.[138] Also, in a sample of MCI patients, Nunes-Silva et al.[139] found that higher cognitive reserve (according to the Cognitive Reserve Index questionnaire) was associated with reduced neural inefficiency measured by EEG. In their study, compared to healthy controls, patients presented smaller P300 amplitudes in central-parietal and parietal electrodes, but cognitive reserve mediated this effect. In a different MEG study, healthy controls with higher cognitive reserve (expressed as a composite of educational attainment and occupation measures) presented lower functional connectivity in the theta, alpha, low beta and high beta bands during the Stenberg's Task than those scoring lower in cognitive reserve, which was interpreted as a sign that the later required a greater effort to perform the same task.[140] Finally, in a recent MEG study on the effect of bilingualism in the brain of healthy older adults, five parietal-occipital clusters were described presenting greater functional connectivity at rest in the theta, low beta and high beta bands in the bilingual sample.[141] Interestingly, posterior brain regions had already been said to present diminished brain functional connectivity at rest in preclinical AD and bilingualism appears to act in the opposite direction.

As it has just been exposed, electrophysiological techniques have a great potential in the study of lifestyle interventions to prevent AD. The studies presented above prove that these techniques are able to track changes in brain function exerted by different proxies of lifestyle and cognitive reserve. Nevertheless, MEG has been widely underused for this purpose. After the huge effort that has been put to describe AD progression employing MEG, it is now time to evaluate if and how such pathways could be altered by improving our daily habits.

References

1. Tripp JH. Physical concepts and mathematical models. In: Williamson SJ, Romani G-L, Kaufman L, Modena I, eds. *Biomagnetism: An Interdisciplinary Approach*. Boston, MA: Springer US; 1983:101–139. https://doi.org/10.1007/978-1-4757-1785-3_6.
2. Murakami S, Okada Y. Contributions of principal neocortical neurons to magnetoencephalography and electroencephalography signals. *J Physiol*. 2006;575(3):925–936. https://doi.org/10.1113/jphysiol.2006.105379.
3. Sarvas J. Basic mathematical and electromagnetic concepts of the biomagnetic inverse problem. *Phys Med Biol*. 1987;32(1):11–22. https://doi.org/10.1088/0031-9155/32/1/004.
4. Oppenheim AV, Willsky AS, Nawab H. *Signals and Systems*. 2nd ed. Prentice-Hall; 1997.
5. Berger H. Uber das Elektroenkephalogramm des Menschen. *Arch Psychiatr Nervenkr*. 1929;87:527–570.
6. Cohen D. Magnetoencephalography: evidence of magnetic fields produced by alpha-rhythm currents. *Science*. 1968;161(3843):784–786. https://doi.org/10.1126/science.161.3843.784[80-].
7. Mallat S. *A Wavelet Tour of Signal Processing*. Elsevier; 2009. https://doi.org/10.1016/B978-0-12-374370-1.X0001-8.
8. Shannon CE. A mathematical theory of communication. *Bell Syst Tech J*. 1948;27(4):623–656. https://doi.org/10.1002/j.1538-7305.1948.tb00917.x.
9. Lempel A, Ziv J. On the complexity of finite sequences. *IEEE Trans Inf Theory*. 1976;22:75–81. https://doi.org/10.1109/TIT.1976.1055501.
10. Richman JS, Moorman JR. Physiological time-series analysis using approximate entropy and sample entropy. *Am J Physiol Heart Circ Physiol*. 2000;278(6): H2039–H2049. https://doi.org/10.1152/ajpheart.2000.278.6.H2039.
11. Tsallis C. Possible generalization of Boltzmann-Gibbs statistics. *J Stat Phys*. 1988;52:479–487. https://doi.org/10.1007/BF01016429.
12. Renyi A. *Probability Theory*. Amsterdam: North-Holland Publishing Company; 1970.
13. Friston KJ. Functional and effective connectivity in neuroimaging: a synthesis. *Hum Brain Mapp*. 1994;2(1–2):56–78. https://doi.org/10.1002/hbm.460020107.
14. Bruns A, Eckhorn R, Jokeit H, Ebner A. Amplitude envelope correlation detects coupling among incoherent brain signals. *Neuroreport*. 2000;11(7):1509–1514. https://doi.org/10.1097/00001756-200005150-00029.
15. Rosenblum MG, Pikovsky AS, Kurths J. Phase synchronization of chaotic oscillators. *Phys Rev Lett*. 1996;76(11):1804–1807. https://doi.org/10.1103/PhysRevLett.76.1804.
16. Nunez PL, Srinivasan R, Westdorp AF, et al. EEG coherency I: statistics, reference electrode, volume conduction, Laplacians, cortical imaging, and interpretation at multiple scales. *Electroencephalogr Clin Neurophysiol*. 1997;103(5):499–515. https://doi.org/10.1016/S0013-4694(97)00066-7.
17. Lachaux J-P, Rodriguez E, Martinerie J, Varela FJ. Measuring phase synchrony in brain signals. *Hum Brain Mapp*. 1999;8:194–208. http://www.ma.utexas.edu/users/davis/reu/ch3/cwt/lachaux.pdf. Accessed 29 September 2017.
18. Stam CJ, Nolte G, Daffertshofer A. Phase lag index: assessment of functional connectivity from multi channel EEG and MEG with diminished bias from common sources. *Hum Brain Mapp*. 2007;28(11):1178–1193. https://doi.org/10.1002/hbm.20346.
19. Paluš M. Detecting nonlinearity in multivariate time series. *Phys Lett A*. 1996;213:138–147. https://doi.org/10.1016/0375-9601(96)00116-8.

20. Stam CJ, Van Dijk BW. Synchronization likelihood: an unbiased measure of generalized synchronization in multivariate data sets. *Physica D*. 2002;163(3–4):236–251. https://doi.org/10.1016/S0167-2789(01)00386-4.
21. Papo D. Why should cognitive neuroscientists study the brain's resting state? *Front Hum Neurosci*. 2013;7:45. https://doi.org/10.3389/fnhum.2013.00045.
22. Ohl FW, Scheich H, Freeman WJ. Change in pattern of ongoing cortical activity with auditory category learning. *Nature*. 2001;412(6848):733–736. https://doi.org/10.1038/35089076.
23. Douw L, Schoonheim MM, Landi D, et al. Cognition is related to resting-state small-world network topology: an magnetoencephalographic study. *Neuroscience*. 2011;175:169–177. https://doi.org/10.1016/j.neuroscience.2010.11.039.
24. van den Heuvel MP, Mandl RCW, Kahn RS, Hulshoff Pol HE. Functionally linked resting-state networks reflect the underlying structural connectivity architecture of the human brain. *Hum Brain Mapp*. 2009;30(10):3127–3141. https://doi.org/10.1002/hbm.20737.
25. Song M, Zhou Y, Li J, et al. Brain spontaneous functional connectivity and intelligence. *Neuroimage*. 2008;41(3):1168–1176. https://doi.org/10.1016/j.neuroimage.2008.02.036.
26. Damoiseaux JS, Rombouts SARB, Barkhof F, et al. Consistent resting-state networks across healthy subjects. *Proc Natl Acad Sci U S A*. 2006;103(37):13848–13853. https://doi.org/10.1073/pnas.0601417103.
27. Fox MD, Snyder AZ, Vincent JL, Corbetta M, Van Essen DC, Raichle ME. The human brain is intrinsically organized into dynamic, anticorrelated functional networks. *Proc Natl Acad Sci U S A*. 2005;102(27):9673–9678. https://doi.org/10.1073/pnas.0504136102.
28. Biswal B, Zerrin Yetkin F, Haughton VM, Hyde JS. Functional connectivity in the motor cortex of resting human brain using echo-planar mri. *Magn Reson Med*. 1995;34(4):537–541. https://doi.org/10.1002/mrm.1910340409.
29. Cordes D, Haughton VM, Arfanakis K, et al. Mapping functionally related regions of brain with functional connectivity MR imaging. *AJNR Am J Neuroradiol*. 2000;21(9):1636–1644. http://www.ajnr.org/content/21/9/1636.short. Accessed 12 September 2017.
30. Zhao S, Li C, Uono S, Yoshimura S, Toichi M. Human cortical activity evoked by contextual processing in attentional orienting. *Sci Rep*. 2017;7(1):2962. https://doi.org/10.1038/s41598-017-03104-1.
31. Mason MF, Norton MI, Van Horn JD, Wegner DM, Grafton ST, Macrae CN. Wandering minds: the default network and stimulus-independent thought. *Science*. 2007;315(5810):393–395. [80-]. http://science.sciencemag.org/content/315/5810/393. Accessed 12 September 2017.
32. Buckner RL, Andrews-Hanna JR, Schacter DL. The brain's default network: anatomy, function, and relevance to disease. *Ann N Y Acad Sci*. 2008;1124(1):1–38. https://doi.org/10.1196/annals.1440.011.
33. Anokhin AP, Birbaumer N, Lutzenberger W, Nikolaev A, Vogel F. Age increases brain complexity. *Electroencephalogr Clin Neurophysiol*. 1996;99(1):63–68. https://doi.org/10.1016/0921-884X(96)95573-3.
34. Van Cappellen Van Walsum AM, YAL P, Berendse HW, et al. A neural complexity measure applied to MEG data in Alzheimer's disease. *Clin Neurophysiol*. 2003;114:1034–1040. https://doi.org/10.1016/S1388-2457(03)00072-5.
35. Abatzoglou I, Anninos P, Adamopoulos A, Koukourakis M. Nonlinear analysis of brain magnetoencephalographic activity in Alzheimer disease patients. *Acta Neurol Belg* 2007;107(2):34–39. http://www.ncbi.nlm.nih.gov/pubmed/17710838. Accessed 2 April 2019.

36. Gomez C, Hornero R, Abasolo D, Fernandez A, Escudero J. Analysis of MEG recordings from Alzheimer's disease patients with sample and multiscale entropies. In: *2007 29th Annual International Conference of the IEEE Engineering in Medicine and Biology Society*, IEEE; 2007:6183–6186. vol 2007. https://doi.org/10.1109/IEMBS.2007.4353767.
37. Hornero R, Abásolo D, Escudero J, Gómez C. Nonlinear analysis of electroencephalogram and magnetoencephalogram recordings in patients with Alzheimer's disease. *Philos Trans R Soc A Math Phys Eng Sci*. 2009;367:317–336. https://doi.org/10.1098/rsta.2008.0197.
38. Gomez C, Abasolo D, Poza J, Fernandez A, Hornero R. MEG analysis in Alzheimer's disease computing approximate entropy for different frequency bands. *Conf Proc IEEE Eng Med Biol Soc*. 2010;2010:2379–2382. https://doi.org/10.1109/IEMBS.2010.5627236.
39. Poza J, Hornero R, Escudero J, Fernández A, Sánchez CI. Regional analysis of spontaneous MEG rhythms in patients with Alzheimer's disease using spectral entropies. *Ann Biomed Eng*. 2008;36(1):141–152. https://doi.org/10.1007/s10439-007-9402-y.
40. Poza J, Hornero R, Escudero J, Fernandez A, Gomez C. Analysis of spontaneous MEG activity in Alzheimer's disease using time-frequency parameters. *Conf Proc IEEE Eng Med Biol Soc*. 2008;2008:5712–5715. https://doi.org/10.1109/IEMBS.2008.4650511.
41. Gomez C, Hornero R, Abasolo D, Fernandez A, Poza J. Study of the MEG background activity in Alzheimer's disease patients with scaling analysis methods. In: *2009 Annual International Conference of the IEEE Engineering in Medicine and Biology Society*, IEEE; 2009:3485–3488. https://doi.org/10.1109/IEMBS.2009.5334569.
42. Gómez C, Mediavilla Á, Hornero R, Abásolo D, Fernández A. Use of the Higuchi's fractal dimension for the analysis of MEG recordings from Alzheimer's disease patients. *Med Eng Phys*. 2009;31(3):306–313. https://doi.org/10.1016/j.medengphy.2008.06.010.
43. Gómez C, Hornero R, Abásolo D, Fernández A, López M. Complexity analysis of the magnetoencephalogram background activity in Alzheimer's disease patients. *Med Eng Phys*. 2006;28(9):851–859. https://doi.org/10.1016/j.medengphy.2006.01.003.
44. Fernández A, Hornero R, Gómez C, et al. Complexity analysis of spontaneous brain activity in Alzheimer disease and mild cognitive impairment. *Alzheimer Dis Assoc Disord*. 2010;24(2):182–189. https://doi.org/10.1097/WAD.0b013e3181c727f7.
45. Bruña R, Poza J, Gómez C, García M, Fernández A, Hornero R. Analysis of spontaneous MEG activity in mild cognitive impairment and Alzheimer's disease using spectral entropies and statistical complexity measures. *J Neural Eng*. 2012;9(3):036007. https://doi.org/10.1088/1741-2560/9/3/036007.
46. Poza J, Gómez C, García M, Corralejo R, Fernández A, Hornero R. Analysis of neural dynamics in mild cognitive impairment and Alzheimer's disease using wavelet turbulence. *J Neural Eng*. 2014;11(2):026010. https://doi.org/10.1088/1741-2560/11/2/026010.
47. Poza J, Gomez C, Garcia M, Bachiller A, Fernandez A, Hornero R. Analysis of spontaneous MEG activity in mild cognitive impairment and Alzheimer's disease using Jensen's divergence. In: *2014 36th Annual International Conference of the IEEE Engineering in Medicine and Biology Society* IEEE; 2014:1501–1504. https://doi.org/10.1109/EMBC.2014.6943886.
48. Montez T, Poil S-S, Jones BF, et al. Altered temporal correlations in parietal alpha and prefrontal theta oscillations in early-stage Alzheimer disease. *Proc Natl Acad Sci U S A*. 2009;106(5):1614–1619. https://doi.org/10.1073/pnas.0811699106.
49. Engels MMA, Hillebrand A, van der Flier WM, Stam CJ, Scheltens P, van Straaten ECW. Slowing of hippocampal activity correlates with cognitive decline in

early onset Alzheimer's disease. An MEG study with virtual electrodes. *Front Hum Neurosci.* 2016;10:238. https://doi.org/10.3389/fnhum.2016.00238.
50. Escudero J, Hornero R, Poza J, Abásolo D, Fernández A. Assessment of classification improvement in patients with Alzheimer's disease based on magnetoencephalogram blind source separation. *Artif Intell Med.* 2008;43(1):75–85. https://doi.org/10.1016/j.artmed.2008.01.001.
51. Escudero J, Hornero R, Abásolo D, Fernández A. Blind source separation to enhance spectral and non-linear features of magnetoencephalogram recordings. Application to Alzheimer's disease. *Med Eng Phys.* 2009;31(7):872–879. https://doi.org/10.1016/j.medengphy.2009.04.003.
52. Poza J, Hornero R, Abásolo D, Fernández A, García M. Extraction of spectral based measures from MEG background oscillations in Alzheimer's disease. *Med Eng Phys.* 2007;29(10):1073–1083. https://doi.org/10.1016/j.medengphy.2006.11.006.
53. Buzsáki G. *Rhythms of the Brain.* Oxford University Press Inc; 2006.
54. Hindriks R, van Putten MJAM. Thalamo-cortical mechanisms underlying changes in amplitude and frequency of human alpha oscillations. *Neuroimage.* 2013;70:150–163. https://doi.org/10.1016/j.neuroimage.2012.12.018.
55. Berendse H, Verbunt JP, Scheltens P, van Dijk B, Jonkman E. Magnetoencephalographic analysis of cortical activity in Alzheimer's disease: a pilot study. *Clin Neurophysiol.* 2000;111(4):604–612. https://doi.org/10.1016/S1388-2457(99)00309-0.
56. Osipova D, Ahveninen J, Jensen O, Ylikoski A, Pekkonen E. Altered generation of spontaneous oscillations in Alzheimer's disease. *Neuroimage.* 2005;27(4):835–841. https://doi.org/10.1016/J.NEUROIMAGE.2005.05.011.
57. Poza J, Hornero R, Abásolo D, Fernández A, Mayo A. Evaluation of spectral ratio measures from spontaneous MEG recordings in patients with Alzheimer's disease. *Comput Methods Programs Biomed.* 2008;90(2):137–147. https://doi.org/10.1016/j.cmpb.2007.12.004.
58. De Haan W, Stam CJ, Jones BF, Zuiderwijk IM, Van Dijk BW, Scheltens P. Resting-state oscillatory brain dynamics in Alzheimer disease. *J Clin Neurophysiol.* 2008;25(4):187–193. https://doi.org/10.1097/WNP.0b013e31817da184.
59. Fernández A, Maestú F, Amo C, et al. Focal temporoparietal slow activity in Alzheimer's disease revealed by magnetoencephalography. *Biol Psychiatry.* 2002;52(7):764–770. http://www.ncbi.nlm.nih.gov/pubmed/12372668. Accessed 10 April 2018.
60. Fernández A, Arrazola J, Maestú F, et al. Correlations of hippocampal atrophy and focal low-frequency magnetic activity in Alzheimer disease: volumetric MR imaging-magnetoencephalographic study. *AJNR Am J Neuroradiol.* 2003;24(3):481–487. http://www.ncbi.nlm.nih.gov/pubmed/12637301. Accessed 23 April 2018.
61. Fernández A, Hornero R, Mayo A, Poza J, Maestú F, Ortiz Alonso T. Quantitative magnetoencephalography of spontaneous brain activity in alzheimer disease: an exhaustive frequency analysis. *Alzheimer Dis Assoc Disord.* 2006;20(3):153–159. https://doi.org/10.1097/00002093-200607000-00006.
62. López ME, Cuesta P, Garcés P, et al. MEG spectral analysis in subtypes of mild cognitive impairment. *Age (Dordr).* 2014;36(3):9624. https://doi.org/10.1007/s11357-014-9624-5.
63. Besga A, Ortiz L, Fernández A, et al. Structural and functional patterns in healthy aging, mild cognitive impairment, and Alzheimer disease. *Alzheimer Dis Assoc Disord.* 2010;24(1):1–10. https://doi.org/10.1097/WAD.0b013e3181aba730.
64. Fernández A, Turrero A, Zuluaga P, et al. MEG Delta mapping along the healthy aging-Alzheimer's disease continuum: diagnostic implications. *J Alzheimers Dis.* 2013;35(3):495–507. https://doi.org/10.3233/JAD-121912.
65. Garcés P, Vicente R, Wibral M, et al. Brain-wide slowing of spontaneous alpha rhythms in mild cognitive impairment. *Front Aging Neurosci.* 2013;5(Dec):1–7. https://doi.org/10.3389/fnagi.2013.00100.

66. Fernández A, Hornero R, Mayo A, Poza J, Gil-Gregorio P, Ortiz T. MEG spectral profile in Alzheimer's disease and mild cognitive impairment. *Clin Neurophysiol.* 2006;117(2):306–314. https://doi.org/10.1016/j.clinph.2005.10.017.
67. López-Sanz D, Bruña R, Garcés P, et al. Alpha band disruption in the AD-continuum starts in the subjective cognitive decline stage: a MEG study. *Sci Rep.* 2016;6:37685. https://doi.org/10.1038/srep37685.
68. Cuesta P, Barabash A, Aurtenetxe S, et al. Source analysis of spontaneous magnetoencephalograpic activity in healthy aging and mild cognitive impairment: influence of apolipoprotein E polymorphism. *J Alzheimers Dis.* 2015;43(1):259–273. https://doi.org/10.3233/JAD-140633.
69. Cole MW, Yarkoni T, Repovs G, Anticevic A, Braver TS. Global connectivity of prefrontal cortex predicts cognitive control and intelligence. *J Neurosci.* 2012;32(26): 8988–8999. https://doi.org/10.1523/JNEUROSCI.0536-12.2012.
70. Bokde ALW, Ewers M, Hampel H. Assessing neuronal networks: understanding Alzheimer's disease. *Prog Neurobiol.* 2009;89(2):125–133. https://doi.org/10.1016/j.pneurobio.2009.06.004.
71. Stam CJ, Jones BF, Manshanden I, et al. Magnetoencephalographic evaluation of resting-state functional connectivity in Alzheimer's disease. *Neuroimage.* 2006;32(3): 1335–1344. https://doi.org/10.1016/j.neuroimage.2006.05.033.
72. Alonso JF, Poza J, Mañanas MA, Romero S, Fernández A, Hornero R. MEG connectivity analysis in patients with Alzheimer's disease using cross mutual information and spectral coherence. *Ann Biomed Eng.* 2011;39(1):524–536. https://doi.org/10.1007/s10439-010-0155-7.
73. Juan-Cruz C, Gomez C, Poza J, Fernandez A, Hornero R. Analysis of magnetoencephalography signals from Alzheimer's disease patients using granger causality. *Conf Proc IEEE Eng Med Biol Soc.* 2016;2016:724–727. https://doi.org/10.1109/EMBC.2016.7590804.
74. Stam CJ, Van Cappellen van Walsum AM, YAL P, et al. Generalized synchronization of MEG recordings in Alzheimer's disease: evidence for involvement of the gamma band. *J Clin Neurophysiol.* 2002;19(6):562–574. http://www.ncbi.nlm.nih.gov/pubmed/12488788. Accessed 30 October 2018.
75. Gomez C, Martinez-Zarzuela M, Poza J, Diaz-Pernas FJ, Fernandez A, Hornero R. Synchrony analysis of spontaneous MEG activity in Alzheimer's disease patients. In: *2012 Annual International Conference of the IEEE Engineering in Medicine and Biology Society,* vol 2012:IEEE; 2012:6188–6191. https://doi.org/10.1109/EMBC.2012.6347407.
76. Stam CJ, de Haan W, Daffertshofer A, et al. Graph theoretical analysis of magnetoencephalographic functional connectivity in Alzheimer's disease. *Brain.* 2008;132(1): 213–224. https://doi.org/10.1093/brain/awn262.
77. Franciotti R, Iacono D, Della PS, et al. Cortical rhythms reactivity in AD, LBD and normal subjects: a quantitative MEG study. *Neurobiol Aging.* 2006;27(8):1100–1109. https://doi.org/10.1016/J.NEUROBIOLAGING.2005.05.027.
78. Koelewijn L, Bompas A, Tales A, et al. Alzheimer's disease disrupts alpha and beta-band resting-state oscillatory network connectivity. *Clin Neurophysiol.* 2017;128(11): 2347–2357. https://doi.org/10.1016/j.clinph.2017.04.018.
79. Engels MMA, Yu M, Stam CJ, et al. Directional information flow in patients with Alzheimer's disease. A source-space resting-state MEG study. *NeuroImage Clin.* 2017;15:673–681. https://doi.org/10.1016/j.nicl.2017.06.025.
80. Andrews-Hanna JR, Reidler JS, Huang C, Buckner RL. Evidence for the default network's role in spontaneous cognition. *J Neurophysiol.* 2010;104(1):322–335. https://doi.org/10.1152/jn.00830.2009.
81. Mormino EC, Smiljic A, Hayenga AO, et al. Relationships between Beta-amyloid and functional connectivity in different components of the default mode network in aging. *Cereb Cortex.* 2011;21(10):2399–2407. https://doi.org/10.1093/cercor/bhr025.

82. Gomez C, Stam CJ, Hornero R, Fernandez A, Maestu F. Disturbed Beta band functional connectivity in patients with mild cognitive impairment: an MEG study. *IEEE Trans Biomed Eng*. 2009;56(6):1683–1690. https://doi.org/10.1109/TBME.2009.2018454.
83. Gómez C, Juan-Cruz C, Poza J, et al. Alterations of effective connectivity patterns in mild cognitive impairment: an MEG study. *J Alzheimers Dis*. 2017;65:1–12. https://doi.org/10.3233/JAD-170475. Preprint(Preprint.
84. Cuesta P, Garcés P, Castellanos NP, et al. Influence of the APOE ε4 allele and mild cognitive impairment diagnosis in the disruption of the MEG resting state functional connectivity in sources space. *J Alzheimers Dis*. 2015;44(2):493–505. https://doi.org/10.3233/JAD-141872.
85. Ranasinghe KG, Hinkley LB, Beagle AJ, et al. Regional functional connectivity predicts distinct cognitive impairments in Alzheimer's disease spectrum. *NeuroImage Clin*. 2014;5:385–395. https://doi.org/10.1016/j.nicl.2014.07.006.
86. López-Sanz D, Bruña R, Garcés P, et al. Functional connectivity disruption in subjective cognitive decline and mild cognitive impairment: a common pattern of alterations. *Front Aging Neurosci*. 2017;9:109. https://doi.org/10.3389/fnagi.2017.00109.
87. Jones D, Knopman DS, Gunter JL, et al. Cascading network failure across the Alzheimer's disease spectrum. *Brain*. 2015;139(2):547–562. https://doi.org/10.1093/brain/awv338.
88. Busche MA, Eichhoff G, Adelsberger H, et al. Clusters of hyperactive neurons near amyloid plaques in a mouse model of Alzheimer's disease. *Science*. 2008;321(5896):1686–1689. https://doi.org/10.1126/science.1162844.
89. Lopez ME, Bruna R, Aurtenetxe S, et al. Alpha-band hypersynchronization in progressive mild cognitive impairment: a magnetoencephalography study. *J Neurosci*. 2014;34(44):14551–14559. https://doi.org/10.1523/JNEUROSCI.0964-14.2014.
90. Sporns O. Contributions and challenges for network models in cognitive neuroscience. *Nat Neurosci*. 2014;17(5):652–660. https://doi.org/10.1038/nn.3690.
91. Yu M, Engels MMA, Hillebrand A, et al. Selective impairment of hippocampus and posterior hub areas in Alzheimer's disease: an MEG-based multiplex network study. *Brain*. 2017;140(5):1466–1485. https://doi.org/10.1093/brain/awx050.
92. de Haan W, van der Flier WM, Wang H, Van Mieghem PFA, Scheltens P, Stam CJ. Disruption of functional brain networks in Alzheimer's disease: what can we learn from graph spectral analysis of resting-state Magnetoencephalography? *Brain Connect*. 2012;2(2):45–55. https://doi.org/10.1089/brain.2011.0043.
93. Sami S, Williams N, Hughes LE, et al. Neurophysiological signatures of Alzheimer's disease and frontotemporal lobar degeneration: pathology versus phenotype. *Brain*. 2018;141(8):2500–2510. https://doi.org/10.1093/brain/awy180.
94. de Haan W, van der Flier WM, Koene T, Smits LL, Scheltens P, Stam CJ. Disrupted modular brain dynamics reflect cognitive dysfunction in Alzheimer's disease. *Neuroimage*. 2012;59(4):3085–3093. https://doi.org/10.1016/j.neuroimage.2011.11.055.
95. López-Sanz D, Garcés P, Álvarez B, Delgado-Losada ML, López-Higes R, Maestú F. Network disruption in the preclinical stages of Alzheimer's disease: from subjective cognitive decline to mild cognitive impairment. *Int J Neural Syst*. 2017;1750041. https://doi.org/10.1142/S0129065717500411.
96. Bajo R, Pusil S, López ME, et al. Scopolamine effects on functional brain connectivity: a pharmacological model of Alzheimer's disease. *Sci Rep*. 2015;5(1):9748. https://doi.org/10.1038/srep09748.
97. López ME, Engels MMA, van Straaten ECW, et al. MEG Beamformer-based reconstructions of functional networks in mild cognitive impairment. *Front Aging Neurosci*. 2017;9:107. https://doi.org/10.3389/fnagi.2017.00107.

98. Amezquita-Sanchez JP, Adeli A, Adeli H. A new methodology for automated diagnosis of mild cognitive impairment (MCI) using magnetoencephalography (MEG). *Behav Brain Res*. 2016;305:174–180. https://doi.org/10.1016/j.bbr.2016.02.035.
99. Pineda-Pardo JA, Bruña R, Woolrich M, et al. Guiding functional connectivity estimation by structural connectivity in MEG: an application to discrimination of conditions of mild cognitive impairment. *Neuroimage*. 2014;101:765–777. https://doi.org/10.1016/J.NEUROIMAGE.2014.08.002.
99a. Maestú F, Peña J-M, Garcés P, et al. A multicenter study of the early detection of synaptic dysfunction in mild cognitive impairment using magnetoencephalography-derived functional connectivity. *NeuroImage Clin*. 2015;9:103–109. https://doi.org/10.1016/j.nicl.2015.07.011.
100. Fernández A, Turrero A, Zuluaga P, et al. Magnetoencephalographic parietal delta dipole density in mild cognitive impairment. *Arch Neurol*. 2006;63:427–430.
101. López ME, Turrero A, Cuesta P, et al. Searching for primary predictors of conversion from mild cognitive impairment to Alzheimer's disease: a multivariate follow-up study. *J Alzheimers Dis*. 2016;52(1):133–143. https://doi.org/10.3233/JAD-151034.
102. Bajo R, Castellanos NP, Cuesta P, et al. Differential patterns of connectivity in progressive mild cognitive impairment. *Brain Connect*. 2012;2(1):21–24. https://doi.org/10.1089/brain.2011.0069.
103. Busche MA, Konnerth A. Impairments of neural circuit function in Alzheimer's disease. *Philos Trans R Soc Lond B Biol Sci*. 2016;371(1700):20150429. https://doi.org/10.1098/rstb.2015.0429.
104. Nakamura A, Cuesta P, Kato T, et al. Early functional network alterations in asymptomatic elders at risk for Alzheimer's disease. *Sci Rep*. 2017;7(1):1–11. https://doi.org/10.1038/s41598-017-06876-8.
105. Canuet L, Pusil S, López ME, et al. Network disruption and cerebrospinal fluid amyloid-beta and phospho-tau levels in mild cognitive impairment. *J Neurosci*. 2015;35(28):10325–10330. https://doi.org/10.1523/JNEUROSCI.0704-15.2015.
106. Nakamura A, Cuesta P, Fernández A, et al. Electromagnetic signatures of the preclinical and prodromal stages of Alzheimer's disease. *Brain*. 2018;141:1470–1485. https://doi.org/10.1093/brain/awy044.
107. Pineda-Pardo JA, Garcés P, López ME, et al. White matter damage disorganizes brain functional networks in amnestic mild cognitive impairment. *Brain Connect*. 2014;4(5):312–322. https://doi.org/10.1089/brain.2013.0208.
108. Morcom AM, Fletcher PC. Does the brain have a baseline? Why we should be resisting a rest. *Neuroimage*. 2007;37(4):1073–1082. https://doi.org/10.1016/J.NEUROIMAGE.2006.09.013.
109. Kurimoto R, Ishii R, Canuet L, et al. Event-related synchronization of alpha activity in early Alzheimer's disease and mild cognitive impairment: an MEG study combining beamformer and group comparison. *Neurosci Lett*. 2008;443(2):86–89. https://doi.org/10.1016/J.NEULET.2008.07.015.
110. Hata M, Kurimoto R, Kazui H, et al. Alpha event-related synchronization after eye closing differs in Alzheimer's disease and dementia with Lewy bodies: a magnetoencephalography study. *Psychogeriatrics*. 2018;18(3):202–208. https://doi.org/10.1111/psyg.12313.
111. Cheng C-H, Wang P-N, Hsu W-Y, Lin Y-Y. Inadequate inhibition of redundant auditory inputs in Alzheimer's disease: an MEG study. *Biol Psychol*. 2012;89(2):365–373. https://doi.org/10.1016/j.biopsycho.2011.11.010.
112. Josef Golubic S, Aine CJ, Stephen JM, Adair JC, Knoefel JE, Supek S. MEG biomarker of Alzheimer's disease: absence of a prefrontal generator during auditory sensory gating. *Hum Brain Mapp*. 2017;38(10):5180–5194. https://doi.org/10.1002/hbm.23724.

113. Püregger E, Walla P, Deecke L, Dal-Bianco P. Magnetoencephalographic—features related to mild cognitive impairment. *Neuroimage*. 2003;20(4):2235–2244. https://doi.org/10.1016/J.NEUROIMAGE.2003.08.015.
114. López ME, Garcés P, Cuesta P, et al. Synchronization during an internally directed cognitive state in healthy aging and mild cognitive impairment: a MEG study. *Age (Dordr)*. 2014;36(3):9643. https://doi.org/10.1007/s11357-014-9643-2.
115. Maestu F, Baykova E, Ruiz JM, et al. Increased biomagnetic activity in healthy elderly with subjective memory complaints. *Clin Neurophysiol*. 2011;122(3):499–505. https://doi.org/10.1016/j.clinph.2010.08.004.
116. Navas A, Papo D, Boccaletti S, et al. Functional hubs in mild cognitive impairment. *Int J Bifurcation Chaos*. 2015;25(03):1550034. https://doi.org/10.1142/S0218127415500340.
117. Ahmadlou M, Adeli A, Bajo R, Adeli H. Complexity of functional connectivity networks in mild cognitive impairment subjects during a working memory task. *Clin Neurophysiol*. 2014;125(4):694–702. https://doi.org/10.1016/j.clinph.2013.08.033.
118. Cummings JL, Morstorf T, Zhong K. Alzheimer' s disease drug-development pipeline: few candidates, frequent failures. *Alzheimers Res Ther*. 2014;6:1–7.
119. Braak H, Braak E. Neuropathological stageing of Alzheimer-related changes. *Acta Neuropathol*. 1991;82(4):239–259. https://doi.org/10.1007/BF00308809.
120. Jack CR, Wiste HJ, Weigand SD, et al. Age-specific population frequencies of cerebral β-amyloidosis and neurodegeneration among people with normal cognitive function aged 50-89 years: a cross-sectional study. *Lancet Neurol*. 2014;13(10):997–1005. https://doi.org/10.1016/S1474-4422(14)70194-2.
121. Huang P, Fang R, Li BY, Di CS. Exercise-related changes of networks in aging and mild cognitive impairment brain. *Front Aging Neurosci*. 2016;8(Mar). https://doi.org/10.3389/fnagi.2016.00047.
122. Erickson KI, Leckie RL, Weinstein AM. Physical activity, fitness, and gray matter volume. *Neurobiol Aging*. 2014;35(Suppl. 2):S20–S28. https://doi.org/10.1016/j.neurobiolaging.2014.03.034.
123. Choi M, Kim J, Yi E. Development and validation of exercise rehabilitation program for cognitive function and activity of daily living improvement in mild dementia elderly. *J Exerc Rehabil*. 2018;14(2):207–212.
124. Kim K, Lee H. The effects of whole body vibration exercise intervention on electroencephalogram activation and cognitive function in women with senile dementia. *J Exerc Rehabil*. 2018;14(4):586–591.
125. Styliadis C, Kartsidis P, Paraskevopoulos E, Ioannides AA, Bamidis PD. Neuroplastic effects of combined computerized physical and cognitive training in elderly individuals at risk for dementia: an eloreta controlled study on resting states. *Neural Plast*. 2015;2015:172192[ii].
126. de Frutos-Lucas J, López-Sanz D, Zuluaga P, et al. Physical activity effects on the individual alpha peak frequency of older adults with and without genetic risk factors for Alzheimer's disease: a MEG study. *Clin Neurophysiol*. 2018;129(9):1981–1989. https://doi.org/10.1016/J.CLINPH.2018.06.026.
127. Deeny SP, Poeppel D, Zimmerman JB, et al. Exercise, APOE, and working memory: MEG and behavioral evidence for benefit of exercise in e4 carriers Sean. *Biol Psychol*. 2008;78(2):179–187. https://doi.org/10.1016/j.biopsycho.2008.02.007. Exercise. 2008.
128. Sanchez-Lopez J, Silva-Pereyra J, Fernández T, et al. High levels of incidental physical activity are positively associated with cognition and EEG activity in aging. *PLoS One*. 2018;13(1):1–18. https://doi.org/10.1371/journal.pone.0191561.

129. Klados MA, Styliadis C, Frantzidis CA. Beta-band functional connectivity is reorganized in mild cognitive impairment after combined computerized physical and cognitive training. *Front Neurosci.* 2016;10(February):1–12. https://doi.org/10.3389/fnins.2016.00055.
130. Douw L, Nieboer D, van Dijk BW, Stam CJ, Twisk JWR. A healthy brain in a healthy body: brain network correlates of physical and mental fitness. *PLoS One.* 2014;9(2). e88202. https://doi.org/10.1371/journal.pone.0088202.
131. Solfrizzi V, Custodero C, Lozupone M, Imbimbo BP. Relationships of dietary patterns, foods, and micro- and macronutrients with Alzheimer' s disease and late-life cognitive disorders: a systematic review. *J Alzheimers Dis.* 2017;59:815–849. https://doi.org/10.3233/JAD-170248.
132. De WMC, Kamphuis PJGH, Sijben JWC, Scheltens P. Utility of imaging for nutritional intervention studies in Alzheimer's disease. *Eur J Pharmacol.* 2011;668:S59–S69. https://doi.org/10.1016/j.ejphar.2011.07.011.
133. Sizonenko V, Babiloni C, De BEA, et al. Brain imaging and human nutrition: which measures to use in intervention studies? *Br J Nutr.* 2013;110:S1–30.
134. Scheltens P, Twisk JWR, Blesa R, Scarpini E, Von ACAF. Efficacy of souvenaid in mild Alzheimer's disease: results from a randomized, controlled trial. *J Alzheimers Dis.* 2014;31:225–236. https://doi.org/10.3233/JAD-2012-121189.
135. De WH, Stam CJ, Lansbergen MM, Wieggers RL. The Effect of souvenaid on functional brain network organisation in patients with mild Alzheimer's disease: a randomised controlled study. *PLoS One.* 2014;9(1):1–11. https://doi.org/10.1371/journal.pone.0086558.
136. van Straaten ECW, de Waal H, Lansbergen MM, et al. Magnetoencephalography for the detection of intervention effects of a specific nutrient combination in patients with mild Alzheimer's disease: results from an exploratory double-blind, randomized, controlled study. *Front Neurol.* 2016;7:161. https://doi.org/10.3389/fneur.2016.00161.
136a. de Frutos J, Toraño FR, Cuesta P, et al. Eat healthier, keep your brain well connected!. *Alzheimers Dement.* 2019.
137. Steffener J, Stern Y. Exploring the neural basis of cognitive reserve in aging. *Biochim Biophys Acta (BBA)—Mol Basis Dis.* 2012;1822(3):467–473. https://doi.org/10.1016/j.bbadis.2011.09.012.
138. Nunes-Silva VM, Castro-caldas A, Del Rio D, Maestu F, Ortiz T. The ex-illiterate brain the critical period, cognitive reserve and HAROLD model. *Dement Neuropsychol* 2009;3(3):222–227.
139. Gu L, Chen J, Gao L, et al. Cognitive reserve modulates attention processes in healthy elderly and amnestic mild cognitive impairment: an event-related potential study. *Clin Neurophysiol.* 2018;129:198–207. https://doi.org/10.1016/j.clinph.2017.10.030.
140. Lopez ME, Aurtenetxe S, Pereda E, et al. Cognitive reserve is associated with the functional organization of the brain in healthy aging: a MEG study. *Front Aging Neurosci.* 2014;6(Jun):1–9. https://doi.org/10.3389/fnagi.2014.00125.
141. de Frutos-, Lucas J, López-Sanz D, Cuesta P, et al. Enhancement of posterior brain functional networks in bilingual older adults. *Biling Lang Cogn.* 2019;1–14. https://doi.org/10.1017/S1366728919000178.

AD molecular: PET amyloid imaging across the Alzheimer's disease spectrum: From disease mechanisms to prevention

Pierre-François Meyer[a,b], Melissa McSweeney[a,b], Julie Gonneaud[a,b], Sylvia Villeneuve[a,b,*]
[a]Centre for Studies on the Prevention of Alzheimer's Disease, Douglas Mental Health University Institute, Montréal, Canada
[b]McGill University, Montréal, Canada
*Corresponding author: e-mail address: sylvia.villeneuve@mcgill.ca

Contents

Abstract

The advent of amyloid-beta (Aβ) positron emission tomography (PET) imaging has transformed the field of Alzheimer's disease (AD) by enabling the quantification of cortical Aβ accumulation and propagation in vivo. This revolutionary tool has made

Progress in Molecular Biology and Translational Science, Volume 165
ISSN 1877-1173
https://doi.org/10.1016/bs.pmbts.2019.05.001

it possible to measure direct associations between Aβ and other AD biomarkers, to identify factors that influence Aβ accumulation and to redefine entry criteria into clinical trials as well as measure drug target engagement. This chapter summarizes the main findings on the associations of Aβ with other biomarkers of disease progression across the AD spectrum. It discusses investigations of the timing at which Aβ pathology starts to accumulate, demonstrates the clinical utility of Aβ PET imaging and discusses some ethical implications. Finally, it presents genetic and potentially modifiable lifestyle factors that might influence Aβ accumulation and therefore be targets for AD prevention.

1. Introduction

Alzheimer's disease (AD) is the leading cause of dementia, a group of brain disorders associated with cognitive decline severe enough to interfere with activities of daily life. While AD is generally known for its devastating effects on one's ability to form and retain memories, the clinical expression of the disease is preceded by a silent phase notably characterized by the pathological accumulation of misfolded amyloid-beta (Aβ) and hyperphosphorylated tau proteins as well as neuronal damage and synaptic failure. The abnormal accumulation of pathogenic proteins was first described well over a century ago. However, it wasn't until the early 1990s that it was suggested that Aβ had a causal role in AD development.[1] This "Amyloid hypothesis" was supported by the observation that single copy mutations in the amyloid precursor protein (APP) and the gamma-secretase pathway (Presenilins) acted as dominant traits provoking an early onset AD phenotype with ~100% penetrance. However, it was challenged by neuropathological findings suggesting that ~30% of cognitively normal older adults have Aβ or tau at autopsy.[2,3] Up until recently, these findings could not be investigated in vivo owing to the absence of a reliable marker of brain Aβ pathology. The introduction of Aβ imaging in the early 2000s has revolutionized AD clinical research. It is now possible to test the amyloid hypothesis using positron emission tomography (PET) imaging of Aβ and longitudinal evaluations of cognitive performance. Over the last two decades the findings resulting from in vivo imaging studies have forced the field to rethink its conception of AD as a clinical entity, helped in differential diagnosis and influenced the clinical management of individuals with neurodegenerative diseases.

In this chapter we discuss the strengths and limitations of molecular imaging as a method for in vivo quantification of Aβ pathology. We summarize the findings of associations between Aβ and other AD biomarkers in the clinical and pre-clinical phases of the disease and how these may have bearing on clinical research and practice. Finally, we discuss the factors that might postpone Aβ appearance, slow down its rate of accumulation and modify its association with other neurodegenerative markers that may be targeted for disease prevention.

2. Molecular imaging of in vivo Aβ pathology

2.1 Amyloid biology

Despite the initial description of senile plaques dating back to Alois Alzheimer's presentation of findings from the first AD patient in 1907, their Aβ component was only elucidated eight decades later.[4,5] The Aβ peptide results from the sequential cleavage of the larger APP protein by β- and γ-secretase. The APP protein is first cleaved by β-secretase and consequently releases a large soluble APP-β fragment. The remaining membrane-bound portion of the protein is then cleaved by γ-secretase. This process is imprecise and thus yields Aβ peptides of varying length of which the most abundant are those ending at amino acid 40 ($A\beta_{40}$) and 42 ($A\beta_{42}$).[6] These Aβ monomers, particularly the $A\beta_{42}$ fragments, can spontaneously aggregate into oligomers, eventually leading to the formation of fibrillar Aβ that makes up senile plaques.[7–9] While senile plaques constitute the main pathological hallmark of the disease, they may only indicate a final, inert, stage of Aβ accumulation since the soluble oligomeric forms are likely those exerting the strongest neurotoxic effects.[10–13]

2.2 Development of Aβ PET tracers and association with other Aβ biomarkers

In the early 2000s, a number of potential probes were investigated for the imaging of Aβ plaques.[14] Derivatives of Congo red and Thioflavin T—two molecules used for staining Aβ plaques in pathology studies—were developed. However, while these probes did bind strongly to Aβ plaques, they failed to effectively cross the blood brain barrier (BBB) thereby resulting in low binding in vivo. Additional efforts to develop antibody-derived tracers were similarly disappointing owing to limited BBB permeability.[14] The first successful Aβ imaging attempt in humans was achieved using the ^{18}F-FDDNP tracer.[15] Unfortunately, this agent lacked specificity and also bound to neurofibrillary tangles, thus limiting its utility. The AD field had to wait until 2004 for the first clinical study using Pittsburgh Compound B (PiB), a tracer with high affinity and high selectivity for fibrillary Aβ in plaques.[16,17] To date, this remains the most successful and widely used tracer in research settings, often considered as the "gold standard" for imaging of Aβ pathology.[18] However, one of its limitations lies in the use of the ^{11}Carbon radioactive isotope. The relatively short half-life of ^{11}C restricts the clinical utility of this tracer to larger hospitals with a nearby cyclotron. Additionally, and like most Aβ tracers, it shows high affinity mainly for fibrillar Aβ rather than its oligomeric forms.[18–21]

The need for an Aβ tracer that can easily be used in a clinical setting has led to the development and the evaluation of a large number PET or SPECT radiotracers[14] among which three have been approved for clinical use ([^{18}F] Florbetaben, [^{18}F]Florbetapir and [^{18}F]Flutemetamol, Fig. 1).[22] These tracers have the significant advantage over PiB that they are bound to an ^{18}F radioactive isotope which has a longer half-life (~109 mn for ^{18}F vs ~20 mn for ^{11}C). All tracers showed good accuracy for the discrimination of AD patients vs healthy older adults in Phase II and III trials.[23–27] Among newer tracers not yet approved by medical agencies, NAV4694 is a promising ^{18}F tracer which has several appealing features, notably very low unspecific binding in white matter.[28]

Aβ PET tracers were not the first attempts to develop markers to measure Aβ pathology *ante-mortem*. In fact, cerebrospinal fluid measures of Aβ preceded the validation of PiB by more than a decade with the $A\beta_{42}$ fragment being the most sensitive to changes in Aβ accrual.[29,30] Contrary to cortical PET binding, which tends to increase with increasing Aβ plaque deposition in the brain, CSF $A\beta_{42}$ levels decrease progressively with this process. It has been hypothesized that this reduction may result from sequestration of Aβ in plaques at the brain level and is therefore a good proxy of the overall brain Aβ burden.[31]

Aβ-PET presents the distinct advantage over CSF that it provides information on the *quantity* and *extent* of cortical Aβ deposition. Thus, it allows for the study of local interactions between brain Aβ deposition and other disease-related biomarker changes (e.g., brain atrophy, glucose metabolism). While CSF and PET measures of global Aβ deposition are highly correlated, it is possible that they provide distinct information regarding the stage of disease progression.[32–34] Increasing evidence suggests that individuals who are

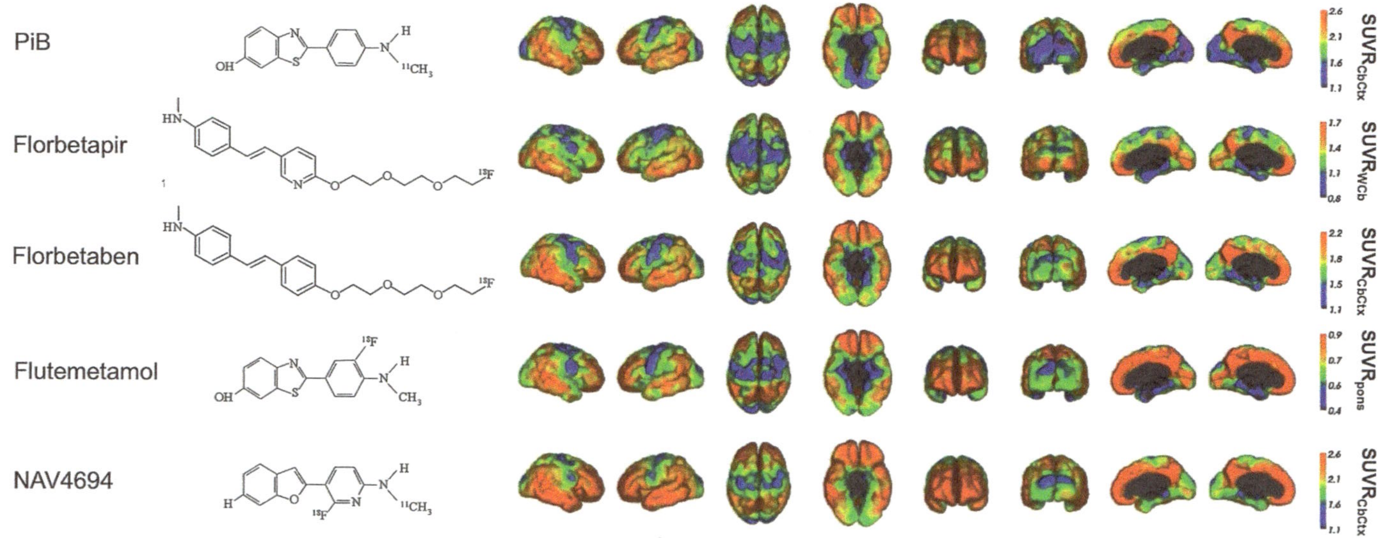

Fig. 1 PET tracers for Aβ imaging. The structure of the first Aβ-specific ligand (PiB), second generation ^{18}F tracers approved for clinical use (Florbetapir, Florbetaben, Flutemetamol) and a third-generation tracer (NAV4694) used for the study of Aβ deposition in humans (left). The stereotypical distribution of Aβ deposition in AD patients is represented for each tracer (right) with high tracer retention in prefrontal, temporal and parietal cortices as well as the cingulate and precuneal areas. *Reproduced with permission from Villemagne VL. Amyloid imaging: past, present and future perspectives.* Ageing Res Rev. *2016;30:95–106.*

positive on CSF Aβ alone (CSF+/PET−) are possibly at an intermediate stage between those who are both negative (CSF−/PET−) or positive (CSF+/PET+) on CSF and PET, although it remains a matter of debate.[35] A study by Palmqvist and colleagues,[34] showed that CSF+/PET− individuals had accruing Aβ PET tracer binding in a restricted set of brain regions thought to be among the first to show Aβ plaques pathology (Fig. 2A). In contrast, CSF+/PET+ persons show widespread accumulation of brain Aβ PET tracer uptake (Fig. 2B). These results suggest that abnormality of CSF Aβ markers may be an earlier event than PET abnormality and thereby allow for earlier detection of AD pathology. Nonetheless, both CSF and PET show similar accuracy for AD diagnosis.[36,37]

More recently, blood tests for the detection of brain Aβ pathology have been developed.[38] The validation of these novel assays uses Aβ PET as a reference to determine Aβ-positivity. These blood tests show good accuracy for the distinction of Aβ PET-negative from -positive individuals and have the advantage of being cheaper and less invasive than PET or CSF collection. Thus, CSF and blood tests for the detection of Aβ pathology may have great potential for future clinical use. Aβ-PET will, however, remain the gold standard to investigate disease mechanisms as it provides information regarding the topography of Aβ lesions.

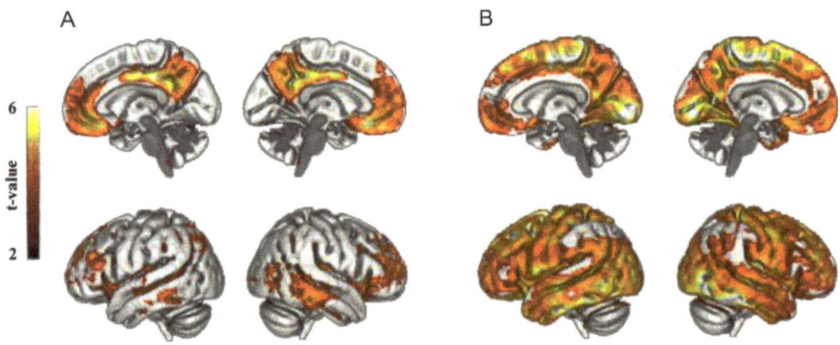

Fig. 2 Aβ accumulation based on CSF/PET status. (A) Individuals showing abnormal CSF Aβ values but subthreshold Aβ PET uptake show increasing PET uptake in medial prefrontal and precuneal regions compared to CSF−/PET− individuals. (B) In contrast CSF+/PET+ participants have a widespread increase in PET binding when compared to CSF−/PET− persons. This suggests that CSF abnormality alone may be an early indicator of the AD pathological process. *Adapted from Palmqvist S, Scholl M, Strandberg O, et al. Earliest accumulation of beta-amyloid occurs within the default-mode network and concurrently affects brain connectivity.* Nat Commun. *2017;8:1214. Springer Nature Limited, CC By 4.0.*

2.3 Aβ-PET quantification and challenges

There is no universal way to quantify cortical Aβ uptake or a widely accepted threshold value to define Aβ positivity, which may be one possible explanation of varying Aβ positivity prevalence across studies, even when they consider older adults in similar age-ranges.[39] It is important to keep in mind that absolute Aβ-PET binding values are extremely difficult to compare across studies because there is substantial differences in PET acquisition parameters, image preprocessing and methods used to define Aβ positivity. While there exists good correlation between different PET Aβ ligands,[40,41] differences in tracer proprieties (e.g., degree and localization of unspecific binding, dynamic range of the tracers) also influence estimations of brain Aβ burden.

Among acquisition parameters that may vary from one study to another, PET scans can be either dynamic or static thereby providing qualitatively different information. While the former corresponds to a full acquisition starting from the time of injection until the tracer activity stabilizes, the latter corresponds to the acquisition when the tracer's activity has reached an equilibrium state. Dynamic scans capture pharmacokinetic and pharmacodynamic properties of the tracer in brain tissue and therefore allow for a truly quantitative measure of tracer uptake.[42] Owing to time constraints, however, static scans are often preferred in clinical and some research settings. Static scan outputs are usually expressed in standardized uptake value ratio (SUVR). SUVRs quantify the amount of tracer uptake using each subject's unspecific binding as its own reference. To do this, a brain region recognized from neuropathology studies as being mainly pathology-free (e.g., cerebellum, white matter) and of having similar biological properties to the regions of interest (e.g., gray matter) is defined as a reference region. The tracer uptake in each brain region of interest is then divided by the uptake in the reference region thereby obtaining an SUVR value theoretically being equal or greater to one. The choice of the optimal reference region for image scaling, however, is subject to debate. The cerebellum has been widely used as reference in cross-sectional studies because it is relatively spared by Aβ accumulation, at least until the most severe stages of amyloidosis. The whole cerebellum is sometimes preferred when using ^{18}F tracers, since they are more prone to high unspecific binding in the white matter.[43] For longitudinal Aβ quantification there seems to be a consensus toward using a white matter reference region rather than a cerebellar one.[44–47] This particular difference may complicate the comparison of results from cross-sectional and longitudinal studies.

Other pre-processing parameters such as the availability of structural magnetic resonance imaging (MRI) scans, correction for partial volume effects, or even the software used for data pre-processing are going to influence PET measures.[47–49] In an effort to improve the comparison of Aβ measurements across pre-processing methods, a working group has proposed a way of standardizing Aβ PET values.[50] The Centiloid project proposes standardized acquisition and pre-processing methods to quantify Aβ PET binding as well as a scaling procedure for PiB and other tracer outputs to a standardized scale. The standardized (or centiloid) scale ranges from 0 to 100, where zero corresponds to values obtained from a group of young individuals and 100 corresponds to values obtained from a group of AD patients. The Centiloid transformation has initially been developed for PiB-PET and has recently been applied to other tracers.[51–54] Using this method, the comparability of multi-site PET scans is improved and may yield common definitions for Aβ positivity.[55,56,56a] Despite these efforts, however, specific tracer kinetics such as the tracer dynamic range may still influence standardized scales and one should always be aware of these.[57]

PET therefore provides a reliable tool to measure Aβ plaque pathology in vivo, but still suffers from a lack of methodological consensus. Despite these difficulties, various Aβ PET tracers and methodologies have been used for imaging of plaque pathology in humans providing important findings which have improved our knowledge of pathological events.

3. Aβ imaging in individuals with cognitive impairment

Dementia refers to a group of syndromes characterized by loss of cognitive functions and inability to perform activities of daily living. In individuals diagnosed with AD dementia, cross-sectional associations between Aβ PET burden, symptoms, and other imaging markers are generally weak. The lack of association of Aβ with other AD biomarkers may owe to the fact that, by the time an individual progresses to the early stages of dementia, the level of cortical Aβ has saturated the neocortex.[58] Nonetheless, in vivo Aβ imaging has enabled the field to recognize the longitudinal dynamics of Aβ accumulation and its associations (or dissociations) with the symptoms and various other biomarkers of AD. Furthermore, Aβ PET used in conjunction with other disease indicators may provide more information about AD progression than Aβ alone. Thus, implementing Aβ PET scan in clinical settings may be an informative tool for patient diagnosis and disease management.

3.1 Associations between Aβ PET and markers of AD progression

Historically, the in vivo diagnosis of AD was based on the nature and the severity of cognitive impairments.[59,60] AD has been classically defined as an amnestic syndrome of the hippocampal type with a signature impairment in episodic memory retrieval that is not rectified by cueing paradigms.[61] In AD, impaired episodic memory can be manifested across a variety of cognitive faculties (free recall, recognition, paired-associate learning) and sensory modalities (auditory, visual, olfactory).[62] AD also affects domains of language ability and semantic knowledge, including object naming, category fluency, semantic categorization, as well as working memory, attention, and visuospatial abilities.[59]

Post mortem assays have shown that the quantity of neurofibrillary tangles, but not Aβ plaques, is correlated with cognitive decline and dementia severity in individuals with AD.[63] Subsequent longitudinal Aβ PET studies have confirmed these findings by demonstrating that the progression of clinical symptoms in late-onset sporadic AD are not coupled with the rate of brain Aβ accumulation,[64] but rather with the rate of neurodegeneration as measured by structural MRI.[65] Other studies have described only weak associations between Aβ PET deposition and late-onset AD dementia symptom severity.[58,66,67] Despite these weak associations, Aβ-positivity is nonetheless associated with an increased risk of conversion to dementia in patients with mild cognitive impairment (MCI).[68] These results might be explained by an observed slowing in rates of Aβ accumulation once symptoms are consistent with a diagnosis of AD dementia.[69] Aβ accumulation is hypothesized to follow a sigmoidal trajectory in which a period of rapid accumulation precedes a quantitative plateau in Aβ deposition, during which dementia symptoms become more severe.[58] This plateau phase may therefore make cross-sectional and longitudinal associations of Aβ with cognition virtually undetectable at this late stage of the disease.

The associations between brain atrophy and Aβ load in AD have also been inconsistent across studies, ranging from absent to showing robust correlations.[70] Assuming that Aβ pathology is no longer a dynamic marker of the disease in the dementia stage,[58] it is possible that observed associations between Aβ and altered brain structural integrity are mediated by associated downstream factors rather than directly caused by Aβ itself. Supporting this idea, the pattern of brain atrophy measured with MRI, or of glucose metabolism measured with Fludeoxyglucose-(FDG) PET does not spatially overlap with brain Aβ deposition.[71,72] Fig. 3 shows AD-related atrophy (orange),

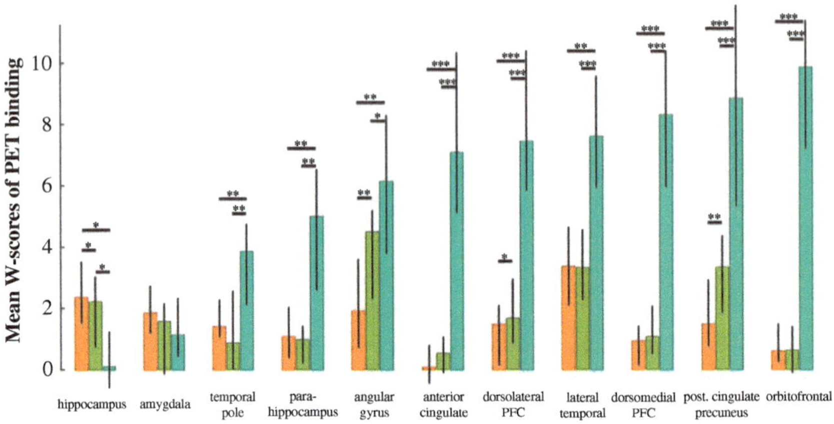

Fig. 3 Aβ deposition pattern does not overlap with atrophy and glucose hypometabolism. The figure shows local degrees of atrophy (orange), hypometabolism (green), and Aβ deposition (blue) expressed as mean W-scores. Increased Aβ deposition does not associate with increased atrophy and glucose hypometabolism. The orbitofrontal cortex for instance has high Aβ deposition but relatively low atrophy and hypometabolism. In contrast, the hippocampus has low Aβ load but nonetheless elevated hypometabolism and atrophy. *$P < 0.05$; **$P < 0.005$; ***$P < 0.0001$. Post. cingulate, Posterior cingulate cortex; PFC, prefrontal cortex. *Reprint from La Joie R, Perrotin A, Barre L, et al. Region-specific hierarchy between atrophy, hypometabolism, and beta-amyloid (Abeta) load in Alzheimer's disease dementia.* J Neurosci. *2012;32:16265–16273, with permission.*

hypometabolism (green), and Aβ deposition (blue) in seven brain regions.[72] One striking feature is that very low Aβ-PET binding is found in the hippocampus and the amygdala, two regions with predominant atrophy and hypometabolism. In contrast, very high Aβ-PET binding is found in the orbitofrontal region which almost no atrophy and mainly preserved brain glucose metabolism.

While the patterns of brain atrophy and neuronal failure do not fully overlap with that of brain Aβ deposition, Aβ seems to accumulate in brain regions which are highly functionally connected.[73] Many of the brain regions affected by Aβ in the disease process are members of the default mode network (DMN) and include regions such as the precuneus, medial orbitofrontal cortex, posterior cingulate cortex, anterior cingulate cortex, and angular gyrus.[34,73–75] The DMN comprises a network of functionally connected brain regions that are co-activated during wakeful rest and inhibited during attention-related cognitive tasks. Additionally, episodic memory retrieval is associated with increased activity of regions in the posterior DMN.[76] While the DMN has been the center of interest of most Aβ-related studies, the Aβ-associated abnormal functional connectivity is not restricted to the DMN[77] and some have therefore argued that the regions expressing strongest Aβ deposition are in fact hubs where multiple networks converge.[70] These hubs could also be the convergence point of multiple pathologies, including tau, the other pathological hallmark of AD.[78,79]

Tau PET imaging is among the most meaningful complementary biomarkers of Aβ PET. Tau radiotracers became available for clinical research in 2012,[80] 8 years after PiB. Much like what is observed for brain atrophy and neuronal death, the topographical pattern and progression of tau deposition is distinct from that of Aβ deposition.[81–84] However, the presence of Aβ seems necessary for tau to spread from the medial temporal lobe to the rest of the neocortex.[85–87] The presence of both proteins is also needed for a definite diagnosis of AD dementia.[88] Thus, using Aβ PET concurrently with tau PET imaging and structural MRI may facilitate the reliable distinction of AD from other diseases in clinical settings.[88] Current tau PET tracers also seem to reliably differentiate topographical patterns between different tauopathies;[89,90] however, they are not presently approved for clinical use.[91]

3.2 Clinical utility and challenges of Aβ PET imaging

3.2.1 Misdiagnosis and the importance of Aβ imaging in clinical settings

The advent of Aβ imaging has created several interesting debates regarding the diagnosis criteria of AD. For instance, the diagnosis of probable AD dementia was initially based on the clinical expression of the disease and pathologically confirmed at autopsy. Autopsy studies have suggested that there is a mismatch between clinical and neuropathological diagnoses for up to 30% of individuals diagnosed with dementia.[92] These misdiagnoses include individuals diagnosed as having AD dementia when they have no AD pathology (false positive) and individuals incorrectly diagnosed with another dementia when they have AD pathology (false negative).[93] It is now possible to support the diagnosis of probable AD with Aβ imaging in vivo. PET studies have shown that ~15% of AD-diagnosed individuals have low Aβ tracer binding.[94] These patients usually have a clinical progression of cognitive and behavioral symptoms that are consistent with a neurodegenerative disease, but most of them do not have the AD-typical neurodegenerative profiles suggesting that, at least in some cases, there was a misdiagnosis.[95,96] These misdiagnoses may have important implications for patients. For instance, a NACC-UDS database study showed that 18%–67% of individuals misdiagnosed with AD during their lifetime were prescribed unnecessary or inappropriate medications.[97] Furthermore, vascular dementia (VD) or Parkinson's disease (PD) patients who were initially misdiagnosed with AD incurred significantly greater annual medical costs (paid to providers through Medicare) than patients who were initially diagnosed correctly.[98] The application of Aβ PET imaging in clinical settings may therefore allow practitioners to make more informed diagnostic decisions since the absence of significant Aβ pathology generally rules out a diagnosis of AD dementia.

3.2.2 Appropriate clinical use

As mentioned above, three Aβ-PET tracers have been FDA-approved for clinical use since 2012: Florbetaben, Florbetapir and Flutemetamol. Florbetapir is currently the most widely used Aβ radiotracer.[99] In 2013, the Amyloid Imaging Taskforce (AIT) developed guidelines to aid clinical translation of Aβ scans.[100] To summarize, according to the AIT's appropriate use criteria, Aβ imaging is appropriate in cases where (a) the patient expresses both subjective and objective cognitive impairment, (b) previous clinical assessment indicated AD as a possible but inconclusive diagnosis, and (c) knowledge of the patient's Aβ status would increase diagnostic certainty and impact patient management. The AIT's examples of appropriate and inappropriate cases are presented in Table 1.

3.2.3 Clinical decision making

Clinical use of Aβ imaging appears to have a significant impact on clinical decision making.[101,102] Across 13 studies on the clinical translation of Aβ imaging conducted between 2012 and 2017, imaging results led to a change in diagnosis in 29% of the cases, an increase in diagnostic confidence in 60% of cases, and prompted an overall change in patient management in 64% of the cases.[101] Additionally, Aβ PET results have a greater impact on practitioners' diagnostic confidence in individuals under 65 years of age, as asymptomatic Aβ deposition is common in older patients and may confound diagnostic accuracy.[100,103]

These results, along with the previously observed rates of misdiagnosis, are in line with the highly anticipated results from the Imaging Dementia-Evidence for Amyloid Scanning (IDEAS) study which evaluates the clinical utility and possible benefits for patient health of Aβ scan use in the clinic.[104] IDEAS includes a sample of 11,409 Medicare recipients with MCI or dementia who meet the AIT's appropriate use criteria for Aβ imaging, making it the largest study assessing Aβ imaging in private clinical settings. In this

Table 1 Appropriate use criteria established by the amyloid imaging taskforce.

Appropriate circumstances for Aβ imaging
Patients with persistent or progressive unexplained mild cognitive impairment
Patients satisfying core clinical criteria for possible Alzheimer's disease because of unclear clinical presentation either atypical clinical course or etiologically mixed presentation
Patients with progressive dementia and atypically early age of onset (<65 years or less in age)
Inappropriate circumstances for Aβ imaging
Patients with core clinical criteria for probable Alzheimer's disease with typical age of onset
To determine dementia severity
Solely based on a positive family history of dementia or presence of *APOE* ε4
Patients with a cognitive complaint that is unconfirmed on clinical examination
In lieu of genotyping for suspected autosomal mutation carriers
In asymptomatic individuals
Non-medical usage (e.g., legal, insurance coverage, or employment screening)

Recapitulated from Johnson KA, Minoshima S, Bohnen NI, et al. Appropriate use criteria for amyloid PET: a report of the Amyloid Imaging Task Force, the Society of Nuclear Medicine and Molecular Imaging, and the Alzheimer's Association. *Alzheimers Dement.* 2013;9:e-1-16.

cohort, post-PET scan changes in patient management were made for 60.2% of patients with MCI and 63.5% of patients with dementia. Physicians reported that 85.2% of changes in patient management were substantially influenced by Aβ PET results. The most frequent change in patient management was the use of AD drugs. This change is crucial because drugs such as cholinesterase inhibitors and memantine can be prescribed to reduce AD symptoms.[105] However, they are also associated with worse outcomes in frontotemporal dementia and other dementias that don't involve Aβ deposition.[106,107]

The IDEAS team is continuing to collect data on long-term patient outcomes following Aβ-PET-supported diagnoses, which may have implications for whether or not Aβ PET will be eligible for insurance coverage in the future.[108] However, IDEAS does not plan to evaluate the clinical utility of the use of Aβ PET in conjunction with other biomarkers or AD-specific clinical assessments. At present, there is limited research on this topic.[103] Future investigations will be needed to evaluate whether diagnostic accuracy, confidence, as well as clinical outcomes, could be improved by use of multi-modal assessments. Furthermore, to our knowledge, none of the discussed studies assessing change in patient diagnosis and clinical decision making following Aβ-PET imaging have reported their results stratified by race. Given that recent investigations suggest that (1) African Americans may have a higher incidence of diagnosis of AD dementia compared with non-Hispanic Caucasian Americans[109,110] and (2) African Americans with a dementia diagnosis are more likely to have AD pathology with another mixed neuropathology,[111,112] it would be clinically important to evaluate whether Aβ imaging increases diagnostic confidence proportionately across racial groups. Thus, future investigations should also consider evaluating whether Aβ PET imaging is equally useful and accessible across demographics.

3.2.4 Disclosure of Aβ-PET results for cognitively impaired individuals

There is an ongoing debate about the ethics of Aβ-positivity status disclosure in clinical and research settings. As there is currently no established disease-modifying treatment for AD, valid concerns about patients' psychological well-being in the face of a positive PET scan disclosure have been raised. While this topic is further elaborated upon in a subsequent chapter, it is worth giving a brief overview of Aβ-PET status disclosure.

Ongoing projects, such as the Risk Evaluation and Education of Alzheimer's Disease—the Study of Communicating Amyloid Neuroimaging (REVEAL-SCAN),[113] are studying the behavioral and psychological impacts of revealing Aβ PET scan results in an ethnically diverse cohort of older adults.[113] One such investigation suggested that individuals with MCI who were informed of their Aβ status following a PET scan focused more on the benefits rather than the negative implications of the disclosure.[114] Participants appreciated the opportunity to plan ahead and most individuals also understood that the scan result was not definitive. However, a previous study found that amnestic MCI patients reported both positive and negative outcomes at 2 weeks and 6 months following the disclosure of their Aβ status.[115] All participants with Aβ-positive results reported advantages of the disclosure at both time points, the majority of them also reported perceived disadvantages at 2 weeks, and half of them continued to report disadvantages 6 months later. Here, perceived advantages included enjoying life more, planning, and medication management and perceived disadvantages included increased worry about symptom onset and fear of stigmatization from relatives.

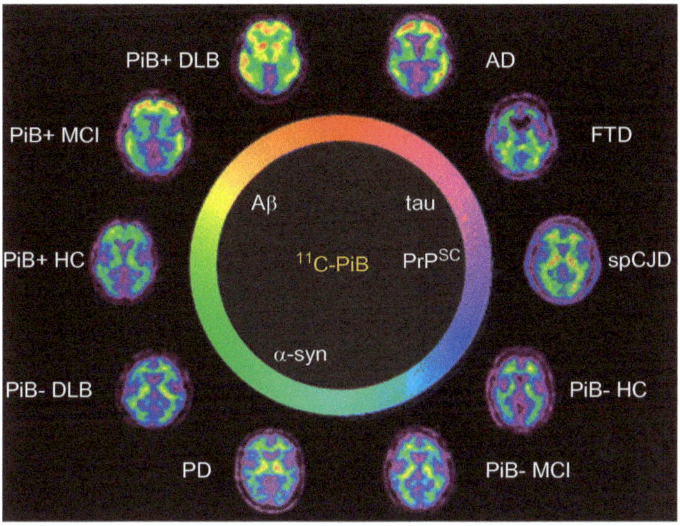

Fig. 4 Aβ positivity is not restricted to AD dementia. Represented are PiB uptake patterns from several Aβ and non-Aβ dementias. Represented clockwise from bottom right corner, Aβ-negative a 73 year-old Aβ-negative control (PiB-HC) subject (MMSE 30), an 83-year-old Aβ-negative subject with mild cognitive impairment (PiB-MCI) (MMSE 28), a 61-year-old Aβ-negative Parkinson's disease (PD) patient (MMSE 27), a 69 year-old Aβ-negative patient with dementia with Lewy Bodies (DLB; MMSE 24), a 77-year-old Aβ-positive healthy control (PiB+HC) subject (MMSE 28), an 82-year-old Aβ-positive subject with mild cognitive impairment (PiB+MCI) (MMSE 28), a 78-year-old Aβ-positive DLB patient (PiB+DLB) (MMSE 19), a 76-year-old Alzheimer's disease (AD) patient (Aβ-positive by definition; MMSE 21), a 59-year-old patient with frontotemporal dementia (FTD; MMSE 20), and a 59-year-old Aβ-positive patient with confirmed sporadic Creutzfeldt-Jakob disease (spCJD). *Reproduced with permission from Fodero-Tavoletti MT, Cappai R, McLean CA, et al. Amyloid imaging in Alzheimer's disease and other dementias.* Brain Imaging Behav. *2009;3:246–261.*

Importantly, a positive Aβ-PET scan alone is not sufficient to rule out other diagnoses or comorbidities. Patients with other age-related neurodegenerative disorders, such as dementia with Lewy bodies (DLB),[116] cerebral Aβ angiopathy,[117] and corticobasal degeneration,[94] also often present with positive Aβ scans (Fig. 4). Nevertheless, as suggested above an Aβ scan could be particularly useful in the evaluation of mildly affected, clinically atypical or early age-at-onset patients.

3.2.5 A β-PET value in disease-modifying interventions

To date, no Aβ-targeting therapy has been approved and the AD field is plagued by large scale failures of these drugs in clinical trials.[118–121] Recent

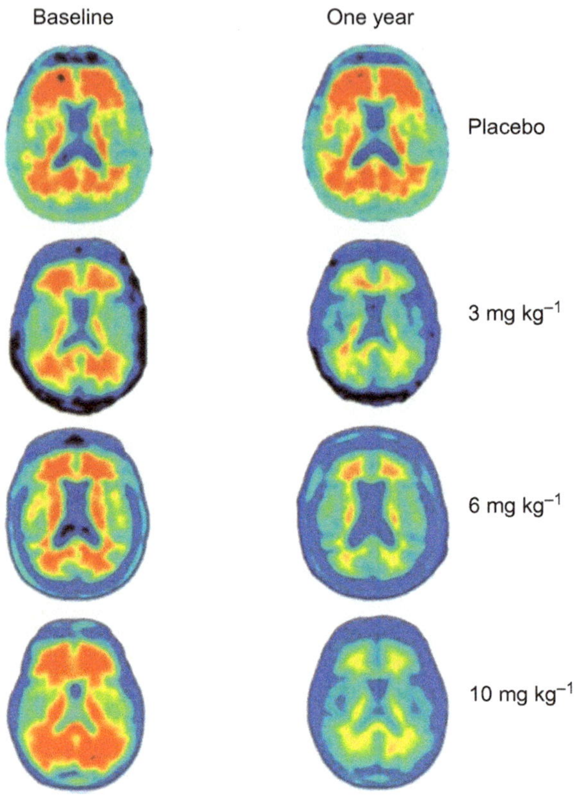

Fig. 5 Use of Aβ PET for evaluation of target engagement. Recent clinical trials have used repeated PET assessments to evaluate whether the drug interacts and clears Aβ deposits. In the phase I trial of the antibody aducanumab, there was an appreciable dose-dependent reduction in Aβ-PET tracer retention after 1 year of treatment. *Reproduced with permission from Sevigny J, Chiao P, Bussiere T, et al. The antibody aducanumab reduces Abeta plaques in Alzheimer's disease.* Nature. *2016;537:50–56.*

compounds have successfully reduced the amount of Aβ in the brain but without slowing of cognitive decline.[122] For example, Fig. 5 shows that the antibody aducanumab successfully decreased fibrillary Aβ deposits as measured with PET in a dose-dependent manner over a 1-year period. However, two phase III trials of this drug were recently terminated following a futility analysis suggesting it would not meet its primary endpoint of reducing cognitive decline in patients with early AD.[123] Many explanations have been brought forward for such treatments' failure to curb cognitive decline, including biochemical properties of the antibodies evaluated and the timing of intervention. However, one important caveat may have been the lack of systematic evaluation of presence of Aβ pathology in enrolled patients. For example, the failed Bapineuzumab trials enrolled >30% of individuals who were PET Aβ-negative thereby treating patients for a pathology they did not have.[124] One important concern that has arisen from these failed Aβ-targeting trials is whether or not Aβ is a valid target for AD treatment. Also, because Aβ positivity in cognitively normal individuals may be a strong indicator of future AD dementia risk, interventions are now moving even earlier, enrolling patients in the pre-symptomatic stages of disease.

3.3 Conclusions

By the time clinical dementia and AD neuropathology are present, one is already in an advanced stage of AD. At this point in the disease, network dysfunction and brain atrophy are already well underway, and the associations between Aβ and dementia symptoms are difficult to detect. Consequently, this clinical disease stage is probably too late for Aβ-targeting therapy. However, the knowledge of Aβ status appears to be valuable in clinical settings, as it has a measurable influence on physicians' diagnostic and treatment decisions.

4. Aβ imaging in cognitively healthy older adults

It is estimated that Aβ starts to accumulate more than two decades before the onset of dementia.[125–127] One of the most important advances provided by Aβ PET imaging is the visualization of Aβ deposition in the brain of cognitively healthy individuals along with the possibility of longitudinal monitoring of other AD biomarkers, including cognitive status. Over the past ~15 years, several studies have tested the association of Aβ with other markers of brain disease in cognitively normal individuals, both in cross-sectional and longitudinal designs, to better understand early disease mechanisms and identify new prevention strategies.

4.1 Associations of Aβ with other markers of AD progression

Much like what is observed in cognitively impaired patients, the association between brain Aβ deposition and objective measures of cognitive performance in cognitively normal older adults is weakly apparent in cross-sectional studies and, may instead be more related to increasing subjective cognitive decline (i.e., a report of perceived, not objectively measurable, worsening of cognitive abilities).[128] Meta-analyses suggest that, if anything, there are very subtle associations between increasing Aβ load and worse objective cognitive performance in cognitively normal older adults.[129,130] Associations between baseline Aβ burden and longitudinal cognitive change are less ambiguous and several studies suggest that higher Aβ-PET binding is accompanied by faster subsequent cognitive decline.[69,131–135] This may be particularly evident in people with subjective cognitive decline.[136] However, accelerated cognitive decline in cognitively normal older adults may only become evident after 3–4 years of follow-up,[137] stressing the need for longitudinal research designs that span over decades. People with a positive Aβ-PET scan have also been shown to have an increased risk of converting to dementia.[68,138] Finally, the association of Aβ with decline seems potentiated by evidence of an ongoing neurodegenerative process.[139–142] In this context, neurodegeneration could indicate the presence of a comorbid pathology or a more advanced disease stage.

Similar to what is observed in individuals with dementia, the cross-sectional findings assessing the link between Aβ load and brain damage are inconclusive. For instance, some cross-sectional evaluations suggest that atrophy in a pattern reminiscent of AD is observable in Aβ PET-positive older adults[143,144] while others report no observable atrophy in Aβ positive vs negative persons.[145,146] As is the case for cognitive findings, longitudinal studies provide more consistent evidence of associations between Aβ status and longitudinal brain atrophy.[147–151] This association is also seemingly more apparent for individuals who already have some degree of brain atrophy.[152] In essence, longitudinal brain atrophy would only be accelerated (or apparent) for individuals with evidence of *both* Aβ and neurodegeneration, while it would occur at similar rates for groups of older individuals without either or both Aβ positivity or brain atrophy.[150]

Regarding FDG-PET findings, some reports suggest that global glucose metabolism may increase (i.e., hypermetabolism) in Aβ-positive cognitively normal older adults, possibly reflecting compensation mechanisms.[153,154] Alternatively, it is also possible that high lifelong metabolism drives Aβ deposition.[155] Consistent with this idea, brain regions with the highest Aβ burdens are also those that have high metabolic activity throughout the lifespan.[156] One study of >600 cognitively normal individuals, however, suggested that there exists a modest, negative association of FDG-PET with Aβ-PET in AD-typical regions with aging.[157] These conflicting results may owe in part to the timing in the pathological process at which studies are performed. In the early stages, increased brain metabolic activity may either be a driver of or a compensatory response to Aβ deposition. At later stages, however, decreased glucose metabolism in the presence of Aβ may reflect early neuronal failure.[158,159]

The timing in the pathological process may also influence findings between Aβ and brain functional connectivity. As previously mentioned, there is a strong overlap between brain Aβ deposition and brain regions that are known to be highly functionally connected.[73] Extensive preclinical work suggests that abnormal neuronal function leads to extracellular release of Aβ[160] and this may modulate the risk of plaque formation.[161] Older adults with childhood-onset epilepsy have, for instance, been found to have increased Aβ burden in late middle age.[162] Increased brain activity throughout the lifespan may therefore facilitate Aβ deposition.[163] Thus, an abnormal increase in neural activity may play an important role in initiating brain Aβ deposition. Similarly to what was proposed for FDG, increased connectivity across different brain networks could also help individuals maintain their cognitive performance while Aβ starts to accumulate.[164,165] Increased Aβ deposition has been associated with greater neural activity in task-positive regions[166,167] and decreased inhibition in task-negative regions (i.e., DMN-like regions) during memory encoding tasks.[168] At later stages, however, Aβ might be related to decreased within-network brain connectivity, possibly reflecting an ongoing neurodegenerative process.[169,170]

Until recently, it was not possible to study the interplay between Aβ and tau in the brain. The combination of Aβ radiotracers with novel tau radioligands now provide an insight on how these two proteins co-occur and interact in vivo. Recent work suggests that Aβ-positivity is accompanied by elevated tau-PET binding in brain regions known to be affected by tau pathology in cognitively normal older adults.[86,87,171] When taken as a continuous variable, global Aβ load is associated with higher regional tau-PET binding even in cognitively unimpaired individuals without clinically significant (i.e., subthreshold) levels of tau pathology.[172] These Aβ-associated elevations in tau-PET binding are especially prevalent in the entorhinal cortex and medial temporal regions, which are among the earliest regions to be affected by *tau* pathology in AD.[173,174]

While the emergence of Aβ pathology may be partly driven by lifelong abnormal neural function or metabolism, the presence of Aβ deposition in older adults relates to future brain atrophy, neuronal failure and cognitive decline.[135,175,176] Longitudinal findings therefore suggest that Aβ accumulation is an early event in the AD pathological process. The limited evidence of immediate biomarker associations, however, raises the possibility that other actors that coexist with Aβ positivity (e.g., tau) may drive brain and cognitive changes.[177] The future combination of Aβ and tau PET tracers in longitudinal studies will likely improve our appreciation of causal mechanisms.

4.2 Clinical implications

4.2.1 New research clinical criteria

With the availability of Aβ-PET imaging, an increasing amount of research has been devoted to improving the characterization of the preclinical phase of AD. Biomarker-based diagnostic criteria have been proposed by the National Institute on Aging and the Alzheimer's Association (NIA-AA) to identify the disease before the occurrence of the first clinical symptoms.[88] According to the NIA-AA guidelines, these preclinical diagnosis criteria should be restricted to research, their purpose being to accelerate AD-related research discoveries. Aβ-PET imaging has greatly contributed to the conceptual shift of considering AD as a continuum rather than a clinical entity. This framework proposes to classify the disease stage in which a person may be based on the presence or absence of the three pathological hallmarks of AD: Aβ, tau and neurodegeneration. More specifically, this framework suggests that the presence of Aβ is sufficient to be on the AD continuum, while any pathological change in the absence of Aβ-positivity relates to non-AD disease processes. While this biomarker-based definition of AD is not universally accepted, it nevertheless brings a framework to study the changes preceding and probably causing cognitive impairment. Considering AD as a continuum, rather than a chronic disease severe enough to interfere with daily activities, also opens new horizons for preventive interventions.

4.2.2 Disclosure of Aβ-positivity results

The question of disclosure is particularly sensitive in research settings where a proportion of healthy older adults are nonetheless Aβ positive. While cognitively normal individuals would not be eligible for a clinical Aβ-PET scan according to the AIT's appropriate use criteria, they are nevertheless at increased risk of AD dementia. Thus, ethical questions about Aβ-PET disclosure in cognitively normal individuals, bridging the disciplines of psychology, medicine and philosophy, have risen to the forefront of translational AD imaging research.

Clinicians and researchers agree that harm reduction is a guiding principle of biomarker status disclosure. Previous reviews have raised concerns about the well-being of individuals presented with an early positive Aβ scan, ranging from negative emotions to risk of self-harm. Thus, it has been argued that only patients who meet the AIT's appropriate use criteria should be eligible for AB-PET scan results disclosure.[178] Individuals in early stages of dementia have preserved awareness of their declining cognition and ability to execute planned actions, putting them at potential risk of suicide following a diagnosis of preclinical dementia.[179] Arguably, this severe risk may translate to cases of individuals who present as cognitively normal but are already worried about self-perceived changes in their own cognitive faculties. Suicide is a rare but possible outcome following preclinical disclosure, especially within the first 3 months, and should be taken as a serious outcome risk on a case-by-case basis.[179]

To combat deleterious disclosure-related outcomes rooted in misunderstanding, it is important for individuals to be well informed about what a positive Aβ scan means: while it portends increased risk for AD dementia, it is not a definitive diagnosis and it does not guarantee that one will progress to AD dementia. A recent study assessed healthy older adults' understanding of positive Aβ PET results following disclosure and distribution of informational materials.[180] Among 50 participants, 62% correctly understood that their results indicated a higher risk of progressing to AD dementia and also recognized that there is still a possibility of not developing the disease, although 12% misinterpreted their Aβ status as either an indication of imminent AD onset or equivalent to an AD diagnosis. While comparisons of psychological outcomes between individuals who understood their results and those who didn't were not reported, it is important to ensure that the people understand that a positive scan alone does not equate to a diagnosis or a certain prognosis of AD dementia.

Guidelines about Aβ-PET results disclosure in research settings assessing cognitively normal populations have been drafted in order to minimize harm and maximize participant well-being.[181] Such guidelines recommend accounting for pre-existing psychiatric risk factors when evaluating whether or not results disclosure is appropriate on a case-by-case basis, and also emphasize the importance of education about the meaning of a positive PET scan. Most importantly, clinical researchers should ensure that individuals are informed about psychosocial support resources following disclosure. Despite concerns that Aβ status disclosure among cognitively normal individuals would lead to adverse outcomes, studies assessing responses to disclosure have suggested that positive responses have generally outweighed the negative ones.[182]

4.2.3 Preventive interventions

While Aβ does not seem like an appropriate therapeutic target at the dementia stage, extensive data linking Aβ-positivity to subsequent brain atrophy and cognitive decline suggests that it may be a better target at the earliest stages of the disease. Thus, it is now thought that prevention trials targeting Aβ accumulation have the best chance of success.[183,184] It seems evident that Aβ PET will hold a key role in such trials. For instance, in novel pharmacoprevention trials such as the A4 study, evidence of Aβ-positivity using either PET or cerebrospinal fluid measures is required for enrollment,[185] and will be necessary to evaluate target engagement. Carefully designed prevention trials, including those in the more predictive autosomal dominant AD, will elucidate *whether* and *when* Aβ can be targeted for prevention.[186] A possibility remains that anti-Aβ therapies may be more efficient even before there is measurable evidence of plaque pathology.

5. "Early" and subthreshold Aβ imaging

5.1 Pitfalls of Aβ-positivity classifications

Most findings discussed thus far emerge from studies that consider Aβ PET as a dichotomous variable, classifying individuals as being Aβ-positive (high binding) vs Aβ-negative (low binding). While this classification simplifies interpretation and is more easily translatable to clinical settings, it also has limitations. As can be seen in Fig. 6, there is not always a clear dichotomy between individuals with (i.e., above the threshold) and without (i.e., below the threshold) Aβ deposition, raising the question of intermediate cases.[74,187] Using a global Aβ index, individuals with "subthreshold" values are thus categorized as Aβ-negative even if most of them probably have some degree of Aβ pathology.[74] Fig. 7, middle row, shows an example of an individual who would be classified as Aβ-negative based on conventional Aβ-PET thresholds but who, nevertheless, shows evidence of moderate Aβ pathology at autopsy according to the Consortium to Establish a Registry for Alzheimer's Disease (CERAD) criteria.

Based on estimated rates of Aβ accumulation,[125] it has been suggested that early increases in PiB-PET signal can be captured about 7 years before an individual reaches a conventional, or conservative, threshold of Aβ positivity.[74] Accordingly, increasing evidence suggests that "subthreshold" individuals accumulate Aβ deposition over time ("Aβ accumulators").[34,188] This "Aβ accumulator" status is additionally associated with memory decline and longitudinal increase in tau-PET binding,[188–190] further suggesting that

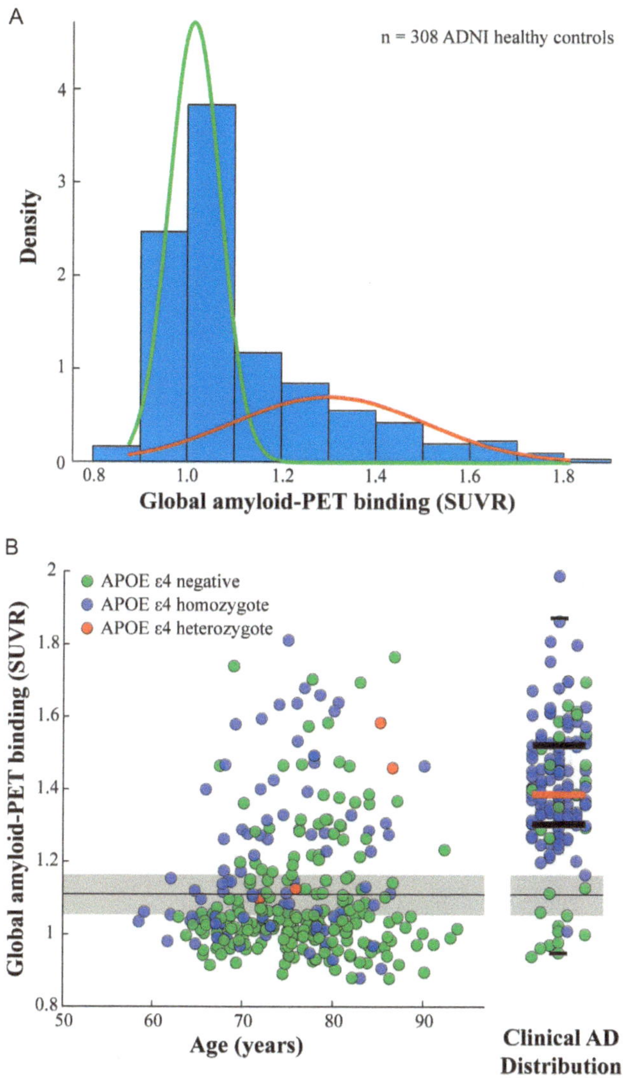

Fig. 6 Characteristic Aβ binding. (A) When plotting the distribution of global cortical Aβ PET tracer retention (SUVR), it often displays a bimodal distribution with a portion of individuals having low binding (usually centered around SUVR = 1.0) and a group with higher binding (here, centered on SUVR = 1.3). Data obtained from Aβ-PET scans (here with AV45) of ADNI participants (http://adni.loni.usc.edu/) who are cognitively healthy. (B) Representation of global cortical Aβ tracer retention as a function of age in ADNI cognitively normal controls (left). The distribution of Aβ tracer binding in clinically diagnosed AD patients is represented on the right. The full black line represents the tracer's Aβ-positivity threshold and gray area highlights a ±5% interval of borderline individuals whose status might change depending on pre-analytical methods or tracer used. The distribution on the right represents the distribution of cortical Aβ binding values in a group of clinically diagnosed AD patients from the ADNI. Clinically diagnosed AD patients with subthreshold Aβ binding may be misdiagnosed and clinicians could reconsider their diagnosis given their negative scan (see Section 3.2.1).

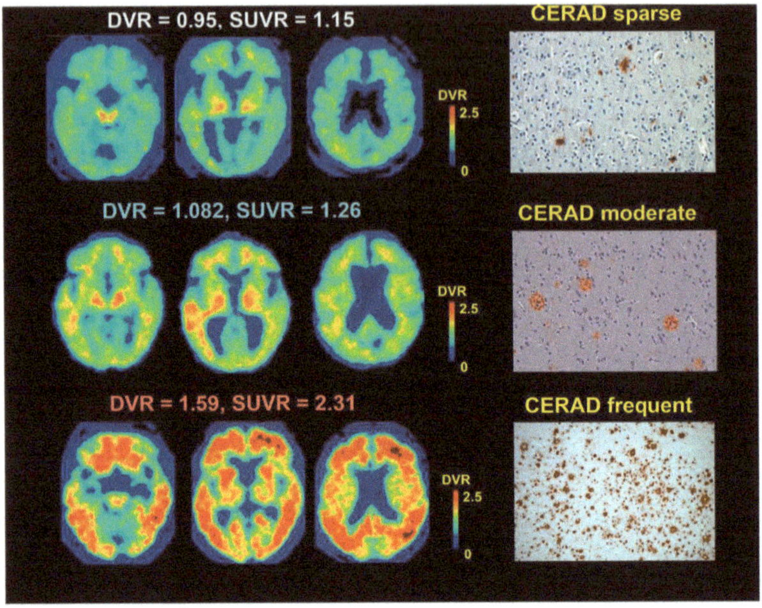

Fig. 7 PiB-PET binding and post-mortem Aβ load. The figure shows trans-axial PiB slices from a patient with CERAD sparse (top row), moderate (middle row) and frequent (bottom row) neuritic plaques. PiB-PET trans-axial slices are presented in neurological orientation. Photomicrographs of Aβ-immunohistochemistry are shown at ×10 (top and bottom rows) or ×20 (middle row) magnification. *Reproduced with permission from Villeneuve S, Rabinovici GD, Cohn-Sheehy BI, et al. Existing Pittsburgh Compound-B positron emission tomography thresholds are too high: statistical and pathological evaluation.* Brain. *2015;138:2020–2033.*

subthreshold levels of Aβ binding are clinically relevant. Considering that (1) Aβ accumulation is supposed to trigger the cascade of brain alterations leading to cognitive decline and AD dementia and (2) it might be more beneficial to stop early Aβ accumulation than to remove it after it has already progressed throughout the cortex, these Aβ-negative individuals at risk of becoming accumulators are probably the most suitable population to target for anti-Aβ prevention trials.

Another disadvantage of classifying individuals as Aβ-positive vs Aβ-negative is that it gets rid of potentially valuable information regarding the spatial distribution of Aβ deposition. Indeed, one key aspect for the early detection of Aβ accumulation may lie in identifying "where" it starts accumulating. While the pattern of tau deposition in AD is described to follow a stereotypical spreading pattern, "ground zero" of Aβ accumulation remains to be determined. Autopsy cases were the first to describe the pattern and spread of Aβ deposits. While Braak and Braak indicate that spreading might start in the basal neocortex (Stage A),[191] Thal and collaborators found evidence of Aβ deposits throughout the neocortex since the very first stage of beta-amyloidosis.[192] In vivo Aβ-PET imaging enables further investigation of this question. Results have been inconsistent, with some studies reporting that the earliest manifestations of amyloidosis involve the temporal and frontotemporal lobes,[75,193,194] the orbitofrontal-amygdala-hippocampus axes,[195] or more generally throughout the neocortex.[196] The most common assumption, however, is that Aβ deposition starts accumulating in the precuneus and frontal medial regions, often referred to as "AD-signature" regions.[34,74] One hypothesis is that the frontotemporal areas are among the first to show increased Aβ binding,[75,193,194] but this may be a nonspecific process. In contrast, Aβ accumulation in the precuneus and frontal medial regions may be more specific to the AD pathogenetic process and may be more predictive of future Aβ accumulation (Fig. 8).

Alternatively, discrepancies might also indicate that Aβ deposition does not necessarily follow a stereotypical progression and that not all individuals will start accumulating Aβ in the same brain regions.[192] As an example Fig. 9A shows that, in individuals with low (Cluster 1) and high (Cluster 2)

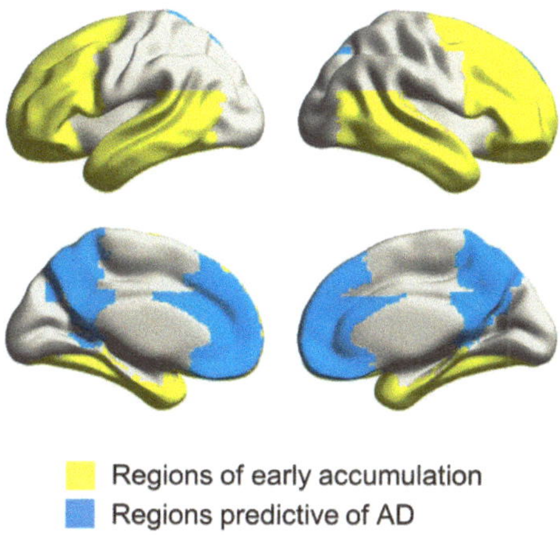

Fig. 8 Preferential regions of Aβ accumulation. Stereotypical patterns of Aβ accumulation when measured using PET. Frontotemporal regions (yellow) may be among the first to show Aβ accumulation but likely represent a process non-specific to AD. In contrast, precuneal and medial frontotemporal regions (blue) are better predictors of future Aβ accumulation and global Aβ burden and are therefore possibly more predictive of an ongoing AD pathological process. *Adapted from Gonneaud J, Arenaza-Urquijo EM, Mezenge F, et al. Increased florbetapir binding in the temporal neocortex from age 20 to 60 years.* Neurology. *2017;89:2438–2446*

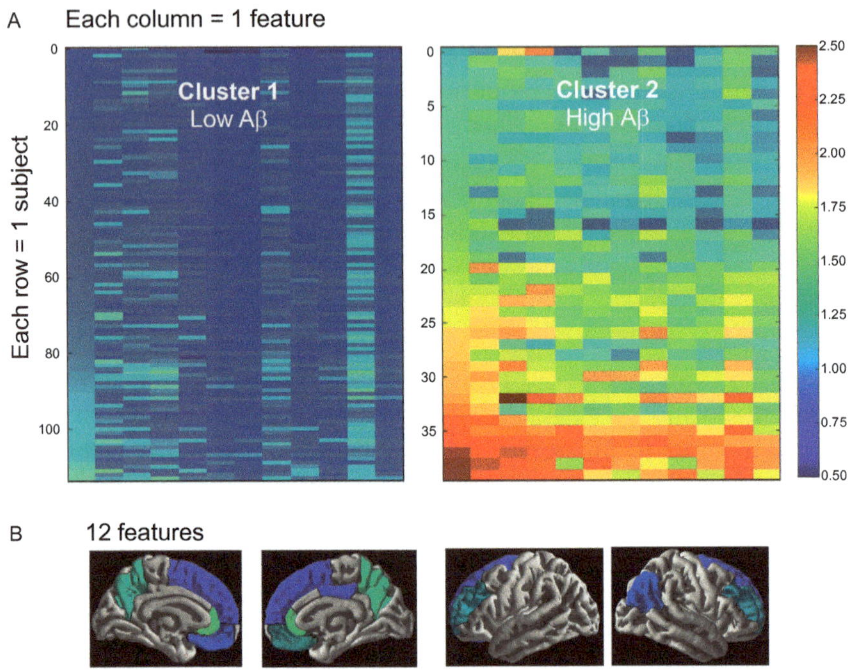

Fig. 9 Non-stereotypical deposition of Aβ. The figure shows the results of a cluster analysis, classifying individuals into two groups based on their Aβ-PET binding values. Panel (A) shows the classification of individuals into low-Aβ (left) or high-Aβ (right) groups as well as intensity of tracer binding (DVR; increasing values indicate more Aβ; color scale ranges from 0.5 to 2.5) in 12 brain regions (features represented in (B) with lighter colors corresponding to the features that have the highest weight in the model). From left to right, the 12 features are: rostral anterior cingulate left hemisphere (lh); rostral anterior cingulate right hemisphere (rh); precuneus rh; precuneus lh; medial orbitofrontal rh; rostral middle frontal lh; rostral middle frontal rh; inferior parietal rh; medial orbitofrontal lh; superior orbitofrontal rh; posterior cingulate rh; and superior orbitofrontal lh. One can notice that, even in the high-Aβ group, the region showing highest tracer uptake may vary from one person to the other, without a stereotypical pattern of tracer distribution. *Reproduced with permission from Villeneuve S, Rabinovici GD, Cohn-Sheehy BI, et al. Existing Pittsburgh Compound-B positron emission tomography thresholds are too high: statistical and pathological evaluation.* Brain. *2015;138:2020–2033.*

Aβ deposition, SUVR values from 12 brain regions (features) which are sensitive to early Aβ PET binding (Fig. 9B) do not follow a stereotypical binding pattern. More particularly, this figure suggests an inter-individual variability in the relative amount of Aβ deposition across key brain regions showing early PET binding in Aβ-positive individuals. For instance, while some individuals (e.g., row 0 of cluster 2) show high PET binging in the precuneus (column 3 and 4), others (e.g., row 16 of cluster 2) have no binding in this same key region despite the fact that they have high brain Aβ burden. Further studies including longitudinal designs and PET-autopsy comparisons are still needed to elucidate this question.

5.2 How early should we go to detect first evidence of Aβ accumulation?

Related to the question of the earliest regional indicators of Aβ accumulation, some studies have investigated the earliest evidence of Aβ binding throughout the age spectrum of adulthood. Most studies investigating the clinical and prognostic relevance of Aβ-PET have principally focused on late-middle aged adults (~50–60 years of age) even when hypothetical models suggest that Aβ accumulation may occur before that. In fact, considering that some individuals can develop AD dementia before the age of 65 and that Aβ starts accumulating more than two decades before disease onset, some individuals should show evidence of Aβ deposition as early as in their 40s. In line with the objective of identifying individuals at the earliest phase of the pathogenetic process, recent investigations have attempted to identify the youngest age at which Aβ aggregation may start. A handful of studies including individuals <50 years old suggest that Aβ accumulation may start early in adulthood.[75,196–198] This early increase may even be associated with cognitive performance in middle-aged individuals.[197] Although these PET findings were initially surprising, they concurred with some neuropathological studies describing the occurrence of Aβ plaques in individuals—principally *APOE* ε4 carriers—as young as 40 years of age, or even before.[191,199,200]

Thus, Aβ-PET imaging has allowed researchers to gather the first in vivo evidence of the earliest protein accumulation. This has strengthened the need of studying Aβ binding in younger populations and taking advantage of the full information provided by PET scans (i.e., both global and regional Aβ load).

6. Factors influencing amyloid accumulation

The prevalence of Aβ deposition in cognitively normal individuals increases with age, going from ~10% in one's 50s to >30% in individuals over 80 years of age.[68,201] Not all individuals age 80+ have Aβ, however, suggesting that inter-individual differences influence Aβ accumulation.

With the emergence of Aβ imaging, it has been shown that several non-modifiable and potentially modifiable factors can influence Aβ deposition.

6.1 Non-modifiable factors

It was suggested quite early on that "sporadic" AD may have a strong hereditary component.[202,203] Having a first-degree family history (FH) of "sporadic" AD is associated with a 1- to 14-fold increased risk of dementia.[204,205] Accordingly, a family history of AD has been associated with an increased risk of Aβ positivity.[206,207] More recently, our group showed that proximity to one's first-degree relative's age at symptom onset correlates with increased Aβ biomarker load, an association that was stronger in women when compared to men.[208] While these observations may be partly driven by the fact that family history and *APOE* ε4 highly co-occur,[209,210] they may also act synergistically to increase Aβ load.[211]

It is now well established that carriers of the ε4 allele have a 4- to 16-fold increased risk of AD dementia.[212,213] One mechanism by which *APOE* might increase AD risk is via clearance of cerebral Aβ.[214,215] When matched for age, *APOE* ε4 carriers usually have more Aβ deposition than non-carriers.[68,201] Fig. 10 shows an example of two 67-year-old cognitively

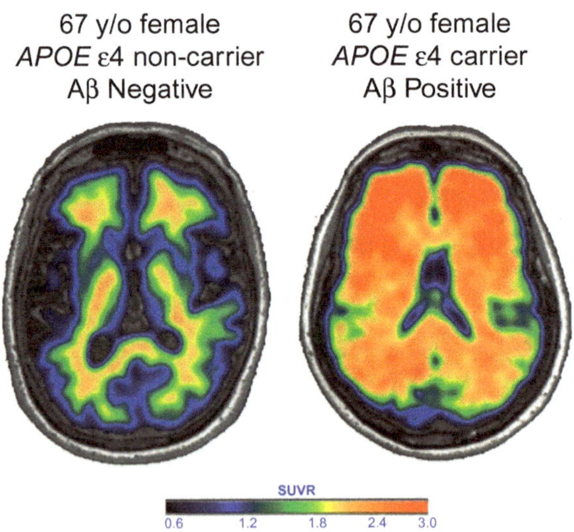

Fig. 10 APOE ε4 and risk of Aβ positivity. Aβ-PET scan of an age and sex matched Aβ-negative APOE ε4 non-carrier (left) and of an Aβ-positive carrier (right) pair from the PREVENT-AD study of cognitively unimpaired older adults at risk of AD.[216]

normal individuals; the one on the left panel is an *APOE* ε4 non-carrier with minimal Aβ-PET binding and the one on the right panel is an *APOE* ε4 carrier with high levels of Aβ-PET binding.

6.2 Modifiable factors

Given the repeated failures of clinical trials involving disease-modifying drugs for AD, a new research field is now investigating modifiable factors that could postpone or slow down Aβ accumulation for disease prevention. Up to 35% of dementia risk is attributable to lifestyle and behavioral factors that are potentially modifiable.[217] Cardiovascular risk factors may be the most recognized among these since higher cardiovascular risk factors in midlife have been associated with increased risk of cognitive decline and dementia.[218–220] More recently, it was suggested that abnormal vascular changes may occur before Aβ abnormality,[221] and that aggressive treatment for hypertension may result in diminished risk of cognitive impairment.[222] The advent of Aβ PET imaging has helped clarify the pathological pathways by which vascular risk factors could influence AD risk. Several studies have shown that midlife cardiovascular risk factors are associated with higher Aβ-PET binding in later life.[223–226] However, these associations were either unobservable or less frequently reported at older ages, suggesting that cardiovascular health earlier in life may have long-term effects on the accrual of AD pathology.[224,226–228]

Cognitive and physical activities have also been identified as potential modifiable factors that influence AD risk,[217,229] and increasing evidence from the PET literature suggest a complex relationship between cerebral Aβ and these modifiable factors.[230] In the preclinical phase of the disease, positive cognitive and physical lifestyle factors, such as higher levels of education, more lifetime cognitive engagement and increased exercise, have been associated with lower Aβ burden.[231–235,235a] Interestingly, most of these lifestyle characteristics have also been found to buffer the detrimental effects of *APOE* ε4 allele on Aβ burden.[236–238] Lifestyle might therefore mitigate Aβ-related genetic predispositions. Fig. 11 gives an example of how higher lifetime cognitive activity, a positive lifestyle factor, can mitigate the risk of Aβ deposition even in the presence of this major genetic risk factor.

An increasing number of neuroimaging studies point toward an important association of sleep with Aβ accumulation. In general, reports suggest that indicators of disturbed sleep such as shorter sleep duration, increased

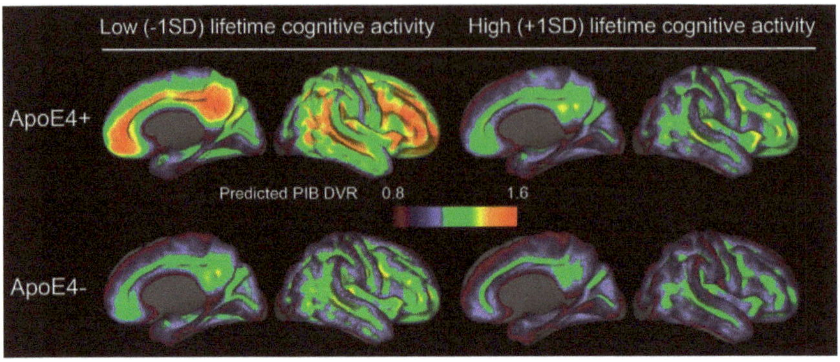

Fig. 11 Lifestyle can mitigate genetic risk. Visualization of the interaction between APOE ε4 carrier status and lifetime cognitive activity. Individuals carrying the APOE ε4 allele are at increased risk of developing Aβ pathology, but this effect is reduced by higher lifetime cognitive activity. *Figure adapted from Wirth M, Villeneuve S, La Joie R, Marks SM, Jagust WJ. Gene-environment interactions: lifetime cognitive activity, APOE genotype, and beta-amyloid burden.* J Neurosci. *2014;34:8612–8617.*

sleep latency or poor sleep quality are associated with increasing Aβ deposition in the brain.[239–242] Other treatable medical conditions such as obstructive sleep apnea and abnormal blood pressure during sleep are also related to brain Aβ levels.[243,244] Thus, sleep may play a critical role in the process of Aβ clearance from the brain.[245] However, increased sleepiness in older adults may not necessarily be beneficial for Aβ clearance, possibly because it may reflect underlying sleep disorders.[246] Furthermore, disturbed sleep may not only affect Aβ deposition but also contribute to the expression of associated cognitive symptoms.[247,248] Increasing sleep quality, even in conditions such as narcolepsy, may reduce Aβ accumulation,[249] so improving sleep in late-life may have important implications for future preventive interventions.

Finally, one major limitation of current AD PET imaging research is the underrepresentation of racial and ethnic minorities.[250,251] Further studies with racially and ethnically diverse samples are necessary to elucidate if the same genetic and modifiable factors influence Aβ-related risk across demographics.

7. Conclusion

The study of AD has been considerably accelerated in the last three decades as the field has developed disease biomarkers. Among these, radiotracers for molecular imaging of Aβ plaques have greatly contributed to our understanding of disease etiology and the sequence of pathophysiological events leading up to AD dementia. PET studies have allowed researchers to posit that Aβ accumulation occurs over several decades without apparent immediate effects on cognitive performance or brain atrophy. Individuals with Aβ deposition, however, show cognitive decline and accelerated atrophy when followed over time. Thus, there seems to be a decades-long pre-symptomatic period that leads to clinical expression of the disease in which Aβ deposition may be an early event, but not necessarily the direct cause of brain and cognitive decline. The repeated failures of clinical trials of disease-modifying drugs may therefore owe, at least in part, to interventions being too little too late as massive brain changes have already occurred in cognitively impaired individuals. Thus, the pre-symptomatic phase holds tremendous potential for interventions aiming to prevent pathological accumulation of Aβ deposition or its associated brain changes even before Aβ occurs. One major hurdle to this approach remains the difficulty to identify Aβ-negative accumulators, a topic of high interest at the moment for the Aβ-imaging community. However, once identified, encouraging preventive lifestyle habits in at-risk individuals could achieve a 10% reduction in risk factors and prevent more than one million cases worldwide.[252]

Aβ-PET imaging will continue to hold an important role in future interventional studies for enrollment, evaluation of target engagement, and efficacy. Combining this modality with more recent imaging markers of disease processes (e.g., tau, neuro-inflammation) will also yield vital information regarding disease mechanisms and progression to identify viable targets and interventions which should, at long last, reduce AD morbidity.

References

1. Hardy JA, Higgins GA. Alzheimer's disease: the amyloid cascade hypothesis. *Science*. 1992;256:184–185.
2. Crystal H, Dickson D, Fuld P, et al. Clinico-pathologic studies in dementia: nondemented subjects with pathologically confirmed Alzheimer's disease. *Neurology*. 1988;38:1682–1687.
3. Katzman R, Terry R, DeTeresa R, et al. Clinical, pathological, and neurochemical changes in dementia: a subgroup with preserved mental status and numerous neocortical plaques. *Ann Neurol*. 1988;23:138–144.
4. Glenner GG, Wong CW. Alzheimer's disease and Down's syndrome: sharing of a unique cerebrovascular amyloid fibril protein. *Biochem Biophys Res Commun*. 1984;122:1131–1135.
5. Glenner GG, Wong CW. Alzheimer's disease: initial report of the purification and characterization of a novel cerebrovascular amyloid protein. *Biochem Biophys Res Commun*. 1984;120:885–890.
6. Murphy MP, LeVine 3rd H. Alzheimer's disease and the amyloid-beta peptide. *J Alzheimer's Dis*. 2010;19:311–323.

7. Suzuki N, Cheung TT, Cai XD, et al. An increased percentage of long amyloid beta protein secreted by familial amyloid beta protein precursor (beta APP717) mutants. *Science*. 1994;264:1336–1340.
8. Bitan G, Kirkitadze MD, Lomakin A, Vollers SS, Benedek GB, Teplow DB. Amyloid beta-protein (Abeta) assembly: Abeta 40 and Abeta 42 oligomerize through distinct pathways. *Proc Natl Acad Sci USA*. 2003;100:330–335.
9. Selkoe DJ. Alzheimer's disease: genes, proteins, and therapy. *Physiol Rev*. 2001;81: 741–766.
10. Hartley DM, Walsh DM, Ye CP, et al. Protofibrillar intermediates of amyloid beta-protein induce acute electrophysiological changes and progressive neurotoxicity in cortical neurons. *J Neurosci*. 1999;19:8876–8884.
11. Mucke L, Masliah E, Yu GQ, et al. High-level neuronal expression of abeta 1-42 in wild-type human amyloid protein precursor transgenic mice: synaptotoxicity without plaque formation. *J Neurosci*. 2000;20:4050–4058.
12. Klein WL, Krafft GA, Finch CE. Targeting small Abeta oligomers: the solution to an Alzheimer's disease conundrum? *Trends Neurosci*. 2001;24:219–224.
13. Yu L, Petyuk VA, Tasaki S, et al. Association of cortical beta-amyloid protein in the absence of insoluble deposits with Alzheimer disease. *JAMA Neurol*. 2019, https://doi.org/10.1001/jamaneurol.2019.0834.
14. Villemagne VL, Rowe CC. Amyloid PET ligands for dementia. *PET Clin*. 2010;5:33–53.
15. Shoghi-Jadid K, Small GW, Agdeppa ED, et al. Localization of neurofibrillary tangles and beta-amyloid plaques in the brains of living patients with Alzheimer disease. *Am J Geriatr Psychiatry*. 2002;10:24–35.
16. Klunk WE, Engler H, Nordberg A, et al. Imaging brain amyloid in Alzheimer's disease with Pittsburgh Compound-B. *Ann Neurol*. 2004;55:306–319.
17. Ikonomovic MD, Klunk WE, Abrahamson EE, et al. Post-mortem correlates of in vivo PiB-PET amyloid imaging in a typical case of Alzheimer's disease. *Brain*. 2008;131:1630–1645.
18. Cohen AD, Rabinovici GD, Mathis CA, Jagust WJ, Klunk WE, Ikonomovic MD. Using Pittsburgh Compound B for in vivo PET imaging of fibrillar amyloid-beta. *Adv Pharmacol*. 2012;64:27–81.
19. Lister-James J, Pontecorvo MJ, Clark C, et al. Florbetapir f-18: a histopathologically validated Beta-amyloid positron emission tomography imaging agent. *Semin Nucl Med*. 2011;41:300–304.
20. Fodero-Tavoletti MT, Brockschnieder D, Villemagne VL, et al. In vitro characterization of [18F]-florbetaben, an Abeta imaging radiotracer. *Nucl Med Biol*. 2012;39: 1042–1048.
21. Serdons K, Verduyckt T, Vanderghinste D, et al. Synthesis of 18F-labelled 2-(4'-fluorophenyl)-1,3-benzothiazole and evaluation as amyloid imaging agent in comparison with [11C]PIB. *Bioorg Med Chem Lett*. 2009;19:602–605.
22. Villemagne VL, Dore V, Burnham SC, Masters CL, Rowe CC. Imaging tau and amyloid-beta proteinopathies in Alzheimer disease and other conditions. *Nat Rev Neurol*. 2018;14:225–236.
23. Clark CM, Schneider JA, Bedell BJ, et al. Use of florbetapir-PET for imaging beta-amyloid pathology. *JAMA*. 2011;305:275–283.
24. Clark CM, Pontecorvo MJ, Beach TG, et al. Cerebral PET with florbetapir compared with neuropathology at autopsy for detection of neuritic amyloid-beta plaques: a prospective cohort study. *Lancet Neurol*. 2012;11:669–678.
25. Barthel H, Gertz HJ, Dresel S, et al. Cerebral amyloid-beta PET with florbetaben (18F) in patients with Alzheimer's disease and healthy controls: a multicentre phase 2 diagnostic study. *Lancet Neurol*. 2011;10:424–435.

26. Sabri O, Sabbagh MN, Seibyl J, et al. Florbetaben PET imaging to detect amyloid beta plaques in Alzheimer's disease: phase 3 study. *Alzheimers Dement.* 2015;11: 964–974.
27. Curtis C, Gamez JE, Singh U, et al. Phase 3 trial of flutemetamol labeled with radioactive fluorine 18 imaging and neuritic plaque density. *JAMA Neurol.* 2015;72: 287–294.
28. Nyberg S, Jonhagen ME, Cselenyi Z, et al. Detection of amyloid in Alzheimer's disease with positron emission tomography using [^{11}C]AZD2184. *Eur J Nucl Med Mol Imaging.* 2009;36:1859–1863.
29. Seubert P, Vigo-Pelfrey C, Esch F, et al. Isolation and quantification of soluble Alzheimer's beta-peptide from biological fluids. *Nature.* 1992;359:325–327.
30. Motter R, Vigo-Pelfrey C, Kholodenko D, et al. Reduction of beta-amyloid peptide42 in the cerebrospinal fluid of patients with Alzheimer's disease. *Ann Neurol.* 1995;38:643–648.
31. Strozyk D, Blennow K, White LR, Launer LJ. CSF Abeta 42 levels correlate with amyloid-neuropathology in a population-based autopsy study. *Neurology.* 2003;60: 652–656.
32. Mattsson N, Insel PS, Donohue M, et al. Independent information from cerebrospinal fluid amyloid-beta and florbetapir imaging in Alzheimer's disease. *Brain.* 2015;138: 772–783.
33. Palmqvist S, Mattsson N, Hansson O. Alzheimer's disease neuroimaging I. Cerebrospinal fluid analysis detects cerebral amyloid-beta accumulation earlier than positron emission tomography. *Brain.* 2016;139:1226–1236.
34. Palmqvist S, Scholl M, Strandberg O, et al. Earliest accumulation of beta-amyloid occurs within the default-mode network and concurrently affects brain connectivity. *Nat Commun.* 2017;8:1214.
35. Landau SM, Lu M, Joshi AD, et al. Comparing positron emission tomography imaging and cerebrospinal fluid measurements of beta-amyloid. *Ann Neurol.* 2013;74: 826–836.
36. Morris E, Chalkidou A, Hammers A, Peacock J, Summers J, Keevil S. Diagnostic accuracy of (18)F amyloid PET tracers for the diagnosis of Alzheimer's disease: a systematic review and meta-analysis. *Eur J Nucl Med Mol Imaging.* 2016;43:374–385.
37. Blennow K. Cerebrospinal fluid protein biomarkers for Alzheimer's disease. *NeuroRx.* 2004;1:213–225.
38. Nakamura A, Kaneko N, Villemagne VL, et al. High performance plasma amyloid-beta biomarkers for Alzheimer's disease. *Nature.* 2018;554:249–254.
39. Chetelat G, La Joie R, Villain N, et al. Amyloid imaging in cognitively normal individuals, at-risk populations and preclinical Alzheimer's disease. *Neuroimage Clin.* 2013;2:356–365.
40. Vandenberghe R, Van Laere K, Ivanoiu A, et al. 18F-flutemetamol amyloid imaging in Alzheimer disease and mild cognitive impairment: a phase 2 trial. *Ann Neurol.* 2010;68:319–329.
41. Villemagne VL, Mulligan RS, Pejoska S, et al. Comparison of 11C-PiB and 18F-florbetaben for Abeta imaging in ageing and Alzheimer's disease. *Eur J Nucl Med Mol Imaging.* 2012;39:983–989.
42. Yoder KK. *Basic PET Data Analysis Techniques. Positron Emission Tomography-Recent Developments in Instrumentation, Research and Clinical Oncological Practice.* IntechOpen; 2013.
43. Villemagne VL. Amyloid imaging: past, present and future perspectives. *Ageing Res Rev.* 2016;30:95–106.
44. Landau SM, Fero A, Baker SL, et al. Measurement of longitudinal beta-amyloid change with 18F-florbetapir PET and standardized uptake value ratios. *J Nucl Med.* 2015;56:567–574.

45. Chen K, Roontiva A, Thiyyagura P, et al. Improved power for characterizing longitudinal amyloid-beta PET changes and evaluating amyloid-modifying treatments with a cerebral white matter reference region. *J Nucl Med*. 2015;56:560–566.
46. Shokouhi S, McKay JW, Baker SL, et al. Reference tissue normalization in longitudinal (18)F-florbetapir positron emission tomography of late mild cognitive impairment. *Alzheimers Res Ther*. 2016;8:2.
47. Schwarz CG, Senjem ML, Gunter JL, et al. Optimizing PiB-PET SUVR change-over-time measurement by a large-scale analysis of longitudinal reliability, plausibility, separability, and correlation with MMSE. *Neuroimage*. 2017;144:113–127.
48. Bourgeat P, Villemagne VL, Dore V, et al. Comparison of MR-less PiB SUVR quantification methods. *Neurobiol Aging*. 2015;36(suppl 1):S159–S166.
49. Schwarz CG, Jones DT, Gunter JL, et al. Contributions of imprecision in PET-MRI rigid registration to imprecision in amyloid PET SUVR measurements. *Hum Brain Mapp*. 2017;38:3323–3336.
50. Klunk WE, Koeppe RA, Price JC, et al. The Centiloid project: standardizing quantitative amyloid plaque estimation by PET. *Alzheimers Dement*. 2015;11. 1-15.e1-4.
51. Rowe CC, Jones G, Dore V, et al. Standardized expression of 18F-NAV4694 and 11C-PiB beta-amyloid PET results with the Centiloid scale. *J Nucl Med*. 2016;57: 1233–1237.
52. Navitsky M, Joshi AD, Kennedy I, et al. Standardization of amyloid quantitation with florbetapir standardized uptake value ratios to the Centiloid scale. *Alzheimers Dement*. 2018;14:1565–1571.
53. Rowe CC, Dore V, Jones G, et al. (18)F-Florbetaben PET beta-amyloid binding expressed in Centiloids. *Eur J Nucl Med Mol Imaging*. 2017;44:2053–2059.
54. Bourgeat P, Dore V, Fripp J, et al. Implementing the Centiloid transformation for (11) C-PiB and beta-amyloid (18)F-PET tracers using CapAIBL. *Neuroimage*. 2018;183: 387–393.
55. La Joie R, Ayakta N, Seeley WW, et al. Multisite study of the relationships between antemortem [(11)C]PIB-PET Centiloid values and postmortem measures of Alzheimer's disease neuropathology. *Alzheimers Dement*. 2019;15:205–216.
56. Salvado G, Molinuevo JL, Brugulat-Serrat A, et al. Centiloid cut-off values for optimal agreement between PET and CSF core AD biomarkers. *Alzheimers Res Ther*. 2019;11:27.
56a. Tudorascu DL, Minhas DS, Lao PJ, et al. The use of centiloids for applying [(11)C]PiB classification cutoffs across region-of-interest delineation methods. *Alzheimers Dement*. 2018;10:332–339.
57. Su Y, Flores S, Wang G, et al. Comparison of Pittsburgh compound B and florbetapir in cross-sectional and longitudinal studies. *Alzheimers Dement*. 2019;11:180–190.
58. Jack Jr CR, Wiste HJ, Lesnick TG, et al. Brain beta-amyloid load approaches a plateau. *Neurology*. 2013;80:890–896.
59. Weintraub S, Wicklund AH, Salmon DP. The neuropsychological profile of Alzheimer disease. *Cold Spring Harb Perspect Med*. 2012;2:a006171.
60. Salmon DP, Bondi MW. Neuropsychological assessment of dementia. *Annu Rev Psychol*. 2009;60:257–282.
61. Dubois B, Albert ML. Amnestic MCI or prodromal Alzheimer's disease? *Lancet Neurol*. 2004;3:246–248.
62. Salmon DP. Disorders of memory in Alzheimer's disease. 2nd ed. Amsterdam, Netherlands: Elsevier Science Publishers B.V.; 2000:155–195. Handbook of Neuropsychology: Memory and Its Disorders; vol 2.
63. Arriagada PV, Growdon JH, Hedley-Whyte ET, Hyman BT. Neurofibrillary tangles but not senile plaques parallel duration and severity of Alzheimer's disease. *Neurology*. 1992;42:631–639.

64. Engler H, Forsberg A, Almkvist O, et al. Two-year follow-up of amyloid deposition in patients with Alzheimer's disease. *Brain*. 2006;129:2856–2866.
65. Jack Jr CR, Lowe VJ, Weigand SD, et al. Serial PIB and MRI in normal, mild cognitive impairment and Alzheimer's disease: implications for sequence of pathological events in Alzheimer's disease. *Brain*. 2009;132:1355–1365.
66. Rowe CC, Ng S, Ackermann U, et al. Imaging beta-amyloid burden in aging and dementia. *Neurology*. 2007;68:1718–1725.
67. Furst AJ, Rabinovici GD, Rostomian AH, et al. Cognition, glucose metabolism and amyloid burden in Alzheimer's disease. *Neurobiol Aging*. 2012;33:215–225.
68. Roberts RO, Aakre JA, Kremers WK, et al. Prevalence and outcomes of amyloid positivity among persons without dementia in a longitudinal, population-based setting. *JAMA Neurol*. 2018;75:970–979.
69. Villain N, Chetelat G, Grassiot B, et al. Regional dynamics of amyloid-beta deposition in healthy elderly, mild cognitive impairment and Alzheimer's disease: a voxelwise PiB-PET longitudinal study. *Brain*. 2012;135:2126–2139.
70. Jagust W. Imaging the evolution and pathophysiology of Alzheimer disease. *Nat Rev Neurosci*. 2018;19:687–700.
71. Lehmann M, Ghosh PM, Madison C, et al. Diverging patterns of amyloid deposition and hypometabolism in clinical variants of probable Alzheimer's disease. *Brain*. 2013;136:844–858.
72. La Joie R, Perrotin A, Barre L, et al. Region-specific hierarchy between atrophy, hypometabolism, and beta-amyloid (Abeta) load in Alzheimer's disease dementia. *J Neurosci*. 2012;32:16265–16273.
73. Buckner RL, Snyder AZ, Shannon BJ, et al. Molecular, structural, and functional characterization of Alzheimer's disease: evidence for a relationship between default activity, amyloid, and memory. *J Neurosci*. 2005;25:7709–7717.
74. Villeneuve S, Rabinovici GD, Cohn-Sheehy BI, et al. Existing Pittsburgh Compound-B positron emission tomography thresholds are too high: statistical and pathological evaluation. *Brain*. 2015;138:2020–2033.
75. Gonneaud J, Arenaza-Urquijo EM, Mezenge F, et al. Increased florbetapir binding in the temporal neocortex from age 20 to 60 years. *Neurology*. 2017;89:2438–2446.
76. Wagner AD, Shannon BJ, Kahn I, Buckner RL. Parietal lobe contributions to episodic memory retrieval. *Trends Cogn Sci*. 2005;9:445–453.
77. Jones DT, Knopman DS, Gunter JL, et al. Cascading network failure across the Alzheimer's disease spectrum. *Brain*. 2016;139:547–562.
78. Villeneuve S, Wirth M, La Joie R. Are AD-typical regions the convergence point of multiple pathologies? *Front Aging Neurosci*. 2015;7:42.
79. Sepulcre J, Schultz AP, Sabuncu M, et al. In vivo tau, amyloid, and gray matter profiles in the aging brain. *J Neurosci*. 2016;36:7364–7374.
80. Zhang W, Arteaga J, Cashion DK, et al. A highly selective and specific PET tracer for imaging of tau pathologies. *J Alzheimers Dis*. 2012;31:601–612.
81. Braak H, Braak E. Neuropathological stageing of Alzheimer-related changes. *Acta Neuropathol*. 1991;82:239–259.
82. Braak H, Alafuzoff I, Arzberger T, Kretzschmar H, Del Tredici K. Staging of Alzheimer disease-associated neurofibrillary pathology using paraffin sections and immunocytochemistry. *Acta Neuropathol*. 2006;112:389–404.
83. Brettschneider J, Del Tredici K, Lee VM, Trojanowski JQ. Spreading of pathology in neurodegenerative diseases: a focus on human studies. *Nat Rev Neurosci*. 2015;16:109–120.
84. Iaccarino L, Tammewar G, Ayakta N, et al. Local and distant relationships between amyloid, tau and neurodegeneration in Alzheimer's Disease. *Neuroimage Clin*. 2018;17:452–464.

85. Musiek ES, Holtzman DM. Three dimensions of the amyloid hypothesis: time, space and 'wingmen'. *Nat Neurosci*. 2015;18:800–806.
86. Scholl M, Lockhart SN, Schonhaut DR, et al. PET imaging of tau deposition in the aging human brain. *Neuron*. 2016;89:971–982.
87. Vemuri P, Lowe VJ, Knopman DS, et al. Tau-PET uptake: regional variation in average SUVR and impact of amyloid deposition. *Alzheimers Dement*. 2017;6: 21–30.
88. Jack Jr CR, Bennett DA, Blennow K, et al. NIA-AA research framework: toward a biological definition of Alzheimer's disease. *Alzheimers Dement*. 2018;14:535–562.
89. Hall B, Mak E, Cervenka S, Aigbirhio FI, Rowe JB, O'Brien JT. In vivo tau PET imaging in dementia: pathophysiology, radiotracer quantification, and a systematic review of clinical findings. *Ageing Res Rev*. 2017;36:50–63.
90. Ossenkoppele R, Rabinovici GD, Smith R, et al. Discriminative accuracy of [18F] flortaucipir positron emission tomography for Alzheimer disease vs other neurodegenerative disorders. *JAMA*. 2018;320:1151–1162.
91. Okamura N, Harada R, Ishiki A, Kikuchi A, Nakamura T, Kudo Y. The development and validation of tau PET tracers: current status and future directions. *Clin Transl Imaging*. 2018;6:305–316.
92. Beach TG, Monsell SE, Phillips LE, Kukull W. Accuracy of the clinical diagnosis of Alzheimer disease at National Institute on Aging Alzheimer Disease Centers, 2005–2010. *J Neuropathol Exp Neurol*. 2012;71:266–273.
93. Fischer CE, Qian W, Schweizer TA, et al. Determining the impact of psychosis on rates of false-positive and false-negative diagnosis in Alzheimer's disease. *Alzheimers Dement (N Y)*. 2017;3:385–392.
94. Ossenkoppele R, Jansen WJ, Rabinovici GD, et al. Prevalence of amyloid PET positivity in dementia syndromes: a meta-analysis. *JAMA*. 2015;313:1939–1949.
95. Chetelat G, Ossenkoppele R, Villemagne VL, et al. Atrophy, hypometabolism and clinical trajectories in patients with amyloid-negative Alzheimer's disease. *Brain*. 2016;139:2528–2539.
96. Landau SM, Horng A, Fero A, Jagust WJ. Alzheimer's disease neuroimaging I. Amyloid negativity in patients with clinically diagnosed Alzheimer disease and MCI. *Neurology*. 2016;86:1377–1385.
97. Gaugler JE, Ascher-Svanum H, Roth DL, Fafowora T, Siderowf A, Beach TG. Characteristics of patients misdiagnosed with Alzheimer's disease and their medication use: an analysis of the NACC-UDS database. *BMC Geriatr*. 2013;13:137.
98. Hunter CA, Kirson NY, Desai U, Cummings AK, Faries DE, Birnbaum HG. Medical costs of Alzheimer's disease misdiagnosis among US Medicare beneficiaries. *Alzheimers Dement*. 2015;11:887–895.
99. Villemagne VL, Dore V, Bourgeat P, et al. Abeta-amyloid and tau imaging in dementia. *Semin Nucl Med*. 2017;47:75–88.
100. Johnson KA, Minoshima S, Bohnen NI, et al. Appropriate use criteria for amyloid PET: a report of the Amyloid Imaging Task Force, the Society of Nuclear Medicine and Molecular Imaging, and the Alzheimer's Association. *Alzheimers Dement*. 2013;9. e-1-16.
101. Barthel H, Sabri O. Clinical use and utility of amyloid imaging. *J Nucl Med*. 2017;58:1711–1717.
102. Fantoni ER, Chalkidou A, JT OB, Farrar G, Hammers A. A systematic review and aggregated analysis on the impact of amyloid PET brain imaging on the diagnosis, diagnostic confidence, and management of patients being evaluated for Alzheimer's disease. *J Alzheimers Dis*. 2018;63:783–796.
103. Spallazzi M, Barocco F, Michelini G, et al. The incremental diagnostic value of [18F] Florbetaben PET and the pivotal role of the neuropsychological assessment in clinical practice. *J Alzheimers Dis*. 2019;67:1235–1244.

104. Rabinovici GD, Gatsonis C, Apgar C, et al. Association of amyloid positron emission tomography with subsequent change in clinical management among medicare beneficiaries with mild cognitive impairment or dementia. *JAMA*. 2019;321:1286–1294.
105. Schneider LS. Alzheimer disease pharmacologic treatment and treatment research. *Continuum (Minneap Minn)*. 2013;19:339–357.
106. Mendez MF, Shapira JS, McMurtray A, Licht E. Preliminary findings: behavioral worsening on donepezil in patients with frontotemporal dementia. *Am J Geriatr Psychiatry*. 2007;15:84–87.
107. Boxer AL, Knopman DS, Kaufer DI, et al. Memantine in patients with frontotemporal lobar degeneration: a multicentre, randomised, double-blind, placebo-controlled trial. *Lancet Neurol*. 2013;12:149–156.
108. (CMS) CfMMS. *Decision Memo for Beta Amyloid Positron Emission Tomography in Dementia and Neurodegenerative Disease (CAG-00431N) [Online]*. Available at: https://www.cms.gov/medicare-coverage-database/details/nca-decision-memo.aspx?NCAId=265; 2019. Accessed 19 April 2019.
109. Steenland K, Goldstein FC, Levey A, Wharton W. A meta-analysis of Alzheimer's disease incidence and prevalence comparing African-Americans and Caucasians. *J Alzheimers Dis*. 2016;50:71–76.
110. McDonough IM. Beta-amyloid and cortical thickness reveal racial disparities in preclinical Alzheimer's disease. *Neuroimage Clin*. 2017;16:659–667.
111. Barnes LL, Leurgans S, Aggarwal NT, et al. Mixed pathology is more likely in black than white decedents with Alzheimer dementia. *Neurology*. 2015;85:528–534.
112. Graff-Radford NR, Besser LM, Crook JE, Kukull WA, Dickson DW. Neuropathologic differences by race from the National Alzheimer's Coordinating Center. *Alzheimers Dement*. 2016;12:669–677.
113. Risk Evaluation and Education for Alzheimer's Disease—the Study of Communicating Amyloid Neuroimaging (REVEAL-SCAN) [Online]. Available at: https://clinicaltrials.gov/show/NCT02959489; 2019. Accessed 19 April 2019.
114. Lingler JH, Roberts JS, Kim H, et al. Amyloid positron emission tomography candidates may focus more on benefits than risks of results disclosure. *Alzheimers Dement*. 2018;10:413–420.
115. Vanderschaeghe G, Schaeverbeke J, Bruffaerts R, Vandenberghe R, Dierickx K. Amnestic MCI patients' experiences after disclosure of their amyloid PET result in a research context. *Alzheimers Res Ther*. 2017;9:92.
116. Gomperts SN, Rentz DM, Moran E, et al. Imaging amyloid deposition in Lewy body diseases. *Neurology*. 2008;71:903–910.
117. Farid K, Charidimou A, Baron JC. Amyloid positron emission tomography in sporadic cerebral amyloid angiopathy: a systematic critical update. *Neuroimage Clin*. 2017;15:247–263.
118. Salloway S, Sperling R, Fox NC, et al. Two phase 3 trials of bapineuzumab in mild-to-moderate Alzheimer's disease. *N Engl J Med*. 2014;370:322–333.
119. Ostrowitzki S, Lasser RA, Dorflinger E, et al. A phase III randomized trial of gantenerumab in prodromal Alzheimer's disease. *Alzheimers Res Ther*. 2017;9:95.
120. Honig LS, Vellas B, Woodward M, et al. Trial of solanezumab for mild dementia due to Alzheimer's disease. *N Engl J Med*. 2018;378:321–330.
121. Egan MF, Kost J, Voss T, et al. Randomized trial of verubecestat for prodromal Alzheimer's disease. *N Engl J Med*. 2019;380:1408–1420.
122. Sevigny J, Chiao P, Bussiere T, et al. The antibody aducanumab reduces Abeta plaques in Alzheimer's disease. *Nature*. 2016;537:50–56.
123. Biogen. *Biogen and Eisai to Discontinue Phase 3 Engage and Emerge Trials of Aducanumab in Alzheimer's Disease [Online]*. Available at: http://investors.biogen.com/news-releases/news-release-details/biogen-and-eisai-discontinue-phase-3-engage-and-emerge-trials; 2019. Accessed 19 April 2019.

124. Liu E, Schmidt ME, Margolin R, et al. Amyloid-beta 11C-PiB-PET imaging results from 2 randomized bapineuzumab phase 3 AD trials. *Neurology*. 2015;85:692–700.
125. Villemagne VL, Burnham S, Bourgeat P, et al. Amyloid beta deposition, neurodegeneration, and cognitive decline in sporadic Alzheimer's disease: a prospective cohort study. *Lancet Neurol*. 2013;12:357–367.
126. Bateman RJ, Xiong C, Benzinger TL, et al. Clinical and biomarker changes in dominantly inherited Alzheimer's disease. *N Engl J Med*. 2012;367:795–804.
127. McDade E, Wang G, Gordon BA, et al. Longitudinal cognitive and biomarker changes in dominantly inherited Alzheimer disease. *Neurology*. 2018;91:e1295–e1306.
128. Perrotin A, Mormino EC, Madison CM, Hayenga AO, Jagust WJ. Subjective cognition and amyloid deposition imaging: a Pittsburgh Compound B positron emission tomography study in normal elderly individuals. *Arch Neurol*. 2012;69:223–229.
129. Hedden T, Oh H, Younger AP, Patel TA. Meta-analysis of amyloid-cognition relations in cognitively normal older adults. *Neurology*. 2013;80:1341–1348.
130. Jansen WJ, Ossenkoppele R, Tijms BM, et al. Association of cerebral amyloid-beta aggregation with cognitive functioning in persons without dementia. *JAMA Psychiat*. 2018;75:84–95.
131. Vos SJ, Xiong C, Visser PJ, et al. Preclinical Alzheimer's disease and its outcome: a longitudinal cohort study. *Lancet Neurol*. 2013;12:957–965.
132. Wirth M, Oh H, Mormino EC, Markley C, Landau SM, Jagust WJ. The effect of amyloid beta on cognitive decline is modulated by neural integrity in cognitively normal elderly. *Alzheimers Dement*. 2013;9. 687-698.e681.
133. Farrell ME, Kennedy KM, Rodrigue KM, et al. Association of longitudinal cognitive decline with amyloid burden in middle-aged and older adults: evidence for a dose-response relationship. *JAMA Neurol*. 2017;74:830–838.
134. Mormino EC, Papp KV, Rentz DM, et al. Early and late change on the preclinical Alzheimer's cognitive composite in clinically normal older individuals with elevated amyloid beta. *Alzheimers Dement*. 2017;13:1004–1012.
135. Soldan A, Pettigrew C, Fagan AM, et al. ATN profiles among cognitively normal individuals and longitudinal cognitive outcomes. *Neurology*. 2019;92:e1567–e1579.
136. Vogel JW, Varga Dolezalova M, La Joie R, et al. Subjective cognitive decline and beta-amyloid burden predict cognitive change in healthy elderly. *Neurology*. 2017;89: 2002–2009.
137. Donohue MC, Sperling RA, Petersen R, et al. Association between elevated brain amyloid and subsequent cognitive decline among cognitively normal persons. *JAMA*. 2017;317:2305–2316.
138. Burnham SC, Bourgeat P, Doré V, et al. Clinical and cognitive trajectories in cognitively healthy elderly individuals with suspected non-Alzheimer's disease pathophysiology (SNAP) or Alzheimer's disease pathology: a longitudinal study. *Lancet Neurol*. 2016;15:1044–1053.
139. Mormino EC, Betensky RA, Hedden T, et al. Synergistic effect of beta-amyloid and neurodegeneration on cognitive decline in clinically normal individuals. *JAMA Neurol*. 2014;71:1379–1385.
140. Soldan A, Pettigrew C, Cai Q, et al. Hypothetical preclinical Alzheimer disease groups and longitudinal cognitive change. *JAMA Neurol*. 2016;73:698–705.
141. Bilgel M, An Y, Helphrey J, et al. Effects of amyloid pathology and neurodegeneration on cognitive change in cognitively normal adults. *Brain*. 2018;141:2475–2485.
142. Knopman DS, Lundt ES, Therneau TM, et al. Entorhinal cortex tau, amyloid-beta, cortical thickness and memory performance in non-demented subjects. *Brain*. 2019;142:1148–1160.
143. Becker JA, Hedden T, Carmasin J, et al. Amyloid-beta associated cortical thinning in clinically normal elderly. *Ann Neurol*. 2011;69:1032–1042.

144. Dore V, Villemagne VL, Bourgeat P, et al. Cross-sectional and longitudinal analysis of the relationship between Abeta deposition, cortical thickness, and memory in cognitively unimpaired individuals and in Alzheimer disease. *JAMA Neurol*. 2013;70: 903–911.
145. Mattsson N, Insel PS, Aisen PS, et al. Brain structure and function as mediators of the effects of amyloid on memory. *Neurology*. 2015;84:1136–1144.
146. Wirth M, Madison CM, Rabinovici GD, Oh H, Landau SM, Jagust WJ. Alzheimer's disease neurodegenerative biomarkers are associated with decreased cognitive function but not beta-amyloid in cognitively normal older individuals. *J Neurosci*. 2013;33: 5553–5563.
147. Andrews KA, Frost C, Modat M, et al. Acceleration of hippocampal atrophy rates in asymptomatic amyloidosis. *Neurobiol Aging*. 2016;39:99–107.
148. Bourgeat P, Chetelat G, Villemagne VL, et al. Beta-amyloid burden in the temporal neocortex is related to hippocampal atrophy in elderly subjects without dementia. *Neurology*. 2010;74:121–127.
149. Chetelat G, Villemagne VL, Villain N, et al. Accelerated cortical atrophy in cognitively normal elderly with high beta-amyloid deposition. *Neurology*. 2012;78:477–484.
150. Gordon BA, Blazey T, Su Y, et al. Longitudinal beta-amyloid deposition and hippocampal volume in preclinical Alzheimer disease and suspected non-Alzheimer disease pathophysiology. *JAMA Neurol*. 2016;73:1192–1200.
151. Hanseeuw BJ, Betensky RA, Mormino EC, et al. PET staging of amyloidosis using striatum. *Alzheimers Dement*. 2018;14:1281–1292.
152. Knopman DS, Jack Jr CR, Wiste HJ, et al. Selective worsening of brain injury biomarker abnormalities in cognitively normal elderly persons with beta-amyloidosis. *JAMA Neurol*. 2013;70:1030–1038.
153. Oh H, Habeck C, Madison C, Jagust W. Covarying alterations in Abeta deposition, glucose metabolism, and gray matter volume in cognitively normal elderly. *Hum Brain Mapp*. 2014;35:297–308.
154. Ossenkoppele R, Zwan MD, Tolboom N, et al. Amyloid burden and metabolic function in early-onset Alzheimer's disease: parietal lobe involvement. *Brain*. 2012;135: 2115–2125.
155. Johnson SC, Christian BT, Okonkwo OC, et al. Amyloid burden and neural function in people at risk for Alzheimer's disease. *Neurobiol Aging*. 2014;35:576–584.
156. Oh H, Madison C, Baker S, Rabinovici G, Jagust W. Dynamic relationships between age, amyloid-beta deposition, and glucose metabolism link to the regional vulnerability to Alzheimer's disease. *Brain*. 2016;139:2275–2289.
157. Lowe VJ, Weigand SD, Senjem ML, et al. Association of hypometabolism and amyloid levels in aging, normal subjects. *Neurology*. 2014;82:1959–1967.
158. Pascoal TA, Mathotaarachchi S, Mohades S, et al. Amyloid-beta and hyperphosphorylated tau synergy drives metabolic decline in preclinical Alzheimer's disease. *Mol Psychiatry*. 2017;22:306–311.
159. Adams JN, Lockhart SN, Li L, Jagust WJ. Relationships between tau and glucose metabolism reflect Alzheimer's disease pathology in cognitively normal older adults. *Cereb Cortex*. 2019;29:1997–2009.
160. Cirrito JR, Yamada KA, Finn MB, et al. Synaptic activity regulates interstitial fluid amyloid-beta levels in vivo. *Neuron*. 2005;48:913–922.
161. Bero AW, Yan P, Roh JH, et al. Neuronal activity regulates the regional vulnerability to amyloid-beta deposition. *Nat Neurosci*. 2011;14:750–756.
162. Joutsa J, Rinne JO, Hermann B, et al. Association between childhood-onset epilepsy and amyloid burden 5 decades later. *JAMA Neurol*. 2017;74:583–590.
163. Jagust WJ, Mormino EC. Lifespan brain activity, beta-amyloid, and Alzheimer's disease. *Trends Cogn Sci*. 2011;15:520–526.

164. Mormino EC, Smiljic A, Hayenga AO, et al. Relationships between beta-amyloid and functional connectivity in different components of the default mode network in aging. *Cereb Cortex*. 2011;21:2399–2407.
165. Hahn A, Strandberg TO, Stomrud E, et al. Association between earliest amyloid uptake and functional connectivity in cognitively unimpaired elderly. *Cereb Cortex*. 2019;29:2173–2182.
166. Mormino EC, Brandel MG, Madison CM, Marks S, Baker SL, Jagust WJ. Abeta deposition in aging is associated with increases in brain activation during successful memory encoding. *Cereb Cortex*. 2012;22:1813–1823.
167. Leal SL, Landau SM, Bell RK, Jagust WJ. Hippocampal activation is associated with longitudinal amyloid accumulation and cognitive decline. *Elife*. 2017;6:e22978.
168. Sperling RA, Laviolette PS, O'Keefe K, et al. Amyloid deposition is associated with impaired default network function in older persons without dementia. *Neuron*. 2009;63:178–188.
169. Hedden T, Van Dijk KR, Becker JA, et al. Disruption of functional connectivity in clinically normal older adults harboring amyloid burden. *J Neurosci*. 2009;29:12686–12694.
170. Sheline YI, Raichle ME, Snyder AZ, et al. Amyloid plaques disrupt resting state default mode network connectivity in cognitively normal elderly. *Biol Psychiatry*. 2010;67: 584–587.
171. Mishra S, Gordon BA, Su Y, et al. AV-1451 PET imaging of tau pathology in preclinical Alzheimer disease: defining a summary measure. *Neuroimage*. 2017;161:171–178.
172. McSweeney M, Binette AP, Meyer P-F, et al. Early increase in tau-PET signal is associated with Aβ burden, CSF P-tau levels and cognition in cognitively normal late-middle-aged adults. *Alzheimers Dement*. 2018;14:P222–P223.
173. Braak H, Thal DR, Ghebremedhin E, Del Tredici K. Stages of the pathologic process in Alzheimer disease: age categories from 1 to 100 years. *J Neuropathol Exp Neurol*. 2011;70:960–969.
174. Kaufman SK, Del Tredici K, Thomas TL, Braak H, Diamond MI. Tau seeding activity begins in the transentorhinal/entorhinal regions and anticipates phospho-tau pathology in Alzheimer's disease and PART. *Acta Neuropathol*. 2018;136:57–67.
175. Jack Jr CR, Wiste HJ, Knopman DS, et al. Rates of beta-amyloid accumulation are independent of hippocampal neurodegeneration. *Neurology*. 2014;82:1605–1612.
176. Jack Jr CR, Therneau TM, Wiste HJ, et al. Transition rates between amyloid and neurodegeneration biomarker states and to dementia: a population-based, longitudinal cohort study. *Lancet Neurol*. 2016;15:56–64.
177. Pontecorvo MJ, Devous MD, Kennedy I, et al. A multicentre longitudinal study of flortaucipir (18F) in normal ageing, mild cognitive impairment and Alzheimer's disease dementia. *Brain*. 2019; https://doi.org/10.1093/brain/awz090.
178. Witte MM, Foster NL, Fleisher AS, et al. Clinical use of amyloid-positron emission tomography neuroimaging: practical and bioethical considerations. *Alzheimers Dement* 2015;1:358–367.
179. Draper B, Peisah C, Snowdon J, Brodaty H. Early dementia diagnosis and the risk of suicide and euthanasia. *Alzheimers Dement*. 2010;6:75–82.
180. Mozersky J, Sankar P, Harkins K, Hachey S, Karlawish J. Comprehension of an elevated amyloid positron emission tomography biomarker result by cognitively normal older adults. *JAMA Neurol*. 2018;75:44–50.
181. Harkins K, Sankar P, Sperling R, et al. Development of a process to disclose amyloid imaging results to cognitively normal older adult research participants. *Alzheimers Res Ther*. 2015;7:26.
182. Burns JM, Johnson DK, Liebmann EP, Bothwell RJ, Morris JK, Vidoni ED. Safety of disclosing amyloid status in cognitively normal older adults. *Alzheimers Dement*. 2017;13:1024–1030.

183. Golde TE, Schneider LS, Koo EH. Anti-abeta therapeutics in Alzheimer's disease: the need for a paradigm shift. *Neuron*. 2011;69:203–213.
184. Sperling RA, Jack Jr CR, Aisen PS. Testing the right target and right drug at the right stage. *Sci Transl Med*. 2011;3:111cm133.
185. Sperling RA, Rentz DM, Johnson KA, et al. The A4 study: stopping AD before symptoms begin? *Sci Transl Med*. 2014;6:228fs213.
186. Bateman RJ, Benzinger TL, Berry S, et al. The DIAN-TU next generation Alzheimer's prevention trial: adaptive design and disease progression model. *Alzheimers Dement*. 2017;13:8–19.
187. Mormino EC, Brandel MG, Madison CM, et al. Not quite PIB-positive, not quite PIB-negative: slight PIB elevations in elderly normal control subjects are biologically relevant. *Neuroimage*. 2012;59:1152–1160.
188. Landau SM, Horng A, Jagust WJ. Alzheimer's disease neuroimaging I. Memory decline accompanies subthreshold amyloid accumulation. *Neurology*. 2018;90:e1452–e1460.
189. Farrell ME, Chen X, Rundle MM, Chan MY, Wig GS, Park DC. Regional amyloid accumulation and cognitive decline in initially amyloid-negative adults. *Neurology*. 2018;91:e1809–e1821.
190. Leal SL, Lockhart SN, Maass A, Bell RK, Jagust WJ. Subthreshold amyloid predicts tau deposition in aging. *J Neurosci*. 2018;38:4482–4489.
191. Braak H, Braak E. Frequency of stages of Alzheimer-related lesions in different age categories. *Neurobiol Aging*. 1997;18:351–357.
192. Thal DR, Rub U, Orantes M, Braak H. Phases of A beta-deposition in the human brain and its relevance for the development of AD. *Neurology*. 2002;58:1791–1800.
193. Cho H, Choi JY, Hwang MS, et al. In vivo cortical spreading pattern of tau and amyloid in the Alzheimer disease spectrum. *Ann Neurol*. 2016;80:247–258.
194. Grothe MJ, Barthel H, Sepulcre J, et al. In vivo staging of regional amyloid deposition. *Neurology*. 2017;89:2031–2038.
195. Sepulcre J, Sabuncu MR, Becker A, Sperling R, Johnson KA. In vivo characterization of the early states of the amyloid-beta network. *Brain*. 2013;136:2239–2252.
196. Rodrigue KM, Kennedy KM, Devous Sr MD, et al. Beta-amyloid burden in healthy aging: regional distribution and cognitive consequences. *Neurology*. 2012;78:387–395.
197. Bischof GN, Rodrigue KM, Kennedy KM, Devous Sr MD, Park DC. Amyloid deposition in younger adults is linked to episodic memory performance. *Neurology*. 2016;87:2562–2566.
198. Bussy A, Snider BJ, Coble D, et al. Effect of apolipoprotein E4 on clinical, neuroimaging, and biomarker measures in noncarrier participants in the dominantly inherited Alzheimer network. *Neurobiol Aging*. 2019;75:42–50.
199. Pletnikova O, Rudow GL, Hyde TM, et al. Alzheimer lesions in the autopsied brains of people 30 to 50 years of age. *Cogn Behav Neurol*. 2015;28:144–152.
200. Pletnikova O, Kageyama Y, Rudow G, et al. The spectrum of preclinical Alzheimer's disease pathology and its modulation by ApoE genotype. *Neurobiol Aging*. 2018;71:72–80.
201. Jansen WJ, Ossenkoppele R, Knol DL, et al. Prevalence of cerebral amyloid pathology in persons without dementia: a meta-analysis. *JAMA*. 2015;313:1924–1938.
202. Breitner JC, Folstein MF. Familial Alzheimer Dementia: a prevalent disorder with specific clinical features. *Psychol Med*. 1984;14:63–80.
203. Mohs RC, Breitner JC, Silverman JM, Davis KL. Alzheimer's disease. Morbid risk among first-degree relatives approximates 50% by 90 years of age. *Arch Gen Psychiatry*. 1987;44:405–408.
204. Fratiglioni L, Ahlbom A, Viitanen M, Winblad B. Risk factors for late-onset Alzheimer's disease: a population-based, case-control study. *Ann Neurol*. 1993;33:258–266.

205. Cannon-Albright LA, Foster NL, Schliep K, et al. Relative risk for Alzheimer disease based on complete family history. *Neurology*. 2019;92:e1745–e1753.
206. Mosconi L, Rinne JO, Tsui WH, et al. Amyloid and metabolic positron emission tomography imaging of cognitively normal adults with Alzheimer's parents. *Neurobiol Aging*. 2013;34:22–34.
207. Mosconi L, Murray J, Tsui WH, et al. Brain imaging of cognitively normal individuals with 2 parents affected by late-onset AD. *Neurology*. 2014;82:752–760.
208. Villeneuve S, Vogel JW, Gonneaud J, et al. Proximity to parental symptom onset and amyloid-beta burden in sporadic Alzheimer disease. *JAMA Neurol*. 2018;75:608–619.
209. Sager MA, Hermann B, La Rue A. Middle-aged children of persons with Alzheimer's disease: APOE genotypes and cognitive function in the Wisconsin Registry for Alzheimer's Prevention. *J Geriatr Psychiatry Neurol*. 2005;18:245–249.
210. Scarabino D, Gambina G, Broggio E, Pelliccia F, Corbo RM. Influence of family history of dementia in the development and progression of late-onset Alzheimer's disease. *Am J Med Genet B Neuropsychiatr Genet*. 2016;171B:250–256.
211. Yi D, Lee Y, Byun MS, et al. Synergistic interaction between APOE and family history of Alzheimer's disease on cerebral amyloid deposition and glucose metabolism. *Alzheimers Res Ther*. 2018;10:84.
212. Poirier J, Davignon J, Bouthillier D, Kogan S, Bertrand P, Gauthier S. Apolipoprotein E polymorphism and Alzheimer's disease. *Lancet*. 1993;342:697–699.
213. van der Lee SJ, Wolters FJ, Ikram MK, et al. The effect of APOE and other common genetic variants on the onset of Alzheimer's disease and dementia: a community-based cohort study. *Lancet Neurol*. 2018;17:434–444.
214. Zlokovic BV. Neurovascular pathways to neurodegeneration in Alzheimer's disease and other disorders. *Nat Rev Neurosci*. 2011;12:723–738.
215. Villeneuve S, Brisson D, Marchant NL, Gaudet D. The potential applications of Apolipoprotein E in personalized medicine. *Front Aging Neurosci*. 2014;6:154.
216. Breitner JCS, Poirier J, Etienne PE, Leoutsakos JM, PREVENT-AD Research G. Rationale and structure for a new center for studies on prevention of Alzheimer's disease (StoP-AD). *J Prev Alzheimers Dis*. 2016;3:236–242.
217. Livingston G, Sommerlad A, Orgeta V, et al. Dementia prevention, intervention, and care. *Lancet*. 2017;390:2673–2734.
218. Kivipelto M, Ngandu T, Fratiglioni L, et al. Obesity and vascular risk factors at midlife and the risk of dementia and Alzheimer disease. *Arch Neurol*. 2005;62:1556–1560.
219. Whitmer RA, Sidney S, Selby J, Johnston SC, Yaffe K. Midlife cardiovascular risk factors and risk of dementia in late life. *Neurology*. 2005;64:277–281.
220. Power MC, Mormino E, Soldan A, et al. Combined neuropathological pathways account for age-related risk of dementia. *Ann Neurol*. 2018;84:10–22.
221. Iturria-Medina Y, Sotero RC, Toussaint PJ, Mateos-Perez JM, Evans AC. Alzheimer's disease neuroimaging I. Early role of vascular dysregulation on late-onset Alzheimer's disease based on multifactorial data-driven analysis. *Nat Commun*. 2016;7:11934.
222. SPRINT MIND Investigators for the SPRINT Research Group, Williamson JD, Pajewski NM, et al. Effect of intensive vs standard blood pressure control on probable dementia: a randomized clinical trial. *JAMA*. 2019;321:553–561.
223. Vemuri P, Knopman DS, Lesnick TG, et al. Evaluation of amyloid protective factors and Alzheimer disease neurodegeneration protective factors in elderly individuals. *JAMA Neurol*. 2017;74:718–726.
224. Gottesman RF, Schneider AL, Zhou Y, et al. Association between midlife vascular risk factors and estimated brain amyloid deposition. *JAMA*. 2017;317:1443–1450.
225. Nagga K, Gustavsson AM, Stomrud E, et al. Increased midlife triglycerides predict brain beta-amyloid and tau pathology 20 years later. *Neurology*. 2018;90:e73–e81.

226. Rabin JS, Schultz AP, Hedden T, et al. Interactive associations of vascular risk and beta-amyloid burden with cognitive decline in clinically normal elderly individuals: findings from the harvard aging brain study. *JAMA Neurol*. 2018;75:1124–1131.
227. Rodrigue KM, Rieck JR, Kennedy KM, Devous Sr MD, Diaz-Arrastia R, Park DC. Risk factors for beta-amyloid deposition in healthy aging: vascular and genetic effects. *JAMA Neurol*. 2013;70:600–606.
228. Reed B, Villeneuve S, Mack W, DeCarli C, Chui HC, Jagust W. Associations between serum cholesterol levels and cerebral amyloidosis. *JAMA Neurol*. 2014;71:195–200.
229. Muller S, Preische O, Sohrabi HR, et al. Relationship between physical activity, cognition, and Alzheimer pathology in autosomal dominant Alzheimer's disease. *Alzheimers Dement*. 2018;14:1427–1437.
230. Arenaza-Urquijo EM, Wirth M, Chetelat G. Cognitive reserve and lifestyle: moving towards preclinical Alzheimer's disease. *Front Aging Neurosci*. 2015;7:134.
231. Landau SM, Marks SM, Mormino EC, et al. Association of lifetime cognitive engagement and low beta-amyloid deposition. *Arch Neurol*. 2012;69:623–629.
232. Liang KY, Mintun MA, Fagan AM, et al. Exercise and Alzheimer's disease biomarkers in cognitively normal older adults. *Ann Neurol*. 2010;68:311–318.
233. Wirth M, Haase CM, Villeneuve S, Vogel J, Jagust WJ. Neuroprotective pathways: lifestyle activity, brain pathology, and cognition in cognitively normal older adults. *Neurobiol Aging*. 2014;35:1873–1882.
234. Arenaza-Urquijo EM, Bejanin A, Gonneaud J, et al. Association between educational attainment and amyloid deposition across the spectrum from normal cognition to dementia: neuroimaging evidence for protection and compensation. *Neurobiol Aging*. 2017;59:72–79.
235. Matthews DC, Davies M, Murray J, et al. Physical activity, mediterranean diet and biomarkers-assessed risk of Alzheimer's: a multi-modality brain imaging study. *Adv J Mol Imaging*. 2014;4:43–57.
235a. Lyons CE, Tudorascu D, Snitz BE, et al. The relationship of current cognitive activity to brain amyloid burden and glucose metabolism. *Am J Geriatr Psychiatry*. 2018;26:977–984.
236. Brown BM, Peiffer JJ, Taddei K, et al. Physical activity and amyloid-beta plasma and brain levels: results from the Australian imaging, biomarkers and lifestyle study of ageing. *Mol Psychiatry*. 2013;18:875–881.
237. Wirth M, Villeneuve S, La Joie R, Marks SM, Jagust WJ. Gene-environment interactions: lifetime cognitive activity, APOE genotype, and beta-amyloid burden. *J Neurosci*. 2014;34:8612–8617.
238. Vemuri P, Lesnick TG, Przybelski SA, et al. Effect of intellectual enrichment on AD biomarker trajectories: longitudinal imaging study. *Neurology*. 2016;86:1128–1135.
239. Spira AP, Gamaldo AA, An Y, et al. Self-reported sleep and beta-amyloid deposition in community-dwelling older adults. *JAMA Neurol*. 2013;70:1537–1543.
240. Spira AP, An Y, Wu MN, et al. Excessive daytime sleepiness and napping in cognitively normal adults: associations with subsequent amyloid deposition measured by PiB PET. *Sleep*. 2018;41. https://academic.oup.com/sleep/article/41/10/zsy152/5088807?casa_token=Gp56s2OG3JEAAAAA:1z_hadFEnypj4-dromJBqJ4OqMMWSjyEK9AABmYCdBHqpHuXiOL3bwHVLYOyUORF7KDuIFaXpx1G.
241. Branger P, Arenaza-Urquijo EM, Tomadesso C, et al. Relationships between sleep quality and brain volume, metabolism, and amyloid deposition in late adulthood. *Neurobiol Aging*. 2016;41:107–114.
242. Shokri-Kojori E, Wang GJ, Wiers CE, et al. Beta-amyloid accumulation in the human brain after one night of sleep deprivation. *Proc Natl Acad Sci USA*. 2018;115:4483–4488.

243. Tarumi T, Harris TS, Hill C, et al. Amyloid burden and sleep blood pressure in amnestic mild cognitive impairment. *Neurology*. 2015;85:1922–1929.
244. Bubu OM, Pirraglia E, Andrade AG, et al. Obstructive sleep apnea and longitudinal Alzheimer's disease biomarker changes. *Sleep*. 2019; https://doi.org/10.1093/sleep/zsz048.
245. Xie L, Kang H, Xu Q, et al. Sleep drives metabolite clearance from the adult brain. *Science*. 2013;342:373–377.
246. Carvalho DZ, St Louis EK, Knopman DS, et al. Association of excessive daytime sleepiness with longitudinal beta-amyloid accumulation in elderly persons without dementia. *JAMA Neurol*. 2018;75:672–680.
247. Mander BA, Marks SM, Vogel JW, et al. Beta-amyloid disrupts human NREM slow waves and related hippocampus-dependent memory consolidation. *Nat Neurosci*. 2015;18:1051–1057.
248. Andre C, Tomadesso C, de Flores R, et al. Brain and cognitive correlates of sleep fragmentation in elderly subjects with and without cognitive deficits. *Alzheimers Dement*. 2019;11:142–150.
249. Gabelle A, Jaussent I, Bouallegue FB, et al. Reduced brain amyloid burden in elderly patients with narcolepsy type 1. *Ann Neurol*. 2019;85:74–83.
250. Williams MM, Scharff DP, Mathews KJ, et al. Barriers and facilitators of African American participation in Alzheimer disease biomarker research. *Alzheimer Dis Assoc Disord*. 2010;24(suppl):S24–S29.
251. Barnes LL. Biomarkers for Alzheimer dementia in diverse racial and ethnic minorities—a public health priority. *JAMA Neurol*. 2019;76:251–253.
252. Barnes DE, Yaffe K. The projected effect of risk factor reduction on Alzheimer's disease prevalence. *Lancet Neurol*. 2011;10:819–828.

1cii

AD molecular: Imaging tau aggregates with positron emissions tomography

Tobey J. Betthauser*
Wisconsin Alzheimer's Disease Research Center, University of Wisconsin School of Medicine and Public Health, Madison, WI, United States
*Corresponding author: e-mail address: tjbetthauser@medicine.wisc.edu

Contents

Abstract

Pathologic aggregates of tau protein are observed in several neurodegenerative diseases and are used to diagnose and stage disease postmortem. Recent advances in positron emission tomography radioligands allow for the detection of aggregated tau proteins in living persons. This chapter describes the development and characterization of several positron emission tomography radioligands used to detect tau pathophysiology in vivo, and how these ligands are being used in clinical aging and neurodegenerative disease research with a focus on imaging tau aggregates in Alzheimer's disease.

1. Introduction

The microtubule associated protein tau plays a role in the stabilization of microtubules and has several pathologic forms observed in neurodegenerative diseases. Recent and rapid progress has been made in the area of

Progress in Molecular Biology and Translational Science, Volume 165
ISSN 1877-1173
https://doi.org/10.1016/bs.pmbts.2019.07.007

molecular imaging of pathologic tau aggregates using positron emission tomography (PET). The focus of tau PET ligand discovery and clinical research has largely centered on imaging neurofibrillary tangles (NFTs) observed in Alzheimer's disease (AD). This chapter gives a brief summary of tau protein aggregates observed in neurodegenerative diseases, features of these aggregates and their implications for PET imaging, and an overview of the development and use of tau PET radioligands in clinical research.

2. Tau protein and tauopathies

In the human brain, tau is an intracellular protein expressed in neurons that plays a role in the assembly and stabilization of microtubules, and belongs to the microtubule-associated protein (MAP) family.[1] Tau expressed in human adults can have six isoforms as a consequence of alternative splicing of exons 2, 3 and 10.[2] These isoforms are frequently subdivided as 3-repeat (3R) or 4-repeat (4R) based on the number of repeat regions in the microtubule binding domain. In neurodegenerative diseases, it is postulated that abnormal phosphorylation of tau results in dissociation of tau from the microtubule after which tau begins to self-aggregate[2–4] forming insoluble aggregates of tau protein. Tau aggregates differ in the expression of 3R and 4R isoforms, ultrastructural, and often have differentiable spatial deposition patterns postmortem.[5–9] The structural differences between tau aggregates likely affect the accessibility of binding sites for PET radioligands, which has consequences for ligand specificity and sensitivity as is discussed in more detail in Section 3.[10,11]

In AD, dissociated 3R and 4R tau form paired-helical filaments[12] (PHF) which make up NFTs, a hallmark proteinopathy of AD (the other hallmark being beta-amyloid plaques, see chapter "AD molecular" by Abeta, Villeneuve, and Sylvia).[7,12–14] Notably for PET ligand development and specificity, tau aggregates observed in AD differ in ultrastructure from tau aggregates observed in many non-AD tauopathies that tend to have predominantly straight-filaments composed of 4R tau (e.g., progressive supranuclear palsy [PSP]; corticobasalar degeneration [CBD]; argyrophilic grain disease [AGD]) or twisted filaments composed of 3R tau (e.g., Pick's disease).[5,15] A recent study also indicated molecular differences in the core structures of PHF and straight filament tau aggregates, and suggested these differences might affect the specificity of PET ligands to one type of tau aggregate over another.[11]

2.1 Neuropathological staging of neurofibrillary tangles

Postmortem neuropathological characterization of tau in aging and AD has been extensively studied in thousands of human autopsy cases spanning the human lifespan.[13,16–19] Neuropathological staging of NFTs (i.e., Braak staging or NFT staging) was initially characterized by Heiko and Eva Braak using Gallyas silver staining[7,20] and was later revised by Braak and colleagues using a more standardized approach that incorporated the anti-tau antibody AT8,[13] which labels tau phosphorylated at serine 202 and threonine 205.[21] These key studies described the neuropathological deposition pattern of pathologic tau, which follows an ordered hierarchical spatial pattern in AD. The pattern of pathologic tau staging (i.e., Braak staging) begins with soluble, intracellular pretangles observed in the locus coeruleus and other brainstem nuclei,[19] followed by pretangles in the medial temporal lobe (MTL) including the transentorhinal cortex. After the pretangle stage, aggregation of insoluble argyrophilic NFTs begins in the transentorhinal cortex, followed by the hippocampus and entorhinal cortex, and continue to spread inferiolaterally along the temporal lobe and ultimately into the greater association cortex (see Braak et al. for a more detailed description of affected regions[13]). Following neuronal death, tau tangles can remain in the extracellular space and are referred to as ghost tangles.[13] The frequency and severity of NFT stages increase with increasing age and correlates more strongly with cognitive decline compared to Aβ plaques at autopsy.[22] In addition to spatial and structural progression, the expression of 3- and 4-repeat isoforms progresses from mostly 4R to mostly 3R over the transition from pretangles to ghost tangles with both isoforms expressed in NFTs in AD.[23] For PET radioligand development, this ordered spatial pattern provides a useful benchmark for characterizing tau PET radioligands in vivo. Currently, postmortem pathological confirmation of beta-amyloid plaques and NFTs provides the only definitive diagnosis of Alzheimer's disease.[24]

3. Tau PET ligand development

3.1 Overview of tau PET ligand development

Radioligands developed for PET neuroimaging have chemicophysical and pharmacokinetic properties that generally make them amenable to quantifying receptor targets in the human brain. Readers are referred to several previous works that discuss these properties, pharmacokinetic modeling and consensus PET nomenclature in greater detail.[10,25–30] To briefly summarize these works, passively diffusible neuroimaging radioligands need to cross the blood-brain barrier in high enough concentrations (max standard uptake value (SUV) $> \sim 1$) to facilitate detection with PET scanners, exhibit a sufficiently rapid pharmacokinetic profile (i.e., fast uptake and washout from the brain in target-free regions), be free of radiometabolites in the brain, have low non-specific binding (e.g., white matter) and have high affinity and selectivity for the physiological target of interest (i.e., tau aggregates in the context of this chapter) over other potential brain targets (e.g., beta-amyloid plaques). Candidate ligands amenable to radiolabeling with fluorine-18 are often preferred over carbon-11 labeled ligands to facilitate distribution of PET tracers to clinical research sites without the means to generate radioisotopes onsite. Ideally for noninvasive quantification, radioligands are reversibly bound (i.e., can be modeled with a reversible two-tissue compartment model) to receptor targets over the duration of the PET scan[31] with stable binding estimates (e.g., binding potential [BP_{ND}][32] or standard uptake value ratio [SUVR]) achieved using reference tissue methods[33–38] within minutes to 1–2 h following bolus tracer injection.

Imaging tau aggregates superimposes additional challenges for PET radioligand development beyond those mentioned in the previous paragraph. As described above, tau aggregates are primarily intracellular until later stages of disease, vary in the expression of 3R and 4R isoforms across different disease states and neurodegenerative diseases, and differ in ultrastructure across various types of tauopathies. The intracellular localization of tau aggregates requires PET radioligands to be able to enter the neurons to access aggregated tau binding sites throughout most of the pathologic process. Ultrastructural differences between various tauopathies also likely affect the accessibility of PET ligands to potential binding sites and therefor have consequences for ligand specificity for different tau aggregates.[11] Tau also belongs to a group of proteins referred to as amyloids (e.g., beta-amyloid, alpha synuclein, Huntington, etc.) that are frequently observed in neurodegenerative diseases and are defined by their tendency to self-aggregate into fibrillary structures with beta-pleated sheets.[10] Neurodegenerative diseases often co-occur in late-life and protein aggregates present in AD (i.e., beta-amyloid plaques) and non-AD neurodegenerative diseases co-localize with tau aggregates.[39,40] In AD, Aβ plaques and NFTs co-localize in neocortical regions and the density of Aβ plaques has been reported to be 4–20 times higher than the density of NFTs.[7,14] Therefore in order to make accurate inference about pathologic tau aggregates from

PET imaging, tau PET radioligands need to have high selectivity for tauopathies over other protein aggregates and high selectivity for specific forms of pathologic tau aggregates (e.g., PHF over straight filaments) if they are to be used for differentiating various tauopathies in vivo. Given the potential for differences in ligand selectivity for the various tau aggregates, this work generally refers to "tau PET" or "tau imaging" as encompassing tau aggregates observed in all tauopathies and makes further distinctions where evidence exists to support selectivity of ligands for particular forms of pathologic tau.

3.2 PET radioligands for imaging tau aggregates

Currently there are several PET radioligands for imaging pathologic tau aggregates in various stages of development. First generation tau PET ligands include FDDNP,[41] Flortaucipir (aka T807, AV-1451)[42] and T808 (aka AV-680,[43] the THK series (THK-523,[44] THK-5105, THK-5117 [and enantiomerically pure (S)THK-5117, aka THK-5317], THK-5317),[45] THK-5351[46]) and PBB3.[47] More recently developed ligands and those in earlier stages of development include MK-6240,[48–50] PI-2620,[51] JNJ-311 (aka JNJ-64349311),[52] RO-948 (aka RO6958948),[53] PM-PBB3,[54] *N*-methyl Lansoprazole (aka NML),[55,56] and GTP-1[57] (Fig. 1).

3.2.1 FDDNP

[^{18}F]FDDNP was the first radioligand to observe tau aggregates in human subjects in vivo.[41] Initial study indicated that FDDNP could differentiate probable AD groups from mild cognitive impairment (MCI) and cognitively unimpaired groups, and showed elevated FDDNP binding in a case with pathologically confirmed AD brought to autopsy 14 months after FDDNP imaging.[58] However, in vitro studies indicated that FDDNP bound to both beta-amyloid plaques and NFTs[59] and therefor FDDNP is unsuitable for selective imaging of tau aggregates over beta-amyloid in AD. Results from a later study directly comparing the selective beta-amyloid tracer [^{11}C]Pittsburgh Compound B (PiB) to [^{18}F]FDDNP in human subjects ranging from controls to AD patients were consistent with the in vitro findings that FDDNP binds to both beta-amyloid and non-beta amyloid targets.[60]

3.2.2 THK series

The THK series was developed from an initial screening that identified quinolone derivatives (BF-158, BF-170, THK-523 [aka BF-242]) as potential ligands for tau and Aβ fibrils.[61] THK-523, and 2-arylquinoline

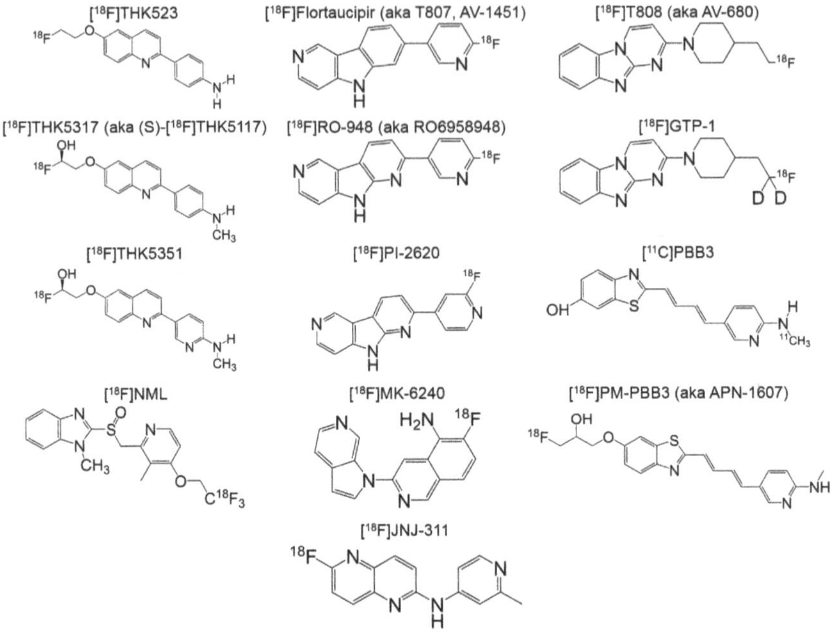

Fig. 1 Chemical structures of candidate PET radioligands for imaging tau protein aggregates.

derivatives THK-5105, THK-5117 and THK-5116 were characterized in preclinical studies[62] and subsequently [^{18}F]THK-523,[63] [^{18}F]THK-5105[62] and [^{18}F]THK-5117[64] proceeded to initial human studies. THK-523 indicated fast brain uptake and clearance, but limited sensitivity for tau aggregates and high white matter binding precluding visual and quantitative assessment of tau aggregates in vivo.[63] Similar characteristics were observed for THK-5105, but with a slower pharmacokinetic profile.[62] In vitro and in vivo human studies identified [^{18}F]THK-5317 (aka (*S*)-[^{18}F]THK-5117) as the most promising of these ligands for in vivo tau imaging. Full pharmacokinetic analysis (i.e., with arterial sampling) of THK-5317 supported reversible 2-tissue compartment kinetics and simplified quantification strategies using the cerebellum as a reference region.[65] Due to high nonspecific white matter binding with THK-5317, a further derivative ([^{18}F]THK-5351, aka GE-216) was developed to be less lipophilic in an effort to reduce nonspecific white matter binding.[46] [^{18}F]THK-5351 initially showed promise for selective imaging of tau aggregates, but also exhibited high binding in the basal ganglia in older controls and AD patients that was not consistent with known distributions of NFTs in AD (demonstrated in Figs. 2 and 3).[46,66,67] It was later revealed that a considerable

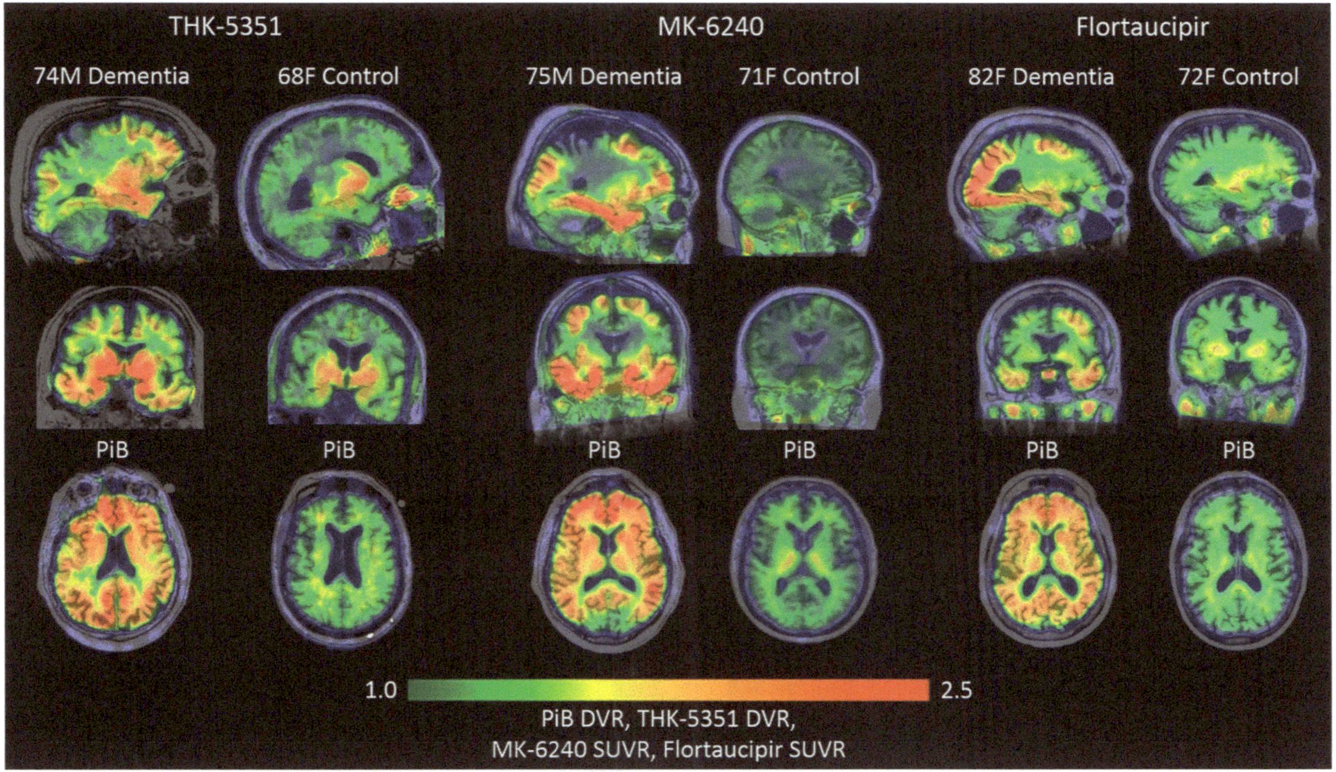

Fig. 2 Sample images for tau PET ligands. Parametric images comparing THK-5351, MK-6240, and Flortaucipir in amyloid negative control and amyloid positive dementia cases. Each column of images represents an individual person.

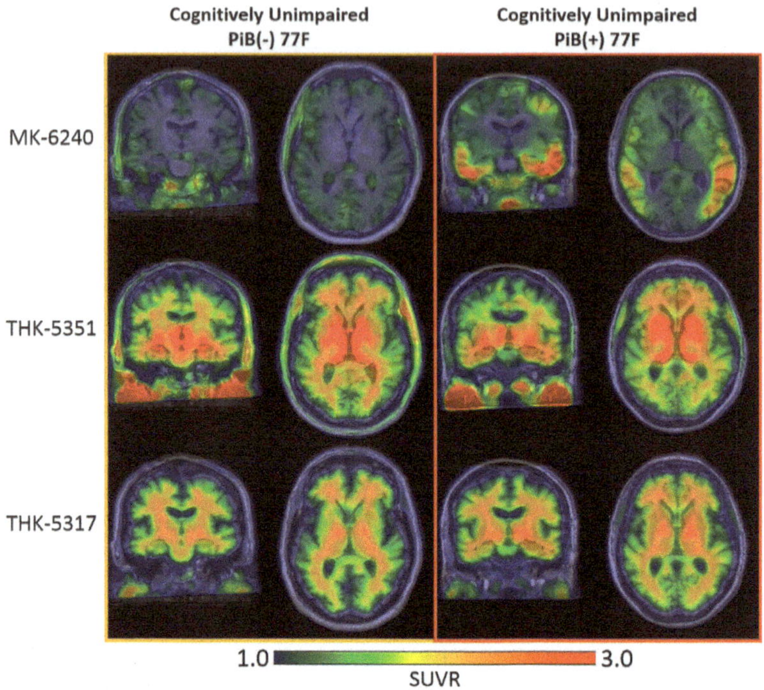

Fig. 3 Comparison of MK-6240, THK-5317 and THK-5351 parametric images. Direct comparisons of MK-6240 (top row), THK-5351 (middle row) and THK-5317 (bottom row) parametric SUVR images in two subjects scanned with all three tracers at the University of Wisconsin-Madison Alzheimer's Disease Research Center. THK-5317 and THK-5351 PET scans were acquired approximately 40 and 30 months prior to MK-6240 PET scans, respectively.

fraction of THK-5351 specific binding was displaced by selegiline, an irreversible monoamine oxidase-B (MAO-B) inhibitor, in non-human primates (cynomolgus)[68] and in humans[69] suggesting a high affinity of THK-5351 for MAO-B. Imaging to postmortem correlates from a human study would affirm the correspondence between THK-5351 in vivo and the postmortem confirmed distribution of MAO-B.[70] This study also identified differences in in vitro tissue preparation and processing that accounted for the discrepancies between previous in vitro studies and between in vitro and in vivo findings. Chemical similarities coupled with reports of similar binding patterns between THK-5351 and its predecessors (e.g., THK-5117) also provide some evidence that these antecedent compounds might also bind to MAO-B, which may explain the shared THK-5351 and THK-5117 binding sites observed in vitro[71] and a conference report that indicated

THK-5351 and THK-5317 were more highly correlated in the basal ganglia than in the cortex in a limited number of humans scanned with both tracers.[72] THK-5317 blocking studies in humans investigating off-target MAO-B binding with MAO-B inhibitors have not yet been performed to validate these observations.

The studies of THK compounds provide some useful insights in the development for tau-specific PET radioligands. Perhaps the most salient point is in regards to methodological considerations and interpretations of in vitro studies and comparisons to in vivo findings. Notably, differences between in vitro tissue preparation and washing, such as fixation and the use of ethanol in washing of in vitro tissues, gave a contradictory results for the affinity of THK-5351 to MAO-B when compared to in vivo imaging findings and in vitro studies using fresh tissue without the use of fixation or ethanol. In a recent review paper written by the developers of THK, they specifically acknowledge that fixation of tissue with formalin, paraformaldehyde or alcohol can denature proteins and inactivate enzymatic activity while not altering aggregated proteins and that this was indeed the case with THK-5351.[73] They also note that they observed differences in results between functional assays and ligand-binding assays wherein the former were less sensitive in evaluating the binding ability of candidate PET ligands to enzymes.[74] These studies also highlight the caution in interpretation of in vitro results that is warranted when in vitro findings fail to recapitulate those observed in vivo, and the importance of in vivo competition studies to identify off-target binding targets when possible. This latter point also draws attention to another challenge for developing tau PET tracers, which is the difficulty in safely performing blocking studies in human subjects to identify reversible tracer binding and the affinity of tau PET tracers for various in vivo receptors sites (i.e., off-target and target binding). The work by Guo and Colleagues[68] provides evidence that drug challenge studies in nonhuman primates might be a viable alternative to human studies for identifying non-tau in vivo targets of candidate tau PET radioligands, which is also supported by the lack of tauopathies observed in wild-type nonhuman primates and the presence of many other shared neurobiological targets.[75]

3.2.3 Flortaucipir (aka T807, AV-1451) and T808 (aka AV-680)

[^{18}F]Flortaucipir is the most widely used and studied tau PET radioligand to date. An initial screening of fluorescent compounds observed that earlier tau ligand candidates T557 and T726 stained PHF tau and to a lesser extent $A\beta_{1-42}$, but ^{18}F-labeled analogs showed minimal brain uptake in rodents.[76,77] [^{18}F]T808 and [^{18}F]Flortaucipir were identified as the lead compounds through an iterative process designed to maximize autoradiographical discrimination between immunohistochemically reactive PHF tau and $A\beta_{1-42}$ in frontotemporal tissue sections from three groups (i.e., high PHF, high Aβ; low PHF, high Aβ; and no PHF or Aβ). Autoradiographic comparisons indicated that T808 had a 27-fold selectivity for PHF tau over $A\beta_{1-42}$ and minimal signal in GM and WM in non-AD tissues. Further in vitro studies of Flortaucipir and T808 indicated selectivity to PHF tau over $A\beta_{1-42}$ in competition binding assays and autoradiographic comparisons in human AD and control tissues, a favorable pharmacokinetic profile in mice and rats, and no off-target in vitro binding to over 72 other central nervous system targets, including MAO-A and -B measured by a functional assay.

First human studies for [^{18}F]Flortaucipir and [^{18}F]T808 in a limited number of AD, MCI and older control cases indicated that both tracers had suitable in vivo pharmacokinetic profiles, low white matter binding, and showed differential binding patterns between AD and control cases.[42,43] T808 also exhibited signal in cortical bone consistent with in vivo defluorination.[43] Later studies with full kinetic modeling (i.e., with arterial sampling) indicated Flortaucipir exhibited a kinetic profile consistent with the reversible two-tissue compartment model in higher binding regions associated with pathological staging of NFTs.[78–80] Early reports of clinical studies using Flortaucipir further demonstrated that Flortaucipir binding differentiated probable AD groups from older controls in regions associated with Braak NFT staging and showed associations between cognitive test outcomes and Flortaucipir in populations that included controls and persons with probable AD dementia (see Section 4 for more details).[81–83] These studies also showed some initial concordance between Flortaucipir binding and the hierarchy of NFTs that is observed in pathology studies.[84] In addition to tau related findings, early reports also commented on Flortaucipir retention in the basal ganglia, substantia nigra, and the choroid plexus, which was inconsistent with pathological staging of NFTs in AD (sample Flortaucipir images in an AD and control case are shown in Fig. 2).

The sources of potential off-target Flortaucipir binding and the potential for Flortaucipir to bind to non-AD tauopathies in humans remains to be fully understood with several conflicting results in the present literature. Autoradiographical studies using fixed tissues and ethanol in washing steps identified neuromelanin- and melanin-containing cells, hemorrhagic lesions, iron, calcifications in the choroid plexus and leptomeningeal melanin as potential sites of off-target binding.[85,86] These studies also indicated that Flortaucipir had higher specific binding in AD cases compared to non-AD tauopathies. In general, these studies did not observe the proportionally high signal in the basal ganglia that is observed in vivo. In vitro autoradiography and binding displacement studies using fresh frozen tissues with non-ethanol washing conditions did observe displaceable binding in the basal ganglia, which was displaced by clorgyline (MAO-A inhibitor) and selegiline (MAO-B inhibitor) with affinities similar to those observed for AD tau aggregates suggesting that MAO-A and -B are sites of off-target Flortaucipir binding.[49,71,87] One of these studies[87] indicated discrepancies for MAO binding between functional (indicated μM affinity to MAO enzymes) and competition assays (indicated nM affinity to MAO A and B), and suggested that differences in the level of observed off-target binding between regions were due to a combination of the differences in Flortaucipir affinity for MAO-A and MAO-B and the known concentrations of these enzymes.[88] A preclinical study in rhesus macaques with arterial sampling investigating another candidate tau PET ligand (MK-6240, see below) showed that the total distribution volume (V_T) of [^{18}F]Flortaucipir was reduced throughout the brain in the presence of 0.5 or 1.5 mg/kg unlabeled Flortaucipir, whereas the total V_T of [^{18}F]MK-6240 was unchanged in the presence of 0.5 mg/kg unlabeled MK-6240, suggesting reversible off-target Flortaucipir binding to non-tau targets in vivo. In humans in vivo, iron-sensitive MR imaging indicated an association between Flortaucipir binding in the putamen and age-related iron deposition.[89] In addition, Flortaucipir basal ganglia SUVR values have been shown to be as high as values in NFT-associated regions of probable AD participants,[81,90] and have been shown to have a moderate positive correlation with age, even when partialing out biomarker measures of pathologic beta-amyloid.[91] The positive basal ganglia age association with Flortaucipir in the absence of pathophysiologic amyloid may suggest that this signal is partially due to MAO-B, which has been shown with [^{11}C]Deprenyl to increase with age between 5% and 10% per decade in adult humans.[92] Contrary to these findings, a retrospective study comparing Flortaucipir SUV ($SUV = \text{activity concentration} \times \text{body mass}/\text{injected activity}$) in Parkinson's disease patients that either were or were not using MAO-B inhibitors at the time of their PET scan indicated no group differences across several brain regions.[93]

There are a growing number of in vitro and in vivo studies investigating Flortaucipir in non-AD tauopathies, including some with imaging to post-mortem correlates. The largest study comparing clinically diagnosed PSP patients to Parkinson's disease patients and controls indicated that the PSP group had significantly higher levels of Flortaucipir binding in the basal ganglia compared to the other groups, but there was considerable overlap between these groups, interindividual differences within the PSP group, and the authors suggest that off-target basal ganglia binding likely confounds the interpretation of tau-related signal in these cases, which may explain the overlap between groups.[94] In vivo studies with smaller numbers of PSP cases have indicated mixed results. Most, but not all, of these studies report higher Flortaucipir binding in the basal ganglia of probable PSP cases relative to controls, but are discrepant as to whether Flortaucipir binding is differentiable between control and PSP groups and whether Flortaucipir binding correlates with symptomology in regions of know tauopathy in PSP.[95–99] Studies with imaging to postmortem correlates from CBD, PSP and *MAPT* mutation carrier cases also indicated discrepant results with some reporting agreement and others reporting minimal or no agreement between in vivo Flortaucipir and postmortem distributions of 4R tau aggregates, whereas in vivo Flortaucipir binding is highly correlated with AT8 immunoreactivity in AD postmortem.[94,100–104]

Despite the broad use of Flortaucipir in clinical research studies, the sources and magnitude of potential off-target binding, and the ability of Flortaucipir to detect non-AD tauopathies still require further characterization. As with THK-5351, there are notable differences between in vitro autoradiography and displacement binding assay methods that preclude direct comparisons of results across studies, and likely explain the discrepant results. For example, in vitro studies using fixed tissues and ethanol in wash conditions[85,86,103] report little to no displaceable binding of Flortaucipir in MAO containing regions (e.g., putamen) whereas studies using fresh frozen tissues without ethanol washes indicate moderate to high displacement of Flortaucipir by self-block and MOA-A and MAO-B inhibitors in these same regions and in in vitro binding assays of other regions.[49,71,87] Another finding from these Flortaucipir characterization studies that is of interest for future ligand screening comes from comparisons of functional and ligand displacement assays for enzyme targets. Consistently, functional assays indicated MAO enzymatic function was intact in the presence of μM concentrations of Flortaucipir but ligand displacement assays indicated nanomolar affinity of Flortaucipir for both MAO-A and MAO-B, which may indicate that functional assays are not suitable for screening candidate PET radioligands. Similar to observations with MAO-A and -B, the discrepancies in the ability of Flortaucipir to detect non-AD tauopathies may be attributable to a combination of lower density (i.e., low B_{max}) of tau aggregates in non-AD tauopathies compared to AD, and a lower affinity (i.e., higher K_D) of Flortaucipir for non-AD tauopathies, which was observed in comparisons of AD and PSP tissues in vitro.[87] As of this writing, there have been no reported in vivo human studies evaluating the effect of MAO-A or MAO-B inhibitors on Flortaucipir binding using a blocking study design despite MAO-A and MAO-B inhibitors being approved for clinical use in humans. These types of studies along with direct comparisons to other tracers thought to be more specific to tau aggregates (e.g., MK-6240, PI-2620, RO-948) are needed to ascertain the potential limitations of Flortaucipir for detecting tau aggregates, and to better interpret Flortaucipir imaging findings in clinical research studies.

3.2.4 PBB3, AM-PBB3, and PM-PBB3

Phenyl/pyridinyl-butadienyl-benzothiazoles/benzothiazoliums (i.e., PBBs) were screened and developed as ligands for tau aggregates based on observations from beta-amyloid tracers, histopathological dyes thioflavin-S, Congo red and FSB, and chemical characteristics that enabled access and binding to β-pleated sheets.[47] Double fluorescent staining in human AD and Pick's disease tissues indicated several PBBs with selectivity for tau aggregates over beta-amyloid, and notably also identified compounds that bound to AD tau aggregates and Pick's bodies,[47] a 3R tauopathy with twisted and straight filament ultrastructures.[105] Based on this initial evaluation, [^{11}C]PBB3 was selected as the lead compound for initial human studies which showed rapid peripheral metabolism, adequate brain penetrance and moderate washout in the brains of controls. In 3 PiB(+) AD cases, [^{11}C]PBB3 showed elevated binding in regions consistent with Braak NFT, but also exhibited high signal in the striatum and thalamus. Brain homogenate and autoradiography studies comparing PBB3 with other potential tau imaging ligands in human AD tissues showed that PBB3 was not displaced by Flortaucipir or THK compounds indicating a separate binding site to AD tau aggregates compared to most other tau PET ligands.[71] However, it was soon revealed that PBB3 experienced photoisomerization under fluorescent light, resulting in rapidly decreasing radiochemical purity of formulated [^{11}C]PBB3 within minutes of formulation,[106] and that metabolism of [^{11}C]PBB3 resulted in a brain-penetrable, non-polar radiometabolite[106,107] in humans complicating quantification of PBB3 binding.[108] Derivatives of PBB3 ([^{18}F]PM-PBB3 (aka APN-1607) and [^{18}F]AM-PBB3) were therefore developed to resolve the metabolite confound.[54] Initial conference reports suggest that unlike most other available PET radioligands, PM-PBB3 binds to tau aggregates in primarily 4R (PSP, CBD) and primarily 3R (Pick's disease) tauopathies in addition to mixed 3R and 4R tau aggregates in AD.[109–111] In addition, these early reports indicated PM-PBB3 resulted in higher binding potential estimates in AD patients and lower basal ganglia signal in healthy controls compared to PBB3.[54] All reports indicate that [^{18}F]PM-PBB3 will be further evaluated in humans in several different tauopathies.

3.2.5 MK-6240

MK-6240 was discovered through a process that initially screened potential ligands using central nervous system multiparameter optimization[112] in the context of minimally modifying chemical properties when radiolabeled with fluorine-18, and emphasizing ligands that minimized nonspecific white matter binding and optimized the pharmacokinetic profile in nonhuman primates.[113] In vitro comparisons of several candidate ligands from the discovery process ultimately led to the selection of [^{18}F]MK-6240 for further in vitro characterization and in vivo preclinical characterization using T808 and Flortaucipir as benchmarks.[49] All in vitro studies were carried out using fresh frozen tissues without ethanol present. Displacement assays and autoradiographical comparisons indicated MK-6240 had a high affinity and selectivity for NFTs over beta-amyloid plaques in immunohistologically confirmed human AD frontal cortex and hippocampus tissues, and had only low non-displaceable binding in the putamen, frontal cortex and entorhinal cortex of control tissue sections devoid of AT8 immunoreactivity (i.e., there was no in vitro evidence of displaceable off-target MK-6240 binding). In NFT containing regions MK-6240 was displaced by T808 and by self-block and indicated a higher affinity for NFTs compared to Flortaucipir under the same in vitro conditions. By comparison, Flortaucipir binding in the hippocampus, frontal cortex and putamen of control slices without AT8 immunoreactivity was displaceable by self-block, but was not displaced by T808.[49] In vitro competition binding assays using tissue devoid of NFT pathology indicated that MK-6240 was unable to displace [^{3}H]Flortaucipir binding at nM concentrations, but that both clorgyline (an MAO-A inhibitor) and Flortaucipir self-block were able to displace [^{3}H]Flortaucipir at subnanomolar concentrations. The in vitro findings suggest Flortaucipir and MK-6240 share common binding sites for NFTs, but not for non-tau molecular targets (i.e., MAO-A in this study). Preclinical study in rhesus macaques indicated rapid uptake and washout in the brain, uptake in the skull suggesting in vivo defluorination, and observed that the total [^{18}F]MK-6240 V_T in the brain was not altered by self-block of 0.5 mg/kg unlabeled MK-6240. Comparatively, [^{18}F]Flortaucipir V_T was displaced throughout the macaque brain in the presence of 0.5 of 1.5 mg/kg unlabeled Flortaucipir, which again suggests that Flortaucipir exhibits displaceable binding to non-tau targets.

Initial human studies with full pharmacokinetic modeling of MK-6240 in controls, MCI and probable AD cases indicated a pharmacokinetic profile consistent with the reversible two-tissue compartment model in regions with suspected neurofibrillary tangle deposition.[50,114] Stable plasma and reference tissue based distribution volume ratio (DVR) estimates were achieved within 90 min of bolus MK-6240 injection and reference tissue DVR estimates using MRTM2 appeared stable with as little as 60 min dynamic scanning.[48,50,114] Studies evaluating quantification methodology indicated nearly 1:1 agreement between DVR estimates and SUVR established 60–90 or 70–90 min post MK-6240 injection, and also observed SUVR increasing for the high binding AD cases as late as 100 min post MK-6240 injection.[48,50,114] A study evaluating MK-6240 in young controls, older controls, MCI and probable AD cases indicated that MK-6240 spatial binding patterns in amyloid positive cases (assessed by PiB PET) were consistent with patterns described in neuropathological staging of NFTs. This study also observed elevated signal in the calvarium, meninges, various sinuses, substantia nigra and between the cerebellum and occipital cortex (tentorium cerebelli), but did not observed elevated binding in the basal ganglia or the choroid plexus with the exception of a single case with moderately elevated choroid signal (sample MK-6240 images are presented for an AD and older control case in Fig. 2, and additional cases in Fig. 3). The extra axial signal in the calvarium and meninges was observed to increase in intensity relative to cerebellar gray matter throughout 120 min post MK-6240 injection,[48] but the extent to which this signal may affect quantification and visual interpretation of tau-related signal has not yet been characterized. There are currently several ongoing clinical trials evaluating MK-6240 for detection of NFTs in AD, and using MK-6240 to characterize NFTs relative to cognitive decline and other biomarkers.

3.2.6 RO-948

The RO compounds were screened using [^{3}H]T808 displacement assays in fresh frozen AD tissues and identifying compounds with lower lipophilicity to minimize nonspecific binding.[115] Six lead compounds were identified from the initial screening and were further tested by labeling these compounds with tritium and conducting autoradiographical comparisons to immunohistochemical stains for beta-amyloid (mAb BAP-2) and tau (mAb pS422). From these lead compounds, three ([^{18}F]RO6958948 [aka RO-948], [^{11}C]RO6931643 [aka RO-643] and [^{11}C]RO6924963 [RO-963]) were selected for further in vitro, preclinical and human studies.[53,116,117] Autoradiographical competition studies in fresh frozen AD tissues indicated that all three compounds displaced [^{18}H]T808 at nanomolar concentrations, and that all three lead compounds did not displace [^{3}H] florbetapir (a PET ligand for imaging beta-amyloid plaques). High concordance was observed between tritiated RO-series compounds and anti-tau (pS422), but not anti-beta-amyloid (BAP-2) immunostaining, and all three compounds indicated low nonspecific binding in non-AD control tissues. Off-target binding was not observed for these compounds in autoradiography, but a selectivity screen (Cerep) of CNS targets did indicate concentration dependent displaceable binding to adenosine A3 receptor with 50% inhibition observed at slightly >30 nM for RO-948. [^{3}H]RO-948 was not inhibited in autoradiographical experiments in the presence of 10 μM concentrations of L-deprenyl (MAO-B inhibitor), clorgyline (MAO-A inhibitor), or lazabemide (reversible MAO-B inhibitor), suggesting no in vitro off-target MAO-A or MAO-B binding. No appreciable [^{3}H] RO-948 binding or T808 displacement was observed in tau immunoreactive PSP, CBD or Pick's disease sections.

Preclinical characterization was conducted in six baboons using PET imaging with arterial sampling.[116] All three lead compounds indicated rapid uptake and clearance from the baboon brain (peak SUV 1.2–2.0) and rapid metabolism of parent radioligands, with RO-963 having slower washout compared to the other compounds. Tissue kinetics in baboons were better described by the two-tissue compartment model compared to the one-tissue compartment model and differences in V_T were observed between the three compounds (RO-963 $V_T = 29.1$ mL/mL; RO-948 $V_T = 6.4$ mL/mL; RO-643 $V_T = 0.7$ mL/mL). Time-activity curves for RO-948 and RO-643 were similar across all brain regions, but were higher in the thalamus and putamen for RO-963 relative to the cerebellum and cortical regions. Minimal white matter binding was observed for all three RO ligands and there was no evidence of in vivo defluorination. No blocking studies were conducted to investigate potential displaceable binding in nonhuman primates.

Initial human studies were conducted for ([^{18}F]RO6958948 [aka RO-948], [^{11}C]RO6931643 [aka RO-643] and [^{11}C]RO6924963 [RO-963]) in young controls, older controls and probable AD patients with PET confirmed beta-amyloid positivity.[53,117] The three tracers were initially compared pairwise in a limited number of young control and AD subjects (each study participant received scans with two of the tracers). All three tracers entered the human brain (peak SUV 2–3.5 for RO-963 and RO-948, 1–2 for RO-643) with RO-948 and RO-643 having faster uptake and washout in young controls compared to RO-963, and the fastest washout occurring with RO-948. In AD participants, pairwise comparisons indicated SUVRs (cerebellar cortex reference region) in NFT associated regions were highest for RO-948, followed by RO-643 and then RO-963. All three tracers were best modeled using the two-tissue compartment model and were free from nonpolar plasma radiometabolites. RO-948 was identified as having an optimal pharmacokinetic profile compared to the other radioligands and was selected for more extensive human characterization.[53]

Initial clinical characterization of RO-948 was performed in 21 participants including PiB and AV45 PET determined beta-amyloid negative young and older controls, and beta-amyloid positive probable AD patients.[53] Full kinetic modeling (i.e., with arterial sampling) was performed in a subset of these participants, and a further subset was used to investigate test-retest variability. The subset with arterial sampling indicated rapid metabolism of parent RO-948 (16% parent fraction at 30 min post injection) free of nonpolar radiometabolites, and a pharmacokinetic profile consistent with the two-tissue compartment model. V_T was observed to be higher in older controls compared to younger controls throughout the brain including the cerebellum. Plasma and reference tissue based graphical analysis (Logan and MRTM2) methods indicated good agreement and stability of DVR estimates for 90-min scan durations (shorter durations were not tested). SUVR determined 60–90 min post injection (cerebellum gray matter reference region) were highly correlated with reference tissue-based DVR estimates, but overestimated DVR and were observed to increase up to 150 min post injection in high binding AD cases. Test-retest variability was $<10\%$ on average, with DVR methods (plasma and reference tissue derived) outperforming SUVR. Binding patterns ascertained from parametric SUVR images were consistent with Flortaucipir in regions associated with pathological staging of NFTs in AD participants. SUVR images showed minimal white matter binding, no off-target binding in the basal ganglia, choroid plexus or the thalamus, and sporadic retention in the cerebellar vermis and extra axially in the meninges. RO-948 is currently being investigated in clinical trials to compare RO-948 to other PET tracers (Flortaucipir and MK-6240) in AD and to characterize tau aggregates in relation to other biomarkers in the context of AD.

3.2.7 Benzimidazole derivatives: astemizole and lansoprazole

Several benzimidazole derivatives were initially screened in vitro as potential ligands for imaging tau aggregates, which showed higher affinity for PHF tau over beta-amyloid plaques and also identified derivatives that bound to tau aggregates in PSP and Pick's disease.[61] Initial preclinical and in vitro study of astemizole and lansoprazole identified these chemicals as candidates for development as PET radioligands for imaging tau aggregates.[118] Radio-labeled derivatives of these compounds were later tested in preclinical models. In vitro, preclinical and initial human studies of ^{11}C- and ^{18}F-labeled N-Methyl Lansoprazole (NML) indicated NML shows promise for tau imaging.[56,119] In vitro studies indicate NML may bind to tau aggregates in both PSP and AD with similar affinity,[56] and with selectivity for tau aggregates over beta-amyloid.[119] In preclinical evaluation, [^{11}C]NML[56] was observed to be a substrate for permeability-glycoprotein 1 in rodents, but did not appear to be a substrate for CNS transporters in rhesus macaques. PET images from rhesus macaques indicated high brain penetrance and rapid washout (peak activity concentration at 3 min post-injection). A study with a limited number of [^{18}F]NML and [^{18}F]Lansoprazole scans in healthy late middle-aged humans indicated that [^{18}F]NML entered the brain and exhibited rapid washout in cerebellar and cortical gray matter (max SUV ~3–4) and had moderate white matter binding following gray matter washout whereas [^{18}F]Lansoprazole uptake was minimal in the human brain throughout the 90-min acquisition.[55] The human study also identified two other candidate Lansoprazole derivatives for further development as tau imaging ligands. Astemizole derivatives have undergone in vitro and preclinical evaluation, which indicated suitable in vitro affinity for tau aggregates but these compounds did not enter rodent brains even under efflux transporter inhibition.[120] Of note for both astemizole and Lansoprazole, these compounds did not greatly displace [^{3}H]T808 binding in AD slices in vitro, which may indicate an alternate tau binding site for these compounds compared to most other tau imaging ligands.[121]

3.2.8 JNJ-311

JNJ-311 (aka JNJ64349311) was identified from a panel of over 4000 small molecule candidates.[52] In a first report, 1 μM JNJ-311 displaced [^{3}H]T808, but not [^{3}H]florbetapir in AD tissue sections with immunohistologically confirmed tau and beta-amyloid aggregates. In vitro work indicated that JNJ-311 competed with T808 with a K_i of 7.9 nM, had >500-fold selectivity for tau over Aβ (determined by comparison of JNJ-311 inhibition of T808 and florbetapir), and only polar metabolites from human, rat, and mouse hepatocytes in vitro. Preclinical evaluation of [^{18}F]JNJ-311 in rats and rhesus macaques indicated good brain penetrance and fast washout of JNJ-311 that was more rapid compared to Flortaucipir and appeared to be consistent across the entire brain.[122] To date, no human studies with JNJ-311 have been reported.

3.2.9 PI-2620

[^{18}F]PI-2620 is an analogue of Flortaucipir (see Figure 1) that has no peer reviewed papers published as of this writing, but has had some human studies reported in conference proceedings which are mentionable. In an undisclosed number of AD, PSP and healthy control participants, conference reports of PI-2620 indicated fast uptake and washout in non-NFT brain regions, higher dynamic range (SUVR 1–4) compared to Flortaucipir, and did not indicate off-target binding in brain regions showing off-target binding with Flortaucipir (e.g., choroid plexus, basal ganglia and meninges). This study also observed focal uptake in the globus pallidus and substantia nigra in PSP patients suggesting PI-2620 might also bind to 4R tauopathies.[123] A later conference report indicated that a group of 17 probable or possible PSP patients had higher PI-2620 binding in the globus pallidus and putamen compared to controls,[124] although SUVR values in these cases were much lower than those reported a separate conference abstract with AD patients.[125]

3.2.10 Genentech tau probe 1 (GTP1)

GTP1 is an analogue of T808 designed to reduce in vivo defluorination.[126,127] As of this writing, characteristics of GTP1 have only been reported at scientific conferences. Conference reports of human studies indicated that GTP demonstrated a pharmacokinetic profile consistent with the two-tissue compartment model in regions of elevated uptake, with SUVR estimates from 60 to 90 min post injection closely matching DVR estimates using the cerebellum gray matter as a reference region.[57] GTP1 had differential binding patterns between healthy controls and AD patients in regions of suspected pathologic tau,[128] did not indicate in vivo defluorination,[126] but also demonstrated elevated signal in the basal ganglia similar to Flortaucipir.[129]

4. Tau PET ligands in clinical Alzheimer's disease research

4.1 Overview of tau PET tracers in clinical research

Advances in developing PET radioligands capable of imaging tau aggregates in vivo spurred massive interest in clinical dementia and aging research. For the first time, it became possible to observe the spatial deposition of tau aggregates over time in living persons enabling large-scale, longitudinal biomarker studies to investigate spatiotemporal patterns of tau deposition in relation to other clinically relevant factors observed in neurodegenerative disease including beta-amyloid plaques, glucose hypometabolism, structural brain changes, and changes in cognition. Additionally, tau aggregate imaging has potential for informing diagnosis, disease staging and prognosis in AD, as well as aiding clinical trials by improving participant screening, providing an indicator for target engagement of anti-tau therapeutics, and providing a measure of pathologic tau for assessing trial efficacy.[130] The potential for new in vivo discovery led to the rapid integration of tau imaging into several previously existing and ongoing AD biomarker studies, despite some features of these PET radiotracers being uncharacterized or not fully understood. In addition to advances in tau imaging, the increasing role of pathologic beta-amyloid and tau biomarkers in stratifying neurodegenerative disease is reflected in the National Institute on Aging and the Alzheimer's Association's recently published research framework.[131] This framework outlines the use of cerebrospinal fluid (CSF) and neuroimaging biomarkers to identify abnormal levels of pathologic beta-amyloid and tau biomarkers (and neurodegeneration)[132] with the intention to move toward an in vivo definition of AD that is based on biological evidence of AD pathophysiology. This section highlights the use of tau aggregate imaging in clinical AD research. This summary is restricted to studies using Flortaucipir due to a lack of peer-reviewed clinical studies for other tracers, and known off-target and radiometabolite issues for THK and PBB3, respectively.

4.2 Tau imaging in clinical research

4.2.1 Pathologic tau cerebrospinal fluid and neuroimaging biomarkers

As mentioned briefly above, inference about the presence of pathologic tau in neurodegenerative disease can be ascertained from evidence of abnormal CSF or neuroimaging biomarkers. Notably, commonly used CSF assays and neuroimaging markers of pathologic tau measure different molecular targets with CSF assays providing information about the amount of phosphorylated (i.e., tau phosphorylated at threonine-181) and total tau in the brain,[133] whereas PET imaging reflects the spatial deposition of insoluble tau aggregates. Given the suggestion in the research framework that measures of pathophysiologic tau can be derived from CSF or PET imaging, a key element to understanding the merits and limits of this approach requires that these biomarkers be compared within the same individuals. Comparisons of CSF and Flortaucipir derived measures of pathophysiologic tau have been reported in some studies.[82,91,134,135] These studies indicated that CSF measures of $pTau_{181}$ were moderately positively correlated with Flortaucipir SUVR overall, and that the strength of these correlations were dependent on disease state (i.e., impaired vs. unimpaired, beta-amyloid positive vs. negative) and spatially differential with the strongest correlations being reported in the inferior and middle temporal gyri and the posterior cingulate gyrus when samples included persons with and without cognitive impairment. One study investigating the agreement between dichotomous measures (i.e., tau +/−) of pathophysiologic tau derived from CSF and Flortaucipir SUVR indicated up to 83% classification agreement.[134] Further study of these relationships in larger samples and studies using other tau PET tracers are needed to better understand when in the AD process and where in the brain CSF and PET neuroimaging markers of pathologic tau are concordant and discordant.

4.2.2 Imaging tau aggregates in cognitively impaired individuals

Several studies from different cohorts have investigated Flortaucipir in samples with clinically impaired and unimpaired persons. Comparisons between clinically classified groups (i.e., cognitively unimpaired controls, MCI and dementia) have shown that these groups have differential spatial deposition patterns of Flortaucipir SUVR.[81–84,135–139] For example, on average beta-amyloid positive cognitively unimpaired persons had higher Flortaucipir SUVR in regions associated with early NFT stages compared to those that were amyloid negative and cognitively unimpaired, whereas beta-amyloid positive cognitively impaired persons had higher Flortaucipir SUVR in regions associated with later NFT stages. Many of these studies have also indicated that a greater spatial extent of elevated Flortaucipir SUVR was associated with lower cognitive performance across the AD continuum. On average, Flortaucipir spatial deposition patterns in regions associated with neuropathological NFT staging follow the deposition pattern of NFTs described by Braak and Braak,[84,136,139] which has also been observed in initial human studies using MK-6240[48] and RO-948.[53] In addition, these studies also indicated that higher levels of pathophysiologic beta-amyloid were needed to observe elevated Flortaucipir SUVR in isocortical regions. Interestingly, a longitudinal study of Flortaucipir indicated that despite the cross-sectionally observed hierarchy in regions associated with NFT staging, longitudinal rates of Flortaucipir increases did not differ regionally in both unimpaired and impaired groups with elevated pathologic beta-amyloid.[140] Comparisons of early onset AD and autosomal dominant (i.e., *preseniln 1, presenilin 2,* or *amyloid precursor protein* genetic mutation carriers) groups with controls and sporadic AD groups showed differential Flortaucipir binding patterns where both the magnitude and spatial extent of Flortaucipir binding was higher in early onset and autosomal dominant groups, and with the autosomal dominant group showing considerably higher levels of Flortaucipir SUVR in the precuneus compared to sporadic AD.[139,141] Studies comparing Flortaucipir SUVR to measures of glucose metabolism (i.e., [^{18}F]fluorodeoxyglucose [FDG] PET imaging) and structural atrophy (e.g., regional volumes or cortical thickness) showed spatial co-localization of elevated Flortaucipir SUVR with lower FDG uptake andlower gray matter volumes and lower cortical thickness.[135,139,141,142]

4.2.3 Imaging tau aggregates in cognitively unimpaired

A growing number of recent imaging studies have begun investigating the roles of pathologic tau and beta-amyloid in cognitively unimpaired older adults to identify the potential contributions of these features to cognitive decline that occurs prior to clinically diagnosed cognitive impairment. The focus on "preclinical AD" (i.e., the stage of AD wherein pathophysiologic changes are occurring in the absence of overt clinical symptomology) is in part due to growing evidence that the accumulation of pathologic beta-amyloid and tau occur several years prior to clinical symptoms onset,[143–145] and the failure of anti-amyloid therapeutics to alter cognitive outcomes when applied during the early dementia stage of AD.[146] Studies in three separate cohorts have indicated that pathologic beta-amyloid (assessed by PET) and tau (Flortaucipir SUVR) exhibit a synergistic effect resulting in steeper cognitive decline in initially cognitively unimpaired persons.[147–149] Notably, most studies investigated this relationship using the entorhinal cortex as a region of interest for tau imaging, which may suggest that elevated Flortaucipir SUVR in the entorhinal cortex is a disease related feature rather than strictly an age-related phenomena. These studies also provide further evidence that PET detectable pathologic beta-amyloid precedes PET detectable pathologic tau, and that these abnormal biomarker levels precede or are concomitant with clinical impairment, which is mostly consistent with the biomarker cascade model proposed by Jack and colleagues.[143,144]

5. Synopsis and future directions

Rapid advances in PET imaging of tau aggregates have enabled clinical researchers to investigate in vivo spatiotemporal patterns of tau deposition in neurodegenerative disease. Initial clinical studies have shown tremendous promise for the utility of tau imaging to stage disease and provide insights into the time course of the pathophysiological processes underlying AD in relation to clinically observed symptomology. These important studies also suggest that tau PET imaging will be a critical component in the development and assessment of novel disease modifying therapeutics and in the differential diagnoses of neurodegenerative diseases during life. However, there is considerable work to be done with regard to the development and characterization of radioligands used for imaging tau aggregates, and to advance tau imaging to clinical use.

Discovery of radioligands for imaging tau aggregates has been a historically arduous task involving the screening of hundreds to thousands of candidate chemicals. Recent advances in the characterization of the molecular structures of protein aggregates[11] may enable a more sophisticated approach utilizing in silico ligand design and discovery of candidate ligands. In silico modeling and docking simulations informed by these molecular structures may help identify binding sites of PET ligands to tau aggregates, which would provide novel insights into the binding mechanisms of tau tracers to tau aggregates, and would help identify unique and shared binding sites between existing and newly developed PET ligands. This idea is currently being explored and has already indicated some proof-of-concept regarding binding mechanisms and competition of currently available tau PET ligands to tau aggregate binding sites.[150] Importantly, more advanced ligand discovery approaches could also help identify ligands that have specificity for non-AD tau aggregates for which there is currently a lack of available PET radioligands.

The specificity of currently available PET radioligands to different tau aggregates and the ability of current PET radioligands to detect non-AD tauopathies in humans has been mostly inconclusive. To this point, most currently available PET ligands for imaging tau aggregates compete for AD tau aggregate binding sites, but there are notable exceptions in PBB3 (and derivative PM-PBB3) and NML. Similar to the discovery of T808, these compounds may provide useful benchmarks for developing tau imaging ligands that are capable of detecting non-AD tauopathies. In addition, these compounds themselves might also be capable of detecting non-AD tauopathies, but further characterization is needed, particularly in human subjects.

Proper characterization of tau imaging radioligands is a key factor for making appropriate inference from in vivo imaging studies. Several studies across multiple candidate radioligands have demonstrated the importance of understanding how the methodological details applied during this process affect in vitro results, and thereby conclusions about ligand specificity and sensitivity to specific types of tau aggregates over other potential CNS targets and other tauopathies. Direct comparisons of various in vitro experimental conditions and their effects on candidate ligand binding would provide useful insights that would enable informed inference when comparing in vitro results across several studies. This will likely be a crucial component for explaining discrepancies observed between in vivo imaging and in vitro autoradiography, and for interpreting imaging to postmortem validation studies. In addition, direct intra-person comparisons of candidate tau tracers both in vitro and in vivo are needed to ascertain the similarities and differences between these radiotracers, which will inform the interpretation of PET imaging findings in clinical research. Relatedly, given the potential for tau PET ligands to bind to MAO enzymes, proper blocking studies in human subjects using clinically approved inhibitory drugs are warranted to conclusively determine whether these radioligands exhibit off-target MAO binding in human subjects.

There are currently no tau imaging radiotracers approved for clinical use, and all compounds in human studies are restricted to investigational use. Larger numbers of imaging to postmortem studies are needed to understand and validate the ability of PET ligands to detect pathologic tau aggregates in pathologically confirmed cases. These types of studies are needed to verify clinical findings from in vivo imaging studies that suggest tau PET imaging with current radioligands is able to inform diagnosis and stage AD. In addition, these studies will inform the detection limits of PET imaging radioligands compared to the higher sensitivity of histopathological detection, which will allow more accurate interpretation of PET imaging findings related to the extent of tau pathology and disease state.

Tau imaging in clinical research has been mostly limited to studies using Flortaucipir. Initial human studies of several newly developed PET ligands are showing considerably less off-target binding and higher binding potentials in the human brain compared to Flortaucipir studies. Direct comparisons of other tau tracers with Flortaucipir in humans is needed to better characterize the differences between these tracers, and help determine whether data from studies using different tau tracers can reasonably be combined in meta-analyses. In addition, there is a need to perform clinical research studies with newly developed tau aggregate radiotracers and to compare these results to similar Flortaucipir studies to validate Flortaucipir imaging findings and inferences about pathologic tau aggregates in relation to elements of aging and neurodegenerative disease.

References

1. Weingarten MD, Lockwood AH, Hwo SY, Kirschner MW. A protein factor essential for microtubule assembly. *Proc Natl Acad Sci USA*. 1975;72(5):1858–1862.
2. Buée L, Bussière T, Buée-Scherrer V, Delacourte A, Hof PR. Tau protein isoforms, phosphorylation and role in neurodegenerative disorders. *Brain Res Rev*. 2000;33(1):95–130.
3. Bancher C, Brunner C, Lassmann H, et al. Accumulation of abnormally phosphorylated tau precedes the formation of neurofibrillary tangles in Alzheimer's disease. *Brain Res*. 1989;477(1–2):90–99.
4. Grundke-Iqbal I, Iqbal K, Tung Y-C, Quinlan M, Wisniewski HM, Binder LI. Abnormal phosphorylation of the microtubule-associated protein tau (tau) in Alzheimer cytoskeletal pathology. *Proc Natl Acad Sci USA*. 1986;83(13):4913–4917.
5. Villemagne VL, Okamura N. Tau imaging in the study of ageing, Alzheimer's disease, and other neurodegenerative conditions. *Curr Opin Neurobiol*. 2016;36:43–51.
6. McKee AC, Stern RA, Nowinski CJ, et al. The spectrum of disease in chronic traumatic encephalopathy. *Brain*. 2013;136(pt 1):43–64.
7. Braak H, Braak E. Neuropathological stageing of Alzheimer-related changes. *Acta Neuropathol*. 1991;82(4):239–259.
8. Forman MS, Zhukareva V, Bergeron C, et al. Signature tau neuropathology in gray and white matter of corticobasal degeneration. *Am J Pathol*. 2002;160(6):2045–2053.
9. Steele JC. Progressive supranuclear palsy. *Arch Neurol*. 1964;10(4):333–359.
10. Mathis CA, Lopresti BJ, Ikonomovic MD, Klunk WE. Small-molecule PET tracers for imaging proteinopathies. *Semin Nucl Med*. 2017;47(5):553–575.
11. Fitzpatrick AWP, Falcon B, He S, et al. Cryo-EM structures of tau filaments from Alzheimer's disease. *Nature*. 2017;547(7662):185–190.
12. Friedhoff P, von Bergen M, Mandelkow E-M, Mandelkow E. Structure of tau protein and assembly into paired helical filaments. *Biochim Biophys Acta*. 2000;1502(1):122–132.
13. Braak H, Alafuzoff I, Arzberger T, Kretzschmar H, Del Tredici K. Staging of Alzheimer disease-associated neurofibrillary pathology using paraffin sections and immunocytochemistry. *Acta Neuropathol*. 2006;112(4):389–404.
14. Thal DR, Rüb U, Orantes M, Braak H. Phases of Aβ-deposition in the human brain and its relevance for the development of AD. *Neurology*. 2002;58(12):1791–1800.
15. Buée L, Delacourte A. Comparative biochemistry of tau in progressive supranuclear palsy, corticobasal degeneration, FTDP-17 and Pick's disease. *Brain Pathol*. 1999;9(4):681–693.
16. Braak H, Thal DR, Ghebremedhin E, Del Tredici K. Stages of the pathologic process in Alzheimer disease: age categories from 1 to 100 years. *J Neuropathol Exp Neurol*. 2011;70(11):960–969.
17. Delacourte A, David JP, Sergeant N, et al. The biochemical pathway of neurofibrillary degeneration in aging and Alzheimer's disease. *Neurology*. 1999;52(6):1158–1165.

18. Braak H, Braak E. Frequency of stages of Alzheimer-related lesions in different age categories. *Neurobiol Aging*. 1997;18(4):351–357.
19. Braak H, Del Tredici K. The preclinical phase of the pathological process underlying sporadic Alzheimer's disease. *Brain*. 2015;138(10):2814–2833.
20. Gallyas F. Silver staining of Alzheimer's neurofibrillary changes by means of physical development. *Acta Morphol Acad Sci Hung*. 1971;19(1):1–8.
21. Goedert M, Jakes R, Vanmechelen E. Monoclonal antibody AT8 recognises tau protein phosphorylated at both serine 202 and threonine 205. *Neurosci Lett*. 1995;189(3):167–170.
22. Nelson PT, Alafuzoff I, Bigio EH, et al. Correlation of Alzheimer disease neuropathologic changes with cognitive status: a review of the literature. *J Neuropathol Exp Neurol*. 2012;71(5):362–381.
23. Uchihara T. Pretangles and neurofibrillary changes: similarities and differences between AD and CBD based on molecular and morphological evolution. *Neuropathology*. 2014;34(6):571–577.
24. Montine TJ, Phelps CH, Beach TG, et al. National Institute on Aging-Alzheimer's Association guidelines for the neuropathologic assessment of Alzheimer's disease: a practical approach. *Acta Neuropathol*. 2012;123(1):1–11.
25. Honer M, Gobbi L, Martarello L, Comley RA. Radioligand development for molecular imaging of the central nervous system with positron emission tomography. *Drug Discov Today*. 2014;19(12):1936–1944.
26. Laruelle M. Relationships between radiotracer properties and image quality in molecular imaging of the brain with positron emission tomography. *Mol Imaging Biol*. 2003;5(6):363–375.
27. Pike VW. PET radiotracers: crossing the blood-brain barrier and surviving metabolism. *Trends Pharmacol Sci*. 2009;30(8):431–440.
28. Pike VW. Considerations in the development of reversibly binding PET radioligands for brain imaging. *Curr Med Chem*. 2016;23(18):1818–1869.
29. Innis RB, Cunningham VJ, Delforge J, et al. Consensus nomenclature for in vivo imaging of reversibly binding radioligands. *J Cereb Blood Flow Metab*. 2007;27(9):1533–1539.
30. Heurling K, Leuzy A, Jonasson M, et al. Quantitative positron emission tomography in brain research. *Brain Res*. 2017;1670:220–234.
31. Logan J, Fowler JS, Volkow ND, Ding YS, Wang GJ, Alexoff DL. A strategy for removing the bias in the graphical analysis method. *J Cereb Blood Flow Metab*. 2001;21(3):307–320.
32. Mintun MA, Raichle ME, Kilbourn MR, Wooten GF, Welch MJ. A quantitative model for the in vivo assessment of drug binding sites with positron emission tomography. *Ann Neurol*. 1984;15(3):217–227.
33. Ichise M, Ballinger JR, Golan H, et al. Noninvasive quantification of dopamine D2 receptors with iodine-123-IBF SPECT. *J Nucl Med*. 1996;37(3):513–520.
34. Logan J, Fowler JS, Volkow ND, Wang G-J, Ding Y-S, Alexoff DL. Distribution volume ratios without blood sampling from graphical analysis of PET data. *J Cereb Blood Flow Metab*. 1996;16(5):834–840.
35. Gunn RN, Lammertsma AA, Hume SP, Cunningham VJ. Parametric imaging of ligand-receptor binding in PET using a simplified reference region model. *Neuroimage*. 1997;6(4):279–287.
36. Cunningham VJ, Hume SP, Price GR, Ahier RG, Cremer JE, Jones AK. Compartmental analysis of diprenorphine binding to opiate receptors in the rat in vivo and its comparison with equilibrium data in vitro. *J Cereb Blood Flow Metab*. 1991;11(1):1–9.
37. Lammertsma AA, Hume SP. Simplified reference tissue model for PET receptor studies. *Neuroimage*. 1996;4(3 pt 1):153–158.

38. Hume SP, Myers R, Bloomfield PM, et al. Quantitation of carbon-11-labeled raclopride in rat striatum using positron emission tomography. *Synapse*. 1992;12(1):47–54.
39. Kapasi A, DeCarli C, Schneider JA. Impact of multiple pathologies on the threshold for clinically overt dementia. *Acta Neuropathol*. 2017;134(2):171–186.
40. James BD, Wilson RS, Boyle PA, Trojanowski JQ, Bennett DA, Schneider JA. TDP-43 stage, mixed pathologies, and clinical Alzheimer's-type dementia. *Brain*. 2016;139(11):2983–2993.
41. Shoghi-Jadid K, Small GW, Agdeppa ED, et al. Localization of neurofibrillary tangles and beta-amyloid plaques in the brains of living patients with Alzheimer disease. *Am J Geriatr Psychiatry*. 2002;10(1):24–35.
42. Chien DT, Bahri S, Szardenings AK, et al. Early clinical PET imaging results with the novel PHF-tau radioligand [F-18]-T807. *J Alzheimers Dis*. 2013;34(2):457–468.
43. Chien DT, Szardenings AK, Bahri S, et al. Early clinical PET imaging results with the novel PHF-tau radioligand [F18]-T808. *J Alzheimers Dis*. 2014;38(1):171–184.
44. Fodero-Tavoletti MT, Okamura N, Furumoto S, et al. 18F-THK523: a novel in vivo tau imaging ligand for Alzheimer's disease. *Brain*. 2011;134(pt 4):1089–1100.
45. Okamura N, Furumoto S, Harada R, et al. Novel 18F-labeled arylquinoline derivatives for noninvasive imaging of tau pathology in Alzheimer disease. *J Nucl Med*. 2013;54(8):1420–1427.
46. Harada R, Okamura N, Furumoto S, et al. 18F-THK5351: a novel PET radiotracer for imaging neurofibrillary pathology in Alzheimer disease. *J Nucl Med*. 2016;57(2):208–214.
47. Maruyama M, Shimada H, Suhara T, et al. Imaging of tau pathology in a tauopathy mouse model and in Alzheimer patients compared to normal controls. *Neuron*. 2013;79(6):1094–1108.
48. Betthauser TJ, Cody KA, Zammit MD, et al. In vivo characterization and quantification of neurofibrillary tau PET radioligand (18)F-MK-6240 in humans from Alzheimer disease dementia to young controls. *J Nucl Med*. 2018;60(1):93–99.
49. Hostetler ED, Walji AM, Zeng Z, et al. Preclinical characterization of 18F-MK-6240, a promising PET tracer for in vivo quantification of human neurofibrillary tangles. *J Nucl Med*. 2016;57(10):1599–1606.
50. Lohith TG, Bennacef I, Vandenberghe R, et al. Brain imaging of Alzheimer dementia patients and elderly controls with (18)F-MK-6240, a PET tracer targeting neurofibrillary tangles. *J Nucl Med*. 2018;60(1):107–114.
51. Barret O, Seibyl J, Stephens A, et al. First-in-human pet studies with the next generation tau agent 18-F pi-2620 in Alzheimer's disease, progressive supranuclear palsy, and controls. In: *Paper presented at Alzheimer's Association International Conference 2017*; 2017.
52. Rombouts FJ, Andres JI, Ariza M, et al. Discovery of N-(pyridin-4-yl)-1,5-naphthyridin-2-amines as potential tau pathology PET tracers for Alzheimer's disease. *J Med Chem*. 2017;60(4):1272–1291.
53. Kuwabara H, Comley RA, Borroni E, et al. Evaluation of (18)F-RO-948 PET for quantitative assessment of tau accumulation in the human brain. *J Nucl Med*. 2018;59(12):1877–1884.
54. Shimada H, Kitamura S, Ono M, et al. First-in-human pet study with 18F-Am-Pbb3 and 18F-Pm-Pbb3. In: *Alzheimer's Association International Conference 2017, London, England*; 2017.
55. Rafique W, Kramer V, Pardo T, et al. Image-guided development of heterocyclic sulfoxides as ligands for tau neurofibrillary tangles: from first-in-man to second-generation ligands. *ACS Omega*. 2018;3(7):7567–7579.
56. Shao X, Carpenter GM, Desmond TJ, et al. Evaluation of [(11)C]N-methyl lansoprazole as a radiopharmaceutical for PET imaging of tau neurofibrillary tangles. *ACS Med Chem Lett*. 2012;3(11):936–941.

57. Bohorquez SS, Barret O, Tamagnan G, et al. Quantification, test-retest and dosimetry of the novel Genentech Tau Probe 1, [18F]GTP1. In: *Human Amyloid Imaging Conference; 2016, Miami Beach, FL*; 2016.
58. Small GW, Kepe V, Ercoli LM, et al. PET of brain amyloid and tau in mild cognitive impairment. *N Engl J Med*. 2006;355(25):2652–2663.
59. Agdeppa ED, Kepe V, Liu J, et al. Binding characteristics of radiofluorinated 6-dialkylamino-2-naphthylethylidene derivatives as positron emission tomography imaging probes for β-amyloid plaques in Alzheimer's disease. *J Neurosci*. 2002;21(24): RC189.
60. Tolboom N, Yaqub M, van der Flier WM, et al. Detection of Alzheimer pathology in vivo using both 11C-PIB and 18F-FDDNP PET. *J Nucl Med*. 2009;50(2):191–197.
61. Okamura N, Suemoto T, Furumoto S, et al. Quinoline and benzimidazole derivatives: candidate probes for in vivo imaging of tau pathology in Alzheimer's disease. *J Neurosci*. 2005;25(47):10857–10862.
62. Okamura N, Furumoto S, Fodero-Tavoletti MT, et al. Non-invasive assessment of Alzheimer's disease neurofibrillary pathology using 18F-THK5105 PET. *Brain*. 2014;137(pt 6):1762–1771.
63. Villemagne VL, Furumoto S, Fodero-Tavoletti MT, et al. In vivo evaluation of a novel tau imaging tracer for Alzheimer's disease. *Eur J Nucl Med Mol Imaging*. 2014;41(5):816–826.
64. Harada R, Okamura N, Furumoto S, et al. [18F] THK-5117 PET for assessing neurofibrillary pathology in Alzheimer's disease. *Eur J Nucl Med Mol Imaging*. 2015;42(7):1052–1061.
65. Jonasson M, Wall A, Chiotis K, et al. Tracer kinetic analysis of (S)-18F-THK5117 as a PET tracer for assessing tau pathology. *J Nucl Med*. 2016;57(4):574–581.
66. Lockhart SN, Baker SL, Okamura N, et al. Dynamic PET measures of tau accumulation in cognitively normal older adults and Alzheimer's disease patients measured using [18F] THK-5351. *PLoS One*. 2016;11(6):e0158460.
67. Betthauser TJ, Lao PJ, Murali D, et al. In vivo comparison of tau radioligands (18) F-THK-5351 and (18)F-THK-5317. *J Nucl Med*. 2017;58(6):996–1002.
68. Guo Q, Ottersatter B, Benninghoff S, et al. Evaluation of the selectivity of Tau PET radioligand THK5351 in AD brain in vitro and nonhuman primate brain in vivo. In: *Human Amyloid Imaging Conference, 2017, Miami Beach, FL*; 2017.
69. Ng KP, Pascoal TA, Mathotaarachchi S, et al. Monoamine oxidase B inhibitor, selegiline, reduces 18F-THK5351 uptake in the human brain. *Alzheimers Res Ther*. 2017;9(1):25.
70. Ishiki A, Harada R, Kai H, et al. Neuroimaging-pathological correlations of [(18)F] THK5351 PET in progressive supranuclear palsy. *Acta Neuropathol Commun*. 2018;6(1):53.
71. Lemoine L, Gillberg PG, Svedberg M, et al. Comparative binding properties of the tau PET tracers THK5117, THK5351, PBB3, and T807 in postmortem Alzheimer brains. *Alzheimers Res Ther*. 2017;9(1):96.
72. Betthauser T, Converse AK, Murali D, et al. Retrospective comparison of [F-18] MK-6240, [F-18]THK5351 and [F-18]THK5317 ((S)-THK-5117) in subjects scanned with all three PET tracers. In: *Human Amyloid Imaging Conference, 2019, Miami Beach, FL*; 2019.
73. Hirvonen J, Kailajarvi M, Haltia T, et al. Assessment of MAO-B occupancy in the brain with PET and [11C]-L-deprenyl-D2: a dose-finding study with a novel MAO-B inhibitor, EVT 301. *Clin Pharmacol Ther*. 2009;85(5):506–512.
74. Harada R, Okamura N, Furumoto S, Yanai K. Imaging protein misfolding in the brain using beta-sheet ligands. *Front Neurosci*. 2018;12:585.

75. Heuer E, Rosen RF, Cintron A, Walker LC. Nonhuman primate models of Alzheimer-like cerebral proteopathy. *Curr Pharm Des*. 2012;18(8):1159–1169.
76. Zhang W, Arteaga J, Cashion DK, et al. A highly selective and specific PET tracer for imaging of tau pathologies. *J Alzheimers Dis*. 2012;31(3):601–612.
77. Xia C-F, Arteaga J, Chen G, et al. [18F]T807, a novel tau positron emission tomography imaging agent for Alzheimer's disease. *Alzheimers Dement*. 2013;9(6):666–676.
78. Barret O, Alagille D, Sanabria S, et al. Kinetic modeling of the tau PET tracer 18F-AV-1451 in human healthy volunteers and Alzheimer's disease subjects. *J Nucl Med*. 2016;58:1124–1131.
79. Wooten DW, Guehl NJ, Verwer EE, et al. Pharmacokinetic evaluation of the tau PET radiotracer (18)F-T807 ((18)F-AV-1451) in human subjects. *J Nucl Med*. 2017;58(3):484–491.
80. Golla SSV, Timmers T, Ossenkoppele R, et al. Quantification of tau load using [(18)F] AV1451 PET. *Mol Imaging Biol*. 2017;19(6):963–971.
81. Johnson KA, Schultz A, Betensky RA, et al. Tau positron emission tomographic imaging in aging and early Alzheimer disease. *Ann Neurol*. 2015;79(1):110–119.
82. Brier MR, Gordon B, Friedrichsen K, et al. Tau and Abeta imaging, CSF measures, and cognition in Alzheimer's disease. *Sci Transl Med*. 2016;8(338):338ra366.
83. Scholl M, Lockhart SN, Schonhaut DR, et al. PET imaging of tau deposition in the aging human brain. *Neuron*. 2016;89(5):971–982.
84. Schwarz AJ, Yu P, Miller BB, et al. Regional profiles of the candidate tau PET ligand 18F-AV-1451 recapitulate key features of Braak histopathological stages. *Brain*. 2016;139(pt 5):1539–1550.
85. Marquié M, Normandin MD, Vanderburg CR, et al. Validating novel tau positron emission tomography tracer [F-18]-AV-1451 (T807) on postmortem brain tissue. *Ann Neurol*. 2015;78(5):787–800.
86. Lowe VJ, Curran G, Fang P, et al. An autoradiographic evaluation of AV-1451 tau PET in dementia. *Acta Neuropathol Commun*. 2016;4(1):58.
87. Vermeiren C, Motte P, Viot D, et al. The tau positron-emission tomography tracer AV-1451 binds with similar affinities to tau fibrils and monoamine oxidases. *Mov Disord* 2018;33(2):273–281.
88. Tong J, Meyer JH, Furukawa Y, et al. Distribution of monoamine oxidase proteins in human brain: implications for brain imaging studies. *J Cereb Blood Flow Metab*. 2013;33(6):863–871.
89. Choi JY, Cho H, Ahn SJ, et al. Off-target (18)F-AV-1451 binding in the basal ganglia correlates with age-related iron accumulation. *J Nucl Med*. 2018;59(1):117–120.
90. Baker SL, Harrison TM, Maass A, La Joie R, Jagust W. Effect of off-target binding on (18)F-Flortaucipir variability in healthy controls across the lifespan. *J Nucl Med*. 2019. http://jnm.snmjournals.org/content/early/2019/03/14/jnumed.118.224113.abstract.
91. Gordon BA, Friedrichsen K, Brier M, et al. The relationship between cerebrospinal fluid markers of Alzheimer pathology and positron emission tomography tau imaging. *Brain*. 2016;139(pt 8):2249–2260.
92. Fowler JS, Volkow ND, Wang GJ, et al. Age-related increases in brain monoamine oxidase B in living healthy human subjects. *Neurobiol Aging*. 1997;18(4):431–435.
93. Hansen AK, Brooks DJ, Borghammer P. MAO-B inhibitors do not block in vivo Flortaucipir([(18)F]-AV-1451) binding. *Mol Imaging Biol*. 2018;20(3):356–360.
94. Schonhaut DR, McMillan CT, Spina S, et al. (18)F-Flortaucipir tau positron emission tomography distinguishes established progressive supranuclear palsy from controls and Parkinson disease: a multicenter study. *Ann Neurol*. 2017;82(4):622–634.
95. Cho H, Choi JY, Hwang MS, et al. Subcortical (18) F-AV-1451 binding patterns in progressive supranuclear palsy. *Mov Disord*. 2017;32(1):134–140.

96. Coakeley S, Cho SS, Koshimori Y, et al. Positron emission tomography imaging of tau pathology in progressive supranuclear palsy. *J Cereb Blood Flow Metab.* 2017;37(9):3150–3160.
97. Smith R, Schain M, Nilsson C, et al. Increased basal ganglia binding of (18)F-AV-1451 in patients with progressive supranuclear palsy. *Mov Disord.* 2017;32(1):108–114.
98. Whitwell JL, Lowe VJ, Tosakulwong N, et al. [(18)F]AV-1451 tau positron emission tomography in progressive supranuclear palsy. *Mov Disord.* 2017;32(1):124–133.
99. Passamonti L, Vazquez Rodriguez P, Hong YT, et al. 18F-AV-1451 positron emission tomography in Alzheimer's disease and progressive supranuclear palsy. *Brain.* 2017;140(3):781–791.
100. Josephs KA, Whitwell JL, Tacik P, et al. [18F]AV-1451 tau-PET uptake does correlate with quantitatively measured 4R-tau burden in autopsy-confirmed corticobasal degeneration. *Acta Neuropathol.* 2016;132(6):931–933.
101. McMillan CT, Irwin DJ, Nasrallah I, et al. Multimodal evaluation demonstrates in vivo (18)F-AV-1451 uptake in autopsy-confirmed corticobasal degeneration. *Acta Neuropathol.* 2016;132(6):935–937.
102. Smith R, Puschmann A, Scholl M, et al. 18F-AV-1451 tau PET imaging correlates strongly with tau neuropathology in MAPT mutation carriers. *Brain.* 2016; 139(pt 9):2372–2379.
103. Marquie M, Normandin MD, Meltzer AC, et al. Pathological correlations of [F-18]-AV-1451 imaging in non-alzheimer tauopathies. *Ann Neurol.* 2017;81(1):117–128.
104. Smith R, Wibom M, Pawlik D, Englund E, Hansson O. Correlation of in vivo [18F]Flortaucipir with postmortem Alzheimer disease tau pathology. *JAMA Neurol.* 2018;76(3):310–317. https://doi.org/10.1001/jamaneurol.2018.3692.
105. Villemagne VL, Fodero-Tavoletti MT, Masters CL, Rowe CC. Tau imaging: Early progress and future directions. *Lancet Neurol.* 2015;14(1):114–124.
106. Hashimoto H, Kawamura K, Igarashi N, et al. Radiosynthesis, photoisomerization, biodistribution, and metabolite analysis of 11C-PBB3 as a clinically useful PET probe for imaging of tau pathology. *J Nucl Med.* 2014;55(9):1532–1538.
107. Hashimoto H, Kawamura K, Takei M, et al. Identification of a major radiometabolite of [11C]PBB3. *Nucl Med Biol.* 2015;42(12):905–910.
108. Kimura Y, Ichise M, Ito H, et al. PET quantification of tau pathology in human brain with 11C-PBB3. *J Nucl Med.* 2015;56(9):1359–1365.
109. Shimada H, Ono M, Tagai K, et al. Preclinical and clinical characterization of 18f-pm-Pbb3, a pet ligand for diverse tau pathologies. In: *Alzheimer's Association International Conference, 2018, Chicago, IL*; 2018.
110. Salinas C, Girmay S, Tempest P, et al. Evidence of differential in vitro and in vivo binding of APN-1607 in progressive supranuclear palsy and Alzheimer's disease. In: *Human Amyloid Imaging Conference, 2019, Miami Beach, FL*; 2019.
111. Shimada H, Tagai K, Kubota M, et al. In vivo distribution pattern of 18F-PM-PBB3 (18F-APN-1607) and its relationship with clinical features in diverse 4-repeat tauopathies. In: *Human Amyloid Imaging Conference, 2019, Miami Beach, FL*; 2019.
112. Wager TT, Hou X, Verhoest PR, Villalobos A. Moving beyond rules: the development of a central nervous system multiparameter optimization (CNS MPO) approach to enable alignment of druglike properties. *ACS Chem Nerosci.* 2010;1(6):435–449.
113. Walji AM, Hostetler ED, Selnick H, et al. Discovery of 6-(fluoro-(18)F)-3-(1H-pyrrolo[2,3-c]pyridin-1-yl)isoquinolin-5-amine ([(18)F]-MK-6240): a positron emission tomography (PET) imaging agent for quantification of neurofibrillary tangles (NFTs). *J Med Chem.* 2016;59(10):4778–4789.
114. Pascoal TA, Shin M, Kang MS, et al. In vivo quantification of neurofibrillary tangles with [(18)F]MK-6240. *Alzheimers Res Ther.* 2018;10(1):74.

115. Gobbi LC, Knust H, Korner M, et al. Identification of three novel radiotracers for imaging aggregated tau in Alzheimer's disease with positron emission tomography. *J Med Chem*. 2017;60(17):7350–7370.
116. Honer M, Gobbi L, Knust H, et al. Preclinical evaluation of (18)F-RO6958948, (11)C-RO6931643 and (11)C-RO6924963 as novel radiotracers for imaging aggregated tau in AD with positron emission tomography. *J Nucl Med*. 2017;59(4):675–681.
117. Wong DF, Comley RA, Kuwabara H, et al. Characterization of 3 novel tau radiopharmaceuticals, (11)C-RO-963, (11)C-RO-643, and (18)F-RO-948, in healthy controls and in Alzheimer subjects. *J Nucl Med*. 2018;59(12):1869–1876.
118. Rojo LE, Alzate-Morales J, Saavedra IN, Davies P, Maccioni RB. Selective interaction of lansoprazole and astemizole with tau polymers: potential new clinical use in diagnosis of Alzheimer's disease. *J Alzheimers Dis*. 2010;19(2):573–589.
119. Fawaz MV, Brooks AF, Rodnick ME, et al. High affinity radiopharmaceuticals based upon lansoprazole for PET imaging of aggregated tau in Alzheimer's disease and progressive supranuclear palsy: synthesis, preclinical evaluation, and lead selection. *ACS Chem Nerosci*. 2014;5(8):718–730.
120. Riss PJ, Brichard L, Ferrari V, et al. Radiosynthesis and characterization of astemizole derivatives as lead compounds toward PET imaging of τ-pathology. *MedChemComm*. 2013;4(5):852–855.
121. Declercq L, Celen S, Lecina J, et al. Comparison of new tau PET-tracer candidates with [18F]T808 and [18F]T807. *Mol Imaging*. 2016;15:1–15.
122. Declercq L, Rombouts F, Koole M, et al. Preclinical evaluation of (18)F-JNJ64349311, a novel PET tracer for tau imaging. *J Nucl Med*. 2017;58(6):975–981.
123. Stephens A, Seibyl JP, Mueller A, et al. Clinical evaluation of 18F-PI-2620, a next generation tau PET agent in subjects with Alzheimer's Disease and progressive supranuclear palsy. In: *Human Amyloid Imaging Conference, 2018, Miami Beach, FL*; 2018.
124. Brendel M, Barthel H, Van Eimeren T, et al. 18F-PI2620 tau-PET in progressive supranuclear palsy—a multi-center evaluation. In: *Society of Nuclear Medicine 2019 Annual Meeting*; 2019.
125. Villemagne VL, Dore V, Mulligan RS, et al. Evaluation of 18F-PI-2620, a novel selective tau tracer for the assessment of Alzheimer's and non-Alzheimer's tauopathies. In: *Human Amyloid Imaging Conference, 2019, Miami Beach, FL*; 2019.
126. Marik J, Tinianow J, Ogasawara A, et al. [18F]GTP1—a tau specific tracer for imaging tau-pathology in AD. In: *Human Amyloid Imaging Conference, 2016, Miami Beach, FL*; 2016.
127. Okamura N, Harada R, Ishiki A, Kikuchi A, Nakamura T, Kudo Y. The development and validation of tau PET tracers: current status and future directions. *Clin Transl Imaging*. 2018;6(4):305–316.
128. Bohorquez SS, Barret O, Tamagnan G, et al. Evaluation of tau burden in a cross-sectional cohort of Alzheimer's disease subjects using [18f]Gtp1 (Genentech tau probe 1). In: *Paper Presented at Alzheimer's Association International Conference 2016*; 2016.
129. Bohorquez SS, Bentsson T, Barret O, et al. Kinetics of [18F]GTP1 (Genentech tau probe 1) in the basal ganglia of Alzheimer's patients and healthy controls. In: *Human Amyloid Imaging Conference, 2017, Miami Beach, FL*; 2017.
130. Villemagne VL, Dore V, Burnham SC, Masters CL, Rowe CC. Imaging tau and amyloid-beta proteinopathies in Alzheimer disease and other conditions. *Nat Rev Neurol*. 2018;14:225–236.
131. Jack Jr CR, Bennett DA, Blennow K, et al. NIA-AA research framework: toward a biological definition of Alzheimer's disease. *Alzheimers Dement*. 2018;14(4):535–562.
132. Jack Jr CR, Bennett DA, Blennow K, et al. A/T/N: an unbiased descriptive classification scheme for Alzheimer disease biomarkers. *Neurology*. 2016;87(5):539–547.

133. Vanmechelen E, Vanderstichele H, Davidsson P, et al. Quantification of tau phosphorylated at threonine 181 in human cerebrospinal fluid: a sandwich ELISA with a synthetic phosphopeptide for standardization. *Neurosci Lett.* 2000;285(1):49–52.
134. La Joie R, Bejanin A, Fagan AM, et al. Associations between [(18)F]AV1451 tau PET and CSF measures of tau pathology in a clinical sample. *Neurology.* 2018;90(4): e282–e290.
135. Mattsson N, Scholl M, Strandberg O, et al. (18)F-AV-1451 and CSF T-tau and P-tau as biomarkers in Alzheimer's disease. *EMBO Mol Med.* 2017;9(9):1212–1223.
136. Lowe VJ, Wiste HJ, Senjem ML, et al. Widespread brain tau and its association with ageing, Braak stage and Alzheimer's dementia. *Brain.* 2018;141(1):271–287.
137. Cho H, Choi JY, Hwang MS, et al. Tau PET in Alzheimer disease and mild cognitive impairment. *Neurology.* 2016;87(4):375–383.
138. Maass A, Landau S, Baker SL, et al. Comparison of multiple tau-PET measures as biomarkers in aging and Alzheimer's disease. *Neuroimage.* 2017;157:448–463.
139. Scholl M, Ossenkoppele R, Strandberg O, et al. Distinct 18F-AV-1451 tau PET retention patterns in early- and late-onset Alzheimer's disease. *Brain.* 2017;140(9):2286–2294.
140. Jack Jr CR, Wiste HJ, Schwarz CG, et al. Longitudinal tau PET in ageing and Alzheimer's disease. *Brain.* 2018;141(5):1517–1528.
141. Gordon BA, Blazey TM, Christensen J, et al. Tau PET in autosomal dominant Alzheimer's disease: relationship with cognition, dementia and other biomarkers. *Brain.* 2019;142(4):1063–1076.
142. Ossenkoppele R, Schonhaut DR, Scholl M, et al. Tau PET patterns mirror clinical and neuroanatomical variability in Alzheimer's disease. *Brain.* 2016;139(pt 5):1551–1567.
143. Jack CR, Knopman DS, Jagust WJ, et al. Tracking pathophysiological processes in Alzheimer's disease: an updated hypothetical model of dynamic biomarkers. *Lancet Neurol.* 2013;12(2):207–216.
144. Jack CR, Knopman DS, Jagust WJ, et al. Hypothetical model of dynamic biomarkers of the Alzheimer's pathological cascade. *Lancet Neurol.* 2010;9(1):119–128.
145. Price JL, Morris JC. Tangles and plaques in nondemented aging and "preclinical" Alzheimer's disease. *Ann Neurol.* 1999;45(3):358–368.
146. Honig LS, Vellas B, Woodward M, et al. Trial of Solanezumab for mild dementia due to Alzheimer's disease. *N Engl J Med.* 2018;378(4):321–330.
147. Aschenbrenner AJ, Gordon BA, Benzinger TLS, Morris JC, Hassenstab JJ. Influence of tau PET, amyloid PET, and hippocampal volume on cognition in Alzheimer disease. *Neurology.* 2018;91(9):e859–e866.
148. Knopman DS, Lundt ES, Therneau TM, et al. Entorhinal cortex tau, amyloid-beta, cortical thickness and memory performance in non-demented subjects. *Brain.* 2019;142(4):1148–1160.
149. Sperling RA, Mormino EC, Schultz AP, et al. The impact of amyloid-beta and tau on prospective cognitive decline in older individuals. *Ann Neurol.* 2019;85(2):181–193.
150. Murugan NA, Chiotis K, Rodriguez-Vieitez E, Lemoine L, Agren H, Nordberg A. Cross-interaction of tau PET tracers with monoamine oxidase B: evidence from in silico modelling and in vivo imaging. *Eur J Nucl Med Mol Imaging.* 2019;46(6):1369–1382.

1ciii

AD molecular: Molecular imaging of Alzheimer's disease: PET imaging of neurotransmitter systems

Adam P. Mecca*
Department of Psychiatry and Alzheimer's Disease Research Unit, Yale University School of Medicine, New Haven, CT, United States
*Corresponding author: e-mail address: adam.mecca@yale.edu

Contents

Abstract

Current understanding of Alzheimer's disease (AD) pathogenesis relies on the observed accumulations of amyloid β and phosphorylated tau aggregates that are thought to play key roles in initiating or propagating disease. However, other processes including changes in synaptic proteins and neurotransmitter loss have been suggested as important etiologies or contributors. Positron emission tomography (PET) imaging allows *in vivo* investigations of molecular changes associated with AD. PET imaging with multiple radiotracers can be used in combination with other modalities such as magnetic resonance imaging (MRI), and with assessments of cognition and neuropsychiatric symptoms to investigate the molecular underpinnings of AD. Studies of synaptic

Progress in Molecular Biology and Translational Science, Volume 165
ISSN 1877-1173
https://doi.org/10.1016/bs.pmbts.2019.04.003

protein changes may improve the understanding of disease mechanisms and provide valuable markers of disease progression and therapeutic efficacy. This chapter will illustrate the importance of *in vivo* molecular imaging in the study of AD with a specific emphasis on PET and radioligands for several non-amyloid targets.

1. Introduction

The pathogenesis of Alzheimer's disease (AD) remains elusive and rests predominantly on the observed accumulations of amyloid β (Aβ) and phosphorylated tau (p-tau) protein aggregates that are thought to play key roles in initiating or propagating the disease. However, other processes including changes in synaptic proteins, neurotransmitter loss, inflammation and cerebrovascular disease have been suggested as important etiologies or contributors. Until recently, much of what is known about AD pathology has been learned from *post mortem* studies and animal models. The advancements of positron emission tomography (PET) imaging allow *in vivo* investigations of molecular changes associated with AD. Molecular imaging can be used to study multiple biomarkers both early in the disease process and longitudinally. PET imaging can be used in combination with other imaging modalities such as magnetic resonance imaging (MRI), as well as assessments of cognition and neuropsychiatric symptoms to investigate the molecular underpinnings of AD. Much of the recent work on PET imaging biomarkers has aided in the diagnosis and understanding of AD through *in vivo* imaging of metabolic changes, Aβ plaque burden, and p-tau accumulation in humans.[1–4] In addition to these important insights, studies of synaptic level protein changes may improve the understanding of disease mechanisms, and provide valuable markers of disease progression and therapeutic efficacy.[5–7] This chapter will illustrate the importance of *in vivo* molecular imaging in the study of AD with a specific emphasis on PET and radioligands for several non-amyloid targets. This will include recent work to develop methods of synapse quantification and recent investigations into metabotropic glutamate receptor subtype 5 (mGluR5) distribution, as well as more developed investigations into serotonergic systems.

2. Synaptic density imaging

Synaptic loss is linked with AD progression and provides a novel target for exploration with PET.[8,9] Synaptic changes are important biological markers of disease and potential therapeutic targets. Our recent PET studies using the small molecule radioligand (*R*)-1-((3-(11C-methyl-11C)pyridin-4-yl)methyl)-4-(3,4,5-trifluorophenyl)pyrrolidin-2-one ([^{11}C]UCB-J) have identified synaptic vesicle glycoprotein 2A (SV2A) as the first *in vivo* marker of synaptic density.[10] Synaptic density is reduced in the medial temporal lobe of individuals with AD compared to healthy participants when assessed with [^{11}C]UCB-J PET.[5] Synaptic density measurements with PET may provide valuable information about spatial patterns of synaptic loss in relation to other AD-related pathology. In addition, future longitudinal studies of synaptic density in AD may contribute to the understanding of an ongoing disease process and lead to the development of both novel treatments and therapeutic biomarkers.

2.1 Synaptic radiotracer development

Synaptic loss is a robust and consistent pathology in AD.[8,9] Synapses are essential for cognitive function, and their loss is well established as the major structural correlate of cognitive impairment in AD.[11–14] Synaptic failure is an early event in AD pathogenesis and individuals with the prodromal stage of mild cognitive impairment (MCI) demonstrate a loss of synapses and several presynaptic proteins.[15–17] For example, *post mortem* studies of the synaptic density marker synaptophysin (SYN) revealed substantial reductions in AD ranging from 25% to 45% in the hippocampus, parietal, temporal, and frontal cortices.[18–20]

One candidate marker for synaptic density is synaptic vesicle glycoprotein 2 (SV2). SV2 has three isoforms (SV2A, SV2B, and SV2C)[21–23] that are expressed in the rat and human brain, but with different distributions. SV2A is expressed in virtually all synapses and is located in synaptic vesicles at presynaptic terminals (Fig. 1). SV2B and SV2C have more restricted distributions.[21–23] The exact role of SV2A in synaptic vesicle cycling and neurotransmitter exocytosis is not fully elucidated, but SV2A may function as a transporter, modulate exocytosis of synaptic vesicles, or modify synaptic function.[24] SV2 is unlikely to have a function in synaptic vesicle formation, since vesicle density and synaptic morphology are unaltered in SV2 knockout mice.[25,26] Instead, SV2 may be part of the synaptic vesicle cycle of regulated exocytosis and endocytosis, which is a continuous process within presynaptic terminals.[27] SV2A may provide a novel approach to study a variety of neurological pathologies, and quantification of SV2A may be a direct measure of synaptic vesicle density. Quantitative immunoblotting

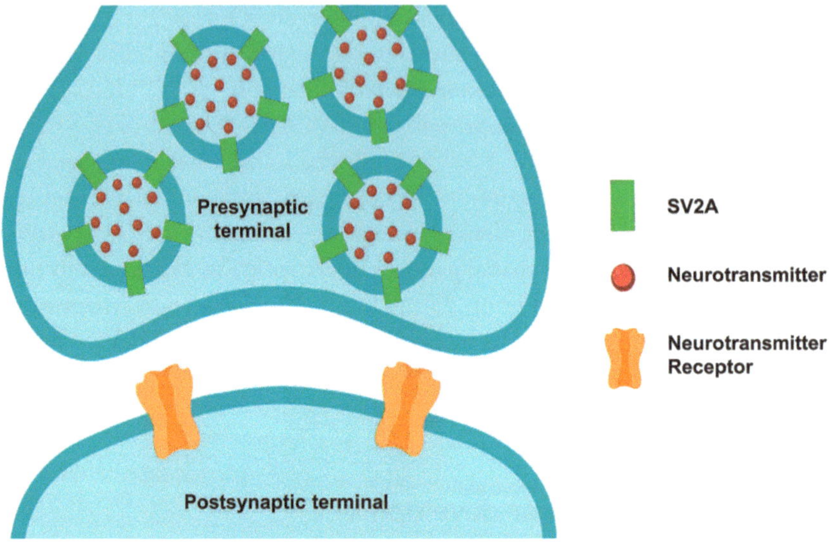

Fig. 1 SV2A cellular localization. SV2A is a transmembrane protein that is ubiquitously expressed in the brain. It is located in synaptic vesicles and is thought to be involved in the regulation of synaptic vesicle release. SV2A = synaptic vesicle glycoprotein 2A.

and mass spectrometry studies have demonstrated about five copies of SV2 per vesicle.[28] This is consistent across vesicles regardless of the neurotransmitter.[28] While immunostaining for SYN has been accepted as a biomarker for synaptic density, SV2A may provide a more accurate measure, due to its greater uniformity in vesicles.

Recently, compounds derived from levetiracetam were developed with appropriate affinity and selectivity for SV2A to be PET tracers.[29] One such compound, [^{11}C]UCB-J (Fig. 2) was ultimately advanced for human studies since it has a high affinity and specificity for SV2A and fast kinetics (Fig. 3 and Video 1 in the online version at https://doi.org/10.1016/bs.pmbts.2019.04.003). Baseline and levetiracetam blocking studies performed in rhesus monkeys showed that brain activity was widely distributed in gray matter with highest uptake in cortex, striatum, and thalamus.[30] Specific binding throughout the gray matter was consistent with SV2A being localized in synaptic terminals.

In additional preclinical baboon studies, the *in vitro* regional distribution of SV2A measured by protein binding assays was highly correlated with PET measured [^{11}C]-UCB-J binding to SV2A volume of distribution (V_T) in all regions (R^2 = 0.72–0.84).[10] There were also strong correlations between SV2A and SYN across all regions (R^2 = 0.88). Both SV2A and SYN

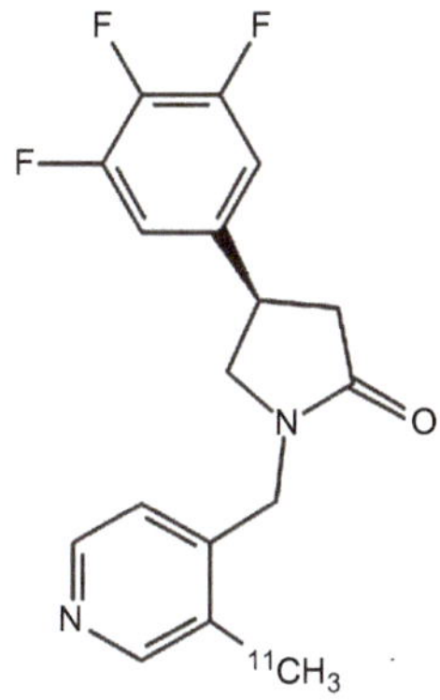

Fig. 2 Chemical structure of [^{11}C]UCB-J. [^{11}C]UCB-J was advanced for human studies because of favorable kinetics, high specificity, and excellent test-retest reliability.

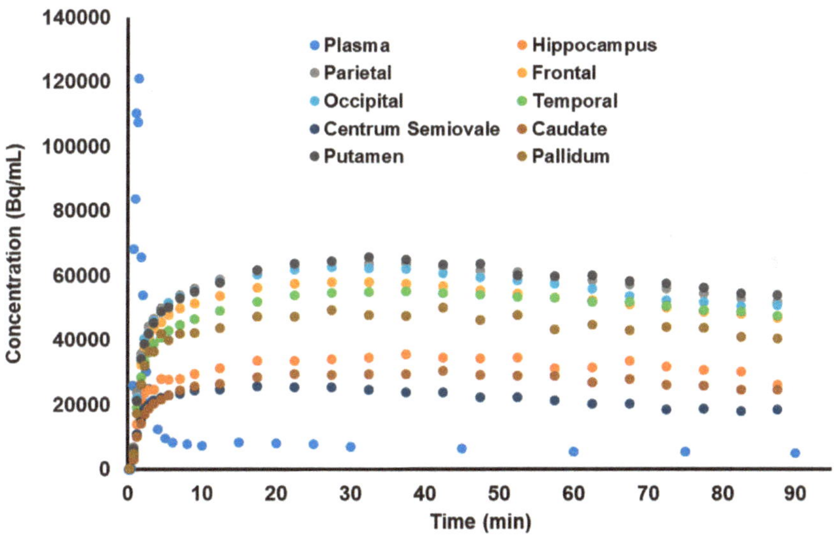

Fig. 3 Time activity curves after [^{11}C]UCB-J injection. Radioactivity concentration is measured over time in the plasma and multiple brain regions of interest. These data are used to quantify receptor concentrations with kinetic modeling.

levels were high in all gray matter regions, but low in white matter. Taken together, these preclinical studies support the utility of SV2A as a biomarker of synaptic density.

Initial human studies were performed to characterize [^{11}C]UCB-J in five cognitively normal participants between the ages of 25 and 55 years old.[31] Each of these participants received two scans on the same day so that test-retest reliability could be assessed. Similar to the results from preclinical animal studies, peak brain activity was high after injection of [^{11}C]UCB-J (standardized uptake value ~10). Uptake was relatively homogenous across gray matter regions, and much lower in white matter regions. Tissue activity curves (TACs) were well described by 1- and 2-tissue compartment models (1TC and 2TC).[32] Therefore, V_T was estimated using 1TC. The test-retest variability of [^{11}C]UCB-J was very low with mean test-retest % differences in V_T ranging from −4% to 2%.[31] Additional levetiracetam displacement studies confirmed that [^{11}C]UCB-J binds specifically to SV2A in humans.[10] Arterial and venous sampling to generate input functions for kinetic modeling is essential for fully quantitative assessments of tracer binding. However, simplified methods of quantitation using a brain reference region are needed for use in clinical research or clinical care. The centrum semiovale is a candidate reference region for [^{11}C]UCB-J PET because it is devoid of specific binding. This is supported by displacement scans performed with the SV2A-selective anticonvulsant levetiracetam.[10]

2.2 Synaptic density measurements in AD

Because [^{11}C]UCB-J binding to SV2A is an excellent candidate biomarker for *in vivo* synaptic density changes, SV2A density was quantified in 10 participants with AD and 11 who were cognitively normal (CN).[5] Participants ranged in age from 61 to 83 years old. AD participants met research diagnostic criteria for amnestic MCI (aMCI)[33] or National Institute on Aging-Alzheimer's Association (NIA-AA) criteria for dementia due to AD (AD-dementia)[34] and were positive for brain Aβ measured by [^{11}C] Pittsburgh Compound B ([^{11}C]PiB) PET. CN participants were negative for brain Aβ measured by [^{11}C]PiB PET. Parametric images of V_T were computed for 8 CN and 9 AD participants using the arterial input function and 1TC modeling (Fig. 4). AD participants demonstrated significantly lower [^{11}C]UCB-J binding to SV2A in the hippocampus compared to CN participants. There was no difference in [^{11}C]UCB-J V_T for the centrum semiovale between CN and AD groups. Using the centrum semiovale reference region, binding potential (BP_{ND}) values were calculated using two step-simplified reference tissue model (SRTM2).[35] Binding potential images calculated using SRTM2 gave regional values with excellent agreement to those computed from 1TC V_T maps.[5] Using SRTM2, hippocampal BP_{ND} was significantly lower in the AD group compared to CN. Exploratory analyses of other brain regions of interest also revealed reductions in

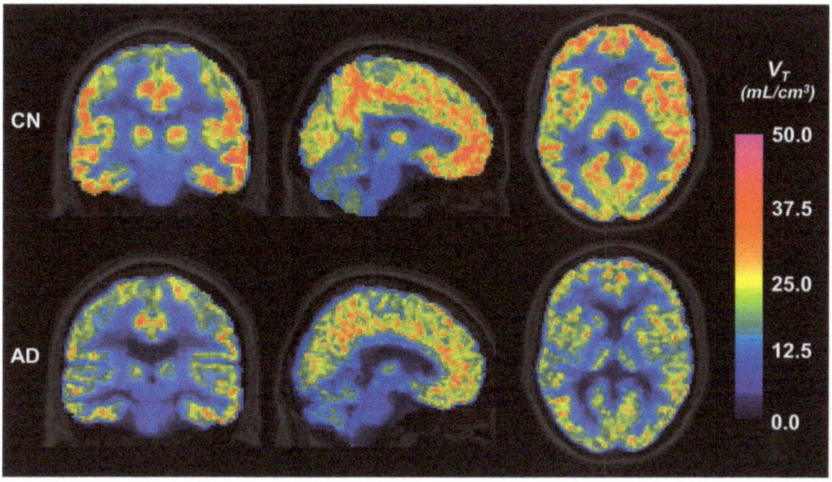

Fig. 4 Parametric image of synaptic density imaging. V_T is calculated for each voxel using 1TC kinetic modeling after [^{11}C]UCB-J injection. SV2A is present throughout the cortex and in subcortical gray matter. AD participants have lower hippocampal and association cortical SV2A. CN = cognitively normal, AD = Alzheimer's disease, 1TC = 1 tissue compartment, V_T = volume of distribution, SV2A = synaptic vesicle glycoprotein 2A.

entorhinal cortex. Hippocampus and entorhinal cortex are among the most atrophied brain regions in early AD,[36,37] raising the question of whether reductions in SV2A binding in these regions were attributable solely to tissue loss. In fact, MRI volumetric analysis[38] demonstrated 25% reduction in entorhinal cortex volume in the AD group, similar to the 27% reduction of SV2A binding (BP_{ND}) in entorhinal cortex. The reduction in SV2A binding in entorhinal cortex was no longer significant after partial volume correction (PVC) to account for gray matter atrophy. By contrast, the 44% reduction of 1TC-derived hippocampal BP_{ND} in AD participants far exceeded the corresponding reduction in hippocampal volume (22%) and persisted after PVC. Whole brain analyses were also performed to verify the identified regional differences in [^{11}C]UCB-J binding to SV2A. Compared to CN participants, AD participants had lower SV2A that was restricted to the medial and inferior temporal regions. However, the magnitude of [^{11}C]UCB-J binding to SV2A was reduced in broader association cortical regions (see Fig. 4 for examples). The K_1 parameter derived from 1TC modeling was used to compare tissue perfusion of AD participants to CN participants. Parametric K_1 images may convey regional metabolic information, similar to metabolism measured with [^{18}F]FDG PET. Analyses

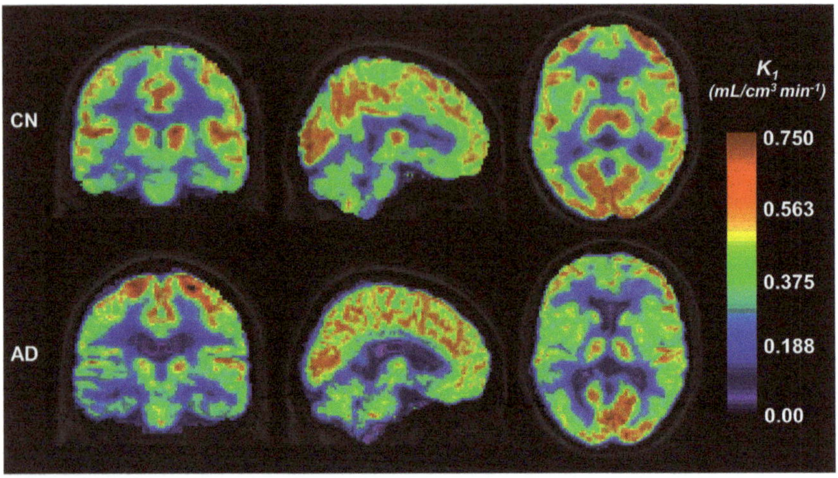

Fig. 5 Parametric image of brain perfusion after [^{11}C]UCB-J injection. K_1 is calculated for each voxel using 1TC kinetic modeling after [^{11}C]UCB-J injection. The pattern of hypoperfusion in AD may be different than synaptic density and closely resemble that of hypometabolism measured with [^{18}F]FDG PET. CN=cognitively normal, AD=Alzheimer's disease, 1TC=1 tissue compartment, PET=positron emission tomography.

of K_1 images revealed that AD participants had lower lateral temporal, lateral parietal, and precuneus perfusion compared to CN participants. This pattern seems to match known AD patterns of hypometabolism and is distinct from the pattern of synaptic density differences (see Fig. 5 for examples). Taken together, these initial studies of synaptic density using [^{11}C]UCB-J PET indicate that measurements of synaptic density may provide valuable insights into AD pathology and serve as powerful tools to track disease progression and therapeutic efficacy.

2.3 New developments in synaptic density imaging

Continued progress in tracer development has yielded candidate second-generation SV2A ligands. These promising molecules are ^{18}F-labeled, providing a compound with a longer radioactive half-life (110 min) compared with the first generation ^{11}C-labeled compounds (20 min half-life). This longer half-life will allow central production and off-site transport for multicenter use. One promising ligand, [^{18}F]SDM-8, has been advanced to human studies based on tracer properties that are similar to [^{11}C]UCB-J. The high brain uptake, fast kinetics, and high specific binding make [^{18}F]SDM-8 a promising radiotracer for SV2A *in vivo* imaging.

Although PET imaging provides valuable spatial information, it can also be expensive, unavailable, and relatively invasive. For that reason, fluid biomarkers of synaptic health are an attractive option as assays for early diagnosis or to detect synaptic damage. Of particular interest is preliminary work to measure SV2A in cerebrospinal fluid (CSF) which suggests that SV2A levels are decreased in AD compared to healthy participants.[39] In addition, neurofilament light chain (NFL), growth-associated protein 43 (GAP-43), neurogranin (Ng), synaptosomal-associated protein 25 (SNAP-25), and synaptotagmin-1 (SYT-1) are potential CSF biomarkers of neuronal damage or synaptic health that are elevated in AD.[40–44] NFL measured in the plasma is also elevated in AD and is highly correlated with CSF NFL.[45] These fluid-based biomarkers may provide important information related to brain neuronal damage and synaptic health. However, the relationship of brain synaptic density to fluid measures of synaptic dysfunction or neuronal damage are yet to be determined. Therefore, future studies will need to investigate the relationships between PET and CSF biomarkers of synaptic and neuronal health.

Finally, multimodal and multi-tracer imaging studies will be essential in attempting to understand both cross-sectional and longitudinal relationships between molecular, structural, and network level changes that occur in AD. For example, various resting state-functional MRI (rs-fMRI) methods have been used to assess intrinsic functional connectivity over a range of spatially-defined extents. Fractional amplitude of low frequency fluctuations (falFF) is a single-voxel indicator of BOLD signal frequency power independent from brain connectivity for which neural activity is thought to be the primary contributor.[46] Regional homogeneity (ReHo) is a measure of local functional connectivity.[23] Group-independent component analysis with dual regression (gICA-DR) can be used to identify common spatiotemporal patterns and project them back to an individual.[47] faLFF, ReHo, and glCA-DR patterns are altered in AD. faLFF is reduced in posterior cingulum, precuneus, and medial superior frontal gyrus in AD compared to CN individuals.[47] ReHo is a measure of local functional connectivity shown to be reduced in the precuneus, supramarginal gyrus, angular gyrus, and inferior parietal lobule in AD compared to CN individuals.[47] gICA-DR within the default mode network (DMN) is reduced in the precuneus in AD compared to CN individuals.[47] In summary, measures of functional connectivity are reduced in AD and the focus of that reduction resides in the posterior cingulate cortex (PCC),[47,48] a hub of intrinsic cortical connectivity. Because the PCC and medial temporal lobe are functionally connected, it has been hypothesized that AD pathology in the PCC can lead to synaptic dysfunction and neuronal loss in the medial temporal lobe.[49] However, the relationship of synaptic density to measures of brain activity and connectivity are yet to be determined.

In summary, synaptic loss is a major structural correlate of cognitive decline due to AD. Therefore, methods to measure changes in synaptic health will be valuable tools for understanding the neurobiology of AD, tracking disease progression, and assessing therapeutic efficacy. PET imaging with specific radioligands that allow the quantification and localization of SV2A, such as [^{11}C]UCB-J and [^{18}F]SDM-2, are a promising method for measurement of synaptic density. SV2A PET imaging methods can be combined with methods to image other pathophysiological changes such as p-tau or Aβ accumulation and changes in functional connectivity, as well as fluid biomarkers and cognitive assessments, to gain a robust understanding of AD neurobiology.

3. Metabotropic glutamate receptor subtype 5 imaging

Metabotropic glutamate subtype 5 receptors (mGluR5) are g-protein coupled receptors that modulate synaptic transmission and may play an important role in AD by mediating the synaptotoxic action of Aβ oligomers (Aβo). Imaging mGluR5 *in vivo* in humans is possible using the PET radioligand 3-[18F]Fluoro-5-[(pyridine-3-yl)ethynyl]benzonitrile ([^{18}F]FPEB) (Fig. 6).[50,51] Recent *in vivo* studies of mGluR5 expression using [^{18}F]FPEB PET show that individuals with AD have significantly less mGluR5 in the hippocampus compared to CN individuals.[6] This reduction in AD may be more widespread, but it is not clearly separable from receptor loss due to gray

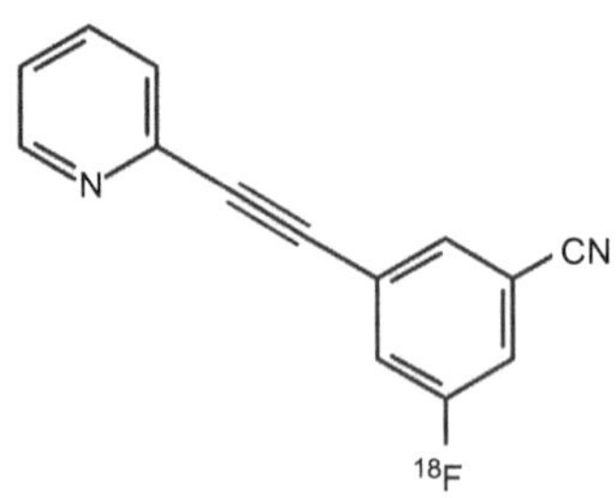

Fig. 6 Chemical structure of [^{18}F]FPEB. [^{18}F]FPEB has been validated in non-human primates and humans. It has high specificity for mGluR5 and good test-retest reliability with bolus-plus-constant-infusion modeling. mGluR5 = metabotropic glutamate receptor subtype 5.

matter atrophy. Further studies to quantify mGluR5 binding in AD may expand our understanding of AD pathogenesis as well as the development of novel biomarkers and treatments.

3.1 Importance of mGluR5 in AD pathophysiology

mGluR5 is located at excitatory synapses[52] and in glial cells (Fig. 7).[53] mGluR5 is expressed throughout the cortex and hippocampus where it modulates synaptic transmission.[54,55] In the rat brain mGluR5 seems to be primarily localized in postsynaptic terminals,[56,57] but it is also present presynaptically.[58] In the primate prefrontal cortex, a substantial proportion of mGluR5 is presynaptic.[59] Preclinical mouse models of AD have been used to explore the role of mGluR5 in AD. Results from these studies indicate that mGluR5 may mediate Aβo toxicity *via* several mechanisms, including promoting the clustering of Aβo as an extracellular scaffold for mGluR5,[60] and serving as a co-receptor for Aβo bound to cellular prion protein (PrP^c) for postsynaptic activation of the tyrosine kinase Fyn.[61,62]

In addition to a more direct involvement in AD pathophysiology, mGluR5 may link Aβ pathology to tau pathology.[63] Complexes of Aβo and PrP^c create a hydrogel phase that recruits mGluR5,[64] leading to activation of the tyrosine kinase Fyn.[61] This activation of Fyn leads to downstream tau phosphorylation.[65] Furthermore, functional tau is required for postsynaptic

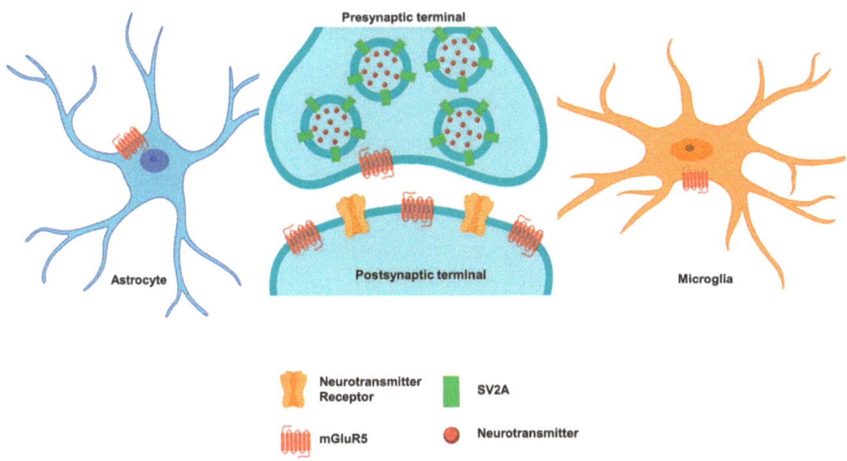

Fig. 7 mGluR5 cellular localization. mGluR5 is predominantly localized at the postsynaptic terminal, but it is also localized in the presynaptic membrane, as well as on astrocytes and microglia. mGluR5 = metabotropic glutamate receptor subtype 5, SV2A = synaptic vesicle glycoprotein 2A.

targeting of Fyn and resulting excitotoxicity mediated by NMDA receptors.[66] The pathways involving Aβo, PrP^c, mGluR5, and Fyn seem to converge on p-tau and the absence of functional tau prevents memory deficits and premature death in transgenic APP23 mice that form Aβ.[66]

3.2 mGluR5 expression in preclinical AD models

Recognition of mGluR5 as a mediator of AD pathology and a potentially important therapeutic target has stimulated the investigation of mGluR5 expression and receptor binding in AD models. Two studies have measured mGluR5 changes in mouse models of AD using PET. Fang et al. investigated mGluR5 in AβPP transgenic mice (tg-ArcSwe) using 3-(6-methyl-pyridin-2-ylethynyl)-cyclohex-2-enone-*O*-[^{11}C]methyl-oxime ([^{11}C]ABP688) PET and reported no difference in binding compared to wild-type mice.[67] However, mGluR5 protein levels were increased in tg-ArcSwe mice when assessed with immunoblot. In a similar study, Lee et al. measured mGluR5 density in 5xFAD mice using [^{18}F]FPEB-PET and immunoblot and observed lower mGluR5 binding and protein levels in the hippocampus and striatum compared to wild-type mice.[68]

3.3 Human mGluR5 imaging

There are several candidate radiotracers for *in vivo* human imaging of mGluR5. Among those studied in humans, [^{11}C]ABP688 may be problematic because of an unexplained significant increase in binding observed in the second scan during test-retest studies.[69] Another candidate, 3-fluoro-5-(2-(2-[^{18}F](fluoromethyl)thiazol-4-yl)ethynyl)benzonitrile ([^{18}F]SP203), is unacceptable for continued studies because it undergoes defluorination in humans.[70] Defluorination is problematic since [^{18}F]fluoride ions bind to bone. Therefore, radioactivity will accumulate in the skull and "spill-in" to adjacent brain tissue, making quantitation of binding in the brain inaccurate. Thus far, [^{18}F]FPEB is the best candidate for continued mGluR5 imaging in humans. [^{18}F]FPEB has high uptake in cortical and subcortical gray matter, low uptake in the cerebellum, and good test-retest reliability with a bolus-plus-constant-infusion method.[50,51] Quantitation with a tissue reference region of whole cerebellum was highly correlated with a method using venous or arterial plasma input functions.[51] For these reasons, [^{18}F]FPEB was used to study mGluR5 distribution in AD compared to CN adults.

Initial investigations of mGluR5 were performed in 16 participants with AD (eight with MCI and eight with dementia) compared with 15 CN participants.[6] Participants ranged in age from 59 to 84 years old. AD participants met research diagnostic criteria for amnestic MCI[33] or NIA-AA criteria for dementia due to AD[34] and were positive for brain Aβ measured by [^{11}C]PiB PET. CN participants were negative for brain Aβ measured by [^{11}C]PiB PET. Parametric images of BP_{ND} were computed using a whole cerebellum reference region (Fig. 8). [^{18}F]FPEB binding to mGluR5 was significantly lower (43%) in the hippocampus of participants with AD compared to CN participants. mGluR5 levels were also lower in the entorhinal cortex and parahippocampal gyrus of participants with AD, but only the lower hippocampal BP_{ND} remained significant after gray matter masking and PVC. Whole brain analyses were also performed to verify the identified regional differences in [^{18}F]FPEB binding to mGluR5. Compared to CN participants, AD participants had lower mGluR5 in the medial temporal lobe. However, the magnitude of [^{18}F]FPEB binding to mGluR5 was reduced in broader association cortical regions (see Fig. 8 for examples).

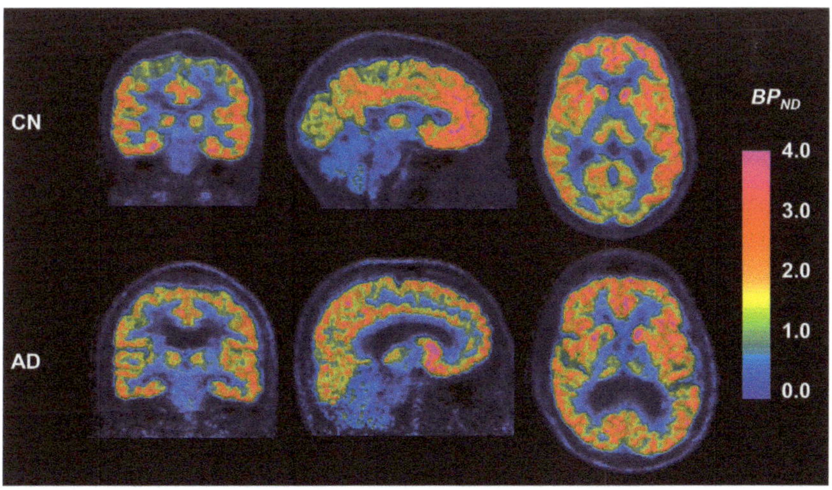

Fig. 8 Parametric images of mGluR5 Binding. BP_{ND} is calculated for each voxel using equilibrium modeling and bolus-plus-constant-infusion modeling of [^{18}F]FPEB. mGluR5 levels are lower in the medial temporal lobe of participants with AD and more subtle declines in mGluR5 may be present throughout the cortex. CN = cognitively normal, AD = Alzheimer's disease, BP_{ND} = binding potential.

3.4 Significance of mGluR5 changes in AD

Human mGluR5 PET results are consistent with *post mortem*[16,71] and *in vivo*[5] evidence of early hippocampal synaptic loss in AD, as well as mGluR5 reductions in AD model mice.[68] Given the evidence for selective vulnerability of association cortex in AD,[72,73] it is somewhat surprising that the distribution of mGluR5 reductions in AD was isolated to the medial temporal lobe. However, this could be due to the relatively larger within group variability and relatively small sample size in this initial study.

The meaning of reduced hippocampal [^{18}F]FPEB binding to mGluR5 in early AD is unclear. Hippocampal reductions in mGluR5 may simply be the product of nonspecific synaptic loss, which would explain the similarity of these results (43% reduction in hippocampal BP_{ND}) with our recent findings using the synaptic PET tracer [^{11}C]UCB-J in a comparable early AD sample.[5] That study demonstrated a 41% reduction in hippocampal BP_{ND}, consistent with postmortem reports of hippocampal synaptic loss in MCI and mild AD.[16,71] Alternatively, the presence of mGluR5 may influence the regional pattern of synaptic loss, given the evidence for involvement of this receptor in AD pathogenesis. mGluR5 has been hypothesized to mediate Aβo synaptotoxicity by a number of mechanisms, including promoting the clustering of Aβo as an extracellular scaffold for mGluR5[60] and serving as a co-receptor for Aβo bound to PrP^c for postsynaptic activation of the tyrosine kinase Fyn.[62,74] If Aβo synaptotoxicity occurs preferentially at mGluR5 sites, then this might also account for the synaptic pattern of mGluR5 reduction. Multi-tracer PET imaging studies with [^{18}F]FPEB and [^{11}C]UCB-J may be able to dissociate the regional pattern of mGluR5 and synaptic losses in early AD.

In summary, [^{18}F]FPEB-PET revealed reductions in hippocampal mGluR5 binding in early AD. mGluR5 modulates synaptic transmission and may play an important role in AD by mediating the synaptotoxic action of Aβo. Therefore, methods to quantify mGluR5 may be valuable tools for understanding the neurobiology of AD. PET imaging with [^{18}F]FPEB allows *in vivo* quantification and localization of mGluR5. mGluR5 PET imaging methods can be combined with methods to image other pathophysiological changes such as p-tau or Aβ accumulation, synaptic density reduction and changes in functional connectivity, as well as fluid biomarkers and cognitive assessments, to gain a robust understanding of AD neurobiology.

4. Serotonergic imaging

The pattern of neuronal loss in AD starts in the entorhinal cortex and hippocampus where glutamatergic cells receive inhibitory serotonin input from the dorsal raphe nucleus. Therefore, imaging components of the serotonergic system may allow investigations of early molecular changes due to AD. Based on the hypothesis that serotonin plays a role in the cognitive and behavioral symptoms of AD, considerable work has been done to investigate the changes in brain serotonergic systems due to AD. Much of this work is based on observations of neuron loss in the raphe nuclei and associated loss of serotonergic cortical projections in AD.[75,76] Specifically, serotonin transporters (SERT), serotonin 2A receptors (5-HT_{2A}) and serotonin 1A receptors (5-HT_{1A}) have been well characterized as sites of action for antidepressant medications and thus have been the focus of radiotracer development.[7,77]

4.1 Serotonergic imaging targets

In vivo molecular imaging studies have focused on several serotonergic protein targets (Fig. 9) using multiple radiotracers (Fig. 10 and Table 1). 5-HT_{2A} receptors are G protein-coupled receptor for serotonin that are predominantly located postsynaptically on pyramidal neurons and interneurons.[77,89] An alternate target for imaging, 5-HT_{1A} receptors, are also G protein-coupled receptors for serotonin. 5-HT_{1A} receptors are located presynaptically in the raphe nuclei as somatodendritic autoreceptors that allow negative feedback of neurotransmitter release. In addition, 5-HT_{1A} receptors are located postsynaptically in the limbic system and neocortex with the highest density in the hippocampus.[77,85] SERT is the transmembrane protein that transports serotonin from the extracellular space across the cell membrane. SERT is abundant on cell bodies, axons, and presynaptic terminals of serotoninergic neurons projecting from the raphe nuclei.[7]

4.2 5-HT_{2A} receptor imaging in AD

[^{18}F]setoperone binding to 5-HT_{2A} was assessed with PET in a group of 9 participants with AD-dementia compared to 37 CN participants. 5-HT_{2A} receptor levels were lower in multiple regions including the temporal cortex, frontal cortex, parietal cortex, temporoparietal junction,

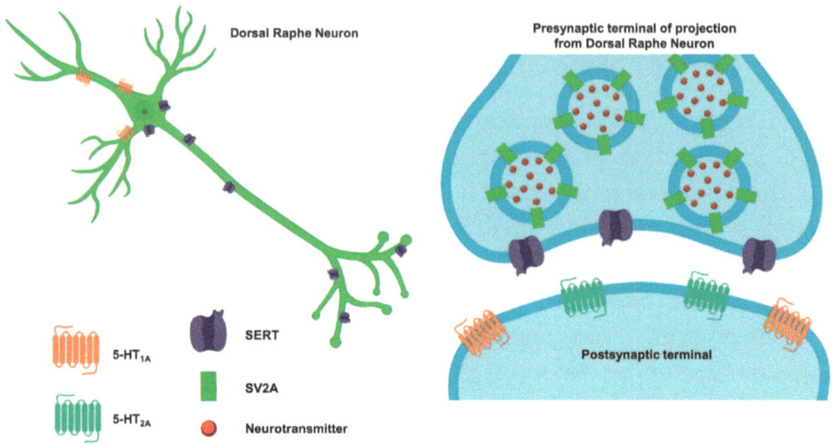

Fig. 9 Serotonergic protein cellular localization. SERT is localized on the cell body, axon, and presynaptically on dendrites of pyramidal cells projecting from the dorsal raphe nucleus. 5-HT$_{1A}$ receptors are located at serotonergic postsynaptic terminals and on the cell bodies and dendrites of dorsal raphe neurons where they serve as autoreceptors to regulate serotonin signaling. 5-HT$_{2A}$ receptors are located at serotonergic postsynaptic terminals. 5-HT$_{1A}$ = 5-hydroxytryptamine-1A receptor, 5-HT$_{2A}$ = 5-Hydroxytryptamine-2A receptor, SERT = serotonin transporter, SV2A = synaptic- vesicle glycoprotein 2A.

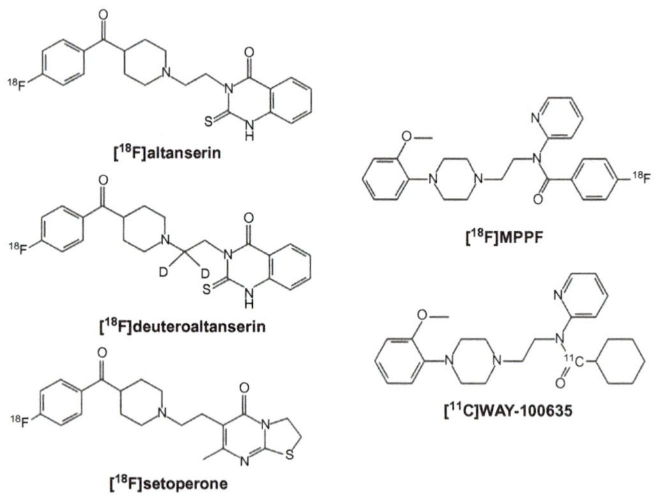

Fig. 10 Serotoninergic radiotracers. Various radioligands have been developed to target serotonergic proteins. 5-HT$_{2A}$ receptor radiotracers include [^{18}F]altanserin, [^{18}F]deuteroaltanserin, and [^{18}F]setoperone. 5-HT$_{1A}$ receptor radiotracers include [^{18}F]MPPF and [^{11}C]WAY-100635. SERT radiotracers include [^{11}C]DASB.

Table 1 Summary of serotonergic *in vivo* molecular imaging studies.

Study	Target	Tracer	Modality	Sample	Significant findings by region
Blin et al.[78]	$5\text{-}HT_{2A}$	[^{18}F]setoperone	PET	12 CN 9 AD	AD < CN—69% temporal, 69% frontal, 55% parietal, 54% temporo-parietal, 35% occipital
Meltzer et al.[79]	$5\text{-}HT_{2A}$	[^{18}F]altanserin	PET	10 CN 9 AD	AD < CN—40% anterior cingulate, 43% prefrontal, 44% sensorimotor
Versijpt et al.[80]	$5\text{-}HT_{2A}$	[^{123}I]5-I-R91150	SPECT	9 CN 9 AD	AD < CN—15% orbitofrontal, 15% prefrontal, 14% lateral frontal, 15% cingulate, 15% sensorimotor, 14% inferior parietal, 8% occipital
Hasselbalch et al.[81]	$5\text{-}HT_{2A}$	[^{18}F]altanserin	PET	17 CN 16 aMCI	aMCI < CN—20–30% posterior cingulate, primary sensorimotor, anterior cingulate, parietal, superior frontal, insula, orbitofrontal, medial temporal, inferior temporal, superior temporal
Santhosh et al.[82]	$5\text{-}HT_{2A}$	[^{18}F] deuteroaltanserin	PET	8 CN 9 AD	AD < CN—43% anterior cingulate
Marner et al.[83, a]	$5\text{-}HT_{2A}$	[^{18}F]altanserin	PET	11 CN 12 AD	AD < CN—28–39% parietal, prefrontal, temporal, occipital, insula, anterior cingulate, mesial temporal
Kepe et al.[84]	$5\text{-}HT_{1A}$	[^{18}F]MPPF	PET	5 CN 6 MCI 8 AD	AD < CN—27% hippocampus, 41% raphe nuclei MCI < CN—13% hippocampus
Lanctot et al.[85]	$5\text{-}HT_{1A}$	[^{11}C]WAY-100635	PET	10 CN 10 AD	AD < CN—16-33% medial temporal, lateral temporal
Truchot et al.[86, a]	$5\text{-}HT_{1A}$	[^{18}F]MPPF	PET	21 CN 11 aMCI 10 AD	AD < CN—35% hippocampus aMCI > CN—59% hippocampus

Continued

Table 1 Summary of serotonergic *in vivo* molecular imaging studies.—cont'd

Study	Target	Tracer	Modality	Sample	Significant findings by region
Truchot et al.[87, a]	$5\text{-}HT_{1A}$	$[^{18}F]$MPPF	PET	21 CN 11 aMCI 10 AD	AD < CN—hippocampus, parahippocampus, inferior temporal, left inferior parietal, left superior parietal, right middle frontal, inferior occipital MCI > CN—inferior occipital, middle occipital
Ouchi et al.[88]	SERT	$[^{11}C]$DASB	PET	10 CN 15 AD	AD < CN—24-33% midbrain, nucleus accumbens, putamen, thalamus
Marner et al.[83, a]	SERT	$[^{11}C]$DASB	PET	11 CN 12 AD	AD < CN—33% medial temporal, 22% occipital, 12% prefrontal
Smith et al.[7]	SERT	$[^{11}C]$DASB	PET	28 CN 28 MCI	MCI < CN—10–38% anterior cingulate, medial frontal, insula, middle temporal, amygdala, hippocampus, precuneus, thalamus, raphe nuclei

[a]Analyses were on performed on the same study sample.

CN = cognitively normal, MCI = mild cognitive impairment, AD = AD dementia, $5\text{-}HT_{1A}$ = serotonin 1A receptor, $5\text{-}HT_{2A}$ = serotonin 2A receptor, SERT = serotonin transporter, PET = positron emission tomography, SPECT = single photon emission computed tomography.

and occipital cortex.[78] A study using [^{18}F]altanserin PET also showed cortical reductions in AD-dementia, but in a less broad distribution.[79] After PVC to account for atrophy, 5-HT$_{2A}$ receptor levels were lower in the cingulate cortex, prefrontal cortex, and sensorimotor cortex of 9 AD participants compared to 10 CN participants. The application of correction for gray matter atrophy and the smaller sample size may explain the more restricted spatial extent of group differences. There have also been several studies using single positron emission computed tomography (SPECT) to quantify AD related changes in 5-HT$_{2A}$. In one such study, [^{123}I]5-I-R91150 binding to 5-HT$_{2A}$ was reduced in the orbitofrontal cortex, prefrontal cortex, lateral frontal cortex, cingulate cortex, sensorimotor cortex, inferior parietal cortex and occipital cortex of 9 AD-dementia participants compared to nine cognitively normal participants.[80] More recently, these studies have been extended to include participants with MCI.[81] 5-HT$_{2A}$ receptor levels were found to be 20–30% lower throughout many cortical and subcortical regions (Table 1). One study using [^{18}F]deuteroaltanserin showed lower 5-HT$_{2A}$ receptor levels in AD dementia, but only in the cingulate cortex.[90] However, the magnitude of 5-HT$_{2A}$ receptor binding was lower in many regions. A broad cortical reduction would be consistent with other studies, but the reason for a more restricted finding using [^{18}F]deuteroaltanserin is not clear. An additional study using [18F]altanserin to quantify 5-HT$_{2A}$ also quantified SERT in the same sample.[83] As in other studies, AD-dementia participants had lower 5-HT$_{2A}$ receptor levels in multiple cortical regions (Table 1). The group differences in SERT were more restricted with the largest magnitude AD decline in the medial temporal cortex.

4.3 5-HT$_{1A}$ receptor imaging in AD

[^{11}C]WAY-100635 binding to 5-HT$_{1A}$ was assessed with PET in a group of 10 participants with AD-dementia compared to 10 CN participants.[85] 5-HT$_{1A}$ receptor levels were lower in the medial and lateral temporal cortex which was restricted to the medial temporal cortex after PVC. In a study by Kepe et al. that included six prodromal AD participants (MCI) as well as eight participants with AD-dementia, 5-HT$_{1A}$ receptor density was significantly lower in the hippocampus and raphe nuclei of the AD-demetia group compared to the CN group ($n = 5$).[84] Furthermore, the MCI group had significantly lower 5-HT$_{1A}$ receptor density only in the hippocampus. Lower 5-HT$_{1A}$ levels in the medial temporal lobe were associated with hypometabiolism measured with [^{18}F]FDG PET. In contrast to these findings, a group of 11 aMCI participants had higher [^{18}F]MPPF binding to 5-HT$_{1A}$ receptors in the hippocampus when compared to 21 CN participants.[86] However, participants at a later stage of disease (AD-dementia, $n=10$) had lower [^{18}F]MPPF binding to 5-HT$_{1A}$ in the hippocampus.[86] These findings are divergent from previous findings by Kepe et al.[84] The discrepancy may be explained by the use of PVC to account for loss of signal due to the relatively low resolution of PET that can lead to "spill-out" of signal in small regions of interest or at boundaries of gray matter-CSF and gray matter-white matter. The larger sample size of Truchot et al.[86] may also provide greater power to detect group differences in early disease stages compared to health. This study was also extended to a whole-brain and limbic-mask analysis. In the whole-brain analysis, there was a significant decrease in 5-HT$_{1A}$ receptor levels in the hippocampus and cortical regions of the AD-dementia group compared to CN group (Table 1).[87] The aMCI group had increased 5-HT$_{1A}$ receptor levels in the inferior occipital cortex, lingual gyrus, and a non-significant increase in the inferior temporal gyrus.[87] The lack of PVC in this analysis may explain the discrepant findings compared to the initial regional analysis. The increase in 5-HT$_{1A}$ receptors in prodromal AD could indicate an early disruption of serotonergic projections from the raphe nuclei. The resulting decrease in serotonin release in the hippocampus would result in less competitive binding and could explain the initial increase in [^{18}F]MPPF binding to 5-HT$_{1A}$ receptors.

4.4 SERT imaging in AD

SERT is present on axons and terminals of serotoninergic neurons projecting from the raphe nuclei. Therefore, it may be a more specific marker of serotonergic terminals compared to 5-HT$_{1A}$ and 5-HT$_{2A}$ receptors that are also located at terminals of non-serotonergic neurons.[91] In a transgenic mouse model of AD, cortical serotonergic degeneration proceeded cortical deposition of amyloid and cortical and hippocampal neuron loss.[92] Thus, it is possible that loss of SERT could be a biomarker of early AD pathophysiology caused by loss of serotonergic neurons and the serotonergic axonal projections.

In agreement with this hypothesis, [^{11}C]DASB binding to SERT was lower in the midbrain, nucleus accumbens, putamen, and thalamus (Table 1) in a group of 15 participants with AD-dementia compared to a group of 10 CN participants.[88] As described above, Marner et al. [83] imaged both SERT and 5-HT$_{2A}$ receptors in 11 AD-dementia participants compared with 11 CN participants. SERT levels were lower in the medial temporal cortex, occipital cortex, and prefrontal cortex (Table 1). The extent of this decline was broader than that of 5-HT_{2A} receptors leading the authors to conclude that there is a relative preservation of serotonergic neurons early in AD that contrasts with the marked loss of 5-HT_{2A} receptors. More recent studies by Smit et al. have imaged SERT in larger samples of participants with MCI.[7] In order to assess the potential confounder of gray matter volume loss, the authors used voxel based morphometry to compare gray matter volume in the MCI group with the CN group. The MCI group had lower gray matter volume in the bilateral middle and inferior temporal gyri, as well as in the right fusiform gyrus. This well characterized sample was also assessed with [^{15}O]water PET to measure regional cerebral blood flow (rCBF). The MCI group had lower rCBF in the left middle temporal gyrus, left posterior cingulate cortex, and left precuneus when compared with the CN group. Compared to these reductions in gray matter volume and rCBF, SERT was reduced more broadly throughout the cortex and subcortex of MCI participants compared to CN participants (Table 1). Lower [^{11}C] DASB binding to SERT was significantly associated with lower episodic memory performance within the MCI group. An extension of this study focused on rs-fMRI connectivity and SERT levels in the dorsal raphe nuclei and precuneus.[93] SERT levels were lower in the MCI group for both regions, but there was no difference in the average rs-fMRI connectivity when comparing MCI to CN participants. Lower SERT density in the dorsal raphe nucleus was associated with lower connectivity between the left hippocampal CA2 and right dentate in MCI, but higher connectivity between these regions in CN participants. Lower SERT density in the dorsal raphe nucleus was also associated with lower connectivity between the left CA2 and right CA3, as well as the right CA2 with bilateral subicula in controls, but there were no significant association in MCI. Furthermore, lower SERT density in the right precuneus was associated with lower connectivity of left retrosplenial cortex and dorsolateral prefrontal cortex in MCI and higher connectivity between these regions in CN participants. Finally, lower SERT density in the right precuneus was associated with higher connectivity of the right retrosplenial cortex and the left dorsal posterior cingulate cortex in MCI but not in CN participants. The meaning of these correlations between SERT and connectivity patterns is not certain, but the authors speculate that serotonergic dysfunction in prodromal AD may lead to alterations in resting state connectivity and ultimately to cognitive dysfunction. This multi-modal imaging study conveys both the strengths and potential pitfalls of correlating multi-tracer or multi-modal data.

On one hand, such a robust study allows analysis of multiple molecular, functional, and cognitive outcomes in the same participant. However, the interpretation of relationships between measurements must be performed with caution and corrections for multiple comparisons should be performed.

In summary, molecular imaging of serotonergic targets may be a viable method for detecting early AD and assessing disease progression. Importantly, studies of serotonergic targets may offer important insights into the differential patterns of receptor/transporter loss and other neuroimaging biomarkers. Furthermore, a multi-modal, multi-tracer approach can be extended to many other molecular targets including synaptic density, mGluR5, Aβ, p-tau, and others. Discerning patterns of longitudinal change in specific molecular biomarkers and the relationships between them could dramatically improve the understanding of AD pathophysiology and assist in the development of an effective therapy.

References

1. Klunk WE, Engler H, Nordberg A, et al. Imaging brain amyloid in Alzheimer's disease with Pittsburgh compound-B. *Ann Neurol.* 2004;55(3):306–319.
2. Wolk DA, Zhang Z, Boudhar S, Clark CM, Pontecorvo MJ, Arnold SE. Amyloid imaging in Alzheimer's disease: comparison of florbetapir and Pittsburgh compound-B positron emission tomography. *J Neurol Neurosurg Psychiatry.* 2012;83(9):923–926.
3. Ishibashi K, Onishi A, Fujiwara Y, Ishiwata K, Ishii K. Relationship between Alzheimer disease-like pattern of 18F-FDG and fasting plasma glucose levels in cognitively normal volunteers. *J Nucl Med.* 2015;56(2):229–233.
4. Mosconi L, Mistur R, Switalski R, et al. FDG-PET changes in brain glucose metabolism from normal cognition to pathologically verified Alzheimer's disease. *Eur J Nucl Med Mol Imaging.* 2009;36(5):811–822.
5. Chen MK, Mecca AP, Naganawa M, et al. Assessing synaptic density in Alzheimer disease with synaptic vesicle glycoprotein 2A positron emission tomographic imaging. *JAMA Neurol.* 2018;75:1215–1224.
6. Mecca AP, McDonald JW, Michalak HR, et al. PET imaging of metabotropic glutamate receptor 5 binding in Alzheimer's disease. *Alzheimers Dement.* 2018;14(7):P501–P503.
7. Smith GS, Barrett FS, Joo JH, et al. Molecular imaging of serotonin degeneration in mild cognitive impairment. *Neurobiol Dis.* 2017;105:33–41.
8. Selkoe DJ. Alzheimer's disease is a synaptic failure. *Science.* 2002;298(5594):789–791.
9. Scheff SW, DeKosky ST, Price DA. Quantitative assessment of cortical synaptic density in Alzheimer's disease. *Neurobiol Aging.* 1990;11(1):29–37.
10. Finnema SJ, Nabulsi NB, Eid T, et al. Imaging synaptic density in the living human brain. *Sci Transl Med.* 2016;8(348):348ra396.
11. Terry RD, Masliah E, Salmon DP, et al. Physical basis of cognitive alterations in Alzheimer's disease: synapse loss is the major correlate of cognitive impairment. *Ann Neurol.* 1991;30(4):572–580.
12. DeKosky ST, Scheff SW. Synapse loss in frontal cortex biopsies in Alzheimer's disease: correlation with cognitive severity. *Ann Neurol.* 1990;27(5):457–464.
13. DeKosky ST, Scheff SW, Styren SD. Structural correlates of cognition in dementia: quantification and assessment of synapse change. *Neurodegeneration.* 1996;5(4):417–421.

14. Hamos JE, DeGennaro LJ, Drachman DA. Synaptic loss in Alzheimer's disease and other dementias. *Neurology*. 1989;39(3):355–361.
15. Masliah E, Mallory M, Alford M, et al. Altered expression of synaptic proteins occurs early during progression of Alzheimer's disease. *Neurology*. 2001;56(1):127–129.
16. Scheff SW, Price DA, Schmitt FA, Mufson EJ. Hippocampal synaptic loss in early Alzheimer's disease and mild cognitive impairment. *Neurobiol Aging*. 2006;27(10): 1372–1384.
17. Pham E, Crews L, Ubhi K, et al. Progressive accumulation of amyloid-beta oligomers in Alzheimer's disease and in amyloid precursor protein transgenic mice is accompanied by selective alterations in synaptic scaffold proteins. *FEBS J*. 2010;277(14):3051–3067.
18. Masliah E, Terry RD, Alford M, DeTeresa R, Hansen LA. Cortical and subcortical patterns of synaptophysinlike immunoreactivity in Alzheimer's disease. *Am J Pathol*. 1991;138(1):235–246.
19. Rao JS, Rapoport SI, Kim HW. Altered neuroinflammatory, arachidonic acid cascade and synaptic markers in postmortem Alzheimer's disease brain. *Transl Psychiatry* 2011;1:e31.
20. Sze CI, Troncoso JC, Kawas C, Mouton P, Price DL, Martin LJ. Loss of the presynaptic vesicle protein synaptophysin in hippocampus correlates with cognitive decline in Alzheimer disease. *J Neuropathol Exp Neurol*. 1997;56(8):933–944.
21. Bajjalieh SM, Peterson K, Linial M, Scheller RH. Brain contains two forms of synaptic vesicle protein 2. *Proc Natl Acad Sci U S A*. 1993;90(6):2150–2154.
22. Bajjalieh SM, Frantz GD, Weimann JM, McConnell SK, Scheller RH. Differential expression of synaptic vesicle protein 2 (SV2) isoforms. *J Neurosci*. 1994;14(9): 5223–5235.
23. Janz R, Sudhof TC. SV2C is a synaptic vesicle protein with an unusually restricted localization: anatomy of a synaptic vesicle protein family. *Neuroscience*. 1999;94(4): 1279–1290.
24. Mendoza-Torreblanca JG, Vanoye-Carlo A, Phillips-Farfan BV, Carmona-Aparicio L, Gomez-Lira G. Synaptic vesicle protein 2A: basic facts and role in synaptic function. *Eur J Neurosci*. 2013;38(11):3529–3539.
25. Crowder KM, Gunther JM, Jones TA, et al. Abnormal neurotransmission in mice lacking synaptic vesicle protein 2A (SV2A). *Proc Natl Acad Sci U S A*. 1999;96(26): 15268–15273.
26. Morgans CW, Kensel-Hammes P, Hurley JB, et al. Loss of the synaptic vesicle protein SV2B results in reduced neurotransmission and altered synaptic vesicle protein expression in the retina. *PLoS One*. 2009;4(4):e5230.
27. Haucke V, Neher E, Sigrist SJ. Protein scaffolds in the coupling of synaptic exocytosis and endocytosis. *Nat Rev Neurosci*. 2011;12(3):127–138.
28. Mutch SA, Kensel-Hammes P, Gadd JC, et al. Protein quantification at the single vesicle level reveals that a subset of synaptic vesicle proteins are trafficked with high precision. *J Neurosci*. 2011;31(4):1461–1470.
29. Mercier J, Archen L, Bollu V, et al. Discovery of heterocyclic nonacetamide synaptic vesicle protein 2A (SV2A) ligands with single-digit nanomolar potency: opening avenues towards the first SV2A positron emission tomography (PET) ligands. *ChemMedChem*. 2014;9(4):693–698.
30. Nabulsi N, Mercier J, Holden D, et al. Synthesis and preclinical evaluation of 11C-UCB-J as a PET tracer for imaging the synaptic vesicle glycoprotein 2A in the brain. *J Nucl Med*. 2016;57:777–784.
31. Finnema SJ, Nabulsi NB, Mercier J, et al. Kinetic evaluation and test-retest reproducibility of [(11)C]UCB-J, a novel radioligand for positron emission tomography imaging of synaptic vesicle glycoprotein 2A in humans. *J Cereb Blood Flow Metab*. 2018;38(11):2041–2052.

32. Innis RB, Cunningham VJ, Delforge J, et al. Consensus nomenclature for in vivo imaging of reversibly binding radioligands. *J Cereb Blood Flow Metab.* 2007;27(9):1533–1539.
33. Petersen RC. Mild cognitive impairment as a diagnostic entity. *J Intern Med.* 2004;256(3):183–194.
34. McKhann GM, Knopman DS, Chertkow H, et al. The diagnosis of dementia due to Alzheimer's disease: recommendations from the National Institute on aging-Alzheimer's association workgroups on diagnostic guidelines for Alzheimer's disease. *Alzheimers Dement.* 2011;7(3):263–269.
35. Mecca AP, Barcelos NM, Wang S, et al. Cortical beta-amyloid burden, gray matter, and memory in adults at varying APOE epsilon4 risk for Alzheimer's disease. *Neurobiol Aging.* 2017;61:207–214.
36. Gomez-Isla T, Price JL, McKeel Jr DW, Morris JC, Growdon JH, Hyman BT. Profound loss of layer II entorhinal cortex neurons occurs in very mild Alzheimer's disease. *J Neurosci.* 1996;16(14):4491–4500.
37. Braak H, Thal DR, Ghebremedhin E, Del Tredici K. Stages of the pathologic process in Alzheimer disease: age categories from 1 to 100 years. *J Neuropathol Exp Neurol.* 2011;70(11):960–969.
38. Fischl B. FreeSurfer. *Neuroimage.* 2012;62(2):774–781.
39. Ashton NJ, Hoglund K, Leuzy A, et al. In: *Cerebrospinal Fluid Synaptic Vesicle Glycoprotein 2A (SV2A) in Alzheimer's Disease. Human Amyloid Imaging 2019 [poster presentation]*; 2019.
40. Ohrfelt A, Brinkmalm A, Dumurgier J, et al. The pre-synaptic vesicle protein synaptotagmin is a novel biomarker for Alzheimer's disease. *Alzheimers Res Ther.* 2016;8(1):41.
41. Brinkmalm A, Brinkmalm G, Honer WG, et al. SNAP-25 is a promising novel cerebrospinal fluid biomarker for synapse degeneration in Alzheimer's disease. *Mol Neurodegener.* 2014;9:53.
42. Portelius E, Olsson B, Hoglund K, et al. Cerebrospinal fluid neurogranin concentration in neurodegeneration: relation to clinical phenotypes and neuropathology. *Acta Neuropathol.* 2018;136(3):363–376.
43. Sandelius A, Portelius E, Kallen A, et al. Elevated CSF GAP-43 is Alzheimer's disease specific and associated with tau and amyloid pathology. *Alzheimers Dement.* 2019;15(1):55–64.
44. Zetterberg H, Skillback T, Mattsson N, et al. Association of cerebrospinal fluid neurofilament light concentration with Alzheimer disease progression. *JAMA Neurol.* 2016;73(1):60–67.
45. Mattsson N, Andreasson U, Zetterberg H, Blennow K. Alzheimer's disease neuroimaging I. Association of plasma neurofilament light with neurodegeneration in patients with Alzheimer disease. *JAMA Neurol.* 2017;74(5):557–566.
46. Zuo XN, Di Martino A, Kelly C, et al. The oscillating brain: complex and reliable. *NeuroImage.* 2010;49(2):1432–1445.
47. Marchitelli R, Aiello M, Cachia A, et al. Simultaneous resting-state FDG-PET/fMRI in Alzheimer disease: relationship between glucose metabolism and intrinsic activity. *NeuroImage.* 2018;176:246–258.
48. Buckner RL, Sepulcre J, Talukdar T, et al. Cortical hubs revealed by intrinsic functional connectivity: mapping, assessment of stability, and relation to Alzheimer's disease. *J Neurosci.* 2009;29(6):1860–1873.
49. Binnewijzend MA, Schoonheim MM, Sanz-Arigita E, et al. Resting-state fMRI changes in Alzheimer's disease and mild cognitive impairment. *Neurobiol Aging.* 2012;33(9):2018–2028.
50. Sullivan JM, Lim K, Labaree D, et al. Kinetic analysis of the metabotropic glutamate subtype 5 tracer [(18)F]FPEB in bolus and bolus-plus-constant-infusion studies in humans. *J Cereb Blood Flow Metab.* 2013;33(4):532–541.

51. Park E, Sullivan JM, Planeta B, et al. Test-retest reproducibility of the metabotropic glutamate receptor 5 ligand [(1)(8)F]FPEB with bolus plus constant infusion in humans. *Eur J Nucl Med Mol Imaging*. 2015;42(10):1530–1541.
52. Awad H, Hubert GW, Smith Y, Levey AI, Conn PJ. Activation of metabotropic glutamate receptor 5 has direct excitatory effects and potentiates NMDA receptor currents in neurons of the subthalamic nucleus. *J Neurosci*. 2000;20(21): 7871–7879.
53. Abushik PA, Niittykoski M, Giniatullina R, et al. The role of NMDA and mGluR5 receptors in calcium mobilization and neurotoxicity of homocysteine in trigeminal and cortical neurons and glial cells. *J Neurochem*. 2014;129(2):264–274.
54. Daggett LP, Sacaan AI, Akong M, et al. Molecular and functional characterization of recombinant human metabotropic glutamate receptor subtype 5. *Neuropharmacology*. 1995;34(8):871–886.
55. Ohnuma T, Augood SJ, Arai H, McKenna PJ, Emson PC. Expression of the human excitatory amino acid transporter 2 and metabotropic glutamate receptors 3 and 5 in the prefrontal cortex from normal individuals and patients with schizophrenia. *Brain Res Mol Brain Res*. 1998;56(1–2):207–217.
56. Shigemoto R, Kinoshita A, Wada E, et al. Differential presynaptic localization of metabotropic glutamate receptor subtypes in the rat hippocampus. *J Neurosci*. 1997;17(19): 7503–7522.
57. Shigemoto R, Nomura S, Ohishi H, Sugihara H, Nakanishi S, Mizuno N. Immunohistochemical localization of a metabotropic glutamate receptor, mGluR5, in the rat brain. *Neurosci Lett*. 1993;163(1):53–57.
58. Gereau RW, Conn PJ. Multiple presynaptic metabotropic glutamate receptors modulate excitatory and inhibitory synaptic transmission in hippocampal area CA1. *J Neurosci*. 1995;15(10):6879–6889.
59. Muly EC, Maddox M, Smith Y. Distribution of mGluR1alpha and mGluR5 immunolabeling in primate prefrontal cortex. *J Comp Neurol*. 2003;467(4):521–535.
60. Renner M, Lacor PN, Velasco PT, et al. Deleterious effects of amyloid beta oligomers acting as an extracellular scaffold for mGluR5. *Neuron*. 2010;66(5):739–754.
61. Um JW, Kaufman AC, Kostylev M, et al. Metabotropic glutamate receptor 5 is a coreceptor for Alzheimer abeta oligomer bound to cellular prion protein. *Neuron*. 2013;79(5):887–902.
62. Haas LT, Salazar SV, Kostylev MA, Um JW, Kaufman AC, Strittmatter SM. Metabotropic glutamate receptor 5 couples cellular prion protein to intracellular signalling in Alzheimer's disease. *Brain*. 2016;139(Pt 2):526–546.
63. Nygaard HB, van Dyck CH, Strittmatter SM. Fyn kinase inhibition as a novel therapy for Alzheimer's disease. *Alzheimers Res Ther*. 2014;6(1):8.
64. Kostylev MA, Tuttle MD, Lee S, et al. Liquid and hydrogel phases of PrP(C) linked to conformation shifts and triggered by Alzheimer's amyloid-beta oligomers. *Mol Cell*. 2018;72(3):426–443.e412.
65. Larson M, Sherman MA, Amar F, et al. The complex PrP(c)-Fyn couples human oligomeric abeta with pathological tau changes in Alzheimer's disease. *J Neurosci*. 2012;32(47):16857–16871a.
66. Ittner LM, Ke YD, Delerue F, et al. Dendritic function of tau mediates amyloid-beta toxicity in Alzheimer's disease mouse models. *Cell*. 2010;142(3):387–397.
67. Fang XT, Eriksson J, Antoni G, et al. Brain mGluR5 in mice with amyloid beta pathology studied with in vivo [11C]ABP688 PET imaging and ex vivo immunoblotting. *Neuropharmacology*. 2017;113(Pt A):293–300.
68. Lee M, Lee HJ, Park IS, et al. Aβ pathology downregulates brain mGluR5 density in a mouse model of Alzheimer. *Neuropharmacology*. 2018;133:512–517.

69. DeLorenzo C, Milak MS, Brennan KG, Kumar JS, Mann JJ, Parsey RV. In vivo positron emission tomography imaging with [^{11}C]ABP688: binding variability and specificity for the metabotropic glutamate receptor subtype 5 in baboons. *Eur J Nucl Med Mol Imaging*. 2011;38(6):1083–1094.
70. Brown AK, Kimura Y, Zoghbi SS, et al. Metabotropic glutamate subtype 5 receptors are quantified in the human brain with a novel radioligand for PET. *J Nucl Med*. 2008;49(12):2042–2048.
71. Scheff SW, Price DA, Schmitt FA, DeKosky ST, Mufson EJ. Synaptic alterations in CA1 in mild Alzheimer disease and mild cognitive impairment. *Neurology*. 2007;68(18): 1501–1508.
72. Braak H, Alafuzoff I, Arzberger T, Kretzschmar H, Del Tredici K. Staging of Alzheimer disease-associated neurofibrillary pathology using paraffin sections and immunocytochemistry. *Acta Neuropathol*. 2006;112(4):389–404.
73. Giannakopoulos P, Hof PR, Bouras C. Selective vulnerability of neocortical association areas in Alzheimer's disease. *Microsc Res Tech*. 1998;43(1):16–23.
74. Um JW, Strittmatter SM. Amyloid-beta induced signaling by cellular prion protein and Fyn kinase in Alzheimer disease. *Prion*. 2013;7(1):37–41.
75. Mann DM, Yates PO. Serotonin nerve cells in Alzheimer's disease. *J Neurol Neurosurg Psychiatry*. 1983;46(1):96.
76. Curcio CA, Kemper T. Nucleus raphe dorsalis in dementia of the Alzheimer type: neurofibrillary changes and neuronal packing density. *J Neuropathol Exp Neurol*. 1984;43(4):359–368.
77. Meltzer CC, Smith G, DeKosky ST, et al. Serotonin in aging, late-life depression, and Alzheimer's disease: the emerging role of functional imaging. *Neuropsychopharmacology*. 1998;18(6):407–430.
78. Blin J, Baron JC, Dubois B, et al. Loss of brain 5-HT2 receptors in Alzheimer's disease. In vivo assessment with positron emission tomography and [18F]setoperone. *Brain*. 1993;116(Pt 3):497–510.
79. Meltzer CC, Price JC, Mathis CA, et al. PET imaging of serotonin type 2A receptors in late-life neuropsychiatric disorders. *Am J Psychiatry*. 1999;156(12):1871–1878.
80. Versijpt J, Van Laere KJ, Dumont F, et al. Imaging of the 5-HT2A system: age-, gender-, and Alzheimer's disease-related findings. *Neurobiol Aging*. 2003;24(4):553–561.
81. Hasselbalch SG, Madsen K, Svarer C, et al. Reduced 5-HT2A receptor binding in patients with mild cognitive impairment. *Neurobiol Aging*. 2008;29(12):1830–1838.
82. Santhosh L, Estok KM, Vogel RS, et al. Regional distribution and behavioral correlates of 5-HT(2A) receptors in Alzheimer's disease with [(18)F]deuteroaltanserin and PET. *Psychiatry Res*. 2009;173(3):212–217.
83. Marner L, Frokjaer VG, Kalbitzer J, et al. Loss of serotonin 2A receptors exceeds loss of serotonergic projections in early Alzheimer's disease: a combined [11C]DASB and [18F] altanserin-PET study. *Neurobiol Aging*. 2012;33(3):479–487.
84. Kepe V, Barrio JR, Huang SC, et al. Serotonin 1A receptors in the living brain of Alzheimer's disease patients. *Proc Natl Acad Sci U S A*. 2006;103(3):702–707.
85. Lanctot KL, Herrmann N, Ganjavi H, et al. Serotonin-1A receptors in frontotemporal dementia compared with controls. *Psychiatry Res*. 2007;156(3):247–250.
86. Truchot L, Costes SN, Zimmer L, et al. Up-regulation of hippocampal serotonin metabolism in mild cognitive impairment. *Neurology*. 2007;69(10):1012–1017.
87. Truchot L, Costes N, Zimmer L, et al. A distinct [18F]MPPF PET profile in amnestic mild cognitive impairment compared to mild Alzheimer's disease. *NeuroImage*. 2008;40(3):1251–1256.
88. Ouchi Y, Yoshikawa E, Futatsubashi M, Yagi S, Ueki T, Nakamura K. Altered brain serotonin transporter and associated glucose metabolism in Alzheimer disease. *J Nucl Med*. 2009;50(8):1260–1266.

89. Miner LA, Backstrom JR, Sanders-Bush E, Sesack SR. Ultrastructural localization of serotonin2A receptors in the middle layers of the rat prelimbic prefrontal cortex. *Neuroscience*. 2003;116(1):107–117.
90. Santhosh L, Estok K, Vogel R, et al. Regional distribution and behavioral correlates of 5-HT2A receptors in Alzheimer's disease with [18F]deuteroaltanserin and PET. *Psychiatry Res Neuroimaging*. 2009;173:212–217.
91. Azmitia EC, Nixon R. Dystrophic serotonergic axons in neurodegenerative diseases. *Brain Res*. 2008;1217:185–194.
92. Liu Y, Yoo MJ, Savonenko A, et al. Amyloid pathology is associated with progressive monoaminergic neurodegeneration in a transgenic mouse model of Alzheimer's disease. *J Neurosci*. 2008;28(51):13805–13814.
93. Barrett FS, Workman CI, Sair HI, et al. Association between serotonin denervation and resting-state functional connectivity in mild cognitive impairment. *Hum Brain Mapp*. 2017;38:3391–3401.

Chapter 2

PD and DLB: Brain imaging in Parkinson's disease and dementia with Lewy bodies

Sarah B. Berman[a,b,c,*], Cameron Miller-Patterson[a]
[a]University of Pittsburgh, Department of Neurology, Pittsburgh, PA, United States
[b]Pittsburgh Institute for Neurodegenerative Diseases, Pittsburgh, PA, United States
[c]University of Pittsburgh, Clinical and Translational Science Institute, Pittsburgh, PA, United States
*Corresponding author: e-mail address: bermans@upmc.edu

Contents

Abstract

Parkinson's disease (PD) and Dementia with Lewy Bodies (DLB) are both neurodegenerative diseases characterized by the abnormal intraneuronal accumulation of misfolded α-Synuclein into Lewy bodies and Lewy neurites. They are distinguished in part by the temporal relationship between the onset of parkinsonism versus cognitive signs, with the former occurring earlier in PD and the latter generally occurring earlier in DLB, along with additional clinical features more prominent in DLB. While clinical criteria help to elucidate diagnosis based on history and exam, definitive diagnosis can only be made at autopsy. More recently, various brain imaging modalities have been utilized to augment diagnostic accuracy made at the bedside and are being studied as possible means to monitor progression or to provide early diagnostic biomarkers. Structural imaging techniques such as structural MRI have some utility in differentiating PD from

Progress in Molecular Biology and Translational Science, Volume 165
ISSN 1877-1173
https://doi.org/10.1016/bs.pmbts.2019.07.009

other causes of parkinsonism, and in differentiating DLB from Alzheimer's disease (AD). In addition, molecular and metabolic brain imaging modalities have shown great promise in elucidating possible diagnostic biomarkers and in providing insight into disease severity and pathogenesis for both PD and DLB. In this review, we summarize the utility of clinically available and research-based brain imaging in diagnosis and monitoring in PD and DLB, which are helping to improve early diagnosis, are providing insight into pathologic mechanisms of disease, and may help to provide biomarker monitoring for disease progression and clinical trials.

1. Introduction

Parkinson's disease (PD) and Dementia with Lewy bodies (DLB) are neurodegenerative diseases classified as Lewy Body diseases, based on the pathognomonic finding of Lewy bodies and Lewy neurites, the abnormal intraneuronal accumulations of misfolded α-Synuclein and other proteins.[1] Clinically, PD is characterized by parkinsonism (bradykinesia with rest tremor and/or rigidity), often with other features such as hyposmia, autonomic insufficiency, sleep, cognitive, and psychiatric dysfunction.[2] DLB is associated with similar clinical signs, but characterized by early cognitive dysfunction and other specific features including visual hallucinations, Rapid Eye Movement sleep behavioral disorder (RBD), and fluctuations in consciousness.[3] Overlapping diagnostic criteria, however, can make clear differential diagnoses difficult: recent DLB consensus diagnostic criteria specify that cognitive symptoms in DLB occur before or within 1 year of parkinsonian motor symptoms, whereas PD dementia only encompasses cognitive symptoms that begin at least 1 year after motor symptoms,[3] but new PD diagnostic criteria do not exclude early cognitive changes from the diagnosis of PD-Mild Cognitive Impairment or PD Dementia.[2] In addition, overlapping features of DLB with Alzheimer's disease (AD), and frequent coexistence of both neuropathologies,[4] further complicate the clinical picture. For much of the early clinical history of diagnosis of PD and DLB, diagnostic utilization of brain imaging modalities was limited. The diagnoses of these neurodegenerative diseases were mainly based on clinical history and exam. But more recently, brain imaging, including structural, metabolic, and molecular imaging modalities, has become an important component of diagnosis and research for these diseases. In this review, we focus separately on PD and DLB, summarizing the utility of clinical and research-based brain imaging.

2. Brain imaging in Parkinson's disease (PD)

2.1 PD structural imaging

Structural imaging modalities that are commonly employed in clinical practice, such as computed tomography (CT) and magnetic resonance imaging (MRI), play a limited role in the diagnosis of Parkinson's disease (PD), since these do not reveal striatal dopaminergic deficits. However, they can often help in the evaluation of other causes of parkinsonism that may also be on the clinician's differential. For example, the "hot cross bun sign" on MRI, resulting from pontine atrophy and gliosis, suggests multiple system atrophy (MSA).[5] Abnormal posterolateral putamen T2 hypointensity and proton density hyperintensity (representing iron deposition) have been shown to be 87.5% and 83.3% sensitive, respectively, and 88.8% and 100% specific, respectively, for a diagnosis of multiple system atrophy-parkinsonism (MSA-P) over PD.[6] The average midbrain-to-pons area ratio on sagittal MRI is significantly smaller in patients with progressive supranuclear palsy (PSP) than in patients with PD, MSA-P, and in asymptomatic patients[7]; the "hummingbird sign" in PSP refers to the appearance of the brainstem on sagittal MRI as a result of this reduced ratio.[5] Excessive white matter disease has been thought to help corroborate a diagnosis of vascular parkinsonism (VP). However, recent literature suggests that striatal MRI abnormalities are neither sensitive nor specific for VP, supported by the fact that most individuals with infarcts in the striatum do not develop parkinsonian signs.[8] Alternatively, lesions of the substantia nigra or nigrostriatal pathway that spare the striatum do seem to correlate with parkinsonism.[8]

While not employed yet in clinical practice, advanced MRI techniques have also been examined as tools to evaluate PD and atypical parkinsonian syndromes (APS) such as MSA and PSP. Compared to patients with PD, apparent diffusion coefficient (ADC) is significantly higher in MSA-P patients in the putamen and in PSP patients in the putamen, caudate, and globus pallidus.[9,10] A meta-analysis of nine studies which evaluated diffusion tensor imaging (DTI) in PD found that fractional anisotropy was significantly lower in the substantia nigra of PD patients compared to healthy controls, suggesting a potential role for DTI in diagnosis.[11]

2.2 PD molecular imaging

Compared to structural imaging modalities, molecular imaging techniques have shown much greater utility in supporting a diagnosis of PD, since these can reveal reductions in presynaptic dopaminergic function. The two most commonly employed modalities for this purpose are single photon emission tomography (SPECT) and positron emission tomography (PET), which indicate reduced striatal dopaminergic input by showing reductions in various functional proteins in the nigrostriatal dopaminergic pathway.[12]

Currently, Dopamine Transporter (DAT) SPECT, commercially available as DaTscan, is the only molecular imaging technique that is approved for clinical practice.[12] DAT SPECT employs the radioligand 123-I-fluoropropyl-CIT, which binds to striatal DAT. Since DAT is a transmembrane protein which transports dopamine from the synaptic cleft back into the presynaptic dopamine neuron, reductions in DAT binding revealed on SPECT correlate with reduced dopaminergic neurons from the substantia nigra to the striatum. In PD, there is typically greater reduction in the posterior putamen than in other parts of the striatum[13] (Fig. 1).

DAT SPECT has been shown to be a useful tool for differentiating neurodegenerative causes of parkinsonism from other non-neurodegenerative causes of parkinsonism, as well as tremor caused by degenerative parkinsonism from other types of tremor.[12] For example, one prospective study found that visual assessment of DAT SPECT images alone had an overall sensitivity and specificity of 96% in identifying parkinsonian (PD, MSA, or PSP) patients over essential tremor (ET) and asymptomatic patients.[14] Jennings et al. compared the diagnostic utility of DAT SPECT to clinical evaluation in differentiating degenerative parkinsonism from various nondegenerative movements

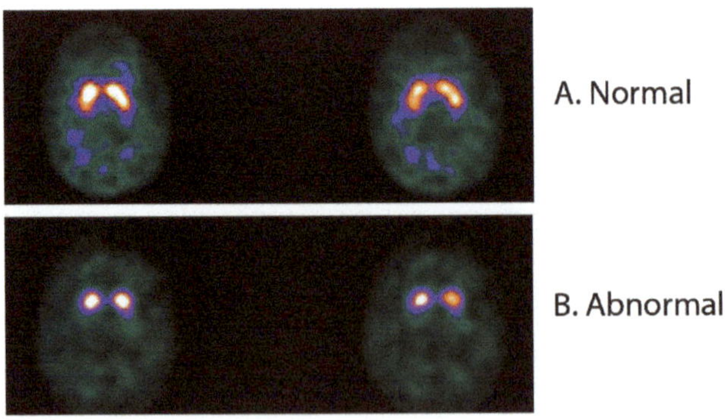

Fig. 1 Dopamine transporter (DAT) single photon emission tomography (SPECT) scan using radioligand 123-I-fluoropropyl-CIT. DAT SPECT images showing presynaptic dopamine terminals in the caudate and putamen in brain slices. The top panel (A) shows normal DAT SPECT imaging with signal in both caudate and putamen. The bottom panel (B) shows asymmetric uptake with reduced uptake in bilateral posterior putamina providing a more circular appearance, as can be seen in both PD and DLB.

disorders, using diagnosis by a blinded movement disorders neurologist at 6-month follow-up as the gold standard. The study found that while both evaluations were equally sensitive for degenerative parkinsonism, DAT SPECT was much more specific.[15] Similarly, a prospective multicenter study found that baseline clinical evaluation was significantly less specific compared to baseline DAT SPECT in identifying tremor due to degenerative parkinsonism versus tremor due to a nondegenerative condition, using blinded clinical evaluation after 3 years as the gold standard.[16] While Lorberboym et al. showed that patients with drug-induced parkinsonism (DIP) can have a positive DAT SPECT, it was postulated that dopamine-blocking agents simply unmasked preclinical PD.[17] However, there is one published case report of a patient with DAT SPECT-positive parkinsonism whose clinical syndrome and DAT binding improved following replacement of lithium with low-dose aripiprazole. The authors suggested that chronic lithium use led to downregulation of DAT by reducing presynaptic dopamine release.[18] Alternatively, patients with a PD diagnosis and a negative DAT SPECT are very unlikely to experience progression of parkinsonism after several years of follow-up, suggesting that they may not in fact have idiopathic PD.[19] These patients also generally do not respond to levodopa and do not have DAT SPECT changes at a later time.[13]

While effective in differentiating degenerative from nondegenerative etiologies of parkinsonism, DAT SPECT has shown poor specificity in differentiating PD from APS such as MSA and PSP, since DAT binding is often reduced in these disorders as well.[12] In a cross-sectional study involving patients with PD, MSA, PSP, and healthy controls, the degree of reduction in DAT binding in the posterior putamen was indistinguishable among patients with PD or APS.[20] However, it has also been shown that DAT reductions tend to be more asymmetric between striata in PD compared to MSA[21,22] as well as PSP.[23] Furthermore, one prospective study found that an anterior-posterior striatal degeneration pattern as opposed to a more global striatal degeneration pattern on DAT SPECT has a 92% positive predictive value for PD over APS or VP.[24] Regardless, there are no clinically-approved standards for differentiating PD from APS on DAT SPECT at this time.

Other binding sites which have been evaluated by molecular imaging as potential diagnostic markers in PD include postsynaptic D2 receptors and dopa decarboxylase (DDC), although these are not typically utilized in clinical practice.[25] D2 SPECT, which involves the radioligand 123-iodobenzamide, has been studied as a diagnostic tool in differentiating PD from APS. One prospective study showed that reduced binding in D2 SPECT in de novo parkinsonism was strongly correlated with APS and poor levodopa responsiveness after several years of follow-up.[26] However, Koch et al. only demonstrated a sensitivity of 87.1% and specificity of 72.5% for D2 SPECT in differentiating PD from APS in a cross-sectional evaluation,[27] while Hellwig et al. showed prospectively that metabolic PET imaging (discussed further below) has much greater diagnostic accuracy in distinguishing PD from APS compared to D2 SPECT.[28] 18-F-fluorodopa uptake using DDC PET has been shown to differ between PD patients and controls. In cases of borderline uptake, healthy controls are more likely to have uniform reduction whereas early PD patients are more likely to have greater reduction in the putamen than caudate.[29] DDC PET has also been evaluated in differentiating between PD, MSA, and PSP. Using discriminant function analysis, PET was able to accurately classify clinically-diagnosed PD patients 64% of the time and PSP patients 70% of the time, although MSA patients were more often miscategorized as having PD or PSP.[30] A more recent publication, however, showed higher diagnostic accuracy in identifying PD and MSA as opposed to PSP.[31] Regardless, DDC PET has not only been shown to correlate well with DAT SPECT in its ability to distinguish PD from control patients,[32] but has also been shown to correlate well in its ability to distinguish early from late PD.[33]

In addition to being evaluated as diagnostic tools in PD, molecular imaging techniques have also been studied as a means of correlating clinical severity with pathological severity. Reduction in uptake on DAT SPECT has been shown to be directly related to severity of motor symptoms, with Marek et al. showing a rate of reduction in striatal uptake of approximately 11.2%/year, compared to a rate in controls of approximately 0.8%/year.[34] Reduction in uptake on DDC PET in the putamen, but not caudate, seems to be directly correlated to motor severity on UPDRS.[35] Similarly, reduction on vesicular monoamine transporter 2 (VMAT2) PET, which uses the radioisotope 18-F-dihydrotetrabenazine, strongly correlated with PD disease severity as quantified by UPDRS motor score and modified Hoehn and Yahr stage.[36] A recent meta-analysis of 142 SPECT and PET studies showed that disease severity is linearly correlated with dopaminergic cell loss greatest in the posterior putamen, which progresses in a negative exponential pattern; this seems to correlate well with the fact that faster clinical progression tends to occur earlier on in PD.[37] In this study, reduction in DDC binding was shown to be consistently less than reduction in binding to DAT or VMAT2, suggesting that DDC is upregulated as PD progresses.[37]

Interestingly, several molecular imaging studies, including DAT SPECT, DDC PET, and dopamine (DA) PET using 11-C-raclopride, have revealed that, unlike overall motor severity, tremor severity does not seem to correlate with nigrostriatal dopaminergic reduction.[38–40] This suggests that development of tremor in PD may be due to pathology outside of the nigrostriatal pathway. Loane et al. employed serotonin transporter (SERT) PET using the radiotracer 11-C-DASB to show that reductions in serotonergic inputs to the striatum and raphe nuclei were significantly associated with worsening postural tremor on the UPDRS, although not rest tremor.[41] This suggests that injury to serotonergic pathways involved in the basal ganglia may be responsible for at least action tremor in PD, unlike bradykinesia and rigidity which seem to be more associated with dopaminergic cell loss.

While molecular imaging has largely been used in PD to evaluate integrity of the nigral dopaminergic nerve terminal, there has also been research attempting to develop effective PET ligands for aggregated α-Synuclein, the major component of Lewy bodies. α-Synuclein-based PET imaging could serve as a useful biomarker for diagnosis, monitoring disease progression, and evaluating response to treatments in clinical trials. At present, however, there are no validated specific PET tracers to detect brain α-Synuclein aggregates. The PET ligand BF227, a benzoxazole compound, has been found to bind to α-Synuclein in the substantia nigra of PD patients at a high concentration. However, it has high affinity for Amyloid beta, which may limit its utility as a selective tracer for PD.[42] Phenothiazine derivatives are also currently being studied as potential α-Synuclein-binding ligands.[42] Tau-binding ligands may also have diagnostic utility in PD, since they may help distinguish PD from an APS such as PSP, although off-target binding and different specificities of the tracers for the multiple conformations of tau isoforms complicate analysis.[43]

Due to their use not only as diagnostic tools but also potentially as biomarkers for trending disease severity, molecular imaging techniques have been incorporated into various clinical trials. For example, Parkinson's Progression Markers Initiative (PPMI), which is an ongoing multicenter observational cohort study to identify biomarkers of disease progression in PD, utilizes DAT SPECT as one of its imaging assessments.[44] The ELLDOPA study found that while UPDRS score was significantly lower in PD patients compared to controls following 40 weeks of levodopa therapy, DAT density on DAT SPECT was also significantly lower, contradicting the clinical results and suggesting the possibility that levodopa could be neurotoxic.[45]

Similarly, the REAL-PET study comparing uptake on DDC PET between patients on either 2 years of levodopa or ropinirole found that reduction in DDC PET uptake in the levodopa group was greater but that clinical outcomes were better.[46] However, the results of both ELLDOPA and REAL-PET are also consistent with the hypothesis that levodopa interferes with radioligand binding as opposed to exacerbating dopaminergic cell loss.[25] Indeed, several subsequent studies including the LEAP study, which compared patients treated with long-term and short-term levodopa, have supported neither a neurotoxic nor neuroprotective effect for levodopa.[47]

2.3 PD metabolic imaging

While not typically used in clinical practice, much research has focused on the use of 18-F-fluorodeoxyglucose PET (FDG PET) to evaluate regional variations of glucose metabolism in the brain in PD. As with molecular imaging, this research may reveal an important diagnostic role for FDG PET in the evaluation of PD. Specific regional patterns of glucose metabolism have been shown to be suggestive of PD as well as specific types of APS. For example, dorsolateral putamen hypermetabolism is a defining characteristic of PD, while bilateral putamen and cerebellar hypometabolism are defining characteristics of MSA.[48] Using predefined imaging features for each disease, Tripathi et al. compared radiographic diagnosis using FDG PET to clinical diagnosis by a movement disorders neurologist during at least 2-year follow-up to attempt to distinguish PD, MSA, PSP, and corticobasal syndrome (CBS).[49] The overall diagnostic accuracy of FDG PET compared to clinical evaluation was 91.7%, while for PD specifically it was 97.6%.[49] A similar study by Brajkovic et al. found a diagnostic accuracy overall of 92% and for PD specifically of 93%.[50]

FDG PET has also been evaluated as a tool to track disease progression. One prospective study of 15 early stage PD patients found that over a 2-year timeframe progression was associated with increasing metabolism in the subthalamic nucleus, globus pallidus, dorsal pons, and primary motor cortex on serial FDG PET.[51] Furthermore, the authors described significant progression of the PD-related motor metabolic covariance pattern (PDRP), which is a characteristic imaging feature seen on FDG PET which indicates general dysfunction of the cortico-striato-thalamocortical pathway in PD; this progression correlated with increasing UPDRS and decreasing DAT binding.[51,52] PD-related cognitive metabolic covariance pattern (PDCP) was also found to increase, but at a slower rate than that of PDRP.[51]

Furthermore, by performing serial FDG PET in initially hemiparkinsonian PD patients, Tang et al. showed that PDRP abnormalities precede motor signs by approximately 2 years.[53]

3. Brain imaging in dementia with Lewy bodies (DLB)

DLB is characterized by cognitive impairment along with parkinsonism, REM sleep behavior disorder (RBD), cognitive fluctuations, and/or visual hallucinations.[3] Similar to PD, it is a synucleinopathy, consisting of intraneuronal accumulations, termed Lewy Bodies, that contain abnormal aggregations of α-Synuclein. Diagnosis of DLB can be more complex than PD, given DLB's similarities to both PD Dementia, and to the more common AD. In addition, concurrent AD and PD neuropathology is often present, leading to clinical signs of both and sometimes complicating diagnosis.[4] Thus, brain imaging in DLB may be helpful in sorting out the clinical picture, although there are fewer studies focused on DLB than on PD with or without dementia. Of note, this review focuses only on *brain* imaging in DLB, and does not include other imaging modalities such as cardiac sympathetic innervation imaging using Iodine-123 metaiodobenzylguanidine cardiac scintigraphy.

3.1 DLB structural imaging

As noted above, structural MRI is generally most helpful in clinical PD assessment for evaluating for other possible causes of parkinsonism. In DLB, given the overlap in pathology and clinical findings with AD, structural MRI may be useful clinically in differential diagnosis. In fact, relative sparing of atrophy in the medial temporal lobe is a supportive biomarker in the updated consensus criteria for DLB.[3] A study by Harper et al. found that medial temporal atrophy could distinguish AD from DLB with a sensitivity of 64% and specificity of 68%.[54] In addition, in a recent large study of 333 patients with DLB, 352 patients with AD, and 233 normal controls from the European DLB Consortium and the AD Neuroimaging Initiative (ADNI), patterns of atrophy by structural MRI were examined.[55] The study found that those with DLB had global atrophy across the frontal, medial temporal, and parietal regions, but overall had less atrophy than in AD, and relatively less medial temporal atrophy. The presence of medial temporal atrophy alone could discriminate between DLB and AD with an accuracy of approximately 65%. In their modeling, this typical pattern of atrophy contributed only marginally for diagnostic utility in those who had the core features of parkinsonism and visual hallucinations, but the absence of medial temporal atrophy had significant utility in distinguishing confirmed DLB from AD when those core clinical signs were *absent*.

3 T MRI susceptibility-weighted imaging (SWI) has also been studied for utility in distinguishing DLB from AD. Shams et al. proposed that the absence of hyperintensity in the substantia nigra "nigrosome," which normally gives a characteristic "swallowtail"-like appearance on SWI imaging in control brains, may be a characteristic feature of DLB.[56] A similar recent study found that the absence of the swallowtail sign on 3 T MRI SWI imaging in the substantia nigra could distinguish DLB from AD and amnestic MCI with similar specificity and sensitivity to DAT SPECT.[57] The authors argue that if validated, this could provide a less expensive alternative to DAT SPECT for imaging dopamine neurons in DLB, but this remains to be determined. Many studies have also evaluated MRI diffusion-tensor imaging (DTI) of white matter tracts in DLB compared to controls or other neurodegenerative diseases.[58–68] These generally report extensive structural connectivity abnormalities in DLB, as measured as reduced fractional anisotropy, but a specific pattern has not been consistently well-defined. Interestingly, several studies report more abnormalities in DTI in visual association areas in DLB, possibly related to the presence of visual hallucinations, and these tend to be more severe than in AD.[62–65]

3.2 DLB molecular brain imaging

3.2.1 DAT SPECT

In DLB, similarly to PD, there is diminished striatal dopamine terminal innervation, as measured by DAT SPECT scans showing reduced striatal 123-I-fluoropropyl-CIT binding, and it has been established as a core biomarker feature of DLB[3] (Fig. 1). This has been demonstrated in numerous studies[69–73] and has been correlated with neuropathology by autopsy.[74] The specificity and sensitivity of DAT SPECT for DLB are estimated to be quite high,[75,76] although many studies are limited by factors such as utilizing high-probability clinical DLB as criteria for scanning, or absence of neuropathological confirmation. Studies that have correlated DAT SPECT with neuropathological confirmation of DLB, although small, do confirm high sensitivity and specificity of this imaging modality in differential diagnosis of DLB,[74,77–79] even though there are reported cases of autopsy-proven DLB pathology lacking abnormal DAT SPECT scans.[74]

There is also much interest in developing imaging biomarkers for detecting prodromal DLB, and DAT SPECT has been studied for its possible utility in early diagnosis. These studies, however, are complicated by overlapping criteria for defining "prodromal DLB" and "cognitive presentations of PD."[2] A recent study utilized DAT SPECT to evaluate striatal DAT binding in "prodromal DLB" subjects ($n = 20$), who did not meet criteria for probable DLB but had mild cognitive impairment (MCI) along with nonmotor features of DLB, such as autonomic symptoms, RBD, and hyposmia, with little or no parkinsonism (which, if present, did not predate the cognitive impairment), and also had occipital hypometabolism on FDG PET scan.[80] These subjects were compared to subjects meeting criteria for probable DLB ($n = 18$) and those with AD ($n = 10$). 123I-FP CIT striatal binding ratios were lower in both prodromal and probable DLB groups compared to AD. The striatal binding ratio was negatively correlated with parkinsonian motor UPDRS score in the DLB group, but not significantly in the prodromal group. In the prodromal group, RBD and hyposmia were negatively correlated with striatal DAT binding. In a longitudinal study by Siepel et al.[81] of seven subjects with mild dementia who ranked low on core symptoms of DLB but who had decreased striatal DAT binding on DAT SPECT scan, all seven eventually developed parkinsonism and cognitive fluctuations over 2–5 years, further suggesting the possibility that abnormal DAT imaging is an early biomarker of prodromal DLB.

3.2.2 α-Synuclein, tau, and amyloid PET

As detailed above, brain imaging methodology to detect α-Synuclein aggregates is an active area of investigation but has yet to be successfully developed and validated. Once developed, this will be of great interest not only in PD imaging but in DLB imaging as well. In addition to α-Synuclein pathology, DLB neuropathology studies show variable degrees of Amyloid and Tau pathology,[4] and thus, many studies have evaluated the utility of Amyloid and Tau PET brain imaging in DLB, since these are available. In general, Amyloid tracer uptake in DLB is less of that in AD and tends to show more variability.[82–85] Increased Amyloid uptake is likely more common in DLB compared to PD with dementia,[84] as might be expected given the more frequent coexistence of AD and DLB pathology.[4] Uptake of the Tau PET tracer AV1451 has also been studied in DLB. Findings have been variable, but in general, those with probable DLB often show higher AV1451 uptake in various regions of the cortex than in controls. In one study, subjects with AD were found to have higher uptake of AV1451, particularly in the medial temporal lobes, compared to those with DLB. Those with DLB, however, had higher uptake in the temporoparietal and occipital regions than in controls, and this was correlated with uptake of the Amyloid tracer Pittsburgh Compound B (PiB).[82] Another study found higher uptake of AV1451 in the inferior temporal gyrus and precuneus in DLB compared to controls.[86] One recent study found increased AV1451 uptake in parietal regions in subjects with DLB compared to both PD and control participants.[87] Other results have been more variable,[88,89] and therefore, specific patterns have not yet been definitively determined.

3.3 DLB metabolic, perfusion, and biochemical imaging

Both FDG PET metabolic imaging and SPECT perfusion brain imaging may be useful in differential diagnosis in DLB. F18-FDG PET studies often reveal hypometabolism in parieto-occipital regions.[90–93] In DLB, FDG PET studies reveal that metabolism is preserved in the posterior cingulate compared to the precuneus and cuneus, termed the "cingulate island" sign.[92,94] In addition, SPECT perfusion studies have validated the cingulate island sign in DLB finding as well, with optimized analysis yielding accuracy, sensitivity, and specificity of 84.6%, 92.3%, and 76.9%, respectively, in distinguishing DLB from AD.[95] However, a study directly comparing the accuracy of FDG PET with SPECT perfusion studies found the cingulate island sign was present with FDG-PET imaging in DLB, but not with SPECT perfusion.[96] The cingulate island sign observed via FDG-PET in DLB has been associated with a lower Braak neurofibrillary tangle stage and was not related to Amyloid pathology.[97] Both occipital hypometabolism or hypoperfusion on FDG-PET or SPECT, and the cingulate island sign on FDG-PET imaging, are considered supportive biomarkers in the diagnostic criteria for DLB.[3]

Proton magnetic resonance spectroscopy (MSR) can give quantitative information on biochemical changes in neurodegenerative diseases. Relatively few studies have examined MRS signatures in DLB, and many are complicated by overlapping AD pathology. Graff-Radford et al. examined specific DLB patients without medial temporal atrophy, as a surrogate for low probability of coexisting AD.[98] Decreased *N*-acetylaspartate (NAA)/Creatine (Cr) ratio is thought to be indicative of neuronal loss, and in these DLB patients, there was preserved NAA/Cr in the posterior cingulate, suggesting normal neuronal volume. Interestingly, this potentially corresponds to the preserved FDG PET uptake that leads to the "cingulate island" sign noted above. In addition, these patients had increased myoinositol/Cr ratio (found in glial cells and thought to be a possible marker of glial proliferation) in the occipital lobe, while having normal NAA/Cr in the same region. Others have reported lower NAA/Cr in the occipital lobe in

DLB, but have not focused on those with lower probability of coexisting AD.[99] Graff-Radford et al. also found elevated choline/Cr ratios in the posterior cingulate in DLB, as they had found in a previous study as well,[100] suggestive of cholinergic dysfunction in this region. While MRS may not yet be of diagnostic utility in DLB, one could hypothesize that the ability to evaluate additional biochemical metabolites using this technique, such as MRS quantification of the antioxidant glutathione,[101] may also prove useful in studies of pathogenesis in DLB.

3.4 Use of imaging to elucidate pathophysiology in DLB

In addition to diagnostic utility, research-based studies may give insight into the underlying pathophysiology of symptoms. For example, the physiological substrate of the visual hallucinations common in DLB are of significant interest, and imaging studies may be useful for trying to understand the underlying pathophysiology. Pezzoli et al. performed a recent comprehensive review of structural and functional (fMRI, PET, and SPECT) imaging studies evaluating visual hallucinations in DLB and PD.[102] Overall, while there were few studies that focused specifically on DLB-associated visual hallucinations, and studies have inconsistent results, the most robust finding was decreased gray matter on structural MRI in the frontal regions in those with either DLB or PDD with visual hallucinations, with some evidence of decreased occipitotemporal metabolic dysfunction. It has been hypothesized that this might represent structural abnormalities of frontal attentional networks, along with hypometabolism in visual processing networks.[102]

Other small studies have provided intriguing suggestions of pathophysiologic changes that may be related to other specific symptoms. For example, the neural dysfunction underlying apathy in DLB was examined in another small study.[103] Here, the presence of apathy was linked to decreased cholinergic innervation in the anterior cingulate cortex by SPECT imaging using the radiotracer ^{123}I-iodobenzovesamicol, a tracer of the vesicular acetylcholine transporter. More general cortical cholinergic dysfunction in DLB, greater than that of AD, was also suggested in another PET imaging study using the radiotracer acetylcholine analogue *N*-[^{11}C]-methyl-4-piperidyl acetate (MP4A).[104] The neural substrate for the classic fluctuations in cognition that characterize DLB is not yet known, but one study found that proton MRS imaging found a correlation between severity of cognitive fluctuations in DLB and increased choline/Cr ratio in the thalamus in DLB compared to AD.[105] Thus, although the underlying functional pathways that lead to specific symptoms in DLB have yet to be fully elucidated, these types of functional and structural imaging studies are helping to clarify possible network substrates for the pathophysiology of specific symptoms.

4. Conclusions

Structural, molecular, and metabolic brain imaging are of increasing utility in PD and DLB, both in differential diagnosis and in understanding pathomechanisms of disease. Continued development of new molecular imaging, such as α-Synuclein or other imaging, and refinement of existing methodologies, are likely to yield improved imaging biomarkers for early diagnosis and monitoring disease progression for clinical studies.

References

1. Spillantini MG, Schmidt ML, Lee VM, Trojanowski JQ, Jakes R, Goedert M. Alpha-synuclein in Lewy bodies. *Nature*. 1997;388(6645):839–840.
2. Postuma RB, Berg D, Stern M, et al. MDS clinical diagnostic criteria for Parkinson's disease. *Mov Disord*. 2015;30(12):1591–1601.
3. McKeith IG, Boeve BF, Dickson DW, et al. Diagnosis and management of dementia with Lewy bodies: fourth consensus report of the DLB consortium. *Neurology*. 2017;89(1):88–100.
4. Jellinger KA. Dementia with Lewy bodies and Parkinson's disease-dementia: current concepts and controversies. *J Neural Transm (Vienna)*. 2018;125(4):615–650.
5. McFarland NR. Diagnostic approach to atypical Parkinsonian syndromes. *Continuum (Minneap Minn)*. 2016;22(4 Movement Disorders):1117–1142.
6. Righini A, Antonini A, Ferrarini M, et al. Thin section MR study of the basal ganglia in the differential diagnosis between striatonigral degeneration and Parkinson disease. *J Comput Assist Tomogr*. 2002;26(2):266–271.
7. Oba H, Yagishita A, Terada H, et al. New and reliable MRI diagnosis for progressive supranuclear palsy. *Neurology*. 2005;64(12):2050–2055.
8. Vizcarra JA, Lang AE, Sethi KD, Espay AJ. Vascular Parkinsonism: deconstructing a syndrome. *Mov Disord*. 2015;30(7):886–894.
9. Schocke MF, Seppi K, Esterhammer R, et al. Diffusion-weighted MRI differentiates the Parkinson variant of multiple system atrophy from PD. *Neurology*. 2002;58(4):575–580.
10. Seppi K, Schocke MF, Esterhammer R, et al. Diffusion-weighted imaging discriminates progressive supranuclear palsy from PD, but not from the parkinson variant of multiple system atrophy. *Neurology*. 2003;60(6):922–927.
11. Cochrane CJ, Ebmeier KP. Diffusion tensor imaging in parkinsonian syndromes: a systematic review and meta-analysis. *Neurology*. 2013;80(9):857–864.
12. Pagano G, Niccolini F, Politis M. Imaging in Parkinson's disease. *Clin Med (Lond)*. 2016;16(4):371–375.
13. Ba F, Martin WR. Dopamine transporter imaging as a diagnostic tool for Parkinsonism and related disorders in clinical practice. *Parkinsonism Relat Disord*. 2015;21(2):87–94.
14. Benamer TS, Patterson J, Grosset DG, et al. Accurate differentiation of parkinsonism and essential tremor using visual assessment of [123I]-FP-CIT SPECT imaging: the [123I]-FP-CIT study group. *Mov Disord*. 2000;15(3):503–510.

15. Jennings DL, Seibyl JP, Oakes D, Eberly S, Murphy J, Marek K. (123I) beta-CIT and single-photon emission computed tomographic imaging vs clinical evaluation in Parkinsonian syndrome: unmasking an early diagnosis. *Arch Neurol.* 2004;61(8):1224–1229.
16. Marshall VL, Reininger CB, Marquardt M, et al. Parkinson's disease is overdiagnosed clinically at baseline in diagnostically uncertain cases: a 3-year European multicenter study with repeat [123I]FP-CIT SPECT. *Mov Disord.* 2009;24(4):500–508.
17. Lorberboym M, Treves TA, Melamed E, Lampl Y, Hellmann M, Djaldetti R. [123I]-FP/CIT SPECT imaging for distinguishing drug-induced parkinsonism from Parkinson's disease. *Mov Disord.* 2006;21(4):510–514.
18. Cilia R, Marotta G, Belletti A, Siri C, Pezzoli G. Reversible dopamine transporter reduction in drug-induced parkinsonism. *Mov Disord.* 2014;29(4):575–577.
19. Marek K, Seibyl J, Eberly S, et al. Longitudinal follow-up of SWEDD subjects in the PRECEPT study. *Neurology.* 2014;82(20):1791–1797.
20. Kim YJ, Ichise M, Ballinger JR, et al. Combination of dopamine transporter and D2 receptor SPECT in the diagnostic evaluation of PD, MSA, and PSP. *Mov Disord.* 2002;17(2):303–312.
21. Varrone A, Marek KL, Jennings D, Innis RB, Seibyl JP. [(123)I]beta-CIT SPECT imaging demonstrates reduced density of striatal dopamine transporters in Parkinson's disease and multiple system atrophy. *Mov Disord.* 2001;16(6):1023–1032.
22. Knudsen GM, Karlsborg M, Thomsen G, et al. Imaging of dopamine transporters and D2 receptors in patients with Parkinson's disease and multiple system atrophy. *Eur J Nucl Med Mol Imaging.* 2004;31(12):1631–1638.
23. Filippi L, Manni C, Pierantozzi M, et al. 123I-FP-CIT in progressive supranuclear palsy and in Parkinson's disease: a SPECT semiquantitative study. *Nucl Med Commun.* 2006;27(4):381–386.
24. Davidsson A, Georgiopoulos C, Dizdar N, Granerus G, Zachrisson H. Comparison between visual assessment of dopaminergic degeneration pattern and semi-quantitative ratio calculations in patients with Parkinson's disease and atypical parkinsonian syndromes using DaTSCAN(R) SPECT. *Ann Nucl Med.* 2014;28(9):851–859.
25. Politis M. Neuroimaging in Parkinson disease: from research setting to clinical practice. *Nat Rev Neurol.* 2014;10(12):708–722.
26. Schwarz J, Tatsch K, Gasser T, et al. 123I-IBZM binding compared with long-term clinical follow up in patients with de novo parkinsonism. *Mov Disord.* 1998;13(1):16–19.
27. Koch W, Hamann C, Radau PE, Tatsch K. Does combined imaging of the pre- and postsynaptic dopaminergic system increase the diagnostic accuracy in the differential diagnosis of parkinsonism? *Eur J Nucl Med Mol Imaging.* 2007;34(8):1265–1273.
28. Hellwig S, Amtage F, Kreft A, et al. [(1)(8)F]FDG-PET is superior to [(1)(2)(3)I]IBZM-SPECT for the differential diagnosis of parkinsonism. *Neurology.* 2012;79(13):1314–1322.
29. Sawle GV, Playford ED, Burn DJ, Cunningham VJ, Brooks DJ. Separating Parkinson's disease from normality. Discriminant function analysis of fluorodopa F 18 positron emission tomography data. *Arch Neurol.* 1994;51(3):237–243.
30. Burn DJ, Sawle GV, Brooks DJ. Differential diagnosis of Parkinson's disease, multiple system atrophy, and Steele-Richardson-Olszewski syndrome: discriminant analysis of striatal 18F-dopa PET data. *J Neurol Neurosurg Psychiatry.* 1994;57(3):278–284.
31. Jin S, Oh M, Oh SJ, et al. Differential diagnosis of parkinsonism using dual-phase F-18 FP-CIT PET imaging. *Nucl Med Mol Imaging.* 2013;47(1):44–51.
32. Ishikawa T, Dhawan V, Kazumata K, et al. Comparative nigrostriatal dopaminergic imaging with iodine-123-beta CIT-FP/SPECT and fluorine-18-FDOPA/PET. *J Nucl Med.* 1996;37(11):1760–1765.
33. Eshuis SA, Maguire RP, Leenders KL, Jonkman S, Jager PL. Comparison of FP-CIT SPECT with F-DOPA PET in patients with de novo and advanced Parkinson's disease. *Eur J Nucl Med Mol Imaging.* 2006;33(2):200–209.

34. Marek K, Innis R, van Dyck C, et al. [123I]beta-CIT SPECT imaging assessment of the rate of Parkinson's disease progression. *Neurology*. 2001;57(11):2089–2094.
35. Broussolle E, Dentresangle C, Landais P, et al. The relation of putamen and caudate nucleus 18F-Dopa uptake to motor and cognitive performances in Parkinson's disease. *J Neurol Sci*. 1999;166(2):141–151.
36. Hsiao IT, Weng YH, Hsieh CJ, et al. Correlation of Parkinson disease severity and 18F-DTBZ positron emission tomography. *JAMA Neurol*. 2014;71(6):758–766.
37. Kaasinen V, Vahlberg T. Striatal dopamine in Parkinson disease: a meta-analysis of imaging studies. *Ann Neurol*. 2017;82(6):873–882.
38. Asenbaum S, Pirker W, Angelberger P, Bencsits G, Pruckmayer M, Brucke T. [123I] beta-CIT and SPECT in essential tremor and Parkinson's disease. *J Neural Transm (Vienna)*. 1998;105(10 – 12):1213–1228.
39. de la Fuente-Fernandez R, Pal PK, Vingerhoets FJ, et al. Evidence for impaired presynaptic dopamine function in parkinsonian patients with motor fluctuations. *J Neural Transm (Vienna)*. 2000;107(1):49–57.
40. Pavese N, Evans AH, Tai YF, et al. Clinical correlates of levodopa-induced dopamine release in Parkinson disease: a PET study. *Neurology*. 2006;67(9):1612–1617.
41. Loane C, Wu K, Bain P, Brooks DJ, Piccini P, Politis M. Serotonergic loss in motor circuitries correlates with severity of action-postural tremor in PD. *Neurology*. 2013;80(20):1850–1855.
42. Mathis CA, Lopresti BJ, Ikonomovic MD, Klunk WE. Small-molecule PET tracers for imaging Proteinopathies. *Semin Nucl Med*. 2017;47(5):553–575.
43. Saeed U, Compagnone J, Aviv RI, et al. Imaging biomarkers in Parkinson's disease and parkinsonian syndromes: current and emerging concepts. *Transl Neurodegener*. 2017;6:8.
44. Marek K, Chowdhury S, Siderowf A, et al. The Parkinson's progression markers initiative (PPMI)—establishing a PD biomarker cohort. *Ann Clin Transl Neurol*. 2018;5(12):1460–1477.
45. Fahn S, Oakes D, Shoulson I, et al. Levodopa and the progression of Parkinson's disease. *N Engl J Med*. 2004;351(24):2498–2508.
46. Whone AL, Watts RL, Stoessl AJ, et al. Slower progression of Parkinson's disease with ropinirole versus levodopa: the REAL-PET study. *Ann Neurol*. 2003;54(1):93–101.
47. Verschuur CVM, Suwijn SR, Boel JA, et al. Randomized delayed-start trial of levodopa in Parkinson's disease. *N Engl J Med*. 2019;380(4):315–324.
48. Eckert T, Barnes A, Dhawan V, et al. FDG PET in the differential diagnosis of parkinsonian disorders. *Neuroimage*. 2005;26(3):912–921.
49. Tripathi M, Dhawan V, Peng S, et al. Differential diagnosis of parkinsonian syndromes using F-18 fluorodeoxyglucose positron emission tomography. *Neuroradiology*. 2013;55(4):483–492.
50. Brajkovic L, Kostic V, Sobic-Saranovic D, et al. The utility of FDG-PET in the differential diagnosis of parkinsonism. *Neurol Res*. 2017;39(8):675–684.
51. Huang C, Tang C, Feigin A, et al. Changes in network activity with the progression of Parkinson's disease. *Brain*. 2007;130(Pt 7):1834–1846.
52. Poston KL, Eidelberg D. FDG PET in the evaluation of Parkinson's disease. *PET Clin*. 2010;5(1):55–64.
53. Tang CC, Poston KL, Dhawan V, Eidelberg D. Abnormalities in metabolic network activity precede the onset of motor symptoms in Parkinson's disease. *J Neurosci*. 2010;30(3):1049–1056.
54. Harper L, Fumagalli GG, Barkhof F, et al. MRI visual rating scales in the diagnosis of dementia: evaluation in 184 post-mortem confirmed cases. *Brain*. 2016;139(Pt 4): 1211–1225.
55. Oppedal K, Ferreira D, Cavallin L, et al. A signature pattern of cortical atrophy in dementia with Lewy bodies: a study on 333 patients from the European DLB consortium. *Alzheimers Dement*. 2019;15(3):400–409.

56. Shams S, Fallmar D, Schwarz S, et al. MRI of the swallow tail sign: a useful marker in the diagnosis of Lewy body dementia? *AJNR Am J Neuroradiol*. 2017;38(9):1737–1741.
57. Kamagata K, Nakatsuka T, Sakakibara R, et al. Diagnostic imaging of dementia with Lewy bodies by susceptibility-weighted imaging of nigrosomes versus striatal dopamine transporter single-photon emission computed tomography: a retrospective observational study. *Neuroradiology*. 2017;59(1):89–98.
58. Bozzali M, Falini A, Cercignani M, et al. Brain tissue damage in dementia with Lewy bodies: an in vivo diffusion tensor MRI study. *Brain*. 2005;128(Pt 7):1595–1604.
59. Firbank MJ, Blamire AM, Krishnan MS, et al. Diffusion tensor imaging in dementia with Lewy bodies and Alzheimer's disease. *Psychiatry Res*. 2007;155(2):135–145.
60. Firbank MJ, Blamire AM, Teodorczuk A, Teper E, Mitra D, O'Brien JT. Diffusion tensor imaging in Alzheimer's disease and dementia with Lewy bodies. *Psychiatry Res*. 2011;194(2):176–183.
61. Ota M, Sato N, Ogawa M, et al. Degeneration of dementia with Lewy bodies measured by diffusion tensor imaging. *NMR Biomed*. 2009;22(3):280–284.
62. Lee JE, Park HJ, Park B, et al. A comparative analysis of cognitive profiles and white-matter alterations using voxel-based diffusion tensor imaging between patients with Parkinson's disease dementia and dementia with Lewy bodies. *J Neurol Neurosurg Psychiatry*. 2010;81(3):320–326.
63. Kantarci K, Avula R, Senjem ML, et al. Dementia with Lewy bodies and Alzheimer disease: neurodegenerative patterns characterized by DTI. *Neurology*. 2010;74(22):1814–1821.
64. Kiuchi K, Morikawa M, Taoka T, et al. White matter changes in dementia with Lewy bodies and Alzheimer's disease: a tractography-based study. *J Psychiatr Res*. 2011;45(8): 1095–1100.
65. Watson R, Blamire AM, Colloby SJ, et al. Characterizing dementia with Lewy bodies by means of diffusion tensor imaging. *Neurology*. 2012;79(9):906–914.
66. O'Donovan J, Watson R, Colloby SJ, Blamire AM, O'Brien JT. Assessment of regional MR diffusion changes in dementia with Lewy bodies and Alzheimer's disease. *Int Psychogeriatr*. 2014;26(4):627–635.
67. Nedelska Z, Schwarz CG, Boeve BF, et al. White matter integrity in dementia with Lewy bodies: a voxel-based analysis of diffusion tensor imaging. *Neurobiol Aging*. 2015;36(6):2010–2017.
68. Delli Pizzi S, Franciotti R, Taylor JP, et al. Structural connectivity is differently altered in dementia with Lewy body and Alzheimer's disease. *Front Aging Neurosci*. 2015;7:208.
69. Ransmayr G, Seppi K, Donnemiller E, et al. Striatal dopamine transporter function in dementia with Lewy bodies and Parkinson's disease. *Eur J Nucl Med*. 2001;28(10):1523–1528.
70. Walker Z, Moreno E, Thomas A, et al. Clinical usefulness of dopamine transporter SPECT imaging with 123I-FP-CIT in patients with possible dementia with Lewy bodies: randomised study. *Br J Psychiatry*. 2015;206(2):145–152.
71. McKeith I, O'Brien J, Walker Z, et al. Sensitivity and specificity of dopamine transporter imaging with 123I-FP-CIT SPECT in dementia with Lewy bodies: a phase III, multicentre study. *Lancet Neurol*. 2007;6(4):305–313.
72. Walker Z, Costa DC, Walker RW, et al. Striatal dopamine transporter in dementia with Lewy bodies and Parkinson disease: a comparison. *Neurology*. 2004;62(9):1568–1572.
73. O'Brien JT, Colloby S, Fenwick J, et al. Dopamine transporter loss visualized with FP-CIT SPECT in the differential diagnosis of dementia with Lewy bodies. *Arch Neurol*. 2004;61(6):919–925.
74. Thomas AJ, Attems J, Colloby SJ, et al. Autopsy validation of 123I-FP-CIT dopaminergic neuroimaging for the diagnosis of DLB. *Neurology*. 2017;88(3):276–283.
75. Papathanasiou ND, Boutsiadis A, Dickson J, Bomanji JB. Diagnostic accuracy of (1)(2)(3)I-FP-CIT (DaTSCAN) in dementia with Lewy bodies: a meta-analysis of published studies. *Parkinsonism Relat Disord*. 2012;18(3):225–229.

76. Brigo F, Turri G, Tinazzi M. 123I-FP-CIT SPECT in the differential diagnosis between dementia with Lewy bodies and other dementias. *J Neurol Sci*. 2015;359(1–2):161–171.
77. Colloby SJ, Firbank MJ, Pakrasi S, et al. A comparison of 99mTc-exametazime and 123I-FP-CIT SPECT imaging in the differential diagnosis of Alzheimer's disease and dementia with Lewy bodies. *Int Psychogeriatr*. 2008;20(6):1124–1140.
78. Morgan S, Kemp P, Booij J, et al. Differentiation of frontotemporal dementia from dementia with Lewy bodies using FP-CIT SPECT. *J Neurol Neurosurg Psychiatry*. 2012;83(11):1063–1070.
79. Walker RW, Walker Z. Dopamine transporter single photon emission computerized tomography in the diagnosis of dementia with Lewy bodies. *Mov Disord*. 2009;24 (Suppl 2):S754–S759.
80. Kasanuki K, Iseki E, Ota K, et al. (123)I-FP-CIT SPECT findings and its clinical relevance in prodromal dementia with Lewy bodies. *Eur J Nucl Med Mol Imaging*. 2017;44(3):358–365.
81. Siepel FJ, Rongve A, Buter TC, et al. (123I)FP-CIT SPECT in suspected dementia with Lewy bodies: a longitudinal case study. *BMJ Open*. 2013;3(4):1–8.
82. Kantarci K, Lowe VJ, Boeve BF, et al. AV-1451 tau and beta-amyloid positron emission tomography imaging in dementia with Lewy bodies. *Ann Neurol*. 2017;81(1):58–67.
83. Donaghy P, Thomas AJ, O'Brien JT. Amyloid PET imaging in Lewy body disorders. *Am J Geriatr Psychiatry*. 2015;23(1):23–37.
84. Edison P, Rowe CC, Rinne JO, et al. Amyloid load in Parkinson's disease dementia and Lewy body dementia measured with [11C]PIB positron emission tomography. *J Neurol Neurosurg Psychiatry*. 2008;79(12):1331–1338.
85. Rowe CC, Ng S, Ackermann U, et al. Imaging beta-amyloid burden in aging and dementia. *Neurology*. 2007;68(20):1718–1725.
86. Gomperts SN, Locascio JJ, Makaretz SJ, et al. Tau positron emission tomographic imaging in the Lewy body diseases. *JAMA Neurol*. 2016;73(11):1334–1341.
87. Smith R, Scholl M, Londos E, Ohlsson T, Hansson O. (18)F-AV-1451 in Parkinson's Disease with and without dementia and in Dementia with Lewy Bodies. *Sci Rep*. 2018;8(1):4717.
88. Lee SH, Cho H, Choi JY, et al. Distinct patterns of amyloid-dependent tau accumulation in Lewy body diseases. *Mov Disord*. 2018;33(2):262–272.
89. Nedelska Z, Josephs KA, Graff-Radford J, et al. (18) F-AV-1451 uptake differs between dementia with lewy bodies and posterior cortical atrophy. *Mov Disord*. 2019;34(3):344–352.
90. Minoshima S, Foster NL, Sima AA, Frey KA, Albin RL, Kuhl DE. Alzheimer's disease versus dementia with Lewy bodies: cerebral metabolic distinction with autopsy confirmation. *Ann Neurol*. 2001;50(3):358–365.
91. Kantarci K, Lowe VJ, Boeve BF, et al. Multimodality imaging characteristics of dementia with Lewy bodies. *Neurobiol Aging*. 2012;33(9):2091–2105.
92. Imamura T, Ishii K, Sasaki M, et al. Regional cerebral glucose metabolism in dementia with Lewy bodies and Alzheimer's disease: a comparative study using positron emission tomography. *Neurosci Lett*. 1997;235(1–2):49–52.
93. Ishii K, Imamura T, Sasaki M, et al. Regional cerebral glucose metabolism in dementia with Lewy bodies and Alzheimer's disease. *Neurology*. 1998;51(1):125–130.
94. Lim SM, Katsifis A, Villemagne VL, et al. The 18F-FDG PET cingulate island sign and comparison to 123I-beta-CIT SPECT for diagnosis of dementia with Lewy bodies. *J Nucl Med*. 2009;50(10):1638–1645.
95. Imabayashi E, Soma T, Sone D, et al. Validation of the cingulate island sign with optimized ratios for discriminating dementia with Lewy bodies from Alzheimer's disease using brain perfusion SPECT. *Ann Nucl Med*. 2017;31(7):536–543.

96. O'Brien JT, Firbank MJ, Davison C, et al. 18F-FDG PET and perfusion SPECT in the diagnosis of Alzheimer and Lewy body dementias. *J Nucl Med*. 2014;55(12):1959–1965.
97. Graff-Radford J, Murray ME, Lowe VJ, et al. Dementia with Lewy bodies: basis of cingulate island sign. *Neurology*. 2014;83(9):801–809.
98. Graff-Radford J, Boeve BF, Murray ME, et al. Regional proton magnetic resonance spectroscopy patterns in dementia with Lewy bodies. *Neurobiol Aging*. 2014;35(6): 1483–1490.
99. Zhong X, Shi H, Shen Z, et al. 1H-proton magnetic resonance spectroscopy differentiates dementia with Lewy bodies from Alzheimer's disease. *J Alzheimers Dis*. 2014;40(4):953–966.
100. Kantarci K, Petersen RC, Boeve BF, et al. 1H MR spectroscopy in common dementias. *Neurology*. 2004;63(8):1393–1398.
101. Rae CD, Williams SR. Glutathione in the human brain: Review of its roles and measurement by magnetic resonance spectroscopy. *Anal Biochem*. 2017;529:127–143.
102. Pezzoli S, Cagnin A, Bandmann O, Venneri A. Structural and functional neuroimaging of visual hallucinations in lewy body disease: a systematic literature review. *Brain Sci*. 2017;7(7):1–30.
103. Mazere J, Lamare F, Allard M, Fernandez P, Mayo W. 123I-Iodobenzovesamicol SPECT imaging of cholinergic Systems in Dementia with Lewy bodies. *J Nucl Med*. 2017;58(1):123–128.
104. Shimada H, Hirano S, Sinotoh H, et al. Dementia with Lewy bodies can be well-differentiated from Alzheimer's disease by measurement of brain acetylcholinesterase activity-a [11C]MP4A PET study. *Int J Geriatr Psychiatry*. 2015;30(11):1105–1113.
105. Delli Pizzi S, Franciotti R, Taylor JP, et al. Thalamic involvement in fluctuating cognition in dementia with Lewy bodies: magnetic resonance evidences. *Cereb Cortex*. 2015;25(10):3682–3689.

Chapter 3

FTD spectrum: Neuroimaging across the FTD spectrum

Jennifer L. Whitwell*
Department of Radiology, Mayo Clinic, Rochester, MN, United States
*Corresponding author: e-mail address: whitwell.jennifer@mayo.edu

Contents

Abstract

Frontotemporal dementia is a complex and heterogeneous neurodegenerative disease that encompasses many clinical syndromes, pathological diseases, and genetic mutations. Neuroimaging has played a critical role in our understanding of the underlying pathophysiology of frontotemporal dementia and provided biomarkers to aid diagnosis. Early studies defined patterns of neurodegeneration and hypometabolism associated with the clinical, pathological and genetic aspects of frontotemporal dementia, with more recent studies highlighting how the breakdown of structural and functional brain networks define frontotemporal dementia. Molecular positron emission tomography ligands allowing the in vivo imaging of tau proteins have also provided important insights, although more work is needed to understand the biology of the currently available ligands.

Progress in Molecular Biology and Translational Science, Volume 165
ISSN 1877-1173
https://doi.org/10.1016/bs.pmbts.2019.05.009

1. Introduction

1.1 Clinical classification of FTD

Frontotemporal dementia (FTD) is a clinical term for a group of neurodegenerative syndromes which are associated with degeneration of the frontal and temporal lobes of the brain. Clinical diagnostic criteria for FTD published in 1998 recognized three clinical FTD syndromes, termed behavioral variant FTD (bvFTD), semantic dementia (SD) and progressive non-fluent aphasia.[1] Behavioral variant FTD was characterized by changes in personality and behavioral abnormalities, such as apathy, disinhibition and obsessive-compulsive behaviors, and loss of insight. Patients with bvFTD can develop language abnormalities as the disease progresses, and patients can also develop parkinsonism and features of motor neuron disease (MND). Updated consensus criteria for bvFTD have since been published.[2] Semantic dementia was characterized by word finding problems (i.e., anomia) and loss of semantic knowledge (such as loss of word, object or face knowledge).[3] Lastly, progressive non-fluent aphasia was characterized predominantly by the presence of non-fluent spontaneous speech output. Since this time, it has been recognized that many patients with progressive non-fluent aphasia display combinations of two clinical features: agrammatic aphasia which is characterized by agrammatic written or spoken output and apraxia of speech which is a motor speech disorder affecting the ability to produce words, typically characterized by sound distortions and segmentations.[4,5] Diagnostic criteria for primary progressive aphasia published in 2011 recognized an agrammatic non-fluent variant of primary progressive aphasia (agPPA) which is defined by the presence of either agrammatism or apraxia of speech,[6] although the language deficits should be the earliest and dominant feature.[7] Furthermore, a clinical entity was described in 2012 in which apraxia of speech is the sole presenting symptom, in the absence of agrammatism or any other evidence or aphasia.[8] This entity was termed primary progressive apraxia of speech (PPAOS).[8] Patients with both agPPA and PPAOS often develop an atypical parkinsonian syndrome over time, characterized by limb apraxia, postural instability and eye movement abnormalities.[9–12] The PPA criteria also defined a semantic variant of PPA which specifically classifies patients with anomia and loss of word knowledge, and hence includes some, although not all, patients that were originally classified as SD.[6]

1.2 Pathological classification of FTD

The pathologies underlying the FTD syndromes are heterogeneous, but are typically defined by the abnormal deposition of one of three proteins: the microtubule-associated protein tau, the TAR DNA binding protein of 43 kDa (TDP-43)[13,14] or, less commonly, the tumor associated protein fused in sarcoma (FUS).[15] Therefore, the majority of cases can be sub-classified into FTLD-tau, FTLD-TDP and FTLD-FUS (Fig. 1), based on the biochemical signature of the abnormally deposited protein.[13,14] Each of these three groups is also sub-classified based on inclusion morphology and lesion distribution. Hence, FTLD-tau includes specific pathologies such as Pick's disease,[16] progressive supranuclear palsy[17] and corticobasal degeneration[18,19]; FTLD-TDP includes four different types (A–D),[20–22] and FTLD-FUS includes the specific pathologies of neuronal intermediate filament inclusion disease,[23,24] basophilic inclusion body disease[25,26] and atypical FTLD with ubiquitin-only immunoreactive changes.[27,28] The syndrome of bvFTD can be associated with many different FTLD pathologies, and can result from FTLD-tau, FTLD-TDP or FTLD-FUS pathology (Fig. 1).[14] There may be some clinical signs that suggest one specific pathology over another, although generally it is very different to predict the underlying pathology in bvFTD. The syndrome of SD has a more homogeneous pathology and typically results from FTLD-TDP pathology (Fig. 1),

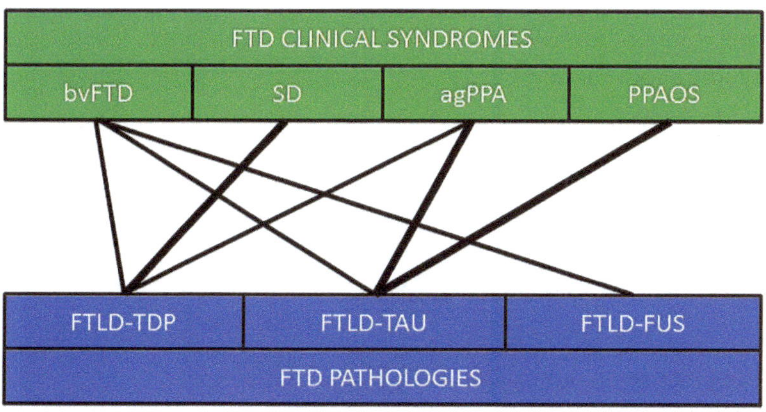

Fig. 1 Schematic illustrating the relationship between FTD clinical syndrome and pathological diagnosis. BvFTD = behavioral variant frontotemporal dementia, SD = semantic dementia; agPPA = agrammatic variant of primary progressive aphasia; PPAOS = primary progressive aphasia.

particularly FTLD-TDP type C[15]. The syndrome PPAOS also has a relatively homogeneous pathology, with most cases resulting from FTLD-tau (Fig. 1) characterized by deposition of 4-repeat tau, i.e., progressive supranuclear palsy or corticobasal degeneration.[10] In fact, the presence of apraxia of speech has been shown to be predictive of an underlying FTLD-tau pathology, even when it co-occurs with agrammatic aphasia.[10] Hence, patients with agPPA, that often have co-occurring apraxia of speech, also often have an underlying FTLD-tau pathology.[10,29] Cases of aPPA have been reported with FTLD-TDP, although this pathology is rarer than FTLD-tau in these patients.[29] Furthermore, patients with the clinical syndromes of FTD have been reported to have the pathology of Alzheimer's disease which is characterized by deposition of 3R + 4R tau and beta-amyloid.[30]

1.3 Neuroimaging in FTD

Neuroimaging research has been critical in increasing our understanding of FTD. Structural magnetic resonance imaging (MRI) and (18F)-fluorodeoxyglucose positron emission tomography (FDG-PET) studies have characterized patterns of neurodegeneration and hypometabolism that define the different clinical and pathological variants of FTD, while diffusion tensor imaging (DTI) and resting-state fMRI have improved our understanding of the structural and functional connectivity breakdowns associated with FTD. The recent availability of molecular PET ligands that bind to beta-amyloid or tau proteins has also allowed the in vivo investigation of brain pathology. The knowledge we have gained from these neuroimaging modalities concerning the pathophysiology of FTD will be discussed in this review.

2. Imaging in the clinical syndromes of FTD

2.1 Behavioral variant FTD

Behavioral variant FTD is typically associated with atrophy on structural MRI and hypometabolism on FDG-PET in the prefrontal cortex and anterior temporal lobes, with relative sparing of more posterior regions of the brain, such as the occipital lobe[31–39] (Fig. 2). Regions in the frontal lobe that are commonly affected in bvFTD include the orbitofrontal cortex, medial and lateral prefrontal cortex and anterior cingulate, along with atrophy of the adjacent insula cortex. Imaging studies have demonstrated that atrophy

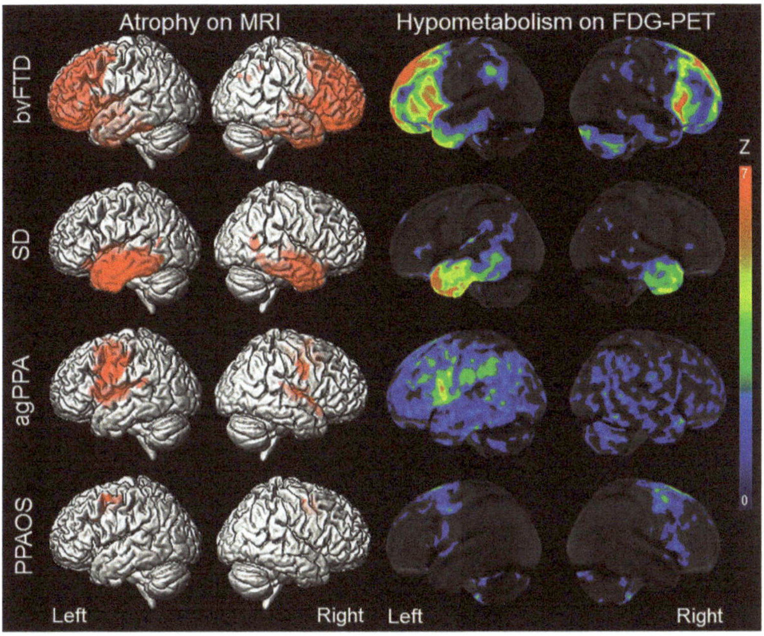

Fig. 2 Patterns of atrophy and hypometabolism associated with the FTD clinical syndromes. Group-level maps depicting gray matter atrophy compared to controls are shown on three dimensional renders of the brain. Hypometabolism on FDG-PET is shown for individual patients, represented as Z score maps depicting abnormalities compared to controls.

of these frontal and anterior temporal regions is related to the behavioral symptoms observed in bvFTD, including apathy, disinhibition, loss of empathy, and aggression.[40–50] Atrophy of subcortical structures, such as caudate, putamen, globus pallidus, thalamus and nucleus accumbens, are also observed, reflecting dysfunction of cortico-striato-thalamic circuits[51] which have also been associated with behavioral abnormalities in bvFTD.[52–61] Neurodegeneration in bvFTD is progressive, with decline in brain volume observed over time. Rates of whole brain volume loss can be as fast as 3% per year,[34,35] with greatest rates of decline observed in the frontal lobes.[62–68] Frontotemporal and subcortical atrophy is observed in the mildest stages of the disease[69] and even before patients meet criteria for bvFTD,[70] and is observed to spread as the disease progresses.[69]

However, a great deal of heterogeneity in patterns of neurodegeneration has been observed within cohorts of bvFTD patients on both MRI and FDG-PET.[71–75] Some patients tend to have more predominant frontal abnormalities,[71,72,75] others have predominant temporal abnormalities,[71,72,75] and some show involvement of temporal and parietal structures.[71,73,75] A predominantly subcortical subtype has also been identified.[71,72] Importantly, clinical characteristics differ across these different anatomical phenotypes, with the frontal phenotype showing worse executive performance and the temporal phenotype showing worse episodic memory impairment.[73,75] Furthermore, the patients with frontal dominant patterns of atrophy tend to show faster clinical decline,[76] while the subcortical subtype shows relatively slow disease progression.[71] Recognizing the different phenotypes of bvFTD could, therefore, be important to help improve prognostic estimates for patients. The phenotypes do not differ in terms of behavioral abnormalities, reflecting the fact that behavioral abnormalities are the key features of a bvFTD diagnosis. Heterogeneity is also observed in the degree of asymmetry in brain atrophy[42,73] with most patients showing symmetric patterns of atrophy, although patients also show right and left-sided patterns of atrophy.[42] In addition, some bvFTD patients have been reported that showed very little atrophy on MRI, despite fulfilling clinical criteria for bvFTD, and show slow rates of clinical decline.[77–79] Some of these patients have been shown to display FTD pathology[80] and are occasionally associated with the presence of a *C9ORF72* genetic mutation[79] (see Section 4). Those patients with additional MND, i.e., FTD-MND, show shorter disease duration than typical bvFTD patients and have been shown to have very similar patterns of atrophy to bvFTD, particularly with involvement of the frontal lobes.[81–83] The degree of atrophy is, however, typically less than that observed in patients with bvFTD,[81,83] although these patients show greater involvement of the frontal and temporal lobes compared to patients with MND in the absence of FTD.[83]

Structural MRI and FDG-PET findings have been shown to be diagnostically useful in differentiating bvFTD from healthy individuals and patients with other neurodegenerative diseases. In particular, the relative pattern of involvement of anterior brain structures nicely differentiates patients with bvFTD from those with typical Alzheimer's dementia which typically targets posterior regions of the brain, including the posterior temporal and parietal lobes.[84] Studies that have utilized pattern classification techniques have found that structural MRI can differentiate bvFTD from controls with an accuracy of 85%,[85,86] and differentiate bvFTD from Alzheimer's dementia with an accuracy of 82%.[86] Diagnostic accuracy has also been shown to be excellent with pattern analysis of FDG-PET, with discriminatory power of 92.2% to differentiate bvFTD from Alzheimer's dementia and 87.6% to differentiate bvFTD from the other language variants of FTD.[87] In fact, simple visual assessments of FDG-PET looking for frontal, anterior cingulate and anterior temporal hypometabolism have excellent specificity (97.6%) and sensitivity (86%) in differentiating bvFTD from Alzheimer's dementia, and can improve differentiation based on clinical features alone.[88] There is also evidence that FDG-PET can improve diagnosis over structural MRI,[89–91] with FDG-PET being more sensitive to underlying dysfunction than structural MRI which can be normal at baseline in bvFTD patients.[92]

Disruption in the structural connectivity of the brain, measured using DTI, is also a feature of bvFTD. Degeneration is observed in white matter tracts that have reciprocal projections into the frontal and temporal lobes, such as the superior longitudinal fasciculus, anterior cingulum, genu of the corpus callosum, uncinate fasciculus and the inferior longitudinal fasciculus[45,93–98] (Fig. 3). Degeneration of these tracts is associated with behavioral severity in bvFTD.[45,96] Degeneration is also observed in fronto-striatal and fronto-thalamic pathways.[51] Less severe degeneration is also observed in tracts that connect into posterior regions of the brain, including the posterior cingulate and posterior aspects of the superior longitudinal fasciculus. Longitudinal studies have observed progressive degeneration of the same set of tracts that show abnormalities in cross-sectional studies, particularly the uncinate fasciculus, genu of the corpus callosum, paracallosal cingulum and frontal white matter,[99–101] with additional progression to involve more posterior white matter tracts located in the posterior temporal and occipital lobe.[102] This demonstrates a spreading of disease from anterior to more posterior brain regions. DTI measures from the anterior corpus callosum are particularly promising in bvFTD since they provide good differentiation of bvFTD patients from other clinical variants of FTD,[93,95] perform well as a longitudinal marker to differentiate bvFTD from controls,[100] and are associated with longitudinal decline in executive function in bvFTD.[103] There is also a suggestion that DTI can pick up abnormalities in patients in which the MRI and FDG-PET scans are normal[104] and DTI findings can show more widespread findings than structural MRI,[98,105] suggesting good sensitivity of DTI. The value of DTI compared to other modalities is less clear. A couple of studies have found that DTI performed better than regional volumes in differentiating bvFTD from controls[106] or from other FTD syndromes,[107] although another found that DTI measures do not perform as well as FDG-PET for diagnosis at the individual-patient level.[104] DTI does, however, appear to add complementary information to structural

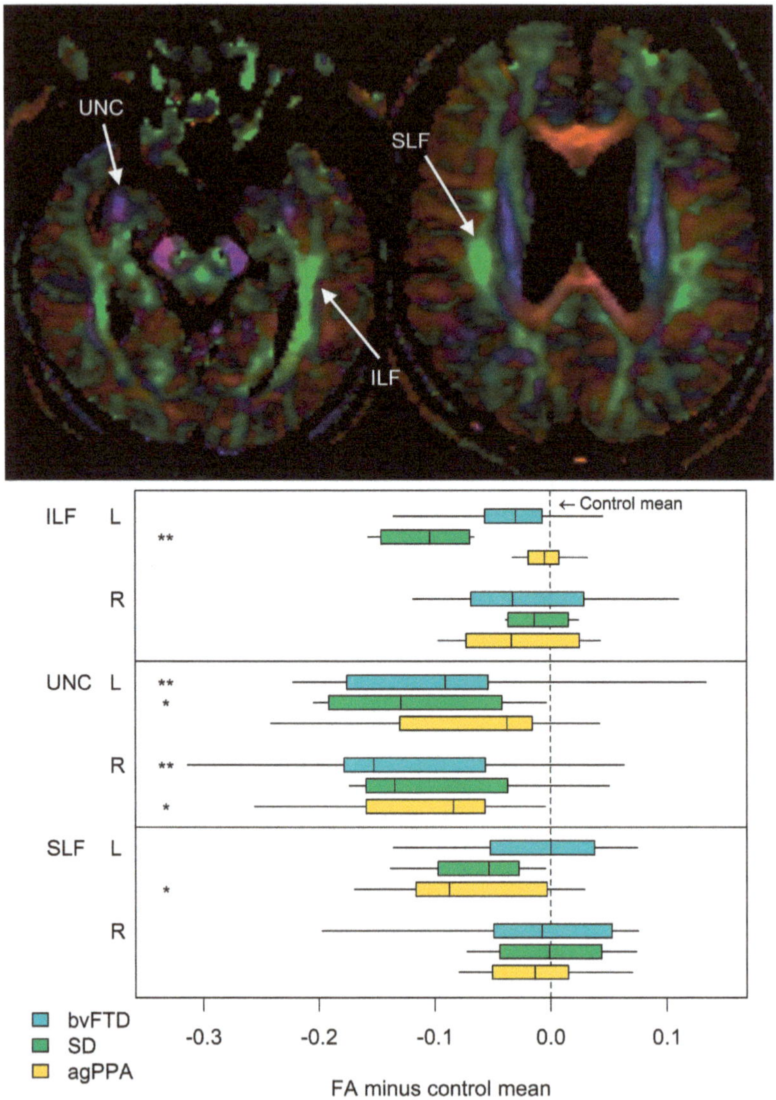

Fig. 3 Diffusion tensor imaging findings in bvFTD, SD and agPPA. The top panel shows color coded fractional anisotropy (FA) maps showing major white matter tracts. The bottom panel shows box-plots of FA values across the three clinical groups. The left inferior longitudinal fasciculus (ILF) is preferentially affected in SD, the uncinate fasciculus (UNC) is affected in both bvFTD and SD, and the left superior longitudinal fasciculus (SLF) is preferentially affected in agPPA. SD = semantic dementia; bvFTD = behavioral variant of FTD; agPPA = agrammatic variant of primary progressive aphasia. *Reproduced with permission from Imaging in Neurodegenerative Disorders (Oxford University Press, 2015, edited by Luca Saba).*

MRI when differentiating bvFTD from controls and Alzheimer's dementia, and hence a combination of markers from different modalities may be the best approach to differential diagnosis.[108]

Functional brain connectivity in bvFTD has also been assessed using resting-state fMRI analyses. Studies that have focused on assessing specific brain networks have shown that one of the characteristic changes observed in bvFTD is decreased functional connectivity within the salience network.[109–113] The salience network is involved in processing socially relevant information and involves anterior cingulate and fronto-insular regions. Similarly, other studies have found bilateral disconnection between frontal and limbic regions of the brain.[111,112,114] Disruptions in connectivity in frontolimbic regions appears to be central to the disease process in bvFTD and has been shown to correlate to the degree of behavioral and executive impairment .[111,115,116] Connectivity from the frontal lobe also appears to decrease with disease progression in bvFTD.[117] However, decreased connectivity has also been observed in the basal ganglia,[114] insula[114] and dorsal attention network,[112] and disconnection between the frontal lobes and posterior regions of the brain, including occipital lobe,[114,118,119] also appear to occur in bvFTD. Increases in connectivity have also been observed in other networks, such as the medial parietal components of the default mode network[109–112] and attention/working memory network,[112] with posterior hubs relatively preserved in bvFTD.[118] This pattern of connectivity changes may be useful to differentiate bvFTD patients from patients with Alzheimer's dementia, which typically show reduced connectivity in the default mode network and posterior brain nodes.[110,112,120] Studies have started to analyze and compare resting state fMRI to other neuroimaging metrics in terms of diagnostic utility. In one study, optimum differentiation of bvFTD and controls was obtained with a combination of structural MRI, DTI and resting state measures.[121] Similarly, optimum differentiation of bvFTD and Alzheimer's dementia was obtained with a combination of DTI and resting state fMRI measures[121] showing that combining functional and structural measures improves differentiation.

Tau PET imaging has been assessed in patients with bvFTD, particularly using the THK-5351 and 18F-flortaucipir ligands. These studies have typically observed elevated uptake in the white matter of the frontal lobe,[122–125] as well as anterior cingulate, temporal lobe, insula, and striatum[123,125] in bvFTD patients compared to controls (Fig. 4). However, results across bvFTD patients are variable, with only 50% of sporadic bvFTD patients in one study demonstrating elevated 18F-flortaucipir uptake compared to

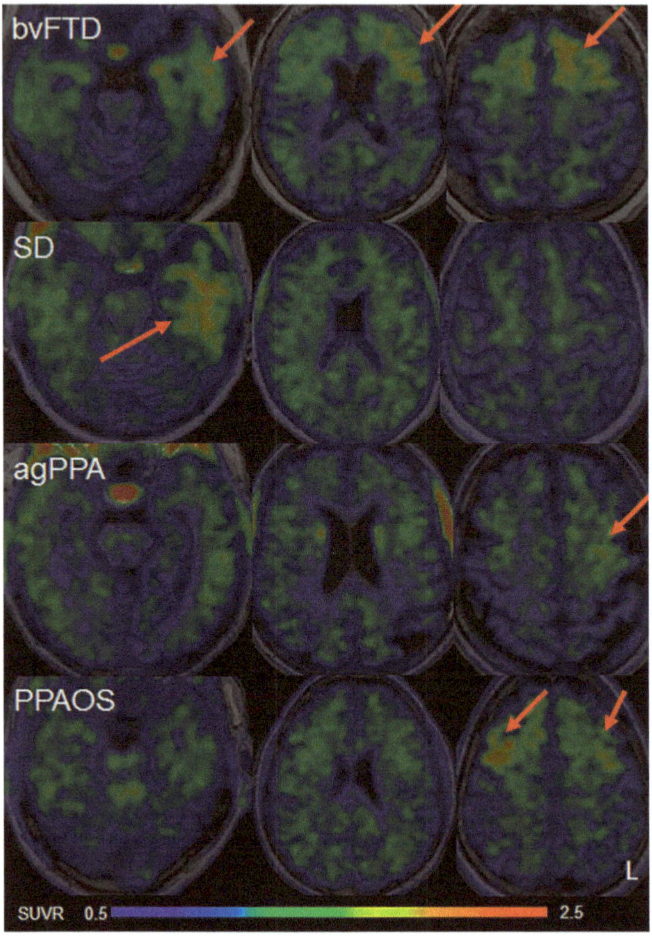

Fig. 4 18F-flortaucipir images from patients with different FTD clinical syndromes. Mild uptake predominantly in the white matter is observed in the frontotemporal lobes in the bvFTD patient, left temporal lobe in the SD patient, left frontal lobe of the agPPA patient and bilateral premotor cortex of the PPAOS patient.

controls.[124] Consistent with structural MRI, the patterns of uptake in bvFTD differ from the more temporoparietal and occipital patterns of uptake observed in Alzheimer's dementia.[122] One study found that frontal white matter uptake using THK-5351 was higher in bvFTD compared to Alzheimer's dementia, although in another study the degree of 18F-flortaucipir uptake in bvFTD was less than the binding range typically seen in Alzheimer's dementia.[124] However, it is very difficult to interpret tau PET findings in bvFTD given the pathological heterogeneity and the fact that a large proportion of cases will likely have an underlying TDP-43 proteinopathy rather than a tauopathy. It is currently unclear what the ligands are binding to in the white matter and basal ganglia and whether it reflects off-target binding to a non-tau target.[126] Furthermore, the tau isoforms present in bvFTD patients with genetic mutations can vary and hence influence tau-PET findings (see Section 3) and many of these bvFTD studies have not screened cases for genetic mutations.[122,123,125] In addition, approximately 50% of bvFTD cases can show amyloid deposition on PET scanning[127] suggesting the possibility of underlying AD pathology which may influence tau-PET findings in bvFTD.[124] The value of tau-PET in bvFTD is therefore currently low until we have a better understanding of the biological underpinnings of ligand binding.

2.2 Semantic dementia

Semantic dementia is associated with characteristic patterns of atrophy on structural MRI and hypometabolism on FDG-PET affecting the temporal lobes.[74,107–110] An anteroposterior gradient of atrophy is typically observed, with most marked atrophy observed in anterior temporal regions (Fig. 2). Degeneration is particularly severe in the fusiform and inferior temporal gyri, although also severely affects the hippocampus, amygdala, entorhinal cortex, and middle temporal gyrus.[128–130] The orbitofrontal cortex, medial frontal lobes, striatum and anterior insula can also be involved.[32,58,130] Asymmetry is also a characteristic feature of SD. Patients which present with anomia and loss of word knowledge (and hence fulfill criteria for the semantic variant of PPA) typically show left-predominant patterns of atrophy and hypometabolism.[32,128–130] However, other patients can show right-predominant patterns of atrophy and may have deficits in other types of semantic knowledge (such as recognizing familiar faces,[131] facial emotions[132]or musical tunes[133,134]) and behavioral abnormalities.[135–138] While the disease typically affects one hemisphere of the anterior temporal lobes early in the disease, atrophy tends to spread within both temporal lobes[62,63,139–142] until patterns are often more symmetrical[136] and the frontal lobes become increasingly involved.[143] The majority of studies assessing neuroimaging features of SD focus on the more common left-sided variant of SD.

The distinctive anterior temporal pattern of atrophy and hypometabolism in SD can be diagnostically useful to differentiate SD from other neurodegenerative diseases. Although anteromedial temporal structures can be involved in both SD and Alzheimer's dementia,[144,145] the anteroposterior gradient of temporal atrophy is not observed in Alzheimer's dementia.[128] One study found that an anteroposterior gradient of hypoperfusion on SPECT can help differentiate patients with SD from those with Alzheimer's dementia.[146] Rates of atrophy of the temporal lobes are also faster in SD compared to Alzheimer's dementia.[63] The patterns of neurodegeneration in SD can overlap with those seen in bvFTD. However, the frontal lobe is usually affected to a greater degree in bvFTD,[65] and a ratio of temporal to frontal atrophy may be useful to distinguish these syndromes.[62] In addition, the temporal pole is more atrophic in SD compared to bvFTD, with left temporal pole volume separating svPPA from bvFTD with 82% sensitivity and 80% specificity in one study.[72] Utilizing both FDG-PET and MRI measures of the temporal pole may in fact improve classification of SD from Alzheimer's dementia and other FTD syndromes better than either modality assessed in isolation.[147]

Consistent with the temporal lobe focus of neurodegeneration, SD also shows reduced structural connectivity within the temporal lobes.[148] White matter tract degeneration is observed in tracts that project from the temporal lobes, including the uncinate fasciculus and the inferior longitudinal fasciculus, particularly in the left hemisphere[94,95,107,145,149–153] (Fig. 3). Degeneration is also observed in frontal lobe white matter tracts, including the genu of the corpus callosum and the arcuate fasciculus, and posterior temporal lobe, although there is usually a relative sparing of tracts in parietal regions of the brain.[94,150,152] While initial white matter degeneration in SD is asymmetric, there is evidence that degeneration spreads to bilateral frontotemporal tracts with disease progression.[102] Compared to bvFTD, SD patients show greater degeneration of the inferior longitudinal fasciculus,[93,94] uncinate fasciculus[154] and inferior fronto-occipital fasciculus,[154] and less degeneration of the genu of the corpus callosum, frontal white matter and parietal regions of the superior longitudinal fasciculus.[105,107] In one study, bvFTD also tended to show greater involvement of tracts in the right hemisphere, while SD showed greater involvement of tracts in the left hemisphere.[154] Semantic dementia appears to involve different white matter projections from the medial temporal lobe compared to Alzheimer's dementia, with SD targeting the uncinate and inferior longitudinal fasciculus and Alzheimer's dementia targeting the cingulum and corpus callosum, demonstrating vulnerability of distinct networks that both involve the medial temporal lobe.[145]

Resting state fMRI studies have shown similarities in how neural networks are disrupted in SD compared to bvFTD, with disruptions in the salience network and frontolimbic connectivity also observed in SD patients.[111] These similarities may reflect the fact that similar behavioral abnormalities can be observed in both SD and bvFTD patients.[111] However, SD appears to be specifically associated with reduced local and distinct connectivity from the anterior temporal lobe. Reduced connectivity has been observed between the left anterior temporal lobe and bilateral frontal and cingulate regions in SD compared to controls and bvFTD.[154] In addition, SD patients show a loss of highly connected hubs in the left inferior temporal lobe, as well as in the orbitofrontal cortex and occipital lobes.[155] These patterns of reduced functional connectivity concur with structural connectivity studies that have shown degeneration of tracts that connect the temporal pole, orbitofrontal cortex and occipital cortex, such as the uncinate fasciculus, inferior longitudinal fasciculus and inferior fronto-occipital fasciculus. The damage of these tracts could be responsible for the functional disconnections between these regions in SD. There is also evidence linking atrophy in SD to disrupted connectivity with the anterior temporal lobe, suggesting that atrophy may follow these connectional pathways.[156]

Recent molecular PET imaging studies have produced some interesting findings in SD. First, amyloid PET studies have shown that a high proportion, in fact up to 30%, of SD patients can have beta-amyloid deposition in their brains.[157–161] The presence of beta-amyloid deposition does not appear to influence the clinical presentation of SD, although the patients with beta-amyloid tend to be older.[161] Although Alzheimer's disease can be the pathological cause of SD in some cases,[29,162–164] this is relatively rare. Hence, the PET findings may reflect age associated beta-amyloid deposition in SD patients where the primary pathology is still a TDP-43 proteinopathy. Second, tau-PET imaging has been performed in patients with SD and, somewhat surprisingly given the underlying pathology, elevated uptake is observed in these patients (Fig. 4). Elevated uptake has been demonstrated in the anteromedial temporal lobe, as well as regions in the basal frontal lobes, insula and putamen, with elevated uptake typically observed in the white matter rather than the gray matter.[124,161,165–168] It remains unclear what the ligand is binding to in these cases. It could reflect off-target binding, reflect low levels of binding to TDP-43 or may be binding to low levels of tau that coexists with TDP-43.[165–167] Tau PET uptake is, in fact, higher in SD patients that show beta-amyloid deposition compared to those without beta-amyloid deposition, perhaps supporting the hypothesis that the ligand may be binding to some tau pathology present in the brain in the beta-amyloid positive patients.[161] Uptake remains elevated in the patients without beta-amyloid deposition, however, suggesting a potential alternative explanation for binding in these patients.[161,168]

2.3 Agrammatic primary progressive aphasia and primary progressive apraxia of speech

The agPPA (referred to as progressive non-fluent aphasia in many studies) is associated with relatively focal patterns of atrophy on MRI and hypometabolism on FDG-PET that target the left posterior frontal lobe, including inferior (Broca's area), middle and superior premotor gyri,[10,153,169–175] which is accompanied by widening of the left perisylvian fissure (Fig. 2). Atrophy and hypometabolism can also be observed in the insula, striatum, superior temporal lobe, and other regions in the left frontal and parietal lobes.[170,171] In contrast to bvFTD and SD, the anterior temporal lobes are typically relatively spared in patients with agPPA. Patients diagnosed with agPPA will most commonly have some combination of agrammatic aphasia and apraxia of speech, and it has been shown that these two clinical features have distinct anatomical correlates.[176] Apraxia of speech is associated with atrophy and hypometabolism in the superior premotor cortex,[8,75] while agrammatic aphasia is associated with abnormalities in the left inferior frontal lobe[176–178] and other temporal and parietal regions[176] within the language network.[179] In fact, patients with PPAOS, which only have apraxia of speech in the absence of aphasia, show very focal patterns of atrophy and hypometabolism involving the bilateral supplementary motor area and lateral superior premotor cortex[8,180] (Fig. 2), and show less involvement of the left inferior frontal lobe and lateral temporal lobe compared to agPPA.[180] As with the other FTD syndromes, atrophy and hypometabolism are progressive in both agPPA and PPAOS. In agPPA, the fastest rates of atrophy are typically observed in the frontal lobes,[141,142,181,182] with progressive volume loss also observed in the motor cortex, temporal and parietal lobes, insula, basal ganglia, and brainstem.[11,142,182,183] Atrophy progresses in the right hemisphere in addition to the left, although left-sided asymmetry is typically maintained over time.[142,183] In PPAOS, bilateral patterns of progressive volume loss over time are observed in the premotor cortex, motor cortex, basal ganglia and midbrain.[9] There is, therefore, overlap in the longitudinal patterns of atrophy observed in agPPA and PPAOS, reflecting progression of common clinical features. Degeneration of the motor cortex and brainstem, for example, are associated with the development and progression of parkinsonism and limb apraxia in these patients.[11] However, agPPA showed faster rates of atrophy in the inferior frontal lobe compared to PPAOS in one study, reflecting greater decline in agrammatism than the PPAOS patients.[11] Agrammatism can develop over time in some PPAOS patients, although not all, and when it does develop it is associated with atrophy of the left inferior frontal lobe, thalamus and putamen, showing that a network of regions is responsible for the development of agrammatic aphasia.[184]

White matter tract degeneration is associated with both agPPA and PPAOS. In agPPA, degeneration has been observed in frontotemporal and frontoparietal connections of the left superior longitudinal fasciculus[94,152,180,185–187] (Fig. 3), particularly involving the arcuate fasciculus that projects into the inferior frontal lobe.[107] The aslant tract that projects from Broca's area to the anterior cingulate and supplementary motor cortex is also affected.[188,189] However, degeneration is also observed in other frontal tracts, including the body of the corpus callosum, anterior cingulate, inferior frontal-occipital fasciculus, and frontostriatal pathways.[107,152,180,186,190] White matter tract findings are typically asymmetric, with greater involvement of the left hemisphere, although tend to be more bilateral and widespead than patterns of atrophy. Degeneration in this network of white matter tracts is associated with the language deficits observed in agPPA.[186,187,190] Temporal lobe white matter tracts can be affected[94,151,186,191] although this is not a consistent finding,[185] and these tracts show greater involvement in SD and bvFTD compared to agPPA.[94,151,185] Generally there is little overlap in patterns of tract degeneration between agPPA and SD.[191] The superior longitudinal fasciculus has been shown to be involved to a greater degree in agPPA compared to bvFTD.[94] Patients with PPAOS show less widespread patterns of white matter tract degeneration than agPPA, with bilateral involvement of the body of the corpus callosum and premotor aspects of the superior longitudinal fasciculus that lie adjacent to regions of gray matter loss in these patients[8,180] (Fig. 5). In fact, it appears as though apraxia of speech severity is particularly associated with degeneration of white matter tracts connecting the supplementary motor area to the inferior frontal cortex, premotor cortex and striatum.[190] White matter tract degeneration worsens over time in agPPA,[11,100,102] with widespread bilateral longitudinal degeneration

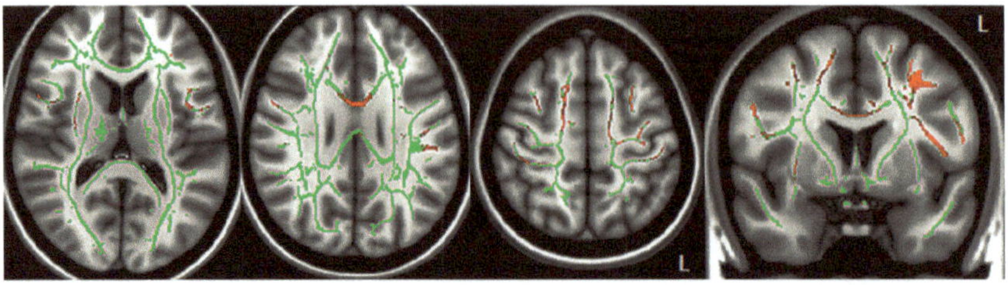

Fig. 5 Diffusion tensor imaging abnormalities in PPAOS. Regions of reduced tract integrity are shown in red on top of a green white matter skeleton of the brain. Results were created using track-based spatial statistics comparing a PPAOS cohort to controls.

observed in the superior longitudinal fasciculus, inferior and superior fronto-occipital fasciculus, cingulum, corticospinal tracts and white matter underlying motor cortex.[11] Longitudinal patterns of white matter tract degeneration are more focal in PPAOS, showing progression in the superior longitudinal fasciculus and body of the corpus callosum and spread to involve the splenium of the corpus callosum and white matter underlying the motor cortex.[9,11] It, therefore, appears that even with disease progression PPAOS remains a more focal disease compared to agPPA, although overlap between the two syndromes is observed.

Consistent with the anatomical findings, disrupted functional connectivity in agPPA has been shown to be asymmetric, predominantly affecting the left hemisphere.[114] Widespread disruptions in connectivity have been observed throughout frontal, temporal and striatal regions in the left hemisphere.[114] Within the speech-language network, agPPA patients have been shown to have a loss of functional hubs in the left fronto-temporal-parietal area and new nodes in more anterior frontal regions and the right frontal lobe compared to controls,[192] suggesting a reorganization of the speech-language network is occurring in agPPA. In fact, longitudinal progression of atrophy in agPPA has been shown to be spatially related to functional connectivity of the speech-language network, supporting the view that neurodegeneration in agPPA spreads through this network.[189] Patients with PPAOS also show disruption of the speech-language network, with disrupted connectivity of the right supplementary motor area and left posterior temporal gyrus to the rest of the speech-language network.[193] The connectivity of the right supplementary motor area correlated with the severity of apraxia of speech.[193] These results fit with MRI/FDG-PET and structural connectivity findings and point toward an important role of the supplementary motor area in apraxia of speech.

Molecular PET imaging using beta-amyloid and tau ligands have been assessed in agPPA and PPAOS patients. Beta-amyloid deposition on PET has been observed in 0–30% of agPPA patients[160,194–196] and 8% of PPAOS patients,[8] although the MRI signatures do not seem to differ according to beta-amyloid status.[197] Elevated 18F-flortaucipir uptake has been observed in both agPPA and PPAOS (Fig. 4). Mildly increased uptake has been observed in frontal white matter and subcortical gray matter structures, incluidng thalamus and basal ganglia, in agPPA, with greater uptake observed in the left hemisphere.[124,166,168] Patterns within agPPA appear to differ depending on the presence of apraxia of speech, with those with apraxia of speech showing greater uptake in the motor cortex.[198] The patterns of tau PET uptake in agPPA differ from those in SD and accounting for frontal versus temporal uptake allows good seperation of these two diseases.[166] Mild tau-PET uptake, predominantly in the white matter, is also observed in PPAOS, with uptake observed in the supplementary motor area and motor cortex.[199] Elevated uptake in the inferior frontal gyrus was also observed in PPAOS patients that develop agramamtic aphasia.[199] Patterns of tau-PET uptake, therefore, match closely with patterns of neurodegeneration observed in both agPPA and PPAOS. Given that agPPA and PPAOS most commonly result from a 4R tauopathy, it is possible that the ligands are detecting the presence of underlying tau proteins. However, autoradiographic studies have not observed good binding of 18F-flortaucipir to 4R tau isoforms,[126,200–202] casting doubt on the biological interpretation of findings in these diseases. It is also currently unclear why the tau-PET findings are located predominantly in the white matter in the agPPA and PPAOS, and more work is needed to understand the mechanisms of binding in these cases.

3. Imaging predicting pathology in FTD

The different pathologies that underlie the FTD syndromes have been shown to be associated with different neuroimaging signatures which can be helpful in allowing the prediction of underlying pathology in FTD. Since future therapeutics will likely target specific proteins or pathologies it will be necessary to provide a pathological diagnosis, rather than just a syndromic diagnosis, for FTD patients. This is particularly important in syndromes where the underlying pathology can be heterogeneous, such as bvFTD and agPPA.

There does not appear to be specific neuroimaging signatures of FTLD-tau or FTLD-TDP, but rather signatures of the specific pathologies within each of these proteinopathies.[203–206] Within the FTLD-tau diseases, Pick's disease has a very different neuroimaging signature compared to PSP and CBD. Pick's disease has a characteristic pattern of knife-edge atrophy of the prefrontal cortex, with additional involvement of the anterior temporal lobes, insula and anterior cingulate[207–209] (Fig. 6). The cortex appears knife-edge due to the dramatic degree of shrinkage of the frontal gyri. In contrast,

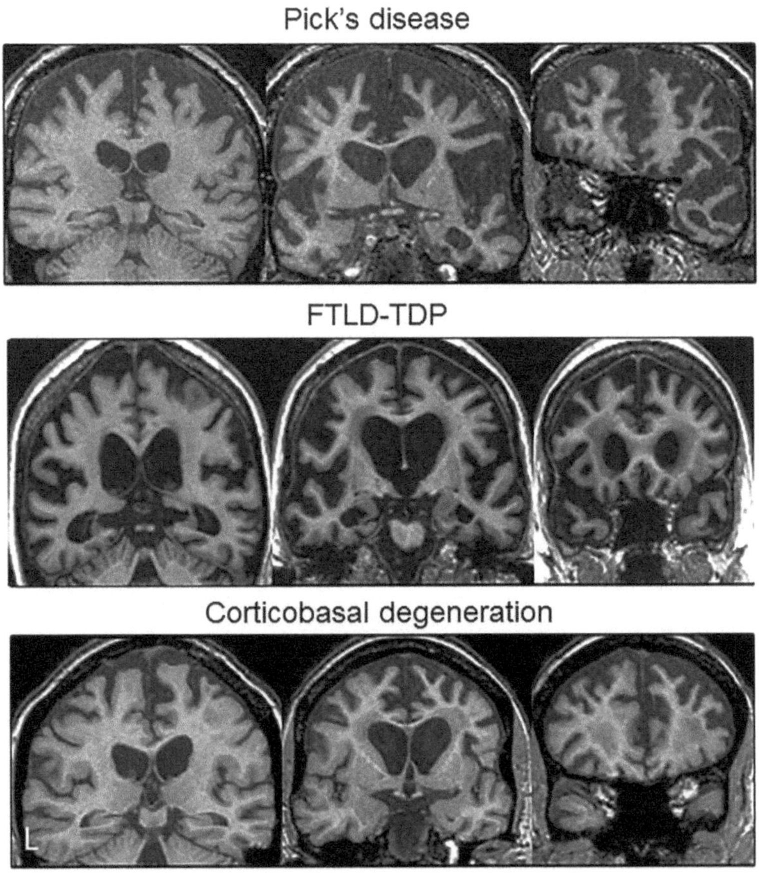

Fig. 6 Individual MRI scans from patients with bvFTD but different pathological diagnoses. The patient with Pick's disease shows knife-edge atrophy of the frontal lobes. The patient with FTLD-TDP shows widespread frontal, temporal and parietal atrophy. The patient with corticobasal degeneration shows atrophy predominantly of the posterior frontal lobes.

PSP and CBD involve more posterior regions of the frontal lobe, including premotor cortex[59] (Fig. 6). Cortical atrophy is typically more widespread in CBD compared to PSP, spreading posteriorly to involve the parietal lobe, and with greater involvement of the basal ganglia.[210,211] Rates of brain atrophy are also typically faster in CBD compared to PSP.[212] Midbrain atrophy is typically considered a diagnostic feature of PSP, although the presence of midbrain atrophy is associated with the clinical syndrome of PSP, rather than the presence of pathological PSP.[213] Hence, midbrain atrophy can be observed in patients that develop the typical clinical features of PSP (postural instability and vertical supranuclear gaze palsy) but have CBD pathology.[213] Cortical atrophy is often asymmetric in CBD, relating to asymmetry that is present clinically, although symmetric CBD cases are also observed[214,215] and asymmetric cases do not always have CBD pathology.[216,217] Different patterns of atrophy are also observed across the FTLD-TDP types. Type A is associated with widespread and asymmetric atrophy of the frontal, temporal and parietal lobes (Fig. 6); type B is associated with atrophy of the frontal and medial temporal lobes; and type C is associated with asymmetric atrophy of the anteromedial temporal lobes and nearly always presents clinically with SD.[218,219] The FTLD-FUS diseases are rarer than FTLD-tau or FTLD-TDP and are associated with frontotemporal atrophy but with additional and severe atrophy of the caudate.[220–222] The caudate has been shown to be involved to a greater degree in FTLD-FUS compared to other FTLD-tau and FTLD-TDP diseases.[220] This caudate signature has been observed across different FTLD-FUS diseases and hence may be a useful biomarker of FTLD-FUS.

Studies have demonstrated that these different FTD pathologies do show different neuroimaging features within groups of patients with the same clinical syndrome. Within patients with bvFTD, patterns of gray matter atrophy differ across the pathologies of Pick's disease, CBD and FTLD-TDP type A (Fig. 6). While all three pathologies showed atrophy of the frontal lobes, likely related to the common clinical presentation, there are differences that may be diagnostically useful. Pick's disease has been shown to have greater atrophy in the anterior frontal lobes, insula and temporal pole compared to CBD,[223] while CBD has greater atrophy in the parietal lobe compared to Pick's disease.[223] Furthermore, FTLD-TDP type has been shown to have greater atrophy in the parietal lobe and posterior temporal lobe compared to both CBD and Pick's disease.[209] Generally, CBD showed milder patterns of atrophy compared to both Pick's disease[223] and FTLD-TDP type A.[209] Imaging has also been shown to be useful in determining the underlying pathology in patients with agPPA. The agPPA patients with CBD pathology showed more widespread gray matter atrophy compared to those with PSP pathology, involving frontal, premotor and motor cortices, insula and striatum, and greater white matter atrophy in anterior parts of the frontal lobe.[224] However, the agPPA patients with PSP pathology developed greater atrophy of the midbrain and pons compared to those with CBD.[224] Furthermore, agPPA patients with FTLD-TDP pathology appear to show less white matter atrophy than agPPA patients with either PSP or CBD pathology.[225] Neuroimaging, therefore, shows promise to be able to aid in the prediction of specific pathologies within bvFTD and agPPA.

4. Imaging in genetic FTD

A relatively large proportion of FTD patients have a family history and an autosomal dominant pattern of inheritance. A number of mutations have now been identified that account for many of these families. The most common mutations occur in the microtubule associated protein tau (*MAPT*)[226] and progranulin (*GRN*)[227,228] genes, along with an expansion of a noncoding GGGGCC hexanucleotide repeat in the *C9ORF72* gene that is genetically linked to chromosome 9p.[229,230] Pathologically, both the *GRN* and *C9ORF72* expansion gene mutations are associated with FTLD-TDP, while *MAPT* mutations form another genetic variant of FTLD-tau. All three mutations are most commonly associated with bvFTD, although more varied clinical presentations are associated with *GRN* mutations, including agPPA,[231–235] and *C9ORF72* repeat expansions are also associated with FTD-MND.[229,230] Neuroimaging has proven incredibly valuable in genetic FTD to define the abnormalities associated with each mutation. This has improved our understanding of heterogeneity in FTD and provided markers that might predict the presence of a genetic mutation. Furthermore, asymptomatic family members with a genetic mutation can be followed before the development of symptoms allowing us to determine the earliest neuroimaging abnormalities in FTD; knowledge that will be crucial to allow early diagnosis.

The three main genetic mutations have been shown to have distinct neuroanatomical correlates on structural MRI. The *MAPT* mutations target the anteromedial temporal lobes, with additional atrophy observed in the frontal lobes, particularly the orbitofrontal cortex.[208,236] Imaging findings are relatively consistent across different mutations, although some mutations tend to show predominant involvement of the medial temporal lobes and others the lateral temporal lobe.[237] The *GRN* mutations can show more variable patterns of atrophy and hypometabolism at the patient level, although they tend to typically show quite widespread and asymmetric involvement[233,238] of the frontal, temporal and parietal lobes.[238,239] Involvement of the parietal lobe is particularly characteristic of *GRN* mutations.[231,238,240] The *C9ORF72* repeat expansion has been shown to be associated with symmetric patterns of atrophy involving the frontal, temporal, parietal and occipital lobes, in addition to the cerebellum[241–244] and thalamus.[242–244] There is evidence that imaging can help differentiate across the three mutations. Studies have found that patients with *GRN* mutations show more asymmetry[242] and greater involvement of the posterior regions of the brain compared to *MAPT* and *C9ORF72* carriers,[231,237,241] while *MAPT* carriers showed greater atrophy of the anterior temporal lobes than both *GRN* and *C9ORF72* carriers[237,241,242] (Fig. 7). Furthermore, the *C9ORF72* repeat expansions showing greater atrophy in parietal and occipital lobes compared to *MAPT* carriers and sporadic bvFTD[241] (Fig. 7). There is some disagreement in the field with the diagnostic utility of the thalamus for detecting *C9ORF72* mutations, with some showing the thalamus differentiates *C9ORF72* patients from sporadic FTD,[245] while others finding that the thalamus does not help to differentiate between the different FTD mutations.[241,246] Differences across FTD mutations have also been identified in longitudinal rates of progression, with *GRN* carriers showing faster rates of atrophy across the brain than both *MAPT*[247,248] and *C9ORF72*,[242,249] as well as sporadic FTD.[249] The *GRN* mutation, therefore, appears to be particularly aggressive.

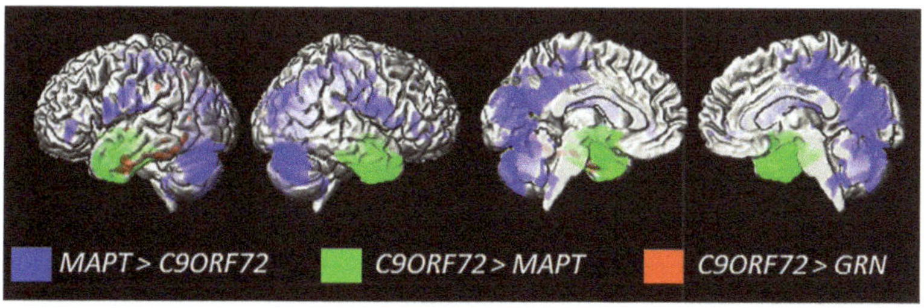

Fig. 7 Three dimensional renders highlighting regional differences in patterns of atrophy among FTD genetic mutations. Blue shows regions with greater atrophy in *C9ORF72* compared to *MAPT*; green shows regions with greater atrophy in *MAPT* compared to *C9ORF72*; and red shows regions with greater atrophy in *GRN* compared to *C9ORF72*.

Studies that have assessed asymptomatic mutation carriers have found that gray matter atrophy can be detected before symptom onset in all three mutations, with regional findings typically matching those observed in symptomatic patients but to a lesser degree.[250–255] Alterations in structural and functional connectivity have also been observed before the onset of clinical symptoms in genetic FTD.[252,256] Degeneration of the uncinate fasciculus has been observed both in symptomatic and asymptomatic *MAPT* mutation carriers[98,253,257]; degeneration of white matter tracts projecting into the frontal, temporal and parietal lobes, such as uncinate fasciculus, anterior corpus callosum, fronto-occipital fasciculus and superior longitudinal fasciculus have been observed in asymptomatic *GRN* mutation carriers[252,258]; and, similarly, degeneration of white matter tracts that project into frontal and posterior brain regions, including the anterior thalamic radiations, have been observed in symptomatic and asymptomatic *C9ORF72*.[98,242,251,254,257] In fact, one study found that white matter tract diffusivity changes in patients with the *C9ORF72* repeat expansion occur 30 years before the onset of symptoms, and occur earlier than diffusivity changes in *MAPT* and *GRN* carriers.[257] Disruptions in functional connectivity in patients with the *C9ORF72* repeat expansion have been observed to overlap with those observed in sporadic bvFTD, with disruptions of the salience network and sensorimotor network in both symptomatic[259] and asymptomatic carriers.[251] This body of work demonstrates that measures from both structural and functional imaging have potential as early disease biomarkers in genetic FTD, although the optimum neuroimaging modality is still unclear.

Elevated uptake has been observed in patients with FTD mutations on tau-PET imaging. The distribution of uptake in the *MAPT* mutations typically mirrors the patterns of neurodegeneration, although the degree of uptake varies across the different *MAPT* mutations due to differences in the characteristic tau isoforms that are deposited in the brain (Fig. 8). Hence, tau mutations, such as V337 M and R406W, which deposit 3R + 4R isoforms have high levels of uptake on tau-PET akin to the levels of uptake observed in Alzheimer's dementia,[124,260–262] whereas mutations that deposit 4R isoforms of tau, such as N279K, S305 N and 10 + 16, show milder uptake.[262–264] Therefore, tau-PET imaging shows promise as a useful neuroimaging modality in some of the *MAPT* mutations. As discussed above, it is unclear what the ligand is binding to in the 4R *MAPT* cases. The underlying mechanism for elevated uptake in *GRN* and *C9ORF72* carriers[124] is also unclear and more work is needed to understand these findings.

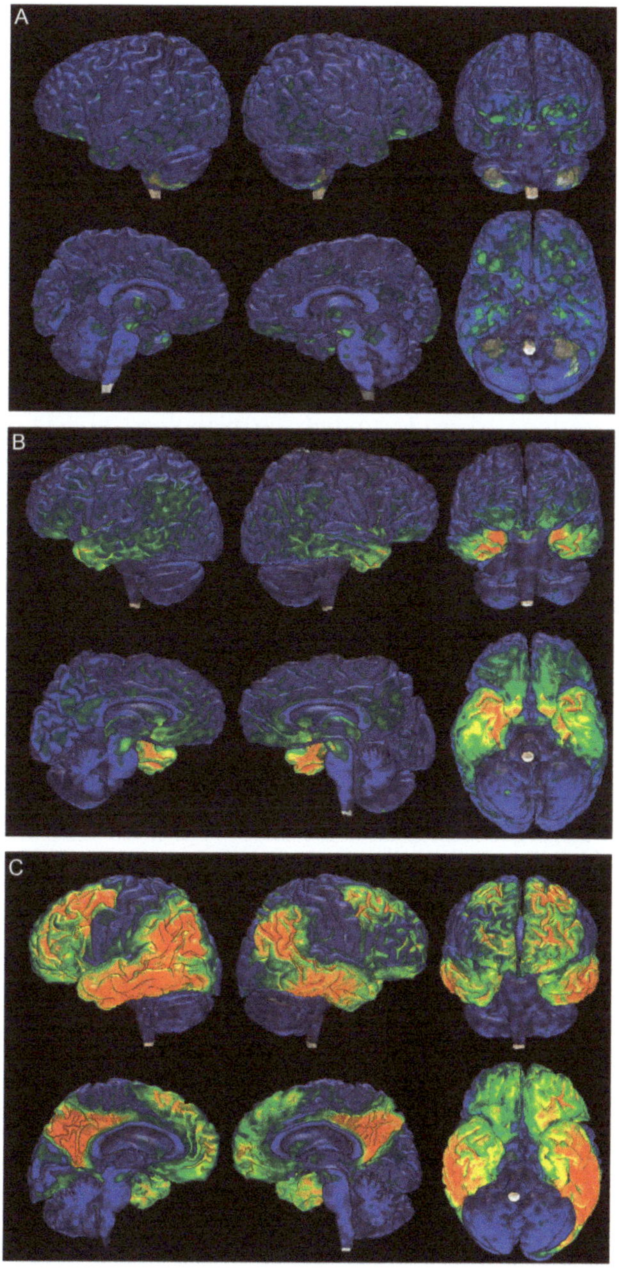

Fig. 8 Representative renderings of tau-PET uptake in *MAPT* mutations relative to Alzheimer's dementia. Tau PET global distributions are displayed on individual participant's brain renderings for 3 participants: (A) patients with 4R *MAPT* with marginally elevated levels of tau PET signal; (B) patients with mixed 3R/4R MAPT, and (C) a representative patient with Alzheimer's dementia. Scale represents standard uptake value ratio from 0.5 to 3. *Figure reproduced with permission from Jones DT, Knopman DS, Graff-Radford J, et al. In vivo (18)F-AV-1451 tau PET signal in MAPT mutation carriers varies by expected tau isoforms.* Neurology. *2018;90(11):e947–e954.*

5. Summary

The past few decades of neuroimaging research in FTD have provided potential disease biomarkers and taught us a lot about the breakdowns in brain structure and function, and the relationship between clinical symptoms and brain abnormalities in bvFTD. This knowledge has improved our ability to inform patients and their families, provide more targeted diagnoses and prognostic estimates, and provided outcomes measures that can be used to track disease progression for clinical treatment trials. However, we still lack definitive biomarkers of underlying pathology in FTD. Molecular PET imaging shows promise but ligands that better target different tau isoforms, and ligands that target FTLD-TDP, will be essential for FTD.

Acknowledgments

Dr. Whitwell is funded by the National Institutes of Health (R01-DC12519, R01-AG50603, R01-NS89757, R21-NS94684 and R01-DC14942).

References

1. Neary D, Snowden JS, Gustafson L, et al. Frontotemporal lobar degeneration: a consensus on clinical diagnostic criteria. *Neurology*. 1998;51(6):1546–1554.
2. Rascovsky K, Hodges JR, Knopman D, et al. Sensitivity of revised diagnostic criteria for the behavioural variant of frontotemporal dementia. *Brain*. 2011;134 (Pt 9):2456–2477.
3. Warrington EK. The selective impairment of semantic memory. *Q J Exp Psychol*. 1975; 27(4):635–657.
4. Duffy J. Apraxia of speech in degenerative neurologic disease. *Aphasiology*. 2006;20 (6):511–527.
5. Jung Y, Duffy JR, Josephs KA. Primary progressive aphasia and apraxia of speech. *Semin Neurol*. 2013;33(4):342–347.
6. Gorno-Tempini ML, Hillis AE, Weintraub S, et al. Classification of primary progressive aphasia and its variants. *Neurology*. 2011;76(11):1006–1014.
7. Mesulam MM. Slowly progressive aphasia without generalized dementia. *Ann Neurol*. 1982;11(6):592–598.
8. Josephs KA, Duffy JR, Strand EA, et al. Characterizing a neurodegenerative syndrome: primary progressive apraxia of speech. *Brain*. 2012;135(Pt 5):1522–1536.
9. Josephs KA, Duffy JR, Strand EA, et al. The evolution of primary progressive apraxia of speech. *Brain*. 2014;137(Pt 10):2783–2795.
10. Josephs KA, Duffy JR, Strand EA, et al. Clinicopathological and imaging correlates of progressive aphasia and apraxia of speech. *Brain*. 2006;129(Pt 6):1385–1398.
11. Tetzloff KA, Duffy JR, Clark HM, et al. Longitudinal structural and molecular neuroimaging in agrammatic primary progressive aphasia. *Brain*. 2018;141(1):302–317.
12. Tetzloff KA, Duffy JR, Strand EA, et al. Clinical and imaging progression over 10 years in a patient with primary progressive apraxia of speech and autopsy-confirmed corticobasal degeneration. *Neurocase*. 2018;24(2):111–120.
13. Josephs KA. Frontotemporal dementia and related disorders: deciphering the enigma. *Ann Neurol*. 2008;64(1):4–14.

14. Josephs KA, Hodges JR, Snowden JS, et al. Neuropathological background of phenotypical variability in frontotemporal dementia. *Acta Neuropathol.* 2011;122(2): 137–153.
15. Neumann M, Rademakers R, Roeber S, Baker M, Kretzschmar HA, Mackenzie IR. A new subtype of frontotemporal lobar degeneration with FUS pathology. *Brain.* 2009;132(Pt 11):2922–2931.
16. Dickson DW. Neuropathology of Pick's disease. *Neurology.* 2001;56(11 Suppl 4): S16–S20.
17. Hauw JJ, Daniel SE, Dickson D, et al. Preliminary NINDS neuropathologic criteria for Steele-Richardson-Olszewski syndrome (progressive supranuclear palsy). *Neurology.* 1994;44(11):2015–2019.
18. Dickson DW, Bergeron C, Chin SS, et al. Office of rare diseases neuropathologic criteria for corticobasal degeneration. *J Neuropathol Exp Neurol.* 2002;61(11): 935–946.
19. Kouri N, Whitwell JL, Josephs KA, Rademakers R, Dickson DW. Corticobasal degeneration: a pathologically distinct 4R tauopathy. *Nat Rev Neurol.* 2011;7(5):263–272.
20. Sampathu DM, Neumann M, Kwong LK, et al. Pathological heterogeneity of frontotemporal lobar degeneration with ubiquitin-positive inclusions delineated by ubiquitin immunohistochemistry and novel monoclonal antibodies. *Am J Pathol.* 2006;169(4):1343–1352.
21. Mackenzie IR, Baborie A, Pickering-Brown S, et al. Heterogeneity of ubiquitin pathology in frontotemporal lobar degeneration: classification and relation to clinical phenotype. *Acta Neuropathol.* 2006;112(5):539–549.
22. Cairns NJ, Bigio EH, Mackenzie IR, et al. Neuropathologic diagnostic and nosologic criteria for frontotemporal lobar degeneration: consensus of the consortium for frontotemporal lobar degeneration. *Acta Neuropathol.* 2007;114(1):5–22.
23. Cairns NJ, Grossman M, Arnold SE, et al. Clinical and neuropathologic variation in neuronal intermediate filament inclusion disease. *Neurology.* 2004;63(8):1376–1384.
24. Josephs KA, Holton JL, Rossor MN, et al. Neurofilament inclusion body disease: a new proteinopathy? *Brain.* 2003;126(Pt 10):2291–2303.
25. Kusaka H, Matsumoto S, Imai T. An adult-onset case of sporadic motor neuron disease with basophilic inclusions. *Acta Neuropathol.* 1990;80(6):660–665.
26. Munoz DG, Neumann M, Kusaka H, et al. FUS pathology in basophilic inclusion body disease. *Acta Neuropathol.* 2009;118(5):617–627.
27. Josephs KA, Lin WL, Ahmed Z, Stroh DA, Graff-Radford NR, Dickson DW. Frontotemporal lobar degeneration with ubiquitin-positive, but TDP-43-negative inclusions. *Acta Neuropathol.* 2008;116(2):159–167.
28. Mackenzie IR, Foti D, Woulfe J, Hurwitz TA. Atypical frontotemporal lobar degeneration with ubiquitin-positive, TDP-43-negative neuronal inclusions. *Brain.* 2008; 131(Pt 5):1282–1293.
29. Spinelli EG, Mandelli ML, Miller ZA, et al. Typical and atypical pathology in primary progressive aphasia variants. *Ann Neurol.* 2017;81(3):430–443.
30. Johnson JK, Head E, Kim R, Starr A, Cotman CW. Clinical and pathological evidence for a frontal variant of Alzheimer disease. *Arch Neurol.* 1999;56(10):1233–1239.
31. Boccardi M, Sabattoli F, Laakso MP, et al. Frontotemporal dementia as a neural system disease. *Neurobiol Aging.* 2005;26(1):37–44.
32. Rosen HJ, Gorno-Tempini ML, Goldman WP, et al. Patterns of brain atrophy in frontotemporal dementia and semantic dementia. *Neurology.* 2002;58(2):198–208.
33. Salmon E, Garraux G, Delbeuck X, et al. Predominant ventromedial frontopolar metabolic impairment in frontotemporal dementia. *Neuroimage.* 2003;20(1):435–440.
34. Grimmer T, Diehl J, Drzezga A, Forstl H, Kurz A. Region-specific decline of cerebral glucose metabolism in patients with frontotemporal dementia: a prospective 18F-FDG-PET study. *Dement Geriatr Cogn Disord.* 2004;18(1):32–36.

35. Diehl J, Grimmer T, Drzezga A, Riemenschneider M, Forstl H, Kurz A. Cerebral metabolic patterns at early stages of frontotemporal dementia and semantic dementia. A PET study. *Neurobiol Aging*. 2004;25(8):1051–1056.
36. Garraux G, Salmon E, Degueldre C, Lemaire C, Laureys S, Franck G. Comparison of impaired subcortico-frontal metabolic networks in normal aging, subcortico-frontal dementia, and cortical frontal dementia. *Neuroimage*. 1999;10(2):149–162.
37. Jeong Y, Cho SS, Park JM, et al. 18F-FDG PET findings in frontotemporal dementia: an SPM analysis of 29 patients. *J Nucl Med*. 2005;46(2):233–239.
38. Ishii K, Sakamoto S, Sasaki M, et al. Cerebral glucose metabolism in patients with frontotemporal dementia. *J Nucl Med*. 1998;39(11):1875–1878.
39. Pan PL, Song W, Yang J, et al. Gray matter atrophy in behavioral variant frontotemporal dementia: a meta-analysis of voxel-based morphometry studies. *Dement Geriatr Cogn Disord*. 2012;33(2–3):141–148.
40. Raczka KA, Becker G, Seese A, et al. Executive and behavioral deficits share common neural substrates in frontotemporal lobar degeneration—a pilot FDG-PET study. *Psychiatry Res*. 2010;182(3):274–280.
41. Schroeter ML, Vogt B, Frisch S, et al. Dissociating behavioral disorders in early dementia-An FDG-PET study. *Psychiatry Res*. 2011;194(3):235–244.
42. Whitwell JL, Xu J, Mandrekar J, et al. Frontal asymmetry in behavioral variant frontotemporal dementia: clinicoimaging and pathogenetic correlates. *Neurobiol Aging*. 2013;34(2):636–639.
43. Rosen HJ, Allison SC, Schauer GF, Gorno-Tempini ML, Weiner MW, Miller BL. Neuroanatomical correlates of behavioural disorders in dementia. *Brain*. 2005;128 (Pt 11):2612–2625.
44. Peters F, Perani D, Herholz K, et al. Orbitofrontal dysfunction related to both apathy and disinhibition in frontotemporal dementia. *Dement Geriatr Cogn Disord*. 2006;21 (5-6):373–379.
45. Hornberger M, Geng J, Hodges JR. Convergent grey and white matter evidence of orbitofrontal cortex changes related to disinhibition in behavioural variant frontotemporal dementia. *Brain*. 2011;134(Pt 9):2502–2512.
46. Krueger CE, Laluz V, Rosen HJ, Neuhaus JM, Miller BL, Kramer JH. Double dissociation in the anatomy of socioemotional disinhibition and executive functioning in dementia. *Neuropsychology*. 2011;25(2):249–259.
47. Massimo L, Powers C, Moore P, et al. Neuroanatomy of apathy and disinhibition in frontotemporal lobar degeneration. *Dement Geriatr Cogn Disord*. 2009;27(1):96–104.
48. Zamboni G, Huey ED, Krueger F, Nichelli PF, Grafman J. Apathy and disinhibition in frontotemporal dementia: insights into their neural correlates. *Neurology*. 2008;71 (10):736–742.
49. Whitwell JL, Sampson EL, Loy CT, et al. VBM signatures of abnormal eating behaviours in frontotemporal lobar degeneration. *Neuroimage*. 2007;35(1):207–213.
50. Woolley JD, Gorno-Tempini ML, Seeley WW, et al. Binge eating is associated with right orbitofrontal-insular-striatal atrophy in frontotemporal dementia. *Neurology*. 2007;69(14):1424–1433.
51. Jakabek D, Power BD, Macfarlane MD, et al. Regional structural hypo- and hyper-connectivity of frontal-striatal and frontal-thalamic pathways in behavioral variant frontotemporal dementia. *Hum Brain Mapp*. 2018;39(10):4083–4093.
52. Bede P, Omer T, Finegan E, et al. Connectivity-based characterisation of subcortical grey matter pathology in frontotemporal dementia and ALS: a multimodal neuroimaging study. *Brain Imaging Behav*. 2018;12(6):1696–1707.
53. Ahmed RM, Goldberg ZL, Kaizik C, et al. Neural correlates of changes in sexual function in frontotemporal dementia: implications for reward and physiological functioning. *J Neurol*. 2018;265(11):2562–2572.

54. Macfarlane MD, Jakabek D, Walterfang M, et al. Striatal atrophy in the behavioural variant of frontotemporal dementia: correlation with diagnosis, negative symptoms and disease severity. *PLoS One*. 2015;10(6): e0129692.
55. Bertoux M, O'Callaghan C, Flanagan E, Hodges JR, Hornberger M. Fronto-striatal atrophy in behavioral variant frontotemporal dementia and Alzheimer's disease. *Front Neurol*. 2015;6:147.
56. Moller C, Dieleman N, van der Flier WM, et al. More atrophy of deep gray matter structures in frontotemporal dementia compared to Alzheimer's disease. *J Alzheimers Dis*. 2015;44(2):635–647.
57. Eslinger PJ, Moore P, Antani S, Anderson C, Grossman M. Apathy in frontotemporal dementia: behavioral and neuroimaging correlates. *Behav Neurol*. 2012;25(2):127–136.
58. Halabi C, Halabi A, Dean DL, et al. Patterns of striatal degeneration in frontotemporal dementia. *Alzheimer Dis Assoc Disord*. 2013;27(1):74–83.
59. Josephs KA, Whitwell JL, Jack Jr CR. Anatomic correlates of stereotypies in frontotemporal lobar degeneration. *Neurobiol Aging*. 2008;29(12):1859–1863.
60. Looi JC, Walterfang M, Styner M, et al. Shape analysis of the neostriatum in subtypes of frontotemporal lobar degeneration: neuroanatomically significant regional morphologic change. *Psychiatry Res*. 2011;191(2):98–111.
61. Perry DC, Whitwell JL, Boeve BF, et al. Voxel-based morphometry in patients with obsessive-compulsive behaviors in behavioral variant frontotemporal dementia. *Eur J Neurol*. 2012;19(6):911–917.
62. Frings L, Mader I, Landwehrmeyer BG, Weiller C, Hull M, Huppertz HJ. Quantifying change in individual subjects affected by frontotemporal lobar degeneration using automated longitudinal MRI volumetry. *Hum Brain Mapp*. 2012;33(7):1526–1535.
63. Krueger CE, Dean DL, Rosen HJ, et al. Longitudinal rates of lobar atrophy in frontotemporal dementia, semantic dementia, and Alzheimer's disease. *Alzheimer Dis Assoc Disord*. 2010;24(1):43–48.
64. Brambati SM, Renda NC, Rankin KP, et al. A tensor based morphometry study of longitudinal gray matter contraction in FTD. *Neuroimage*. 2007;35(3):998–1003.
65. Lu PH, Mendez MF, Lee GJ, et al. Patterns of brain atrophy in clinical variants of frontotemporal lobar degeneration. *Dement Geriatr Cogn Disord*. 2013;35(1-2):34–50.
66. Gordon E, Rohrer JD, Kim LG, et al. Measuring disease progression in frontotemporal lobar degeneration: a clinical and MRI study. *Neurology*. 2010;74(8):666–673.
67. Frings L, Yew B, Flanagan E, et al. Longitudinal grey and white matter changes in frontotemporal dementia and Alzheimer's disease. *PLoS One*. 2014;9(3): e90814.
68. Moller C, Hafkemeijer A, Pijnenburg YAL, et al. Different patterns of cortical gray matter loss over time in behavioral variant frontotemporal dementia and Alzheimer's disease. *Neurobiol Aging*. 2016;38:21–31.
69. Seeley WW, Crawford R, Rascovsky K, et al. Frontal paralimbic network atrophy in very mild behavioral variant frontotemporal dementia. *Arch Neurol*. 2008;65(2): 249–255.
70. Borroni B, Cosseddu M, Pilotto A, et al. Early stage of behavioral variant frontotemporal dementia: clinical and neuroimaging correlates. *Neurobiol Aging*. 2015; 36(11):3108–3115.
71. Ranasinghe KG, Rankin KP, Pressman PS, et al. Distinct subtypes of behavioral variant frontotemporal dementia based on patterns of network degeneration. *JAMA Neurol*. 2016;73(9):1078–1088.
72. Bruun M, Koikkalainen J, Rhodius-Meester HFM, et al. Detecting frontotemporal dementia syndromes using MRI biomarkers. *Neuroimage Clin*. 2019;22: 101711.
73. Cerami C, Dodich A, Lettieri G, et al. Different FDG-PET metabolic patterns at single-subject level in the behavioral variant of fronto-temporal dementia. *Cortex*. 2016;83:101–112.

74. Salmon E, Kerrouche N, Herholz K, et al. Decomposition of metabolic brain clusters in the frontal variant of frontotemporal dementia. *Neuroimage*. 2006;30(3): 871–878.
75. Whitwell JL, Przybelski SA, Weigand SD, et al. Distinct anatomical subtypes of the behavioural variant of frontotemporal dementia: a cluster analysis study. *Brain*. 2009;132(Pt 11):2932–2946.
76. Josephs Jr KA, Whitwell JL, Weigand SD, et al. Predicting functional decline in behavioural variant frontotemporal dementia. *Brain*. 2011;134(Pt 2):432–448.
77. Davies RR, Kipps CM, Mitchell J, Kril JJ, Halliday GM, Hodges JR. Progression in frontotemporal dementia: identifying a benign behavioral variant by magnetic resonance imaging. *Arch Neurol*. 2006;63(11):1627–1631.
78. Kipps CM, Davies RR, Mitchell J, Kril JJ, Halliday GM, Hodges JR. Clinical significance of lobar atrophy in frontotemporal dementia: application of an MRI visual rating scale. *Dement Geriatr Cogn Disord*. 2007;23(5):334–342.
79. Devenney E, Swinn T, Mioshi E, et al. The behavioural variant frontotemporal dementia phenocopy syndrome is a distinct entity—evidence from a longitudinal study. *BMC Neurol*. 2018;18(1):56.
80. Josephs KA, Whitwell JL, Jack CR, Parisi JE, Dickson DW. Frontotemporal lobar degeneration without lobar atrophy. *Arch Neurol*. 2006;63(11):1632–1638.
81. Whitwell JL, Jack Jr CR, Senjem ML, Josephs KA. Patterns of atrophy in pathologically confirmed FTLD with and without motor neuron degeneration. *Neurology*. 2006;66 (1):102–104.
82. Chang JL, Lomen-Hoerth C, Murphy J, et al. A voxel-based morphometry study of patterns of brain atrophy in ALS and ALS/FTLD. *Neurology*. 2005;65(1):75–80.
83. Lillo P, Mioshi E, Burrell JR, Kiernan MC, Hodges JR, Hornberger M. Grey and white matter changes across the amyotrophic lateral sclerosis-frontotemporal dementia continuum. *PLoS One*. 2012;7(8): e43993.
84. Cajanus A, Hall A, Koikkalainen J, et al. Automatic MRI quantifying methods in behavioral-variant frontotemporal dementia diagnosis. *Dement Geriatr Cogn Dis Extra*. 2018;8(1):51–59.
85. Meyer S, Mueller K, Stuke K, et al. Predicting behavioral variant frontotemporal dementia with pattern classification in multi-center structural MRI data. *Neuroimage Clin*. 2017;14:656–662.
86. Moller C, Pijnenburg YA, van der Flier WM, et al. Alzheimer disease and behavioral variant frontotemporal dementia: automatic classification based on cortical atrophy for single-subject diagnosis. *Radiology*. 2016;279(3):838–848.
87. Nazem A, Tang CC, Spetsieris P, et al. A multivariate metabolic imaging marker for behavioral variant frontotemporal dementia. *Alzheimers Dement (Amst)*. 2018;10: 583–594.
88. Foster NL, Heidebrink JL, Clark CM, et al. FDG-PET improves accuracy in distinguishing frontotemporal dementia and Alzheimer's disease. *Brain*. 2007;130(Pt 10):2616–2635.
89. Kerklaan BJ, van Berckel BN, Herholz K, et al. The added value of 18-fluorodeoxyglucose-positron emission tomography in the diagnosis of the behavioral variant of frontotemporal dementia. *Am J Alzheimers Dis Other Demen*. 2014;29(7): 607–613.
90. Vijverberg EG, Wattjes MP, Dols A, et al. Diagnostic accuracy of MRI and additional [18F]FDG-PET for behavioral variant frontotemporal dementia in patients with late onset behavioral changes. *J Alzheimers Dis*. 2016;53(4):1287–1297.
91. Dukart J, Mueller K, Horstmann A, et al. Combined evaluation of FDG-PET and MRI improves detection and differentiation of dementia. *PLoS One*. 2011;6(3): e18111.

92. Pijnenburg YA, Mulder JL, van Swieten JC, et al. Diagnostic accuracy of consensus diagnostic criteria for frontotemporal dementia in a memory clinic population. *Dement Geriatr Cogn Disord*. 2008;25(2):157–164.
93. Agosta F, Scola E, Canu E, et al. White matter damage in frontotemporal lobar degeneration spectrum. *Cereb Cortex*. 2012;22:2705–2714.
94. Whitwell JL, Avula R, Senjem ML, et al. Gray and white matter water diffusion in the syndromic variants of frontotemporal dementia. *Neurology*. 2010;74(16):1279–1287.
95. Zhang Y, Schuff N, Du AT, et al. White matter damage in frontotemporal dementia and Alzheimer's disease measured by diffusion MRI. *Brain*. 2009;132(Pt 9):2579–2592.
96. Borroni B, Brambati SM, Agosti C, et al. Evidence of white matter changes on diffusion tensor imaging in frontotemporal dementia. *Arch Neurol*. 2007;64(2):246–251.
97. Daianu M, Mendez MF, Baboyan VG, et al. An advanced white matter tract analysis in frontotemporal dementia and early-onset Alzheimer's disease. *Brain Imaging Behav*. 2016;10(4):1038–1053.
98. Mahoney CJ, Ridgway GR, Malone IB, et al. Profiles of white matter tract pathology in frontotemporal dementia. *Hum Brain Mapp*. 2014;35(8):4163–4179.
99. Kassubek J, Muller HP, Del Tredici K, et al. Longitudinal diffusion tensor imaging resembles patterns of pathology progression in behavioral variant frontotemporal dementia (bvFTD). *Front Aging Neurosci*. 2018;10:47.
100. Elahi FM, Marx G, Cobigo Y, et al. Longitudinal white matter change in frontotemporal dementia subtypes and sporadic late onset Alzheimer's disease. *Neuroimage Clin*. 2017;16:595–603.
101. Mahoney CJ, Simpson IJ, Nicholas JM, et al. Longitudinal diffusion tensor imaging in frontotemporal dementia. *Ann Neurol*. 2015;77(1):33–46.
102. Lam BY, Halliday GM, Irish M, Hodges JR, Piguet O. Longitudinal white matter changes in frontotemporal dementia subtypes. *Hum Brain Mapp*. 2014;35(7): 3547–3557.
103. Yu J, Lee TMC. Frontotemporal lobar degeneration neuroimaging I. The longitudinal decline of white matter microstructural integrity in behavioral variant frontotemporal dementia and its association with executive function. *Neurobiol Aging*. 2019;76:62–70.
104. Kramer J, Lueg G, Schiffler P, et al. Diagnostic value of diffusion tensor imaging and positron emission tomography in early stages of frontotemporal dementia. *J Alzheimers Dis*. 2018;63(1):239–253.
105. Agosta F, Galantucci S, Magnani G, et al. MRI signatures of the frontotemporal lobar degeneration continuum. *Hum Brain Mapp*. 2015;36(7):2602–2614.
106. Santillo AF, Martensson J, Lindberg O, et al. Diffusion tensor tractography versus volumetric imaging in the diagnosis of behavioral variant frontotemporal dementia. *PLoS One*. 2013;8(7): e66932.
107. Zhang Y, Tartaglia MC, Schuff N, et al. MRI signatures of brain macrostructural atrophy and microstructural degradation in frontotemporal lobar degeneration subtypes. *J Alzheimers Dis*. 2013;33(2):431–444.
108. Moller C, Hafkemeijer A, Pijnenburg YA, et al. Joint assessment of white matter integrity, cortical and subcortical atrophy to distinguish AD from behavioral variant FTD: a two-center study. *Neuroimage Clin*. 2015;9:418–429.
109. Whitwell JL, Josephs KA, Avula R, et al. Altered functional connectivity in asymptomatic MAPT subjects: a comparison to bvFTD. *Neurology*. 2011;77(9):866–874.
110. Zhou J, Greicius MD, Gennatas ED, et al. Divergent network connectivity changes in behavioural variant frontotemporal dementia and Alzheimer's disease. *Brain*. 2010;133(Pt 5):1352–1367.
111. Farb NA, Grady CL, Strother S, et al. Abnormal network connectivity in frontotemporal dementia: evidence for prefrontal isolation. *Cortex*. 2013;49(7): 1856–1873.

112. Filippi M, Agosta F, Scola E, et al. Functional network connectivity in the behavioral variant of frontotemporal dementia. *Cortex*. 2013;49(9):2389–2401.
113. Tuovinen T, Rytty R, Moilanen V, et al. The effect of gray matter ICA and coefficient of variation mapping of BOLD data on the detection of functional connectivity changes in Alzheimer's disease and bvFTD. *Front Hum Neurosci*. 2016;10:680.
114. Reyes P, Ortega-Merchan MP, Rueda A, et al. Functional connectivity changes in behavioral, semantic, and nonfluent variants of frontotemporal dementia. *Behav Neurol*. 2018;2018: 9684129.
115. Sedeno L, Couto B, Garcia-Cordero I, et al. Brain network organization and social executive performance in frontotemporal dementia. *J Int Neuropsychol Soc*. 2016;22(2):250–262.
116. Day GS, Farb NA, Tang-Wai DF, et al. Salience network resting-state activity: prediction of frontotemporal dementia progression. *JAMA Neurol*. 2013;70(10): 1249–1253.
117. Hafkemeijer A, Moller C, Dopper EG, et al. A longitudinal study on resting state functional connectivity in behavioral variant frontotemporal dementia and Alzheimer's disease. *J Alzheimers Dis*. 2017;55(2):521–537.
118. Agosta F, Sala S, Valsasina P, et al. Brain network connectivity assessed using graph theory in frontotemporal dementia. *Neurology*. 2013;81(2):134–143.
119. Hafkemeijer A, Moller C, Dopper EG, et al. Resting state functional connectivity differences between behavioral variant frontotemporal dementia and Alzheimer's disease. *Front Hum Neurosci*. 2015;9:474.
120. Filippi M, Basaia S, Canu E, et al. Brain network connectivity differs in early-onset neurodegenerative dementia. *Neurology*. 2017;89(17):1764–1772.
121. Bouts M, Moller C, Hafkemeijer A, et al. Single subject classification of Alzheimer's disease and behavioral variant frontotemporal dementia using anatomical, diffusion tensor, and resting-state functional magnetic resonance imaging. *J Alzheimers Dis*. 2018;62(4):1827–1839.
122. Son HJ, Oh JS, Roh JH, et al. Differences in gray and white matter (18)F-THK5351 uptake between behavioral-variant frontotemporal dementia and other dementias. *Eur J Nucl Med Mol Imaging*. 2019;46(2):357–366.
123. Nam G, Jeong HJ, Kang JM, et al. (18)F-THK5351 PET imaging in the behavioral variant of frontotemporal dementia. *Dement Neurocogn Disord*. 2018;17(4): 163–173.
124. Tsai RM, Bejanin A, Lesman-Segev O, et al. (18)F-flortaucipir (AV-1451) tau PET in frontotemporal dementia syndromes. *Alzheimers Res Ther*. 2019;11(1):13.
125. Cho H, Seo SW, Choi JY, et al. Predominant subcortical accumulation of (18) F-flortaucipir binding in behavioral variant frontotemporal dementia. *Neurobiol Aging*. 2018;66:112–121.
126. Lowe VJ, Curran G, Fang P, et al. An autoradiographic evaluation of AV-1451 Tau PET in dementia. *Acta Neuropathol Commun*. 2016;4(1):58.
127. Ryan KA, Hammers D, DeLeon A, et al. Agreement among neuropsychological and behavioral data and PiB findings in diagnosing Frontotemporal Dementia. *J Clin Neurosci*. 2017;44:128–132.
128. Chan D, Fox NC, Scahill RI, et al. Patterns of temporal lobe atrophy in semantic dementia and Alzheimer's disease. *Ann Neurol*. 2001;49(4):433–442.
129. Galton CJ, Patterson K, Graham K, et al. Differing patterns of temporal atrophy in Alzheimer's disease and semantic dementia. *Neurology*. 2001;57(2):216–225.
130. Mummery CJ, Patterson K, Price CJ, Ashburner J, Frackowiak RS, Hodges JR. A voxel-based morphometry study of semantic dementia: relationship between temporal lobe atrophy and semantic memory. *Ann Neurol*. 2000;47(1):36–45.
131. Josephs KA, Whitwell JL, Vemuri P, et al. The anatomic correlate of prosopagnosia in semantic dementia. *Neurology*. 2008;71(20):1628–1633.

132. Omar R, Rohrer JD, Hailstone JC, Warren JD. Structural neuroanatomy of face processing in frontotemporal lobar degeneration. *J Neurol Neurosurg Psychiatry*. 2011;82(12):1341–1343.
133. Hsieh S, Hornberger M, Piguet O, Hodges JR. Neural basis of music knowledge: evidence from the dementias. *Brain*. 2011;134(Pt 9):2523–2534.
134. Johnson JK, Chang CC, Brambati SM, et al. Music recognition in frontotemporal lobar degeneration and Alzheimer disease. *Cogn Behav Neurol*. 2011;24(2):74–84.
135. Chan D, Anderson V, Pijnenburg Y, et al. The clinical profile of right temporal lobe atrophy. *Brain*. 2009;132(Pt 5):1287–1298.
136. Edwards-Lee T, Miller BL, Benson DF, et al. The temporal variant of frontotemporal dementia. *Brain*. 1997;120(Pt 6):1027–1040.
137. Thompson SA, Patterson K, Hodges JR. Left/right asymmetry of atrophy in semantic dementia: behavioral-cognitive implications. *Neurology*. 2003;61(9):1196–1203.
138. Josephs KA, Whitwell JL, Knopman DS, et al. Two distinct subtypes of right temporal variant frontotemporal dementia. *Neurology*. 2009;73(18):1443–1450.
139. Brambati SM, Rankin KP, Narvid J, et al. Atrophy progression in semantic dementia with asymmetric temporal involvement: a tensor-based morphometry study. *Neurobiol Aging*. 2009;30(1):103–111.
140. Rohrer JD, McNaught E, Foster J, et al. Tracking progression in frontotemporal lobar degeneration: serial MRI in semantic dementia. *Neurology*. 2008;71(18):1445–1451.
141. Whitwell JL, Anderson VM, Scahill RI, Rossor MN, Fox NC. Longitudinal patterns of regional change on volumetric MRI in frontotemporal lobar degeneration. *Dement Geriatr Cogn Disord*. 2004;17(4):307–310.
142. Rogalski E, Cobia D, Martersteck A, et al. Asymmetry of cortical decline in subtypes of primary progressive aphasia. *Neurology*. 2014;83(13):1184–1191.
143. Staffaroni AM, Ljubenkov PA, Kornak J, et al. Longitudinal multimodal imaging and clinical endpoints for frontotemporal dementia clinical trials. *Brain*. 2019;142(2): 443–459.
144. Schroeter ML, Neumann J. Combined imaging markers dissociate Alzheimer's disease and frontotemporal lobar degeneration—an ALE meta-analysis. *Front Aging Neurosci*. 2011;3:10.
145. Bejanin A, Desgranges B, La Joie R, et al. Distinct white matter injury associated with medial temporal lobe atrophy in Alzheimer's versus semantic dementia. *Hum Brain Mapp*. 2017;38(4):1791–1800.
146. Talbot PR, Snowden JS, Lloyd JJ, Neary D, Testa HJ. The contribution of single photon emission tomography to the clinical differentiation of degenerative cortical brain disorders. *J Neurol*. 1995;242(9):579–586.
147. Tahmasian M, Shao J, Meng C, et al. Based on the network degeneration hypothesis: separating individual patients with different neurodegenerative syndromes in a preliminary hybrid PET/MR study. *J Nucl Med*. 2016;57(3):410–415.
148. Andreotti J, Dierks T, Wahlund LO, Grieder M. Diverging progression of network disruption and atrophy in Alzheimer's disease and semantic dementia. *J Alzheimers Dis*. 2017;55(3):981–993.
149. Acosta-Cabronero J, Patterson K, Fryer TD, et al. Atrophy, hypometabolism and white matter abnormalities in semantic dementia tell a coherent story. *Brain*. 2012;134 (Pt 7):2025–2035.
150. Agosta F, Henry RG, Migliaccio R, et al. Language networks in semantic dementia. *Brain*. 2010;133(Pt 1):286–299.
151. Mahoney CJ, Malone IB, Ridgway GR, et al. White matter tract signatures of the progressive aphasias. *Neurobiol Aging*. 2013;34(6):1687–1699.
152. Schwindt GC, Graham NL, Rochon E, et al. Whole-brain white matter disruption in semantic and nonfluent variants of primary progressive aphasia. *Hum Brain Mapp*. 2013;34(4):973–984.

153. Botha H, Duffy JR, Whitwell JL, et al. Classification and clinicoradiologic features of primary progressive aphasia (PPA) and apraxia of speech. *Cortex.* 2015;69:220–236.
154. Meijboom R, Steketee RM, Ham LS, van der Lugt A, van Swieten JC, Smits M. Differential Hemispheric predilection of microstructural white matter and functional connectivity abnormalities between respectively semantic and behavioral variant frontotemporal dementia. *J Alzheimers Dis.* 2017;56(2):789–804.
155. Agosta F, Galantucci S, Valsasina P, et al. Disrupted brain connectome in semantic variant of primary progressive aphasia. *Neurobiol Aging.* 2014;35(11):2646–2655.
156. Collins JA, Montal V, Hochberg D, et al. Focal temporal pole atrophy and network degeneration in semantic variant primary progressive aphasia. *Brain.* 2017;140(2):457–471.
157. Brown EE, Graff-Guerrero A, Houle S, et al. Amyloid deposition in semantic dementia: a positron emission tomography study. *Int J Geriatr Psychiatry.* 2016;31(9): 1064–1074.
158. Santos-Santos MA, Rabinovici GD, Iaccarino L, et al. Rates of amyloid imaging positivity in patients with primary progressive aphasia. *JAMA Neurol.* 2018;75:342–352.
159. Villarejo-Galende A, Llamas-Velasco S, Gomez-Grande A, et al. Amyloid pet in primary progressive aphasia: case series and systematic review of the literature. *J Neurol.* 2017;264(1):121–130.
160. Josephs KA, Duffy JR, Strand EA, et al. APOE epsilon4 influences beta-amyloid deposition in primary progressive aphasia and speech apraxia. *Alzheimers Dement.* 2014;10(6):630–636.
161. Whitwell JL, Martin PR, Duffy JR, et al. The influence of beta-amyloid on [(18)F]AV-1451 in semantic variant of primary progressive aphasia. *Neurology.* 2019;92(7): e710–e722.
162. Davies RR, Hodges JR, Kril JJ, Patterson K, Halliday GM, Xuereb JH. The pathological basis of semantic dementia. *Brain.* 2005;128(Pt 9):1984–1995.
163. Harris JM, Gall C, Thompson JC, et al. Classification and pathology of primary progressive aphasia. *Neurology.* 2013;81(21):1832–1839.
164. Mesulam M, Wicklund A, Johnson N, et al. Alzheimer and frontotemporal pathology in subsets of primary progressive aphasia. *Ann Neurol.* 2008;63(6):709–719.
165. Bevan-Jones WR, Cope TE, Jones PS, et al. [18F]AV-1451 binding in vivo mirrors the expected distribution of TDP-43 pathology in the semantic variant of primary progressive aphasia. *J Neurol Neurosurg Psychiatry.* 2018;89:1032–1037.
166. Josephs KA, Martin PR, Botha H, et al. [18F]AV-1451 tau-PET and primary progressive aphasia. *Ann Neurol.* 2018;83:599–611.
167. Makaretz SJ, Quimby M, Collins J, et al. Flortaucipir tau PET imaging in semantic variant primary progressive aphasia. *J Neurol Neurosurg Psychiatry.* 2018;89:1024–1031.
168. Cho H, Kim HJ, Choi JY, et al. (18)F-flortaucipir uptake patterns in clinical subtypes of primary progressive aphasia. *Neurobiol Aging.* 2019;75:187–197.
169. Gorno-Tempini ML, Dronkers NF, Rankin KP, et al. Cognition and anatomy in three variants of primary progressive aphasia. *Ann Neurol.* 2004;55(3):335–346.
170. Grossman M, McMillan C, Moore P, et al. What's in a name: voxel-based morphometric analyses of MRI and naming difficulty in Alzheimer's disease, frontotemporal dementia and corticobasal degeneration. *Brain.* 2004;127(Pt 3):628–649.
171. Rohrer JD, Warren JD, Modat M, et al. Patterns of cortical thinning in the language variants of frontotemporal lobar degeneration. *Neurology.* 2009;72(18):1562–1569.
172. Josephs KA, Duffy JR, Fossett TR, et al. Fluorodeoxyglucose F18 positron emission tomography in progressive apraxia of speech and primary progressive aphasia variants. *Arch Neurol.* 2010;67(5):596–605.

173. Perneczky R, Diehl-Schmid J, Pohl C, Drzezga A, Kurz A. Non-fluent progressive aphasia: cerebral metabolic patterns and brain reserve. *Brain Res*. 2007;1133(1): 178–185.
174. Nestor PJ, Graham NL, Fryer TD, Williams GB, Patterson K, Hodges JR. Progressive non-fluent aphasia is associated with hypometabolism centred on the left anterior insula. *Brain*. 2003;126(Pt 11):2406–2418.
175. Abe K, Ukita H, Yanagihara T. Imaging in primary progressive aphasia. *Neuroradiology*. 1997;39(8):556–559.
176. Whitwell JL, Duffy JR, Strand EA, et al. Distinct regional anatomic and functional correlates of neurodegenerative apraxia of speech and aphasia: an MRI and FDG-PET study. *Brain Lang*. 2013;125(3):245–252.
177. Amici S, Brambati SM, Wilkins DP, et al. Anatomical correlates of sentence comprehension and verbal working memory in neurodegenerative disease. *J Neurosci*. 2007;27(23):6282–6290.
178. Peelle JE, Troiani V, Gee J, et al. Sentence comprehension and voxel-based morphometry in progressive nonfluent aphasia, semantic dementia, and nonaphasic frontotemporal dementia. *J Neurolinguistics*. 2008;21(5):418–432.
179. Turken AU, Dronkers NF. The neural architecture of the language comprehension network: converging evidence from lesion and connectivity analyses. *Front Syst Neurosci*. 2011;5:1.
180. Josephs KA, Duffy JR, Strand EA, et al. Syndromes dominated by apraxia of speech show distinct characteristics from agrammatic PPA. *Neurology*. 2013;81(4):337–345.
181. Rohrer JD, Clarkson MJ, Kittus R, et al. Rates of hemispheric and lobar atrophy in the language variants of frontotemporal lobar degeneration. *J Alzheimers Dis*. 2012;30(2):407–411.
182. Brambati SM, Amici S, Racine CA, et al. Longitudinal gray matter contraction in three variants of primary progressive aphasia: a tenser-based morphometry study. *Neuroimage Clin*. 2015;8:345–355.
183. Rogalski E, Cobia D, Harrison TM, Wieneke C, Weintraub S, Mesulam MM. Progression of language decline and cortical atrophy in subtypes of primary progressive aphasia. *Neurology*. 2011;76(21):1804–1810.
184. Whitwell JL, Duffy JR, Machulda MM, et al. Tracking the development of agrammatic aphasia: a tensor-based morphometry study. *Cortex*. 2017;90:138–148.
185. Galantucci S, Tartaglia MC, Wilson SM, et al. White matter damage in primary progressive aphasias: a diffusion tensor tractography study. *Brain*. 2011;134(Pt 10):3011–3029.
186. Grossman M, Powers J, Ash S, et al. Disruption of large-scale neural networks in non-fluent/agrammatic variant primary progressive aphasia associated with frontotemporal degeneration pathology. *Brain Lang*. 2013;127:106–120.
187. Marcotte K, Graham NL, Fraser KC, et al. White matter disruption and connected speech in non-fluent and semantic variants of primary progressive aphasia. *Dement Geriatr Cogn Dis Extra*. 2017;7(1):52–73.
188. Catani M, Mesulam MM, Jakobsen E, et al. A novel frontal pathway underlies verbal fluency in primary progressive aphasia. *Brain*. 2013;136(Pt 8):2619–2628.
189. Mandelli ML, Vilaplana E, Brown JA, et al. Healthy brain connectivity predicts atrophy progression in non-fluent variant of primary progressive aphasia. *Brain*. 2016;139(Pt 10):2778–2791.
190. Mandelli ML, Caverzasi E, Binney RJ, et al. Frontal white matter tracts sustaining speech production in primary progressive aphasia. *J Neurosci*. 2014;34(29):9754–9767.
191. Routier A, Habert MO, Bertrand A, et al. Structural, microstructural, and metabolic alterations in primary progressive aphasia variants. *Front Neurol*. 2018;9:766.

192. Mandelli ML, Welch AE, Vilaplana E, et al. Altered topology of the functional speech production network in non-fluent/agrammatic variant of PPA. *Cortex*. 2018;108:252–264.
193. Botha H, Utianski RL, Whitwell JL, et al. Disrupted functional connectivity in primary progressive apraxia of speech. *Neuroimage Clin*. 2018;18:617–629.
194. Rabinovici GD, Jagust WJ, Furst AJ, et al. Abeta amyloid and glucose metabolism in three variants of primary progressive aphasia. *Ann Neurol*. 2008;64(4):388–401.
195. Leyton CE, Villemagne VL, Savage S, et al. Subtypes of progressive aphasia: application of the International consensus criteria and validation using beta-amyloid imaging. *Brain*. 2011;134(Pt 10):3030–3043.
196. Chare L, Hodges JR, Leyton CE, et al. New criteria for frontotemporal dementia syndromes: clinical and pathological diagnostic implications. *J Neurol Neurosurg Psychiatry*. 2014;85(8):865–870.
197. Whitwell JL, Weigand SD, Duffy JR, et al. Clinical and MRI models predicting amyloid deposition in progressive aphasia and apraxia of speech. *Neuroimage Clin*. 2016;11:90–98.
198. Utianski RL, Whitwell JL, Schwarz CG, et al. Tau uptake in agrammatic primary progressive aphasia with and without apraxia of speech. *Eur J Neurol*. 2018;25(11):1352–1357.
199. Utianski RL, Whitwell JL, Schwarz CG, et al. Tau-PET imaging with [18F]AV-1451 in primary progressive apraxia of speech. *Cortex*. 2018;99:358–374.
200. Marquie M, Normandin MD, Vanderburg CR, et al. Validating novel tau positron emission tomography tracer [F-18]-AV-1451 (T807) on postmortem brain tissue. *Ann Neurol*. 2015;78(5):787–800.
201. Sander K, Lashley T, Gami P, et al. Characterization of tau positron emission tomography tracer [18F]AV-1451 binding to postmortem tissue in Alzheimer's disease, primary tauopathies, and other dementias. *Alzheimers Dement*. 2016;12(11):1116–1124.
202. Ono M, Sahara N, Kumata K, et al. Distinct binding of PET ligands PBB3 and AV-1451 to tau fibril strains in neurodegenerative tauopathies. *Brain*. 2017;140(3):764–780.
203. Whitwell JL, Warren JD, Josephs KA, et al. Voxel-based morphometry in tau-positive and tau-negative frontotemporal lobar degenerations. *Neurodegener Dis*. 2004;1 (4-5):225–230.
204. Pereira JM, Williams GB, Acosta-Cabronero J, et al. Atrophy patterns in histologic vs clinical groupings of frontotemporal lobar degeneration. *Neurology*. 2009;72(19):1653–1660.
205. Whitwell JL, Jack Jr CR, Senjem ML, et al. MRI correlates of protein deposition and disease severity in postmortem frontotemporal lobar degeneration. *Neurodegener Dis*. 2009;6(3):106–117.
206. Whitwell JL, Josephs KA. Neuroimaging in frontotemporal lobar degeneration–predicting molecular pathology. *Nat Rev Neurol*. 2012;8(3):131–142.
207. Lieberman AP, Trojanowski JQ, Lee VM, et al. Cognitive, neuroimaging, and pathological studies in a patient with Pick's disease. *Ann Neurol*. 1998;43(2):259–265.
208. Whitwell JL, Josephs KA, Rossor MN, et al. Magnetic resonance imaging signatures of tissue pathology in frontotemporal dementia. *Arch Neurol*. 2005;62(9):1402–1408.
209. Whitwell JL, Jack Jr CR, Parisi JE, et al. Imaging signatures of molecular pathology in behavioral variant frontotemporal dementia. *J Mol Neurosci*. 2011;45(3):372–378.
210. Josephs KA, Whitwell JL, Dickson DW, et al. Voxel-based morphometry in autopsy proven PSP and CBD. *Neurobiol Aging*. 2008;29(2):280–289.
211. Rohrer JD, Lashley T, Schott JM, et al. Clinical and neuroanatomical signatures of tissue pathology in frontotemporal lobar degeneration. *Brain*. 2011;134(Pt 9):2565–2581.

212. Whitwell JL, Jack Jr CR, Parisi JE, et al. Rates of cerebral atrophy differ in different degenerative pathologies. *Brain*. 2007;130(Pt 4):1148–1158.
213. Whitwell JL, Jack Jr CR, Parisi JE, et al. Midbrain atrophy is not a biomarker of progressive supranuclear palsy pathology. *Eur J Neurol*. 2013;20(10):1417–1422.
214. Hassan A, Whitwell JL, Boeve BF, et al. Symmetric corticobasal degeneration (S-CBD). *Parkinsonism Relat Disord*. 2010;16(3):208–214.
215. Schrag A, Good CD, Miszkiel K, et al. Differentiation of atypical parkinsonian syndromes with routine MRI. *Neurology*. 2000;54(3):697–702.
216. Josephs KA, Tang-Wai DF, Edland SD, et al. Correlation between antemortem magnetic resonance imaging findings and pathologically confirmed corticobasal degeneration. *Arch Neurol*. 2004;61(12):1881–1884.
217. Whitwell JL, Jack Jr CR, Boeve BF, et al. Imaging correlates of pathology in corticobasal syndrome. *Neurology*. 2010;75(21):1879–1887.
218. Rohrer JD, Geser F, Zhou J, et al. TDP-43 subtypes are associated with distinct atrophy patterns in frontotemporal dementia. *Neurology*. 2010;75(24):2204–2211.
219. Whitwell JL, Jack Jr CR, Parisi JE, et al. Does TDP-43 type confer a distinct pattern of atrophy in frontotemporal lobar degeneration? *Neurology*. 2010;75(24):2212–2220.
220. Josephs KA, Whitwell JL, Parisi JE, et al. Caudate atrophy on MRI is a characteristic feature of FTLD-FUS. *Eur J Neurol*. 2010;17(7):969–975.
221. Seelaar H, Klijnsma KY, de Koning I, et al. Frequency of ubiquitin and FUS-positive, TDP-43-negative frontotemporal lobar degeneration. *J Neurol*. 2010;257(5):747–753.
222. Rohrer JD, Lashley T, Holton J, et al. The clinical and neuroanatomical phenotype of FUS associated frontotemporal lobar degeneration. *J Neurol Neurosurg Psychiatry*. 2011;82:1405–1407.
223. Rankin KP, Mayo MC, Seeley WW, et al. Behavioral variant frontotemporal dementia with corticobasal degeneration pathology: phenotypic comparison to bvFTD with Pick's disease. *J Mol Neurosci*. 2011;45(3):594–608.
224. Santos-Santos MA, Mandelli ML, Binney RJ, et al. Features of patients with nonfluent/agrammatic primary progressive aphasia with underlying progressive supranuclear palsy pathology or corticobasal degeneration. *JAMA Neurol*. 2016;73(6):733–742.
225. Caso F, Mandelli ML, Henry M, et al. In vivo signatures of nonfluent/agrammatic primary progressive aphasia caused by FTLD pathology. *Neurology*. 2014;82(3): 239–247.
226. Hutton M, Lendon CL, Rizzu P, et al. Association of missense and 5'-splice-site mutations in tau with the inherited dementia FTDP-17. *Nature*. 1998;393(6686): 702–705.
227. Baker M, Mackenzie IR, Pickering-Brown SM, et al. Mutations in progranulin cause tau-negative frontotemporal dementia linked to chromosome 17. *Nature*. 2006;442(7105):916–919.
228. Cruts M, Gijselinck I, van der Zee J, et al. Null mutations in progranulin cause ubiquitin-positive frontotemporal dementia linked to chromosome 17q21. *Nature*. 2006;442(7105):920–924.
229. Dejesus-Hernandez M, Mackenzie IR, Boeve BF, et al. Expanded GGGGCC hexanucleotide repeat in noncoding region of C9ORF72 causes chromosome 9p-linked FTD and ALS. *Neuron*. 2011;72:245–256.
230. Renton AE, Majounie E, Waite A, et al. A hexanucleotide repeat expansion in C9ORF72 is the cause of chromosome 9p21-linked ALS-FTD. *Neuron*. 2011;72:257–268.
231. Beck J, Rohrer JD, Campbell T, et al. A distinct clinical, neuropsychological and radiological phenotype is associated with progranulin gene mutations in a large UK series. *Brain*. 2008;131(Pt 3):706–720.

232. Pickering-Brown SM, Rollinson S, Du Plessis D, et al. Frequency and clinical characteristics of progranulin mutation carriers in the Manchester frontotemporal lobar degeneration cohort: comparison with patients with MAPT and no known mutations. *Brain*. 2008;131(Pt 3):721–731.
233. Kelley BJ, Haidar W, Boeve BF, et al. Prominent phenotypic variability associated with mutations in Progranulin. *Neurobiol Aging*. 2009;30(5):739–751.
234. Masellis M, Momeni P, Meschino W, et al. Novel splicing mutation in the progranulin gene causing familial corticobasal syndrome. *Brain*. 2006;129(Pt 11):3115–3123.
235. Mesulam M, Johnson N, Krefft TA, et al. Progranulin mutations in primary progressive aphasia: the PPA1 and PPA3 families. *Arch Neurol*. 2007;64(1):43–47.
236. Spina S, Farlow MR, Unverzagt FW, et al. The tauopathy associated with mutation +3 in intron 10 of Tau: characterization of the MSTD family. *Brain*. 2008;131(Pt 1):72–89.
237. Whitwell JL, Jack Jr CR, Boeve BF, et al. Voxel-based morphometry patterns of atrophy in FTLD with mutations in MAPT or PGRN. *Neurology*. 2009;72(9):813.
238. Le Ber I, Camuzat A, Hannequin D, et al. Phenotype variability in progranulin mutation carriers: a clinical, neuropsychological, imaging and genetic study. *Brain*. 2008;131(Pt 3):732–746.
239. Whitwell JL, Jack Jr CR, Baker M, et al. Voxel-based morphometry in frontotemporal lobar degeneration with ubiquitin-positive inclusions with and without progranulin mutations. *Arch Neurol*. 2007;64(3):371–376.
240. Rohrer JD, Warren JD, Omar R, et al. Parietal lobe deficits in frontotemporal lobar degeneration caused by a mutation in the progranulin gene. *Arch Neurol*. 2008;65(4):506–513.
241. Whitwell JL, Weigand SD, Boeve BF, et al. Neuroimaging signatures of frontotemporal dementia genetics: C9ORF72, tau, progranulin and sporadics. *Brain*. 2012;135(Pt 3):794–806.
242. Mahoney CJ, Beck J, Rohrer JD, et al. Frontotemporal dementia with the C9ORF72 hexanucleotide repeat expansion: clinical, neuroanatomical and neuropathological features. *Brain*. 2012;135(Pt 3):736–750.
243. Agosta F, Ferraro PM, Riva N, et al. Structural and functional brain signatures of C9orf72 in motor neuron disease. *Neurobiol Aging*. 2017;57:206–219.
244. Devenney EM, Landin-Romero R, Irish M, et al. The neural correlates and clinical characteristics of psychosis in the frontotemporal dementia continuum and the C9orf72 expansion. *Neuroimage Clin*. 2017;13:439–445.
245. Schonecker S, Neuhofer C, Otto M, et al. Atrophy in the thalamus but not cerebellum is specific for C9orf72 FTD and ALS patients—an atlas-based volumetric MRI study. *Front Aging Neurosci*. 2018;10:45.
246. Bocchetta M, Gordon E, Cardoso MJ, et al. Thalamic atrophy in frontotemporal dementia—not just a C9orf72 problem. *Neuroimage Clin*. 2018;18:675–681.
247. Rohrer JD, Ridgway GR, Modat M, et al. Distinct profiles of brain atrophy in frontotemporal lobar degeneration caused by progranulin and tau mutations. *Neuroimage*. 2010;53(3):1070–1076.
248. Whitwell JL, Weigand SD, Gunter JL, et al. Trajectories of brain and hippocampal atrophy in FTD with mutations in MAPT or GRN. *Neurology*. 2011;77(4):393–398.
249. Whitwell JL, Boeve BF, Weigand SD, et al. Brain atrophy over time in genetic and sporadic frontotemporal dementia: a study of 198 serial magnetic resonance images. *Eur J Neurol*. 2015;22(5):745–752.
250. Bertrand A, Wen J, Rinaldi D, et al. Early cognitive, structural, and microstructural changes in presymptomatic C9orf72 carriers younger than 40 years. *JAMA Neurol*. 2018;75(2):236–245.
251. Lee SE, Sias AC, Mandelli ML, et al. Network degeneration and dysfunction in presymptomatic C9ORF72 expansion carriers. *Neuroimage Clin*. 2017;14:286–297.

252. Olm CA, McMillan CT, Irwin DJ, et al. Longitudinal structural gray matter and white matter MRI changes in presymptomatic progranulin mutation carriers. *Neuroimage Clin.* 2018;19:497–506.
253. Panman JL, Jiskoot LC, Bouts M, et al. Gray and white matter changes in presymptomatic genetic frontotemporal dementia: a longitudinal MRI study. *Neurobiol Aging.* 2019;76:115–124.
254. Papma JM, Jiskoot LC, Panman JL, et al. Cognition and gray and white matter characteristics of presymptomatic C9orf72 repeat expansion. *Neurology.* 2017;89 (12):1256–1264.
255. Pievani M, Paternico D, Benussi L, et al. Pattern of structural and functional brain abnormalities in asymptomatic granulin mutation carriers. *Alzheimers Dement.* 2014;10 (Suppl 5):S354–S363 e351.
256. Dopper EG, Rombouts SA, Jiskoot LC, et al. Structural and functional brain connectivity in presymptomatic familial frontotemporal dementia. *Neurology.* 2014;83(2): e19–e26.
257. Jiskoot LC, Bocchetta M, Nicholas JM, et al. Presymptomatic white matter integrity loss in familial frontotemporal dementia in the GENFI cohort: a cross-sectional diffusion tensor imaging study. *Ann Clin Transl Neurol.* 2018;5(9):1025–1036.
258. Borroni B, Alberici A, Premi E, et al. Brain magnetic resonance imaging structural changes in a pedigree of asymptomatic progranulin mutation carriers. *Rejuvenation Res.* 2008;11(3):585–595.
259. Lee SE, Khazenzon AM, Trujillo AJ, et al. Altered network connectivity in frontotemporal dementia with C9orf72 hexanucleotide repeat expansion. *Brain.* 2014;137(Pt 11):3047–3060.
260. Spina S, Schonhaut DR, Boeve BF, et al. Frontotemporal dementia with the V337M MAPT mutation: tau-PET and pathology correlations. *Neurology.* 2017;88(8):758–766.
261. Smith R, Puschmann A, Scholl M, et al. 18F-AV-1451 tau PET imaging correlates strongly with tau neuropathology in MAPT mutation carriers. *Brain.* 2016;139(Pt 9):2372–2379.
262. Jones DT, Knopman DS, Graff-Radford J, et al. In vivo (18)F-AV-1451 tau PET signal in MAPT mutation carriers varies by expected tau isoforms. *Neurology.* 2018;90(11): e947–e954.
263. Ikeda A, Shimada H, Nishioka K, et al. Clinical heterogeneity of frontotemporal dementia and Parkinsonism linked to chromosome 17 caused by MAPT N279K mutation in relation to tau positron emission tomography features. *Mov Disord.* 2019;34:568–574.
264. Bevan Jones WR, Cope TE, Passamonti L, et al. [(18)F]AV-1451 PET in behavioral variant frontotemporal dementia due to MAPT mutation. *Ann Clin Transl Neurol.* 2016;3(12):940–947.

Chapter 4

Cerebrovascular disease: Neuroimaging of cerebral small vessel disease

C. Elizabeth Shaaban[a,b], Dana R. Jorgensen[c], Peter J. Gianaros[b,d], Joseph Mettenburg[e], Caterina Rosano[a,b,*]

[a]Department of Epidemiology, Graduate School of Public Health, University of Pittsburgh, Pittsburgh, PA, United States
[b]Center for the Neural Basis of Cognition, University of Pittsburgh, Pittsburgh, PA, United States
[c]University of Pittsburgh Medical Center, Pittsburgh, PA, United States
[d]Department of Psychology, University of Pittsburgh, Pittsburgh, PA, United States
[e]Department of Radiology, University of Pittsburgh, Pittsburgh, PA, United States
*Corresponding author: e-mail address: rosanoc@edc.pitt.edu

Contents

Abstract

The cerebral microvasculature is exceptionally vulnerable to changes due to aging. Both the radiological and clinical manifestations of cerebral small vessel disease (cSVD) in

Progress in Molecular Biology and Translational Science, Volume 165
ISSN 1877-1173
https://doi.org/10.1016/bs.pmbts.2019.07.008

older age are highly heterogeneous, ranging from no symptomatology to devastating neurocognitive complications, including stroke, dementia, and depression. To date, the exact pathogenesis of cSVD is unknown; neither prevention nor treatments are currently available for this potentially very disabling condition. This chapter reviews recent advances in neuroimaging methodologies that have improved our understanding of the appearance of cSVD and the underlying mechanisms. We will discuss new venues for future research focusing on: (a) methodologies to measure early stages of cSVD, examining damage primarily in small vessels and secondarily in parenchyma; and (b) lifespan study designs to capture the onset and evolution of cSVD over time, as well as the temporality of exposure to risk factors.

1. Introduction

Small vessels in the brain undergo remodeling and exhibit signs of damage that accumulate with older age. Although cerebral small vessel disease (cSVD) is common after age 75 in the general population, it is detectable as early as middle age in individuals with vascular conditions, such as hypertension or diabetes.[1–3] The clinical manifestations of cSVD are remarkably variable, ranging from no symptomatology to devastating neurocognitive complications, including stroke, dementia, and depression.

Population longitudinal studies of older adults without clinically overt neurological conditions show that cSVD strongly predicts development of such neurocognitive complications over time, suggesting that the clinical manifestations of cSVD have long "incubation" times. The implication of this observation is that cSVD is potentially detectable years before its complications appear. Thus, capturing the earliest stages of cSVD could offer valuable strategies to characterize this condition and its underlying pathogenesis and therefore to intervene. Strategies to prevent and/or delay cSVD-related neurocognitive complications could generate significant cost savings at the personal and public health levels.

Understanding the mechanisms underlying the development of cSVD is a critical, but currently difficult, step for the prevention of cSVD-related complications.[4,5] There seems to be consensus in the field that the major barrier to the elucidation of cSVD is a lack of diagnostic biomarkers of early stages of cSVD (see National Institute of Neurological Disorders and Stroke-sponsored endeavor MARK VCID[6] and a parallel European effort known as HARNESS[7]). Another important limitation of our understanding of cSVD is that, unlike other pathologies that progress over time, several of the neuroradiological measures of cSVD can appear stable over time, and occasionally even "spontaneously" regress. Current neuroimaging methods do not directly measure small vessels' damage, and instead predominantly capture secondary late-stage effects of cSVD on brain parenchyma, such as white matter disruption, radiologically visible as white matter hyperintensities.

The factors predisposing to the development of cSVD are also not very clear. There is a high variability in the risk factors for this common cerebrovascular condition. Blood pressure and other cardiovascular parameters are associated with at least some of the many radiological manifestations of cSVD; however, older adults with unremarkable cardiovascular history also present with cSVD. This dissociation indicates the existence of multifactorial causal pathways that are more complex than of purely cardiovascular origin. It is perhaps for this reason that treatments reducing cardiovascular risk factors have yielded only modest and inconsistent reductions in radiological markers of cSVD for vascular patient populations.[8] Overall, there is no clear evidence of time trajectories of risk factors or of change in cerebral microvasculature across the lifespan. Studies to date have focused for the most part on samples of old to very-old adults, capturing small vessel changes over narrow windows of time and toward the end of life. Moreover, the temporality of study designs has typically been limited to cross-sectional measures.

This chapter reviews both established and recently developed neuroimaging methods capturing cSVD. We specifically focus on technological advancements in neuroradiology that can capture changes directly affecting the small vessels earlier in the evolution of the pathological processes. Next, we examine the relations between these radiological markers and putative risk factors of cSVD. Lastly, we discuss novel research frameworks to advance the future study of cSVD with a focus on study designs and sampling strategies.

2. Neuroradiological appearances and histopathological correlates of cSVD

Common findings on histopathological examination of postmortem brains of older individuals (i.e., those 60 years and older) include tortuosity of venules and arterioles, capillary rarefaction, small infarcts, microhemorrhages and enlarged perivascular spaces. These features are taken as manifestations of aging microcirculation. With the advent of magnetic resonance imaging (MRI), a wider range of cerebral age-related changes have been identified. These include parenchymal changes, such as infarcts, microhemorrhages, lacunes, periventricular spaces and also abnormalities of cerebral blood flow (CBF), white matter hyperintensities (WMH) and reduced fiber alignment (via diffusion tensor imaging). Although studies relating MRI and histopathological features of age-related cerebral microcirculation are sparse, the above MRI measures are nonetheless considered lesions of presumed vascular origin and are grouped together and referred to as "neuroradiological markers of cerebral small vessel disease (cSVD)," a convention that we hereby adopt for simplicity and for consistency with the STRIVE definition.[9] By this definition, cSVD comprises a syndrome of radiological manifestations affecting brain areas with predominantly poor collateral vascularization (e.g., watershed areas, fronto-subcortical regions, etc.). According to this scheme, radiological markers of cSVD include: white matter hyperintensities (WMH), cerebral microbleeds (CMBs), silent brain infarcts (SBI) and lacunes. The following parenchymal markers of cSVD included in the STRIVE definition will not be discussed: gray matter atrophy and diffusion related indices (e.g., fractional anisotropy and mean diffusivity) because they appear removed from actual vascular pathology.

Commonly used neuroradiological markers of cSVD

- White matter hyperintensities (WMH).
- Silent brain infarcts (SBI).
- Lacunes.
- Cerebral microbleeds (CMB).

Novel markers of cSVD

- Cerebral blood flow (CBF) measured at rest.
- Cerebrovascular reactivity (CVR): changes in CBF after stimulation.
- Morphology of small arteries and veins (e.g., tortuosity, density).

To account for the most recent methodological developments to characterize brain circulation, we include more novel markers of cSVD, which appear to be more direct measures of vasculature integrity. For example, endothelial dysfunction, an early stage marker of cSVD, can be measured by CBF at rest and changes in response to specific challenges (e.g., breath-holding or hypercapnia).[10] We also discuss morphological measures (e.g., tortuosity, density) of arterioles via time-of-flight MRI and of small veins via ultra-high field susceptibility-weighted MRI.

2.1 White matter hyperintensities (WMH)

WMH are broadly defined as areas that appear as bright or "hyperintense" on T2, T2*, proton density and fluid-attenuated inversion recovery

(FLAIR) MRI. They may also appear as hypointense on volumetric T1-weighted MRI. It is generally accepted that the variations in MR contrast that contribute to the visualization of WMH correspond to vastly different parenchymal changes as captured by histopathology.

White matter abnormalities were detected in early CT scans of aged human brains. However, when MRI was developed a few years later, it quickly proved to be more sensitive. Once MRI became economically tractable, its routine use substantially heightened our awareness of these age-associated lesions. But what are they? Beginning in the late 1980s and proceeding up to today, there have been numerous studies attempting to define the neuropathological correlates of WMH.[11–32] These correlative studies used either pre-mortem or postmortem MRI or some combination of techniques. To a reasonable degree of consensus, we can draw the following conclusions:

2.1.1 "Periventricular hyperintense signal" (aka "rims" or "caps")

These lesions that outline the edge of ventricles tend to be non-progressive and have the greatest consensus regarding their neuropathological correlates. They consist of a denudation of ependyma from ventricular lining with variable subependymal gliosis (Fig. 1). They may also reflect periventricular necrosis and axonal degeneration with or without increased intraventricular pressure. Some have found reactive astrocytosis with remarkably well-preserved underlying myelin.

2.1.2 "Punctate and confluent deep WMH"

These lesions can progress over time from punctate to clustered to confluent. Punctate forms show a variety of small vessel pathologies (Fig. 1). Accentuation of perivascular (Virchow-Robin) spaces is associated with a variety of vascular pathologies including: convoluted vessels (arterioles, capillaries or venules), sclerotic vessels (with or without amyloidosis), arteriolosclerosis, hyalinosis and collagenosis.

When WMH are associated with susceptibility artifacts, as detected on sequences sensitive to the magnetic properties of the iron in hemoglobin (e.g., gradient-recalled echo or susceptibility-weighted MRI), these lesions are thought to involve micro-hemorrhage with extravasation of red blood cells, pericyte erythrophagocytosis, hemoglobin degradation and hemosiderin deposition (Fig. 1C and D).[33] How such changes might reflect "vascular integrity" or the blood-brain barrier remains to be fully elucidated.

Some WMH are associated with complete or incomplete infarctions (Fig. 2), including conspicuous lacunar type infarcts or less obvious

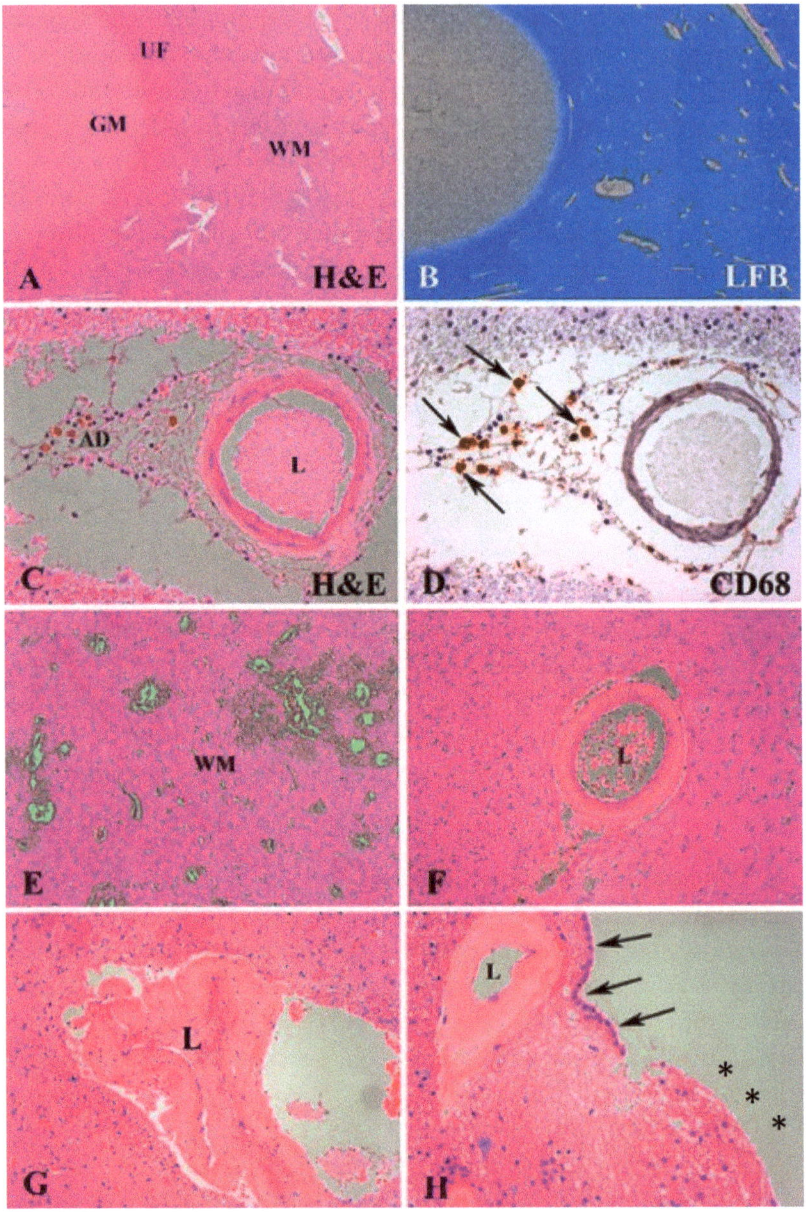

Fig. 1 Deep white matter pathology of age associated with white matter hyperintensities. (A and B) Low power paraffin section of midfrontal cortex from the brain of an 80 year old. (A) Hematoxylin and eosin-stained section shows unremarkable gray matter (GM) with adjacent white matter (WM) containing numerous dilated vascular spaces. Overall the deep white matter staining is more pale than the U-fibers (UF) immediately underlying the gray matter. (B) Successive section stained with luxol fast blue (LFB), where the gray

microinfarcts. Given limitations of sampling microinfarcts (i.e., infarctions not visible to the naked eye), it has been suggested that histopathology grossly underestimates the frequency of microinfarctions (e.g., for every microinfarct noted during routine brain sampling 500 microinfarcts are present in vivo[34]). Reactive gliosis (astrocytic and microglial cellular response) is the sine qua non of CNS damage and repair (Fig. 2). Its associations with WMH and perivascular changes are definitive proof that the histological changes are not fixation or processing artifacts, but are instead representative of in vivo pathology.

Myelin pallor (rarefaction) is a frequently described pathological finding associated with WMH. Despite the pathological term being vague and non-specific, it is frequently misquoted as "demyelination" or other more specific, but inaccurate, descriptors. This has resulted in substantial confusion and inappropriate assumptions about the pathogenesis of myelin pallor. Histologically it is manifest as decreased staining of white matter using any of a number of stains (e.g., H&E, luxol fast blue, staining for myelin basic protein, etc.). Like the MRI finding of WMH, this histological finding of myelin pallor can be broadly interpreted as altered tissue water content with numerous potential etiologies. Simple brain edema (vascular or cytotoxic) can expand the extra- and intracellular space leading to decreased stain density per square mm. Decreased numbers of axons (e.g., loss of fibers projecting through a region of rarefaction secondary to distal stroke) can decrease the tissue density and be associated with staining pallor. Demyelination could lead to decreased staining, but despite frequent use of this term in the literature there is no convincing evidence that primary demyelination (that is selective loss of myelin sheaths as opposed to loss of myelin sheaths secondary to loss of axons) is a true feature of WMH. Based upon measurements of select myelin proteins and alterations in vasoconstrictive

matter appears pale and white matter a deep blue. Numerous dilated perivascular spaces are noted in the white matter. (C and D) High power micrograph of individual blood vessels within deep white matter. (C) The blood vessel lumen (L) is surrounded by a mildly thickened media. The adventitia (AD) surrounding the vessel contains macrophages filled with golden hemosiderin. (D) Special stain for CD68 identifies the hemosiderin laden cells as macrophages (brown cells labeled with arrows). (E) Medium power micrograph of deep white matter showing a "moth eaten" pattern imparted by the numerous dilated perivascular spaces. (F) The lumen of blood vessels is surrounded by a thickly hyalinized vessel wall. (G) Some blood vessels have convoluted and collapsed lumina. (H) Ependymal surface (arrows) of ventricular wall overlay sclerotic vessels. The ventricular surface focally demonstrates ependymal denudation (below asterisks). *Images were provided by Dr. Clayton Wiley (University of Pittsburgh).*

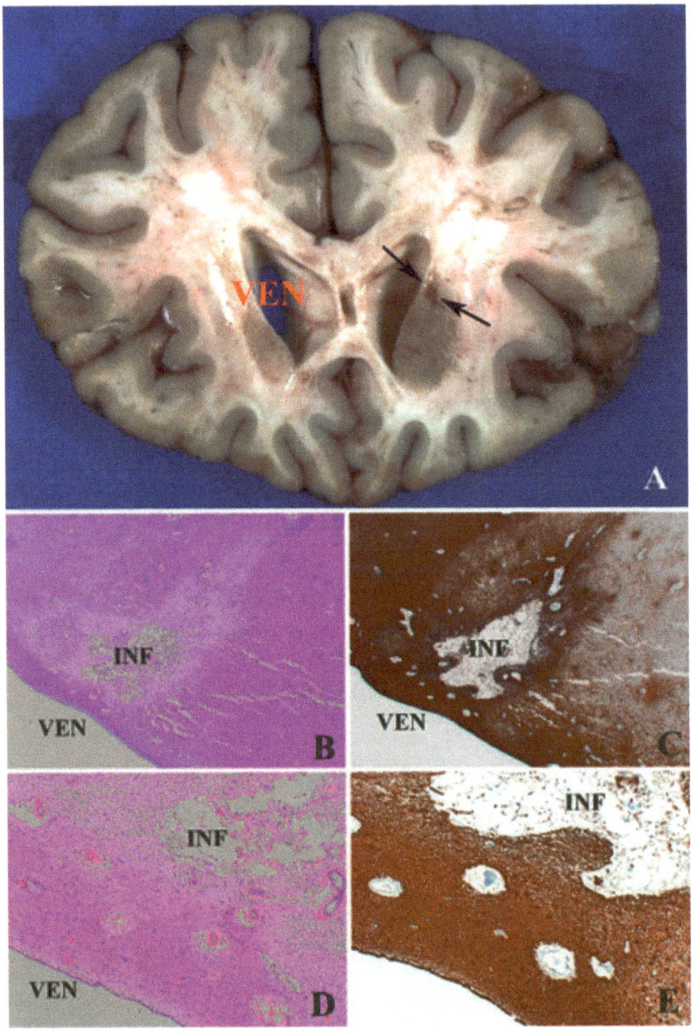

Fig. 2 Lacunar infarction of basal ganglia. (A) Coronal section through the frontal lobe of the brain of an older adult patient. Ventricles (VEN) are mildly dilated, and there is a lacunar infarct in the head of the caudate on the right side (between arrows). (B) Hematoxylin and eosin-stained section of the lacune demonstrates a central region of infarction (INF) with tissue loss. (C) Immunostain for glial fibrillary acidic protein (GFAP) (brown) shows prominent peri-infarct astrocytosis. (D) Higher power micrograph of lacunar infarct showing intact ependyma overlying ventricle (VEN). Tissue surrounding the infarct contains numerous sclerotic blood vessels with dilated perivascular spaces. (E) Immunostain for GFAP demonstrates pronounced astrocytosis. *Images were provided by Dr. Clayton Wiley (University of Pittsburgh).*

endothelin-1 (EDN1) and vascular growth factors (VEGF), Love and colleagues have recently put forward the hypothesis that white matter rarefaction is the result of decreased perfusion.[35] While an enticing explanation of white matter dysfunction, there is no formal proof for demyelination in aged white matter associated with WMH.

Testing the hypothesis that WMH (or at least some of its components) is the result of demyelination is technically quite difficult. Love and colleagues[35] have attempted to compare ratios of myelin associated glycoprotein (MAG) to proteolipid protein (PLP) as a surrogate marker of demyelination. While this is a convenient approach, there are multiple reasons other than demyelination that could lead to changes in the MAG/PLP ratio. Axons in the central nervous system become myelinated once they achieve a diameter of 0.5 μm. Since that is at the resolution limit of light microscopy, ultrastructural analysis (i.e., electron microscopy (EM)) is required to sensitively assess myelination. Most EM studies require perfusion fixed tissue, which of course is essentially impossible to get in humans. Thus, given the ultrastructural artifacts introduced by postmortem studies, the optimal human study would require brain biopsy followed by glutaraldehyde fixation, a perhaps unattainable goal in humans. Therefore, testing the hypothesized role of demyelination in WMH awaits the development of a good animal model that would permit in vivo perfusion and EM confirmation.

2.2 Lacunes and silent brain infarcts

Lacunes are small sharply defined regions of tissue loss most frequently associated with deep gray and white matter. Penetrating vessels supplying the deep white matter and basal ganglia lack collateral networks. Vasculopathy related to uncontrolled hypertension and diabetes, systemic inflammatory and/or hypercoagulable states, and other cardio- and neurovascular risk factors, may result in spontaneous thrombosis of these terminal vessels, which supply small patchwork regions of parenchyma, resulting in lacunar type infarcts. These infarcts differ from territorial infarcts more typically associated with atherosclerosis and cardioembolic disease and manifest as small lake-like regions of parenchymal volume loss, or lacunes. Lacunes are typically visible on FLAIR imaging as 3–15 mm ovoid areas of hypointense signal often (though not always) surrounded by a hyperintense rim.[9] Although these infarcts can frequently result in substantial neurologic deficit, typically due to involvement of the corticospinal tracts and deep gray nuclei, smaller insults may be subclinical or only manifest as the overall burden increases. These silent brain infarcts of the white matter are frequently associated with WMH typically described as chronic small vessel ischemic changes.

2.3 Microbleeds

Cerebral microbleeds (CMBs) are representative of small hemorrhages in the brain stemming from damage to the cerebral small arteries. They are visible on T2* weighted gradient recalled echo (GRE) and susceptibility-weighted imaging (SWI) as ovoid, hypointense signal owing to the paramagnetic properties of the blood breakdown product hemosiderin.[36] These lesions may be up to 10 mm in diameter, but most commonly are 2–5 mm.[9] While CMBs of this size are detectable at typical MRI magnet strengths (1.5 and 3 T), advances in ultra-high field imaging allow for the possibility of imaging even smaller CMBs as well as gains in inter- and intra-rater reliability.[37] Accordingly, the prevalence of CMBs goes up with magnet strength.[37]

CMBs are found either at the cortico-subcortical boundary or in the deep gray matter nuclei and deep white matter. Cortical CMBs are noted in the context of cerebral amyloid angiopathy (CAA), deposits of beta amyloid (aβ) in the small arteries, commonly seen in Alzheimer's disease (AD). Aβ has generally been argued to be a causative agent in these CMBs, although this has been disputed.[38] Recent work suggests that blood-brain barrier disruption is associated with CMBs.[39] Deep CMBs are commonly caused by hypertension. Increased vascular stiffness due to chronic hypertension allows pulsatile flow into the cerebral small vessels, damaging them and leading to CMBs. This pathway may operate through reactive oxygen species-related matrix metalloproteinase increases and damage to the collagen and ultimately the vessels themselves.[40] See Ungvari et al. for a review of CMB physiology.[40]

2.4 Small vessel morphological characteristics

In postmortem specimens with cSVD, small vessels appear as tortuous pencil like structures with a thick basal lamina. Tortuosity of cerebral small arteries and veins is considered a marker of pathology and has been noted in regions of WMH.[41–43] The differentiation of small arteries from small veins is challenging in postmortem specimens, and it is based on a combination of methods, including: (a) negative staining of smooth muscle cells, which are more abundant on the arteriolar than on the venular side, where they acquire distinct morphological characteristics with long cytoplasmic elongations wrapped around the small veins; (b) non-inflammatory, periventricular venulopathy with concentric collagen deposition which causes intramural thickening, stenosis and ultimately luminal occlusion; (c) localization: small veins are more visible in the posterior horns of the ventricle, where arteries are less present due to the watershed characteristics of these regions; and (d) number of branches and amount of capillary-free space surrounding the vessels.[13,42,44–47]

In vivo, small penetrating arteries of the white matter and basal ganglia and small veins in the periventricular areas are typically poorly visualized and below the resolution of most non-invasive techniques such as CT and MRI. However, with ultra-high field time-of-flight MR angiography and susceptibility-weighted imaging, these penetrating small arteries and veins become more conspicuous. Small vessels usually manifest as thin, smooth and relatively straight vasculature with moderate branching. In frank vasculopathies, such as distal internal carotid artery stenosis/occlusion, these penetrating vessels may be recruited as collateral pathways and greatly increase in number, with a more tortuous and aggressively branching pattern, and are prone to hemorrhage. Time-of-flight MR angiography can be used to visualize small arteries as thin thread-like areas of flow-related contrast (Fig. 3). Using this technique, features of cerebrovascular diseasing include reduced vessel number and increases in tortuosity can be examined.[48] The morphology of small veins in vivo is best visualized using susceptibility-weighted imaging (Fig. 3); this approach is very sensitive to motion artifacts, hence utilization of ultra-high field magnets that reduce time of acquisition typically yield images with higher spatial resolution. Number of veins, vein length, tortuosity, and vein number or length per area (density), are measures that can be obtained from this modality.[49–51]

2.4.1 Spatial distribution of morphological abnormalities across segments of the microcirculation

Cerebral small arteries measured via 7 T time-of-flight angiography demonstrate fewer stems and branches in individuals with subcortical vascular dementia and young hypertensive individuals vs. controls.[52,53] In addition, those who are older and individuals who have subcortical vascular dementia have greater arterial tortuosity on 3T and 7 T time-of-flight, respectively.[52,54]

Animal work has suggested that even few "stalls" of blood within the capillaries due to temporary neutrophil adhesion can have a large impact in reducing overall CBF,[55] and human postmortem exam demonstrates

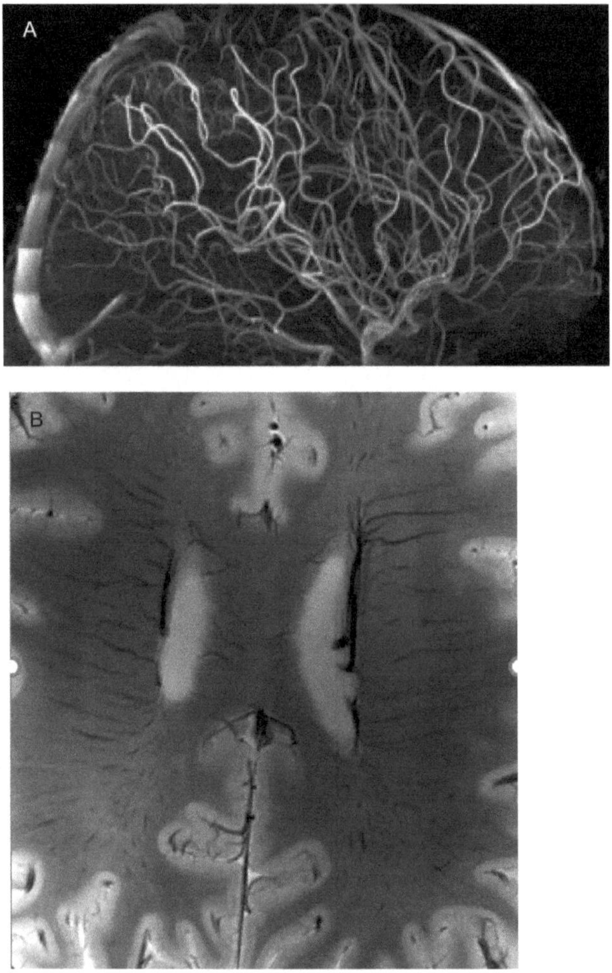

Fig. 3 Time-of-flight angiography of arterial tree and susceptibility-weighted image of medullary small veins. Time-of-flight Angiography (A): Resolution is $0.32 \times 0.32 \times 0.32\ mm^3$. This method can be used to visualize small arteries, which appear as thin thread-like areas of flow-related contrast. (B) Susceptibility-weighted image. This method can be used to visualize small veins. They are shown here as thin, smooth and relatively straight vasculature with moderate branching, darker as compared to surrounding parenchyma. Resolution is $0.2 \times 0.2 \times 1.5\ mm^3$. All images are acquired without the use of any contrast agents at 7T. Images are acquired using the Tic Tac Toe Radiofrequency Coil System (http://rf-research-facility.engineering.pitt.edu/). *Images were provided by Dr. Tamer Ibrahim (University of Pittsburgh).*

the degradation of capillaries into string vessels, which can no longer carry blood flow.[45] However, capillaries are beyond the resolution of human neuroimaging methods, and therefore the pathophysiology and risk and protective factors relating to capillary integrity in humans are unclear.

It is generally assumed that damage in the small vessels is confined to small arteries and capillary beds. The engagement of the venular side of the microcirculation has received less attention in relationship to cSVD. Venules are vulnerable to damage to a greater extent than arterioles because they have thinner, more permeable walls. Venules are also exposed to low pulsatile and slow flow (about half of what is observed in the arteriolar/ capillary side). The slow flow and low pressure in the venules, combined with their thin and permeable walls, facilitate adhesion of inflammatory cells to venular endothelium, and their extravasation on the parenchymal side. Higher pulsatile flow, as observed in hypertension, can damage the venular wall, causing collagenosis, and both venular dilation and lumen narrowing. These changes can cause micro-ischemic events, increase vascular resistance and disrupt upstream hemodynamics, with damage propagating upstream to capillaries and arterioles, leading to abnormalities in blood flow. Another implication of venular damage is venous insufficiency and consequent vessel leakage (i.e., vasogenic edema) compromising interstitial fluid circulation, and causing perivenular space abnormalities. Damage to perivenular spaces has been implicated in disruption of perivascular clearance of β-amyloid. According to this pathway, venular damage may be more closely related to stiffness than to vessel wall's structural damage. For a review of the physiology of age-related changes in the cerebral small venous circulation see Fulop et al.[56] A small 7 T neuroimaging study of older adults found a trend for an association of venous tortuosity with pulse pressure, although it did not find strong associations with other blood pressure parameters.[51] Tortuosity was not found to be significantly associated with age in this restricted age sample of community-dwelling older adults (age 70–89).[51] Relationships of age with venous metrics remain to be fully examined. To date, evidence for a role of venular damage in cSVD is primarily derived from animal studies[57,58] and postmortem studies in humans,[13,42,44,59] thus evidence for associations with risk factors is currently too sparse to reach conclusions.

2.5 Cerebral blood flow (CBF)

Cerebral blood flow and blood volume are substantially different between the gray and white matter of the brain, with 5- to 10-fold greater vasculature evident in the cortical and deep gray matter. Measures of perfusion have historically focused on cortical supply; white matter perfusion has received less attention. Methods of measuring blood flow include PET ($H_2^{15}O$), considered the gold standard, contrast CT perfusion, contrast MRI methods including dynamic susceptibility contrast, and non-contrast MRI arterial spin labeling (ASL).

Global and regional blood flow can be altered by systemic or acute disease states as well as by normal cerebral function through neurovascular coupling. Compensation for systemic disease such as carotid stenosis or heart failure may result in expanded blood volume and reduced blood flow to optimize oxygen extraction and delivery to tissue. Atherosclerosis, or hardening of the arteries, may impair this compensation and decrease neurovascular reserve. A loss of neurovascular reserve may render the supplied parenchyma vulnerable to both acute and chronic ischemic injury. Although ischemic injury to cortical gray matter may directly affect neurons, ischemic injury to white matter may primarily affect supportive glial cells such as oligodendrocytes, and may result in impaired axonal function and/or axonal loss. It is unclear how the physiology of neurovascular compensation differentially impacts cortical gray and white matter supply.

2.5.1 Cerebrovascular reactivity (CVR)

A well-known function of the microcirculation is to readily respond to changes in metabolic needs by modifying its lumen and ultimately the blood flow.[60] Such hemodynamic changes are critical to maintain the needed flow and perfusion even in the presence of varying metabolic needs (lower glucose or lower oxygen), or varying levels of upstream pressure (e.g., occlusions). Poorer vascular reactivity in response to varying demands is considered an abnormality of this regulatory system, and it is defined as endothelial dysfunction (ED). ED is considered to appear early in the course of cSVD, preceding radiologically overt signs of cSVD. Most evidence on ED as an early stage of cSVD is from studies of vasculature in non-CNS districts, primarily because of the challenges in measuring ED in the brain. Cerebral ED can be measured via transcranial doppler as changes in cerebral blood flow in response to hypercapnia.[61] These methods overlook small abnormalities and do not provide information on the spatial distribution of ED. More recently, a combination of hypercapnic stimulus and ASL has enabled more precise measurements of CVR. ED is considered an early marker of the vasculopathic process,[60] preceding structural vascular changes, and parenchymal damage. ED is detectable early in the course of cSVD.[10,61–63] Using CVR, cerebral ED has been observed in studies of older adults, as well as in patients with a history of vascular events.[64] Associations between cerebral ED and poor neurocognitive outcomes have been documented.[65,66]

2.6 Spatial distribution of cSVD across different brain areas and clinical manifestations

The neuroradiological abnormalities described in the previous section are not distributed uniformly through the brain, and instead follow a fronto-subcortical spatial distribution. This distribution has implications for understanding the pathogenesis of cSVD, as well as for its clinical manifestations. The regions at greatest risk of displaying neuroradiological and histopathological manifestations of cSVD are likely to be those where energy utilization is highest, collateral circulation is least, and in circumstances when opportunity to recover is limited (e.g., the watershed areas where collateral vascularization is intrinsically lower or areas where there is β-amyloid accumulation or other pre-existing pathologies). In these regions, microinfarcts would appear first, and neuronal and synaptic losses would be greatest. Disturbances of cerebral perfusion preferentially affect regions at the outskirts of individual nutritional areas, such as inter-territorial borders of non-anastomosing watershed arterial systems in the frontal lobes, the deep striato-capsular areas and deep WM, with subsequent ischemia, infarcts and lacunes in these areas. The particularly low cerebral perfusion pressure of the distal fields between the deep and the superficial arterial system and in the watershed areas make the frontal cortical regions, the basal ganglia GM and the deep WM particularly susceptible to repeated episodes of hypotension, and/or to other toxic/inflammatory insults. Longitudinal studies following changes in radiological markers of cSVD would be critical to clarify the spatio-temporal evolution of damage and provide further clues into the pathogenesis of cSVD.

The spatial distribution of cSVD can also explain its clinical manifestations. Occurrence of SBI and lacunes in the fronto-subcortical areas is of concern because of the role of these areas in executive function[10,67,68] and decision making,[69] which are both impaired in older age. Likewise, WMH in the periventricular areas, where long connecting cortico-subcortical tracts gather, can explain reductions in speed of processing. Executive dysfunction is a hallmark cognitive feature of aging, comprising deficits in information processing speed, psychomotor efficiency, attention, cognitive flexibility, and visuospatial perception. Dysfunction in these domains can impair planning, and problem-solving,[70,71] which are of concern in this population because they can affect independence with activities of daily living. Moreover, executive dysfunction is a well-established risk factor for dementia, depression, disability and mortality.[68,72–78]

In summary, the in vivo neuroradiological methods to characterize cSVD in older age have mostly relied on parenchymal damage, such as WMH, rather than on the pathological events directly affecting the small vessels. A reason for this limitation is that most in vivo brain imaging measures prevent differentiation between the cellular vs. non-cellular volumetric components. Postmortem approaches can distinguish the cellular components of cSVD, but they typically occur in cases of advanced pathology. The advancement of MRI technology holds promise to further our understanding of cSVD, because it captures a greater range of physical properties of the tissue; however, histopathological correlates of these most recent radiological methods have not yet been studied. Research endeavors to capture both in vivo characteristics of early cSVD and its histopathological correlates would involve population-level neuroimaging and brain donation efforts.

3. Causes and risk factors of cSVD: Cardiovascular, metabolic and inflammatory factors

We have recently reviewed causes as well as risk and protective factors of cSVD in detail, and therefore these topics are only briefly summarized here. See Jorgensen/Shaaban et al., for an in-depth review.[79]

3.1 Cardiovascular risk factors

Evidence for an association of risk factors with WMH, SBI/lacunes, and CMBs appears most consistent for hypertension and blood pressure parameters.[40,80–87] In contrast, descriptive studies examining the associations of antihypertensive medications with these outcomes yield mixed findings. Measures of large vessel disease (stiffness, distensibility, intima media thickness, plaque and heart disease) appeared more strongly related to SBI/lacunes than to WMH. This is consistent with a prior report[88] suggesting different risk profile patterns for WMH as compared to SBI/lacunes. A recent study of the Lothian Birth Cohort found that vascular risk factors contributed $<2\%$ to the prevalence of WMH,[89] underscoring the need to examine non-vascular pathways leading to cSVD. Associations of cSVD with inflammation, dyslipidemia, or obesity were not consistent across studies. The state of the evidence is strongest for a positive association of inflammatory factors with WMH. Both CRP and IL-6 have been found to be related to WMH,[90,91] but overall study results are mixed.[90–96] Overall, the evidence does not support a strong link between diabetes and cSVD.[83,85,86,94,97–104]

Diabetes may interact with other risk factors such as sleep duration[104] and SBP,[85] and these interactions should be evaluated in future studies. It has been suggested insulin resistance may be a more sensitive risk factor for SBI/lacunes than diabetes, as both cross-sectional[99] and longitudinal associations[100] have been reported. Among the newer measures of cSVD, Jennings et al. found reduced CBF was associated with increased cardiometabolic risk and worse intima media thickness.[105]

Cumulative exposure, longer duration of exposure, and variability over time in the levels of these factors appear to be important, but few studies examined these characteristics concurrently.

3.2 Sex- and race-related differences in cSVD

Studies have found both a higher prevalence of SBI/lacunes and faster progression of WMH in women compared to men.[97,101,102] Interestingly, sex-related effects on SBI appear stronger for older as compared to younger subgroups. This would appear to mirror the age*sex interaction found in stroke patients, whereby younger women are at lower risk of stroke than men, but older post-menopausal women are at greater risk of stroke than men.[106] The reasons for such sex-related differences and the interaction with age are not clear, but they could be due to differences in hormonal profiles and their changes over time, and/or in overall disease burden in vascular districts outside the brain. Of note, few studies were designed with the primary aim of evaluating cSVD by sex or examined sex by age interactions; hence they may have been under powered to examine sex-related differences. Thus, this possible interaction remains to be formally tested.

Similar to sex differences, the association of race with probability of having cSVD is not clear, with some studies showing strong and others weak/no associations. If a seemingly premature appearance of cSVD among blacks existed, it could be due to the higher burden of cardiovascular risk factors and diseases in blacks as compared to whites. However, substantial evidence in support of this theory is lacking, as few reports of racial differences in neuroradiological markers of cSVD also had comprehensive measures of health-related characteristics. Among the studies examining the role of health-related characteristics as explanatory factors of racial differences, some have found that adjustment for cardiovascular risk factors (smoking, hypertension, diabetes, hypercholesterolemia, and cardiac disease) attenuated racial differences, whereas others did not. As noted above, time of exposure and of measurement of the outcomes, as well as duration of exposure all need to be taken into account. We noticed that those studies that failed to find explanatory factors of racial differences in cSVD were conducted in older subgroups.[98,107] Perhaps the explanatory effect of cardiovascular risk factors is stronger earlier in age, as compared to later in life, when the process of cSVD accumulation has already been unfolding for a long time. Another possibility is that the older subgroups of blacks who are participating in MRI studies are exceptionally resilient. Blacks are known to have a higher risk of death and morbidity as compared to whites; hence, studies with older subgroups of blacks may be vulnerable to survival bias.

3.3 Potential convergence of amyloid-related pathology and cSVD

In the past few years, β-amyloid, the protein associated with Alzheimer's disease plaques, has emerged as a potential risk factor for cSVD. Lower levels of plasma β-amyloid have been reported to be associated with greater WMH progression[108] and lower CSF levels of β-amyloid have been associated cross-sectionally with WMH.[109,110] A recent systematic review found that 10/13 studies reported no association between β-amyloid measured via PET imaging and WMH, while 2 reported that increased β-amyloid was associated with increased WMH and 1/13 reported the opposite.[111]

A recent meta-analysis[112] indicates that overall *APOE4* carrier status is not associated with WMH cross-sectionally or longitudinally, but that these associations differ depending on the presence of other co-morbidities. One study included in this meta-analysis indicates the association is stronger in the presence of hypertension[81]; a related study published after the meta-analysis, also indicates the association between *APOE4* and WMH is stronger in the presence of total cholesterol <200 mg/dL.[113] We found only one study assessing the relationship of *APOE4* carrier status with incident SBI/lacunes,[97] with non-significant results. *APOE4* is associated with CAA, which is in turn associated with CMBs.[114] In our study of small veins,[51] we found a strong association between *APOE4* carrier status and higher tortuosity of small veins. We did not find studies reporting associations with other novel markers of cSVD.

3.4 Genetics of cSVD

Studies evaluating the genetic risk for cSVD found associations with genes important in glial and inflammatory pathways,[115] oxidative phosphorylation pathways,[116] vascular endothelial cell orientation in response to mechanical stress,[117] chromosome 6p25.3,[118] the gene encoding transforming growth factor-beta 1,[119] four single nucleotide polymorphisms in the *NOTCH3* gene (in hypertensives only),[120] and the *MTHFR* genotype.[121] The CHARGE consortium found several novel loci that were suggestive of associations with cSVD, but did not reach genome-wide significance. Specific fibrinogen haplotypes have been related to SBI/lacunes and periventricular WMH,[122] while CRP haplotypes were not associated with WMH or SBI/Lacunes.[123] Overall, there seems to be agreement that genes explain little of the variance in late-life cSVD.[115]

In summary, there is agreement that blood pressure parameters are related with neuroradiological manifestations of cSVD, both cross-sectionally and longitudinally. However, blood pressure and related vascular risk factors appear to explain only a very modest proportion of the variance of cSVD, thus raising the possibility that non-vascular pathways are at play. Moreover, relations between risk factors and neuroradiological manifestations of cSVD appear to follow distinct patterns, with interactions among them that need further study.

4. Discussion: Unanswered questions and research priorities

Based on the evidence reviewed, cSVD appears to be a constellation of heterogeneous risk factors, signs, and symptoms. How can we best proceed to advance our understanding of cSVD, identify its early stages, causes, and temporal evolution? We propose that future studies of cSVD should take a population neuroscience approach that integrates: (a) novel neuroscience methodologies, to characterize cerebrovascular pathology directly, rather than indirectly, via its late stage parenchymal manifestations; (b) population science, to capture the populations (and subgroups) at greater risk of developing cSVD; and (c) geroscience, to carefully characterize the molecular mechanisms underlying cSVD in translational models that bridge basic and clinical models. This population neuroscience approach could potentially actualize the goals of precision medicine, and its foundational principle that different interventions are needed for different people.

4.1 Directly characterize small vessel pathology and mixed-disease states

In vivo and postmortem findings of age-related small vessel disease cover mixed pathologies, including vascular, parenchymal and amyloid-related pathologies. Different combinations of such mixed pathologies may have similar clinical manifestations. To explain the variability of cSVD, it is of paramount importance to accurately identify mixed lesion states. An important question is whether neuroimaging can distinguish and quantify mixed pathogenic features with such accuracy that it could produce biomarkers of onset, evolution and response of cSVD to therapeutic interventions. As described in Section 2, neuroimaging techniques that appear most readily applicable to the study of small vessel pathology include time-of-flight and ultra-high field susceptibility-weighted neuroimaging to discern structural characteristics of small vessels. Accurate studies of cerebral blood flow, especially in the presence of chemical hypoxic stimuli are also very valuable to capture endothelial dysfunction. Further study of the characteristics of the small vessels can have implications for treatment. For example, studying the primary pathogenic events of vascular tortuosity can uncover novel targets for primary prevention, such as determinants of protocollagen and protoelastin generation and transit as extracellular matrix proteins to sites of molecular fractures of the intima and internal elastic lamina and efficiency of elastin repair.

4.2 Sampling strategies and temporality of study designs

Our review found important interactions with age, sex, and race on prevalence and severity of cSVD. We found that, while each factor alone is an important contributor to cSVD, the interactions among these factors are likely more complex than expected. Examining such interactions is important, as they suggest different mechanisms by which cSVD may occur; interactions among risk factors are helpful to inform subgroup-specific analyses (e.g., with high/low levels of a given risk factor) and inform precision medicine approaches—both for risk stratification and intervention. Thus, future studies should include ethnically diverse populations of older adults, adequately powered to test for sex-related difference.

Another aspect of sample selection that requires attention is the inclusion of individuals with multi-morbidities. Although cSVD becomes more common with older age, it is difficult to completely isolate the effects of chronological age on cSVD from that of age-related conditions, including multi-morbidities and dysregulation of biological factors, such as genetic and epigenetic factors.[124–126] Thus, accurate analysis of the age effects on cSVD must also have a comprehensive characterization of risk associated with these other factors, test for interactions of age by morbidities, and be powered to conduct age-stratified analyses. The effects of risk factors on cSVD could also be time-dependent, that is the effects could vary depending on the age at time of exposure (when), not only the duration (for how long). For example, exposure to high blood pressure during middle age has been repeatedly linked to increases in WMH[80,86,87] and risk of vascular dementia.[127] However, with advanced age, high blood pressure often has smaller or non-significant associations with WMH.[128] The reasons for such differences are not entirely clear, but if confirmed in other studies, these findings would have repercussions on the success of intervention strategies, as they may be more effective for younger but not older adults. There are numerous longitudinal descriptive studies examining exposure to blood pressure parameters and cSVD, but few had reliable measures of duration of exposure and/or longitudinal changes in both predictors and cSVD. Length of follow-up was mostly limited to 5 years, with few extending over a period of time long enough to accurately estimate age-related changes. The majority of studies examined populations age 70 and older, with fewer capturing earlier stages of the aging process.

4.2.1 Evolution of cSVD over time remains a key question

We found a wide variability in the rate of accrual of WMH per year, ranging from 5% to 50% per year with some studies also showing apparent reduction of lesions over time. Reasons for such discrepancies may include an over-reliance on cross-sectional designs of adults with a wide age range, rather than repeated measurement of within-person changes over time. Among the studies reporting within-person changes, most had short (<5 years) follow-up times, which may not be long enough to capture changes. We also noted smaller (~5%) accruals were found for the oldest old (>85 years) as compared to adults 65–75 years old, averaging at around 12% across studies; such discrepancies raise the possibility that the oldest adults more likely to return for repeated serial scans are also more likely to be healthier, hence biasing the results toward lower accrual.

Thus, there are many novel questions that cSVD researchers must answer. How rapidly does cSVD evolve? Will rapidity of decline vary for different age groups, sexes, ethnic subgroups, or patterns of comorbid conditions? Will these factors affect decline in different aspects of brain integrity (e.g., blood flow, atrophy) or behavior (e.g., attention, memory)? What factors could delay or accelerate the progression of brain abnormalities, and will they differ across age, racial, or patients' subgroups? If different underlying mechanisms are at play for different age groups, different prevention/intervention strategies may be needed in different age groups.

4.3 Integration of clinical, translational, and basic science methods

There is a need to understand the molecular mechanisms linking risk factors, small vessel pathology, and radiological markers. However, to date, animal models of cSVD have been reductionist, recapitulating cSVD via a single causative factor (see review[129]). Such models fail to capture the full spectrum of phenomena observed in human cSVD,[130] thus reducing the likelihood that identified interventions will effectively translate to humans. Conversely, work in humans has been hampered by limited molecular characterization and tests of mechanism. Neither current animal nor human work sufficiently and consistently addresses questions of critical periods vs. duration of exposure. Moving the field toward effective interventions will require (a) incorporation of more complex models of cSVD in basic science; (b) designing human observation and intervention studies to test mechanistic hypotheses; (c) incorporation of life course approaches in both animal and human research; and (d) consultation early and often in study conduct between basic scientists and human participants researchers to allow for both forward and back-translation.

4.4 What would an ideal research approach look like?

To address the above questions, there is a need to build a program of parallel human and basic science studies with both types of researchers involved in the design and conduct of each study. Specifically, in humans we recommend recruiting a racially diverse cohort of men and women in their early adulthood or perhaps in childhood, when exposure to early life determinants of mid- and late-life health first occurs and is modifiable. A life course approach would increase our understanding of the natural history of cSVD pathophysiology and allow parsing of time-dependent processes including critical periods and exposure duration. Participants would agree to donate their brains and be eligible and willing to undergo serial exams, until as close as possible to death, for multimodal neuroimaging scans, medical examination, interviews on lifestyle, blood draws, and behavioral assessment.

Primary measures of interest would include neuroimaging; molecular markers including multi-omics and genotype; lifestyle, behavior and function; and postmortem ex vivo neuroimaging and histopathology. Neuroimaging measures should be integrative and include those listed in Section 2. Integrating modalities could yield new insights, and such integration is beginning to occur (see Bangen et al.[131] for an example of such a study). Assessing the functional and clinical relevance of such changes would be achieved by including measures of cognitive and physical functioning and following participants for critical endpoints such as cognitive decline, dementia, mobility disability, and death. Sample sizes must be large enough to conduct well-powered sex- and race-stratified analysis of the data, with a focus on interactions between multi-morbidities. A parallel basic science approach would be critical to conduct mechanistic studies of cSVD. Such a study should follow animals across the lifespan and incorporate sufficient numbers of males and females to assess sex differences in cSVD manifestations. One example of an integrated approach would be to study the causal link between cSVD and neuroinflammation both in animal models and in humans using selected radioligands via PET and nanomolecular imaging approaches.[132,133]

Analysis of data should aim to identify patterns of risk factor profiles and explain the variance of the neuroradiological, clinical and postmortem manifestations of cSVD. Specifically, statistical analysis could discriminate among the multiple combinations of events that ultimately lead to cSVD. For example, pattern analyses correlating risk factors with characteristics of cSVD that account for different methods of cSVD ascertainment and population characteristics may uncover distinct patterns of associations or pathways. Tools to handle large amounts of data exist. For example, machine learning and decision tree algorithms can combine high dimensional neuroimaging data, demographics, and clinical variables obtained at repeated time points to produce predictive models of clinically relevant outcomes.[134] We propose that this approach is effective when integrated with an understanding of the underlying conditions and principles of rigorous study design.

Adding some of these exams to existing neuroimaging studies and/or harmonizing and combining human cohorts for analysis could be the most feasible approach. For example, the current human connectome study,[135] existing developmental studies and new NIH developmental imaging initiatives[136] could be feasible options and open new venues for collaborative efforts. The METACOHORTS Consortium has begun a cohort harmonization and pooling initiative, identifying >90 studies for possible inclusion and large total participant number (>600,000).[137] The UK Biobank would also be a promising study on which to build due to its prospective nature, large sample size ($N=500{,}000$), and wealth of variables measured. However, METACOHORTS and UK Biobank would need to be supplemented by other younger and racially/ethnically diverse samples due to their small proportions of non-white participants and those <60 years of age.

In addition to providing novel knowledge of the pathophysiology, descriptive mechanistic studies are needed to plan and design randomized controlled trials targeting cSVD. Such studies could have rippling effects on reducing distal consequences of cSVD, and could potentially prevent future strokes,[138] Alzheimer's disease and related dementias.[139]

Acknowledgments

Lon White, MD, MPH, Senior neuroepidemiologist, Pacific Health Research and Education Institute, Honolulu, Hawaii and Clayton A. Wiley, Director of the Division of Neuropathology, Department of Pathology, University of Pittsburgh.

References

1. Poels MM, Vernooij MW, Ikram MA, et al. Prevalence and risk factors of cerebral microbleeds: an update of the Rotterdam scan study. *Stroke*. 2010;41(10 suppl): S103–S106.
2. Smith EE, O'Donnell M, Dagenais G, et al. Early cerebral small vessel disease and brain volume, cognition, and gait. *Ann Neurol*. 2015;77(2):251–261.
3. Sachdev P, Chen X, Wen W. White matter hyperintensities in mid-adult life. *Curr Opin Psychiatry*. 2008;21(3):268–274.
4. Abbasi J. Nih consortium to study biomarkers for dementia. *JAMA*. 2017; 317(16):1614.
5. NINDS R-N. *Small Vessel Vascular Contributions to Cognitive Impairment and Dementia (VCID) Biomarkers Development Projects (UH2/UH3)*. 2016. https://grants.nih.gov/grants/guide/rfa-files/RFA-NS-16-020.html. Accessed March 22, 2019.
6. https://markvcid.partners.org/. n.d. Accessed March 22, 2019.
7. Smith EE, Biessels GJ, De Guio F, et al. Harmonizing brain magnetic resonance imaging methods for vascular contributions to neurodegeneration. *Alzheimers Dement (Amst)*. 2019;11:191–204.
8. Jennings JR, Mendelson DN, Muldoon MF, et al. Regional grey matter shrinks in hypertensive individuals despite successful lowering of blood pressure. *J Hum Hypertens*. 2012;26(5):295–305.
9. Wardlaw JM, Smith EE, Biessels GJ, et al. Neuroimaging standards for research into small vessel disease and its contribution to ageing and neurodegeneration. *Lancet Neurol*. 2013;12(8):822–838.
10. Wardlaw JM, Smith C, Dichgans M. Mechanisms of sporadic cerebral small vessel disease: insights from neuroimaging. *Lancet Neurol*. 2013;12(5):483–497.
11. Auriel E, Bornstein NM, Berenyi E, et al. Clinical, radiological and pathological correlates of leukoaraiosis. *Acta Neurol Scand*. 2011;123(1):41–47.

12. Awad IA, Johnson PC, Spetzler RF, Hodak JA. Incidental subcortical lesions identified on magnetic resonance imaging in the elderly. II. Postmortem pathological correlations. *Stroke*. 1986;17(6):1090–1097.
13. Black S, Gao F, Bilbao J. Understanding white matter disease: imaging-pathological correlations in vascular cognitive impairment. *Stroke*. 2009;40(3 suppl): S48–S52.
14. Braffman BH, Zimmerman RA, Trojanowski JQ, et al. pathologic correlation with gross and histopathology. 1. Lacunar infarction and Virchow-Robin spaces. *AJR Am J Roentgenol*. 1988;151(3):551–558.
15. Bronge L, Bogdanovic N, Wahlund LO. Postmortem MRI and histopathology of white matter changes in Alzheimer brains. A quantitative, comparative study. *Dement Geriatr Cogn Disord*. 2002;13(4):205–212.
16. Fazekas F, Kleinert R, Offenbacher H, et al. The morphologic correlate of incidental punctate white matter hyperintensities on MR images. *AJNR Am J Neuroradiol*. 1991;12(5):915–921.
17. Fernando MS, O'Brien JT, Perry RH, et al. Comparison of the pathology of cerebral white matter with post-mortem magnetic resonance imaging (MRI) in the elderly brain. *Neuropathol Appl Neurobiol*. 2004;30(4):385–395.
18. Grinberg LT, Amaro Junior E, da Silva AV, et al. Improved detection of incipient vascular changes by a biotechnological platform combining post mortem MRI in situ with neuropathology. *J Neurol Sci*. 2009;283(1–2):2–8.
19. Jouvent E, Poupon C, Gray F, et al. Intracortical infarcts in small vessel disease: a combined 7-T postmortem MRI and neuropathological case study in cerebral autosomal-dominant arteriopathy with subcortical infarcts and leukoencephalopathy. *Stroke*. 2011;42(3):e27–e30.
20. Marshall VG, Bradley Jr WG, Marshall CE, Bhoopat T, Rhodes RH. Deep white matter infarction: correlation of MR imaging and histopathologic findings. *Radiology*. 1988;167(2):517–522.
21. McAleese KE, Firbank M, Hunter D, et al. Magnetic resonance imaging of fixed post mortem brains reliably reflects subcortical vascular pathology of frontal, parietal and occipital white matter. *Neuropathol Appl Neurobiol*. 2013;39(5):485–497.
22. Munoz DG, Hastak SM, Harper B, Lee D, Hachinski VC. Pathologic correlates of increased signals of the centrum ovale on magnetic resonance imaging. *Arch Neurol*. 1993;50(5):492–497.
23. Polvikoski TM, van Straaten EC, Barkhof F, et al. Frontal lobe white matter hyperintensities and neurofibrillary pathology in the oldest old. *Neurology*. 2010;75(23): 2071–2078.
24. Revesz T, Hawkins CP, du Boulay EP, Barnard RO, McDonald WI. Pathological findings correlated with magnetic resonance imaging in subcortical arteriosclerotic encephalopathy (Binswanger's disease). *J Neurol Neurosurg Psychiatry*. 1989;52(12): 1337–1344.
25. Scarpelli M, Salvolini U, Diamanti L, Montironi R, Chiaromoni L, Maricotti M. MRI and pathological examination of post-mortem brains: the problem of white matter high signal areas. *Neuroradiology*. 1994;36(5):393–398.
26. Schmidt R, Schmidt H, Haybaeck J, et al. Heterogeneity in age-related white matter changes. *Acta Neuropathol*. 2011;122(2):171–185.
27. Shim YS, Yang DW, Roe CM, et al. Pathological correlates of white matter hyperintensities on magnetic resonance imaging. *Dement Geriatr Cogn Disord*. 2015;39(1–2): 92–104.
28. Simpson JE, Fernando MS, Clark L, et al. White matter lesions in an unselected cohort of the elderly: astrocytic, microglial and oligodendrocyte precursor cell responses. *Neuropathol Appl Neurobiol*. 2007;33(4):410–419.

29. Simpson JE, Hosny O, Wharton SB, et al. Microarray RNA expression analysis of cerebral white matter lesions reveals changes in multiple functional pathways. *Stroke.* 2009;40(2):369–375.
30. Simpson JE, Ince PG, Higham CE, et al. Microglial activation in white matter lesions and nonlesional white matter of ageing brains. *Neuropathol Appl Neurobiol.* 2007; 33(6):670–683.
31. Smith CD, Snowdon D, Markesbery WR. Periventricular white matter hyperintensities on MRI: correlation with neuropathologic findings. *J Neuroimaging.* 2000;10(1):13–16.
32. Young VG, Halliday GM, Kril JJ. Neuropathologic correlates of white matter hyperintensities. *Neurology.* 2008;71(11):804–811.
33. Janaway BM, Simpson JE, Hoggard N, et al. Brain haemosiderin in older people: pathological evidence for an ischaemic origin of magnetic resonance imaging (MRI) microbleeds. *Neuropathol Appl Neurobiol.* 2014;40(3):258–269.
34. Westover MB, Bianchi MT, Yang C, Schneider JA, Greenberg SM. Estimating cerebral microinfarct burden from autopsy samples. *Neurology.* 2013;80(15): 1365–1369.
35. Love S, Miners JS. Cerebrovascular disease in ageing and Alzheimer's disease. *Acta Neuropathol.* 2016;131(5):645–658.
36. Greenberg SM, Vernooij MW, Cordonnier C, et al. Cerebral microbleeds: a guide to detection and interpretation. *Lancet Neurol.* 2009;8(2):165–174.
37. Conijn MM, Geerlings MI, Biessels GJ, et al. Cerebral microbleeds on MR imaging: comparison between 1.5 and 7T. *AJNR Am J Neuroradiol.* 2011;32(6):1043–1049.
38. van Veluw SJ, Kuijf HJ, Charidimou A, et al. Reduced vascular amyloid burden at microhemorrhage sites in cerebral amyloid angiopathy. *Acta Neuropathol.* 2017;133(3):409–415.
39. Freeze WM, Bacskai BJ, Frosch MP, et al. Blood-brain barrier leakage and microvascular lesions in cerebral amyloid angiopathy. *Stroke.* 2019;50(2):328–335.
40. Ungvari Z, Tarantini S, Kirkpatrick AC, Csiszar A, Prodan CI. Cerebral microhemorrhages: mechanisms, consequences, and prevention. *Am J Physiol Heart Circ Physiol.* 2017;312(6):H1128–H1143.
41. Hooshmand I, Rosenbaum AE, Stein RL. Radiographic anatomy of normal cerebral deep medullary veins: criteria for distinguishing them from their abnormal counterparts. *Neuroradiology.* 1974;7(2):75–84.
42. Fazekas F, Kleinert R, Offenbacher H, et al. Pathologic correlates of incidental MRI white matter signal hyperintensities. *Neurology.* 1993;43(9):1683–1689.
43. Brown WR, Moody DM, Challa VR, Thore CR, Anstrom JA. Venous collagenosis and arteriolar tortuosity in leukoaraiosis. *J Neurol Sci.* 2002;203–204:159–163.
44. Moody DM, Brown WR, Challa VR, Anderson RL. Periventricular venous collagenosis: association with leukoaraiosis. *Radiology.* 1995;194(2):469–476.
45. Brown WR. A review of string vessels or collapsed, empty basement membrane tubes. *J Alzheimers Dis.* 2010;21(3):725–739.
46. Brown WR, Thore CR. Review: cerebral microvascular pathology in ageing and neurodegeneration. *Neuropathol Appl Neurobiol.* 2011;37(1):56–74.
47. Moody DM, Thore CR, Anstrom JA, Challa VR, Langefeld CD, Brown WR. Quantification of afferent vessels shows reduced brain vascular density in subjects with leukoaraiosis. *Radiology.* 2004;233(3):883–890.
48. Bullitt E, Rahman FN, Smith JK, et al. The effect of exercise on the cerebral vasculature of healthy aged subjects as visualized by MR angiography. *AJNR Am J Neuroradiol.* 2009;30(10):1857–1863.
49. Kuijf HJ, Bouvy WH, Zwanenburg JJ, et al. Quantification of deep medullary veins at 7 T brain MRI. *Eur Radiol.* 2016;26(10):3412–3418.

50. Bouvy WH, Kuijf HJ, Zwanenburg JJ, et al. Abnormalities of cerebral deep medullary veins on 7 Tesla MRI in amnestic mild cognitive impairment and early Alzheimer's disease: a pilot study. *J Alzheimers Dis*. 2017;57(3):705–710.
51. Shaaban CE, Aizenstein HJ, Jorgensen DR, et al. In vivo imaging of venous side cerebral small-vessel disease in older adults: an MRI method at 7T. *AJNR Am J Neuroradiol*. 2017;38(10):1923–1928.
52. Seo SW, Kang CK, Kim SH, et al. Measurements of lenticulostriate arteries using 7T MRI: new imaging markers for subcortical vascular dementia. *J Neurol Sci*. 2012;322(1–2):200–205.
53. Kang CK, Park CA, Lee H, et al. Hypertension correlates with lenticulostriate arteries visualized by 7T magnetic resonance angiography. *Hypertension*. 2009;54(5): 1050–1056.
54. Bullitt E, Zeng D, Mortamet B, et al. The effects of healthy aging on intracerebral blood vessels visualized by magnetic resonance angiography. *Neurobiol Aging*. 2010;31(2): 290–300.
55. Cruz Hernandez JC, Bracko O, Kersbergen CJ, et al. Neutrophil adhesion in brain capillaries reduces cortical blood flow and impairs memory function in Alzheimer's disease mouse models. *Nat Neurosci*. 2019;22(3):413–420.
56. Fulop G, Tarantini S, Yabluchanskiy A, et al. Role of age-related alterations of the cerebral venous circulation in the pathogenesis of vascular cognitive impairment. *Am J Physiol Heart Circ Physiol*. 2019;316(5):H1124–H1140.
57. Lai AY, Dorr A, Thomason LA, et al. Venular degeneration leads to vascular dysfunction in a transgenic model of Alzheimer's disease. *Brain*. 2015;138(pt 4): 1046–1058.
58. Nguyen J, Nishimura N, Fetcho RN, Iadecola C, Schaffer CB. Occlusion of cortical ascending venules causes blood flow decreases, reversals in flow direction, and vessel dilation in upstream capillaries. *J Cereb Blood Flow Metab*. 2011;31(11):2243–2254.
59. Pettersen JA, Keith J, Gao F, Spence JD, Black SE. CADASIL accelerated by acute hypotension: arterial and venous contribution to leukoaraiosis. *Neurology*. 2017;88(11):1077–1080.
60. Deanfield JE, Halcox JP, Rabelink TJ. Endothelial function and dysfunction: testing and clinical relevance. *Circulation*. 2007;115(10):1285–1295.
61. Stevenson SF, Doubal FN, Shuler K, Wardlaw JM. A systematic review of dynamic cerebral and peripheral endothelial function in lacunar stroke versus controls. *Stroke*. 2010;41(6):e434–e442.
62. Gouw AA, Seewann A, van der Flier WM, et al. Heterogeneity of small vessel disease: a systematic review of MRI and histopathology correlations. *J Neurol Neurosurg Psychiatry*. 2011;82(2):126–135.
63. Hainsworth AH, Oommen AT, Bridges LR. Endothelial cells and human cerebral small vessel disease. *Brain Pathol*. 2015;25(1):44–50.
64. Blair GW, Doubal FN, Thrippleton MJ, Marshall I, Wardlaw JM. Magnetic resonance imaging for assessment of cerebrovascular reactivity in cerebral small vessel disease: a systematic review. *J Cereb Blood Flow Metab*. 2016;36(5):833–841.
65. Nelson AR, Sweeney MD, Sagare AP, Zlokovic BV. Neurovascular dysfunction and neurodegeneration in dementia and Alzheimer's disease. *Biochim Biophys Acta*. 2016;1862(5):887–900.
66. De Silva TM, Faraci FM. Microvascular dysfunction and cognitive impairment. *Cell Mol Neurobiol*. 2016;36(2):241–258.
67. de Groot JC, de Leeuw FE, Oudkerk M, et al. Cerebral white matter lesions and cognitive function: the rotterdam scan study. *Ann Neurol*. 2000;47(2):145–151.
68. Kuo HK, Lipsitz LA. Cerebral white matter changes and geriatric syndromes: is there a link? *J Gerontol A Biol Sci Med Sci*. 2004;59(8):818–826.

69. Mogenson GJ, Jones DL, Yim CY. From motivation to action: functional interface between the limbic system and the motor system. *Prog Neurobiol*. 1980;14(2–3):69–97.
70. Glisky E. Changes in cognitive function in human aging. In: Riddle DR, ed. *Brain Aging: Models, Methods, and Mechanisms*. CRC Press/Taylor & Francis Boca Raton (FL); 2007 Chapter 1(Available from: https://www.ncbi.nlm.nih.gov/books/NBK3885/?report=classic).
71. Salthouse TA. The processing-speed theory of adult age differences in cognition. *Psychol Rev*. 1996;103(3):403–428.
72. Kuller LH, Arnold AM, Longstreth Jr WT, et al. White matter grade and ventricular volume on brain MRI as markers of longevity in the cardiovascular health study. *Neurobiol Aging*. 2007;28(9):1307–1315.
73. Rosano C, Newman AB. Cardiovascular disease and risk of Alzheimer's disease. *Neurol Res*. 2006;28(6):612–620.
74. Rosano C, Kuller LH, Chung H, Arnold AM, Longstreth Jr WT, Newman AB. Subclinical brain magnetic resonance imaging abnormalities predict physical functional decline in high-functioning older adults. *J Am Geriatr Soc*. 2005;53(4):649–654.
75. Debette S, Markus HS. The clinical importance of white matter hyperintensities on brain magnetic resonance imaging: systematic review and meta-analysis. *BMJ*. 2010;341:c3666.
76. Backman L, Jones S, Berger AK, Laukka EJ, Small BJ. Cognitive impairment in preclinical Alzheimer's disease: a meta-analysis. *Neuropsychology*. 2005;19(4):520–531.
77. Saxton J, Lopez OL, Ratcliff G, et al. Preclinical Alzheimer disease: neuropsychological test performance 1.5 to 8 years prior to onset. *Neurology*. 2004;63(12):2341–2347.
78. Rapp MA, Reischies FM. Attention and executive control predict Alzheimer disease in late life: results from the Berlin Aging Study (BASE). *Am J Geriatr Psychiatry*. 2005;13(2):134–141.
79. Jorgensen DR, Shaaban CE, Wiley CA, Gianaros PJ, Mettenburg J, Rosano C. A population neuroscience approach to the study of cerebral small vessel disease in mid- and late-life: an Invited Review. *Am J Physiol Heart Circ Physiol*. 2018;314(6): H1117–H1136.
80. de Leeuw FE, de Groot JC, Oudkerk M, et al. Hypertension and cerebral white matter lesions in a prospective cohort study. *Brain*. 2002;125(pt 4):765–772.
81. de Leeuw FE, Richard F, de Groot JC, et al. Interaction between hypertension, apoE, and cerebral white matter lesions. *Stroke*. 2004;35(5):1057–1060.
82. Liao D, Cooper L, Cai J, et al. Presence and severity of cerebral white matter lesions and hypertension, its treatment, and its control. The ARIC Study. Atherosclerosis risk in communities study. *Stroke*. 1996;27(12):2262–2270.
83. Knopman DS, Mosley TH, Catellier DJ, Sharrett AR. Cardiovascular risk factors and cerebral atrophy in a middle-aged cohort. *Neurology*. 2005;65(6):876–881.
84. Prabhakaran S, Wright CB, Yoshita M, et al. Prevalence and determinants of subclinical brain infarction: the Northern Manhattan Study. *Neurology*. 2008;70(6):425–430.
85. Knopman DS, Penman A, Catellier D, et al. Vascular risk factors and longitudinal changes on brain MRI The ARIC study. *Neurology*. 2011;76(22):1879–1885.
86. Debette S, Seshadri S, Beiser A, et al. Midlife vascular risk factor exposure accelerates structural brain aging and cognitive decline. *Neurology*. 2011;77(5):461–468.
87. Dufouil C, de Kersaint Gilly A, Besancon V, et al. Longitudinal study of blood pressure and white matter hyperintensities The EVA MRI Cohort. *Neurology*. 2001;56(7):921–926.
88. Khan U, Porteous L, Hassan A, Markus HS. Risk factor profile of cerebral small vessel disease and its subtypes. *J Neurol Neurosurg Psychiatry*. 2007;78(7):702–706.
89. Wardlaw JM, Allerhand M, Doubal FN, et al. Vascular risk factors, large-artery atheroma, and brain white matter hyperintensities. *Neurology*. 2014;82(15): 1331–1338.

90. Satizabal C, Zhu Y, Mazoyer B, Dufouil C, Tzourio C. Circulating IL-6 and CRP are associated with MRI findings in the elderly The 3C-Dijon Study. *Neurology*. 2012;78(10):720–727.
91. Nadkarni NK, Boudreau RM, Studenski SA, et al. Slow gait, white matter characteristics, and prior 10-year interleukin-6 levels in older adults. *Neurology*. 2016;87(19): 1993–1999.
92. Markus HS, Hunt B, Palmer K, Enzinger C, Schmidt H, Schmidt R. Markers of endothelial and hemostatic activation and progression of cerebral white matter hyperintensities. *Stroke*. 2005;36(7):1410–1414.
93. Van Dijk E, Prins N, Vermeer S, et al. C-reactive protein and cerebral small-vessel disease The Rotterdam Scan Study. *Circulation*. 2005;112(6):900–905.
94. Schmidt R, Schmidt H, Fazekas F, et al. Apolipoprotein E polymorphism and silent microangiopathy-related cerebral damage. Results of the Austrian Stroke Prevention Study. *Stroke*. 1997;28(5):951–956.
95. Schmidt R, Schmidt H, Pichler M, et al. C-reactive protein, carotid atherosclerosis, and cerebral small-vessel disease: results of the Austrian Stroke Prevention Study. *Stroke*. 2006;37(12):2910–2916.
96. Aribisala BS, Wiseman S, Morris Z, et al. Circulating inflammatory markers are associated with magnetic resonance imaging-visible perivascular spaces but not directly with white matter hyperintensities. *Stroke*. 2014;45(2):605–607.
97. van Dijk EJ, Prins ND, Vrooman HA, Hofman A, Koudstaal PJ, Breteler MM. Progression of cerebral small vessel disease in relation to risk factors and cognitive consequences: Rotterdam Scan study. *Stroke*. 2008;39(10):2712–2719.
98. Nyquist PA, Bilgel MS, Gottesman R, et al. Extreme deep white matter hyperintensity volumes are associated with African American race. *Cerebrovasc Dis*. 2014;37(4): 244–250.
99. Lee JE, Shin DW, Yun JM, et al. Insulin resistance is a risk factor for silent lacunar infarction. *Stroke*. 2016;47(12):2938–2944.
100. Dearborn JL, Schneider AL, Sharrett AR, et al. Obesity, insulin resistance, and incident small vessel disease on magnetic resonance imaging: atherosclerosis risk in communities study. *Stroke*. 2015;46(11):3131–3136.
101. Longstreth Jr WT, Bernick C, Manolio TA, Bryan N, Jungreis CA, Price TR. Lacunar infarcts defined by magnetic resonance imaging of 3660 elderly people: the Cardiovascular Health Study. *Arch Neurol*. 1998;55(9):1217–1225.
102. Vermeer SE, Koudstaal PJ, Oudkerk M, Hofman A, Breteler MM. Prevalence and risk factors of silent brain infarcts in the population-based Rotterdam Scan Study. *Stroke*. 2002;33(1):21–25.
103. Howard G, Wagenknecht LE, Cai J, Cooper L, Kraut MA, Toole JF. Cigarette smoking and other risk factors for silent cerebral infarction in the general population. *Stroke*. 1998;29(5):913–917.
104. Ramos AR, Dong C, Rundek T, et al. Sleep duration is associated with white matter hyperintensity volume in older adults: the Northern Manhattan Study. *J Sleep Res*. 2014;23(5):524–530.
105. Jennings JR, Heim AF, Kuan DC-H, Gianaros PJ, Muldoon MF, Manuck SB. Use of total cerebral blood flow as an imaging biomarker of known cardiovascular risks. *Stroke*. 2013;44(9):2480–2485.
106. Lisabeth L, Bushnell C. Stroke risk in women: the role of menopause and hormone therapy. *Lancet Neurol*. 2012;11(1):82–91.
107. Gottesman RF, Coresh J, Catellier DJ, et al. Blood pressure and white-matter disease progression in a biethnic cohort. *Stroke*. 2010;41(1):3–8.
108. Kaffashian S, Tzourio C, Soumare A, et al. Plasma beta-amyloid and MRI markers of cerebral small vessel disease: Three-City Dijon study. *Neurology*. 2014;83(22): 2038–2045.

109. Stenset V, Johnsen L, Kocot D, et al. Associations between white matter lesions, cerebrovascular risk factors, and low CSF Abeta42. *Neurology*. 2006;67(5):830–833.
110. Wright CB, Paik MC, Brown TR, et al. Total homocysteine is associated with white matter hyperintensity volume: the Northern Manhattan Study. *Stroke*. 2005;36(6): 1207–1211.
111. Roseborough A, Ramirez J, Black SE, Edwards JD. Associations between amyloid beta and white matter hyperintensities: a systematic review. *Alzheimers Dement*. 2017;13(10):1154–1167.
112. Schilling S, DeStefano AL, Sachdev PS, et al. APOE genotype and MRI markers of cerebrovascular disease: systematic review and meta-analysis. *Neurology*. 2013;81(3): 292–300.
113. Willey JZ, Gardener H, Moon YP, et al. Lipid profile components and subclinical cerebrovascular disease in the northern Manhattan study. *Cerebrovasc Dis*. 2014;37(6):423–430.
114. Auriel E, Greenberg SM. The pathophysiology and clinical presentation of cerebral amyloid angiopathy. *Curr Atheroscler Rep*. 2012;14(4):343–350.
115. Verhaaren BF, Debette S, Bis JC, et al. Multiethnic genome-wide association study of cerebral white matter hyperintensities on MRI. *Circ Cardiovasc Genet*. 2015; 8(2):398–409.
116. Traylor M, Anderson CD, Hurford R, Bevan S, Markus HS. Oxidative phosphorylation and lacunar stroke: genome-wide enrichment analysis of common variants. *Neurology*. 2016;86(2):141–145.
117. Traylor M, Tozer DJ, Croall ID, et al. Genetic variation in PLEKHG1 is associated with white matter hyperintensities (n = 11,226). *Neurology*. 2019;92(8):e749–e757.
118. Neurology Working Group of the Cohorts for Heart and Aging Research in Genomic Epidemiology (CHARGE) Consortium, the Stroke Genetics Network (SiGN), and the International Stroke Genetics Consortium (ISGC). Identification of additional risk loci for stroke and small vessel disease: a meta-analysis of genome-wide association studies. *Lancet Neurol*. 2016;15(7):695–707.
119. Tao HM, Chen GZ, Lu XD, Hu XG, Chen GP, Shao B. TGF-beta 1 codon 10 polymorphism is associated with cerebral SVD. *Can J Neurol Sci*. 2011;38(6):869–873.
120. Schmidt H, Zeginigg M, Wiltgen M, et al. Genetic variants of the NOTCH3 gene in the elderly and magnetic resonance imaging correlates of age-related cerebral small vessel disease. *Brain*. 2011;134(11):3384–3397.
121. Kohara K, Fujisawa M, Ando F, et al. MTHFR gene polymorphism as a risk factor for silent brain infarcts and white matter lesions in the japanese general population The NILS-LSA Study. *Stroke*. 2003;34(5):1130–1135.
122. van Oijen M, Cheung EY, Geluk CE, et al. Haplotypes of the fibrinogen gene and cerebral small vessel disease: the Rotterdam scan study. *J Neurol Neurosurg Psychiatry*. 2008;79(7):799–803.
123. Reitz C, Berger K, de Maat MP, et al. CRP gene haplotypes, serum CRP, and cerebral small-vessel disease: the Rotterdam Scan Study and the MEMO Study. *Stroke*. 2007;38(8):2356–2359.
124. Austad S. *The Geroscience Hypothesis: Is it Possible to Change the Rate of Aging?* Springer International Publishing; 2016.
125. Kennedy BK, Berger SL, Brunet A, et al. Geroscience: linking aging to chronic disease. *Cell*. 2014;159(4):709–713.
126. Lopez-Otin C, Blasco MA, Partridge L, Serrano M, Kroemer G. The hallmarks of aging. *Cell*. 2013;153(6):1194–1217.
127. Yamada M, Kasagi F, Sasaki H, Masunari N, Mimori Y, Suzuki G. Association between dementia and midlife risk factors: the radiation effects research foundation adult health study. *J Am Geriatr Soc*. 2003;51(3):410–414.

128. Dickie DA, Ritchie SJ, Cox SR, et al. Vascular risk factors and progression of white matter hyperintensities in the Lothian Birth Cohort 1936. *Neurobiol Aging.* 2016;42:116–123.
129. Hainsworth AH, Allan SM, Boltze J, et al. Translational models for vascular cognitive impairment: a review including larger species. *BMC Med.* 2017;15(1):16.
130. Hainsworth AH, Markus HS. Do in vivo experimental models reflect human cerebral small vessel disease? A systematic review. *J Cereb Blood Flow Metab.* 2008;28(12): 1877–1891.
131. Bangen KJ, Clark AL, Edmonds EC, et al. Cerebral blood flow and amyloid-β interact to affect memory performance in cognitively normal older adults. *Front Aging Neurosci.* 2017;9:181.
132. Jacobs AH, Tavitian B. Noninvasive molecular imaging of neuroinflammation. *J Cereb Blood Flow Metab.* 2012;32(7):1393–1415.
133. Cerami C, Iaccarino L, Perani D. Molecular imaging of neuroinflammation in neurodegenerative dementias: the role of in vivo PET imaging. *Int J Mol Sci.* 2017;18(5):993.
134. Cole JH, Ritchie SJ, Bastin ME, et al. Brain age predicts mortality. *Mol Psychiatry.* 2017;23(5):1385–1392.
135. http://www.humanconnectomeproject.org/.
136. https://abcdstudy.org/.
137. METACOHORTS Consortium. METACOHORTS for the study of vascular disease and its contribution to cognitive decline and neurodegeneration: an initiative of the Joint Programme for Neurodegenerative Disease Research. *Alzheimers Dement.* 2016;12(12):1235–1249.
138. Smith EE, Saposnik G, Biessels GJ, et al. Prevention of stroke in patients with silent cerebrovascular disease: a scientific statement for healthcare professionals from the American Heart Association/American Stroke Association. *Stroke.* 2017;48(2): e44–e71.
139. Corriveau RA, Koroshetz WJ, Gladman JT, et al. Alzheimer's disease-related dementias summit 2016: national research priorities. *Neurology.* 2017;89(23):2381–2391.

Chapter 5 HIV

5a

Is treated HIV infection still toxic to the brain?

Erin O'Connor*, Thomas Zeffiro*
University of Maryland School of Medicine, Baltimore, MD, United States
*Corresponding authors: e-mail address: erin.oconnor@umm.edu; zeffiro@neurometrika.org

Contents

Abstract

Clinically apparent HIV infection, accompanied by CNS opportunistic infections and HIV encephalopathy, was often associated with profound structural and functional brain effects prior to the introduction of anti-retroviral therapy (ART). With treatment, HIV structural and functional brain effects are smaller and have not been as easily detected. With near complete elimination of CNS opportunistic infections, the HIV neuroimaging research community now grapples with the problem of detecting subtler structural and functional changes against a background of persisting confounds, such as comorbidities and clinical features common in the HIV infected population. This situation also raises the question of whether imaging measure changes that are reported as HIV brain effects are purely related to viral infection, rather than originating from confounding effects that might include age, substance use, hepatitis C coinfection,

Progress in Molecular Biology and Translational Science, Volume 165
ISSN 1877-1173
https://doi.org/10.1016/bs.pmbts.2019.04.001

cerebrovascular risk factors, ART, premorbid cognitive skills and illness duration. In addition to cohort characteristics, variation in image acquisition and analysis techniques may also contribute to study outcome heterogeneity. We review the potential effects of these confounds on detection of HIV infection effects and discuss strategies to avoid or mitigate the effects of these confounds. We then present a systematic approach to measurement, design and analysis in HIV neuroimaging studies, combining both experimental and statistical control techniques to determine if HIV infection effects persist, fluctuate or worsen in groups achieving viral suppression from ART.

1. Introduction

Prior to the introduction of effective HIV treatments, neuroimaging revealed extensive structural brain changes, including the characteristic appearance of opportunistic infections and HIV encephalitis[1] and volumetric loss accompanying HIV encephalopathy.[2] With the widespread availability of anti-retroviral therapy and evolving imaging methods that allow more sensitive characterization of subtle structural and functional changes, HIV neuroimaging now allows characterization of macroarchitectural, microstructural and physiological brain properties in the setting of treated chronic infection.

Nevertheless, studying chronic HIV infection effects on brain structure and function is made difficult by a number of associated measurement, design and analysis complexities. Measurement problems include large variations in regional brain volume and functional connectivity estimates related to differing image acquisition systems, imaging protocols and processing techniques. Design issues stem from the fact that nearly all studies of HIV infection brain effects are observational, necessitating construction of appropriate comparison groups matched retrospectively on participant demographic and clinical characteristics. In addition, HIV infection is often accompanied by a range of comorbid conditions that affect brain structure and function, necessitating exclusion of participants in numbers that may reduce power to detect serostatus effects or limit study generalizability. Finally, collinearity of covariates introduced to isolate nuisance effects may result in statistical modeling concerns, including instability of resulting parameter estimates.

Of these concerns, the association of HIV infection with a range of other medical conditions affecting the brain may be of greatest potential interest. Since the beginning of the AIDS epidemic, changes in brain structure and function in HIV infected individuals have been often attributed to viral effects. Prior to the development of effective treatments, reported serostatus brain effects primarily included opportunistic infections and HIV encephalopathy.[1] The widespread implementation of combination anti-retroviral therapy was followed by a marked decrease in the incidence of CNS opportunistic infections.[3,4] Despite the immune reconstitution and peripheral viral suppression that occurs in HIV infected individuals receiving ART, viral replication may still occur in CNS reservoirs.[5] In addition, several ART-era studies have demonstrated persisting structural associations with cognitive deficits during treatment.[6–9] For these, and possibly other, reasons, the mechanisms responsible for neurological impairment in treated HIV infection are still debated and identification of patients at risk for progressive cognitive impairment remains a diagnostic dilemma.

Numerous studies have focused on HIV associated effects on brain structure and function, with comprehensive reviews providing in depth summaries.[10–17] While macroarchitectural brain MRI changes related to HIV infection in the ART era have grown smaller over time, HIV associated brain structural differences can still be detected in chronic, treated HIV infection using voxel-wise volumetric estimates,[18–20] cortical surface shape analysis,[21–24] and cortical thickness estimation.[18,25] In a meta-analysis of HIV infection effects on total brain volume (TBV), gray matter volume (GMV), white matter volume (WMV) and cerebrospinal fluid volume (CSFV), HIV infection was associated with lower GMV, lower WMV and higher CSFV in studies carried out between 1995 and 2016.[26] An update of these meta-analyses, also including studies from 2017 and 2018, found similar results (Fig. 1).

It is of particular interest that the serostatus effects in all tissue compartments diminished with time. While the source of these temporal trends is not certain, it may be related to changing clinical characteristics of participants in HIV neuroimaging studies, suggesting that widespread use of ART may have decreased macroarchitectural structural changes.[26]

Notably, many HIV neuroimaging studies have been confounded by multiple comorbidities associated with HIV infection, including substance use, hepatitis C co-infection, cerebrovascular risk factors such as diabetes and hypertension, and heterogeneity in ART treatment. In addition, premorbid cognitive abilities may interact with the effects of HIV infection. The Women's Interagency HIV Study reported that reading level, age, years of education, and race were more strongly associated with cognitive deficits than serostatus.[27] These findings raise questions regarding whether some of

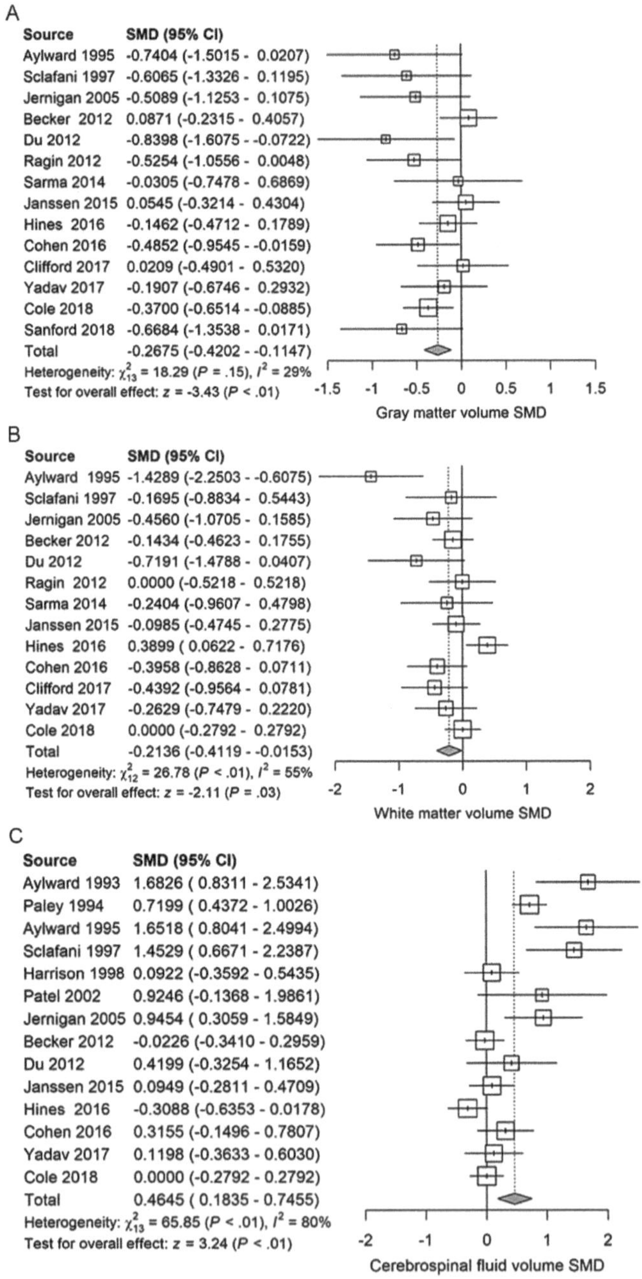

Fig. 1 Meta-analysis of brain volumetric effects of HIV infection showed decreasing (A) GMV, (B) WMV, and increasing (C) CSFV over time with decreasing effect sizes over time. Details can be found in O'Connor et al.[26] Additional studies from 2017 to 2018 have been added.

the reported associations of HIV infection with changes in brain structure and function are more strongly related to comorbid and socioeconomic confounds rather than HIV infection. Accordingly, this chapter will discuss the effects of the major HIV-associated demographic characteristics and comorbidities on brain structure and function, in the context of different approaches taken to control or mitigate these confounds.

2. Confounds in HIV neuroimaging

2.1 Aging effects on brain structure and function

Although aging has effects on brain structure and function that can be observed following HIV infection, there is ongoing debate as to whether HIV infection adds to typical aging effects to cause premature aging or interacts to cause accelerated aging. Multiple cross-sectional studies have shown independent effects of age and serostatus on subcortical gray matter volume.[9,20]

Longitudinal studies that agree with this result have demonstrated HIV-associated, age-related reductions in both GMV and cortical thickness, without showing differences in their rate of change related to serostatus.[18,28] Nevertheless, there is contradictory evidence that supports an HIV by age interaction in brain macroarchitecture using both cross-sectional[29–31] and longitudinal designs.[7,23,32] Whether HIV effects on brain structure and function are halted when effective ART is started, interact with typical age-related changes, or are related to both these mechanisms, has implications for how HIV infection may result in progressive cognitive dysfunction (Fig. 2).

Using MRI to assess WM microstructure, cross-sectional diffusion tensor imaging (DTI) studies in HIV infection generally find fractional anisotropy (FA) decreases and mean diffusivity (MD) increases in WM tracts, with conflicting evidence regarding interactions between age and serostatus effects. Several studies have found independent effects of age and serostatus on DTI measures,[33] even in populations over the age of 60.[34] Longitudinal studies confirming this result show no changes in DTI metrics over approximate 2 year intervals in virally suppressed participants.[35,36] Conflicting evidence from a cross-sectional DTI study reporting an HIV by age interaction in diffusion metrics[37] was confounded by significantly older age in seropositive participants diagnosed prior to the ART era, compared to those diagnosed afterward, who may have had longer durations of viremia.[37] Yet, an HIV by age interaction was also found in subset analysis of aviremic and

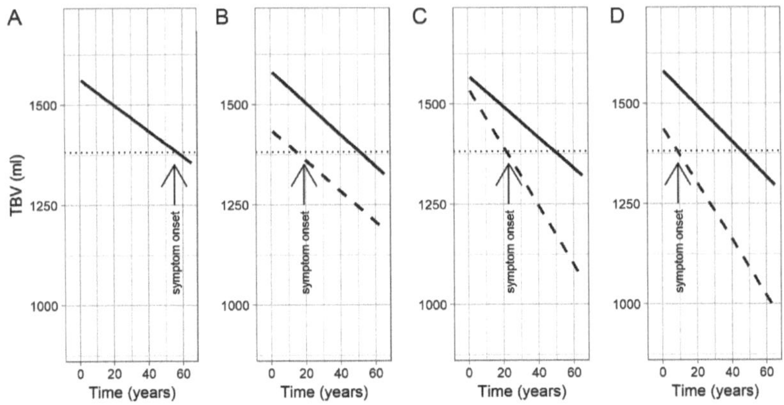

Fig. 2 Simulated demonstration of how additive and interactive effects of age and serostatus on total brain volume (TBV) might interact in the years following HIV infection. In all plots, infection occurs in year 1. (A) In the absence of HIV infection, age is associated with a linear decline in gray matter volume (solid line). (B) HIV infection might have an additive effect with age (dashed line), representing premature aging or legacy effects. (C) HIV infection interacts with age effects (dashed line), representing accelerated aging. (D) HIV infection is associated with both premature and accelerated aging effects (dashed line). Horizontal dotted lines in each condition show the variation in time after initial infection that threshold cognitive or other behavioral effects might become manifest.

HCV negative individuals.[29] In addition, a longitudinal DTI study that found accelerated aging was confounded by heterogeneity in viral suppression.[38]

Functional MRI studies (fMRI) incorporating visual tasks found independent effects of age and serostatus on BOLD-contrast signal changes.[39] Resting state fMRI (rs-fMRI) studies confirm these findings using cross-sectional designs in a mixed treated and untreated cohort of HIV infected participants,[40] and with a longitudinal design in an aviremic cohort studied over 2 years.[28]

Using positron emission tomography[41] and arterial spin labeling[5] in asymptomatic HIV infected samples with undetectable viral load, small reductions in brain glucose uptake and cerebral blood flow were reported in the anterior cingulate cortex in association with HIV infection, but no age by serostatus interaction was observed.[42]

The debate regarding premature or accelerated brain age in HIV is particularly germane to studies of treated, chronically HIV infected individuals who now approach normal life expectancies. Many of them exhibit cognitive and motor fluctuations or decline earlier than would be expected based on their chronological age. A considerable amount of the HIV-neuroimaging literature is consistent with an account stating that HIV infection results in premature brain age, particularly affecting the basal ganglia, and that the structural effects of infection cease with ART initiation.

2.2 Substance use effects on brain structure and function

Much of the HIV neuroimaging literature is confounded by participants' substance use, known to have effects on both brain structure and function. There are a broad range of macrostructural effects reported in association with substance use in HIV negative participants, including opiate, stimulant, cannabis, hallucinogen and alcohol associated GMV reductions.[43–50] Reported WM substance use effects in seronegative populations include lower WMV,[51] as well as altered WM microstructure in normal appearing cerebral WM.[52,53] Functional connectivity changes related to substance use often include altered connectivity in salience network regions such as the amygdala, insula, prefrontal cortex and anterior cingulate cortex.[54–57]

Reported structural and metabolic effects associated with substance use complicate efforts to isolate serostatus effects in HIV infected populations in whom substance use is prevalent.[58–61] In a study reporting reduced FA and increased MD in HIV infected psychostimulant users, the control group consisted of HIV negative participants who did not use psychostimulants, and both groups used other drug classes.[62] Such study designs permit neither clean isolation of serostatus effects nor investigation of potential serostatus and substance use interactions. One possible approach to this dilemma is experimental control using factorial design with cohorts grouped by serostatus and/or specific substance dependence in the absence of other substance use confounds. Studies that exemplify this approach demonstrate synergistic effects of serostatus and alcoholism on white matter microstructure[63] and macroarchitecture, particularly in those with AIDS defining events.[64] A similar design, used to study effects of chronic HIV infection and tobacco smoking, found an HIV by smoking interaction on brain diffusivity, with a much larger effect size related to tobacco smoking.[65]

Volumetric measures of regional GM changes associated with substance use have also been explored using MRI. Global and regional GMV changes in HIV participants are associated smaller mean cortical volumes in rostral middle frontal gyir and inferior and superior parietal lobes, and cortical thinning observed in the posterior cingulate gyrus related to methamphetamine dependence. Nevertheless, no modulating effects of methamphetamine dependence were seen on HIV associated brain structural reductions.[66]

With factorial designs, moderating effects of serostatus on cocaine dependence were found in prefrontal functional connectivity, potentially affecting decision making, and in cerebellar connectivity.[67]

Substance use of varying degree is prevalent in HIV infected populations and can affect both brain structure and function. Various strategies to contend with this challenge are presented in later sections.

2.3 Treatment effects on brain structure and function

Across modalities, ART treatment influences HIV infection effects on both brain structure and function. A longitudinal, multimodality study of chronically infected, virally suppressed HIV participants demonstrated HIV associated lower whole brain FA, higher whole brain MD and lower GMV at baseline, without differences in rates of change for these measures over a 2 year time period, suggesting that ART arrests or minimizes subsequent brain injury.[28] This same study reported no changes in cortical thickness, cerebral perfusion, neural metabolites or functional connectivity in seronegative participants over the same time period.[28]

Higher radial diffusivity has been reported in WM tracts of untreated compared to treated participants with chronic HIV infection[68] and some studies report no alterations in DTI metrics in sustained ART treated HIV infected cohorts[33] relative to controls.

There is substantial evidence that ART treatment may prevent progressive macrostructural brain changes. Circumstantial evidence in support of this idea comes from a meta-analysis that found HIV associated brain volume reductions, with smaller effects in more recently published papers.[26] The chronologically diminishing effect sizes in all tissue compartments suggest that widespread use of ART may lessen macroarchitectural changes after treatment is initiated. With a longitudinal design, cortical thickness and subcortical volume reductions at baseline in aviremic, treated HIV positive individuals had similar rates of changes in brain volume over time when compared to controls.[18]

While the above studies postulate ART effects through investigating virally suppressed and chronically infected participants, studies following cohorts before and after ART initiation allow more causal inferences. Using Magnetic Resonance Spectroscopy,[69] elevated basal ganglia and occipital cortex choline/creatine (Cho/Cr) levels, suggestive of ongoing inflammation found in acute HIV infection, resolved 6 months following ART initiation.[70] Using a longitudinal design, increasing Cho/Cr and myoinositol/creatine (MI/Cr) in the frontal white matter and parietal gray matter in primary HIV infection were reduced after ART initiation.[41] While brain metabolite alterations do occur after ART initiation, brain macroarchitecture and white matter microstructure seem to stabilize early after ART initiation.[20,25,71]

Several of the above studies are consistent with HIV infection having a "legacy effect," whereby HIV associated CNS damage is sustained before ART initiation and then ceases, but may not be reversible,[72] emphasizing the importance of instituting ART treatment early in the disease course.

2.4 Hepatitis C effects on brain structure and function

In HIV negative participants, Hepatitis C (HCV) infection has been associated with reduced cortical thickness in frontal and occipital lobes,[73] cortical hypoperfusion,[74] lower anisotropy and higher diffusivity in WM tracts,[74,75] lower WM NAA/Cr and higher WM myoinositol.[74,75]

For these reasons, HCV effects on brain structure and function in association with HIV needs further investigation, particularly since approximately 25% of HIV infected individuals are co-infected with HCV[76] and reported global cognitive impairment rates are higher among co-infected individuals than those with HIV alone.[77] HCV infection and substance use are typically associated, making isolation of these individual comorbidity effects challenging. In one study, trends for decreased FA and increased MD reported in HCV/HIV co-infection were confounded by substance use history in all HCV participants.[78] One experimental approach is to exclude individuals with recent substance use or dependence history. A DTI study utilizing this experimental approach in a virally suppressed cohort found no additive effects of HCV co-infection on diffusion measures in the corpus callosum.[79] Using this same experimental approach, there is conflicting evidence contributed from a study finding lower anisotropy and higher diffusivity in cerebral WM in association with HCV co-infection in HIV seropositives.[80] Nevertheless, this study was confounded by variable viral suppression of HIV+ individuals.

Macroarchitectural and metabolic effects in HCV/HIV co-infection require further exploration. HCV co-infection was a strong predictor of T2 hyperintense white matter lesions in a study of HIV infected participants studied without a control group.[81] A study reporting higher basal ganglia MI/Cr metabolites in HIV infected participants with acute hepatitis C co-infection excluded individuals with recreational substance use in the past month, but was confounded by fewer ART treated participants in the HIV/HCV co-infected cohort.

The question of whether HIV/HCV co-infection has additive, synergistic or any effects on brain structure and function is relatively underexplored. Given the high rate of coinfection, further study is needed.

2.5 Cardiovascular risk effects on brain structure and function

As the HIV population ages, co-morbid cardiovascular risk factors have increasing importance in the natural course of chronic disease. The MAC Study has thoroughly explored cardiovascular, and more recently neurovascular, risk factors in HIV infection.[82–85] Not only are cerebrovascular disease risk factors in seropositives associated with cognitive impairment,[20,85–87] but individuals with HIV have an increased risk of cerebrovascular events compared with controls,[88] emphasizing the need for reliable risk stratification biomarkers.

In seronegative populations, diabetes, smoking and hypertension are associated with regional GMV reductions.[89–92] Thus, HIV neuroimaging structural studies can be confounded by cardiovascular risk factors.

T2 hyperintense WM lesions are strongly associated with cerebrovascular disease in seronegative populations.[93,94] While the etiology of WM lesions in the HIV population is less clear, there is substantial evidence that cerebrovascular risk factors contribute to WM lesion development more than serostatus. One retrospective study found no associations between diffuse WM lesions and HIV clinical variables or cardiovascular risk factors,[95] but was limited by variability in cardiovascular risk factors recorded in the medical record. Another found that while WM lesion burden was predicted by age, diabetes, and coronary artery disease, with an HIV by diabetes interaction, WM lesion load was not associated with HIV clinical variables.[96] In this study, psychomotor speed correlated with WM lesion volume, but not with serostatus.[96] Neither study, however, discusses the approach used to control for substance use in their cohorts. In virally suppressed HIV infected men, greater WM lesion burden was independently associated with age, higher diastolic blood pressure, duration of immune compromise and cognitive deficits.[97] While ecstasy use was more common in controls, and seropositives had greater lifetime tobacco exposure,[97] there were no other substance use differences between cohorts. Finally, WM lesion load has been associated with age and systolic blood pressure, but not HIV infection clinical parameters.[98]

Microstructural WM changes have also been observed in association with cerebrovascular disease risk factors in HIV infection. In one DTI study, abnormal glucose metabolism was associated with higher MD and lower FA in the caudate and lower FA in the hippocampus in HIV infected participants.[99]

With PET imaging, cardiovascular risk factors predict global and regional caudate and thalamic FDG metabolism in both seropositive and seronegative participants.[100] Given that differences in seropositive and seronegative groups were not reported in this study, it is not clear that these findings were related to serostatus.

Effects of cardiovascular risk factors on brain structure and function in HIV are of increasing importance as the HIV population ages. The availability of older HIV cohorts will allow further exploration of this theme.

2.6 Duration of untreated illness effects on brain structure and function

Length of HIV infection before ART initiation is another confounding variable in the HIV literature stemming from difficulty associated with its accurate reporting. While clinical features of primary HIV infection were described over three decades ago,[101,102] there remain substantial numbers of both HIV infected individuals who are unaware of when their primary HIV infection occurred and also newly diagnosed HIV infected individuals who progress to AIDS within 1 year after diagnosis.[103,104] Although a multimodality MRI study found no association among brain volume, WM microstructure, WM hyperintensity load, cerebral perfusion and duration of infection,[28] it is unclear how illness duration was determined in this study. Contradictory evidence from another study found that longer duration of untreated infection was associated with volume loss in the thalamus, caudate, and cerebellum, and cortical thinning in frontal and temporal lobes and cingulate cortex.[25] In this study, however, there were no significant differences between primary HIV infected and uninfected cohorts. A strong correlation between reduced putamen volume and duration of infection and worsening psychomotor performance has been reported in primary HIV infection.[105]

While assays are available to confirm recent seroconversion in participants with acute HIV infection,[106] accurately determining infection duration is more difficult in chronically infected patients who are uncertain of when infection occurred.

3. Cohort or technique?

Substantial variation in image analysis methods used across studies, regardless of modality, raises the question of whether reported differences related to serostatus result from differences in cohort demographic properties, participant clinical characteristics or image quantification techniques.

3.1 Gray matter structural measures

Region of interest (ROI) volumetric approaches, voxel-wise volumetric approaches, and surface property quantification approaches all provide macrostructural measures of brain structure. While the oldest studies of regional GM volume changes following HIV infection used laborious manual delineation of structure boundaries,[2,107] most recent work uses computational neuroanatomy techniques. In these methods, T1-weighted MR images are first normalized to a template space and then segmented into GM, WM or CSF compartments. Values at each location represent the probability that the voxel is comprised of GM, WM or CSF. The GM tissue probability maps can then be multiplied by the linear and non-linear components of the deformation from native to template space to generate maps of regional GM volume, providing estimates of the absolute amount of GM tissue in each voxel. These voxel-wise GM volume maps can then be directly used for statistical analysis, yielding maps of the effects of interest, including group or covariate effects. Alternatively, GM volume maps may be masked with predefined anatomical or functional parcellations, resulting in ROI measures in the template space that can then be subject to statistical analysis.

Voxel-wise and ROI quantification approaches each have advantages. Voxel-wise approaches are better suited for whole-brain exploratory analysis, but are limited by the need to subsequently use multiple test corrections to adequately control false positives. ROI approaches are better suited to confirmatory designs, being more spatially specific and, if used judiciously, not requiring severe multiple test correction. ROI analysis has been used by investigators studying HIV effects on brain structure.[2] More modern ROI approaches using voxel wise segmentation followed by atlas-based ROI application to extract mean tissue property values have become increasingly popular.[7,105]

Complementary methods to assess gray matter characteristics use surface-based morphometry to estimate cortical thickness, cortical folding and derivative measures such as fractal dimension. Surface modeling has several advantages. For example, is has been shown that using brain surface meshes for spatial registration increases the accuracy of brain registration compared to volume-based registration.[108] Brain surface mesh determination allows estimation of gyrification indices which measure surface complexity[109] or cortical thickness.[110] In addition, inflation or spherical mapping of the cortical surface mesh exposes buried sulci to better visualize coregistered functional activity maps. While surface-based analysis appears to provide added utility in measuring HIV infection effects,[4] the relationship of surface-based to volumetric measures in HIV infection has not been thoroughly examined.

3.2 Functional connectivity measures

Inter-regional functional connectivity methods are of increasing interest in HIV neuroimaging research. Nevertheless, a number of technical and conceptual considerations make it difficult to use resting-state fMRI and have slowed its introduction into routine clinical research.[111]

Hemodynamic signal artifacts resulting from physiological noise, including head motion, cardiac pulsation, and respiratory effects can severely compromise efforts to detect regional modulations in neural activity. Participant head motion is particularly problematic, as it can bias estimated activity correlations between regions. Visual examination of a participant's scan immediately after completion using a movie loop allows repetition of scans when excessive head motion is detected. Nevertheless, even small inter-scan head movements (<0.5 mm) can bias correlation estimates, influencing between-group effect estimates.[112] For this reason, motion correction using rigid body realignment is an obligate part of the rs-fMRI preprocessing pipeline, followed by inclusion of motion estimates in subsequent single-subject statistical modeling.[113] Even images from cooperative participants will have physiologic confounds that need to be addressed. Cardiac pulsation and respiration can cause spurious spatial connectivity patterns.[114] Band-pass filtering to remove fluctuations outside the frequency range of interest mitigates cardiac and respiratory effects and does not require external physiological recordings. In this approach, filtering frequencies lower than $\sim$0.01 Hz and greater than $\sim$0.2 Hz, reduces the effects of non-neuronal physiologic processes.[115]

Global signal regression[116] has often been used for physiologic noise reduction.[117] Nevertheless, it has been shown to spuriously decrease inter-regional correlation estimates, generating false negative correlations that may confound attempts to distinguish regions whose neural activity are positively or negatively associated.[118] For this reason, noise reduction using techniques like anatomical CompCorr are better denoising alternatives.[119]

Systemic CO_2 fluctuations alter BOLD-contrast and contribute to respiratory induced signal variation.[120] To reduce CO_2 fluctuation effects, end-tidal CO_2 can be measured with a face-mask or nasal cannula and the measurements incorporated into the denoising pipeline.[120]

Temporal signal-to-noise ratio (tSNR), the ratio of the mean signal over its temporal standard deviation (SD), reflects the ability to detect BOLD-contrast signal changes,[121] and thus can be used in quality assurance.

ROI-based correlation analysis, independent component analysis detection of canonical networks, and graph theory metrics used to quantify local and global network properties, are likely to be sensitive to very different aspects of inter-regional functional connectivity.[122]

While functional connectivity provides a convenient means to acquire imaging data reflecting inter-regional activity coupling in clinical research settings, its sensitivity to physiological noise from a variety of sources complicates its processing and statistical analysis.

4. Strategies for managing confounds

When considering how confounds are managed in HIV neuroimaging studies, three different approaches have resulted: (1) studies where investigators choose to ignore comorbid brain effects, (2) studies where investigators choose to control for the confounding influence of comorbidities and (3) studies where investigators choose to restrict the sample to participants not contaminated by such comorbidities. Each of these strategies has associated advantages and disadvantages.

4.1 Ignore comorbid brain effects

In the older HIV neuroimaging literature, particularly prior to widespread ART use, investigators primarily focused on answering "*how* HIV affects the brain" rather than "*does* HIV affect the brain?". As a result, comorbid brain effects were not emphasized. Such examples include PET imaging studies that found basal ganglia and thalamus hypermetabolism in AIDS patients[123,124] followed by cortical and subcortical hypometabolism in AIDS dementia complex later stages.[124] Structural studies using this approach reported smaller total brain volumes and smaller basal ganglia volumes associated with HIV.[2] Examination of white matter microstructure using this approach found decreased callosal anisotropy[125]. While many older HIV neuroimaging studies used age matched control groups, examining the influence of education, substance use, hepatitis C co-infection, cardiovascular risk and duration of infection were not central to the study designs.

4.2 Control for the confounding influence of comorbidities

In an era where reliability and reproducibility command great attention, the more recent HIV neuroimaging literature often includes experimental or statistical control to account for potential confounds of serostatus related neuroimaging findings. Experimental control includes matching seropositive and seronegative cohorts on demographic variables and comorbidities. A study using this design reported basal ganglia volumetric reductions in virally suppressed participants, although the HIV infected cohort was slightly younger and had fewer years of education.[4] Statistical control involves treating potential confounds as nuisance variables in multiple regression, since serostatus is the main effect of interest. A study using both experimental and statistical control in a relatively large sample confirmed lower striatal volume in virally suppressed chronic HIV infection[9]. Including covariates known to affect brain structure, even if the serostatus groups were matched on many of those variables, allowed isolation of nuisance variable effects not completely controlled experimentally, serving to reduce the error variance terms (Fig. 3).

There are multiple advantages in study designs using experimental or statistical control of comorbidity confounds. One, the techniques increase the likelihood that potential study participants will meet screening criteria. Second, participants with comorbidities are more representative of the HIV population. Finally, this study design type permits examination of serostatus by comorbidity interactions. The major disadvantage of this study type is that if nuisance variables exhibit collinearity, their individual contribution to the measure of interest cannot be determined and associated parameter estimates may be unstable.

4.3 Restrict the sample to participants not affected by comorbidities

To more specifically study HIV effects on brain structure and function, some investigators have chosen an experimental design that largely eliminates confounds. For example, a study of brain macrostructure and white matter microstructure in acute or primary HIV infection, excluding participants

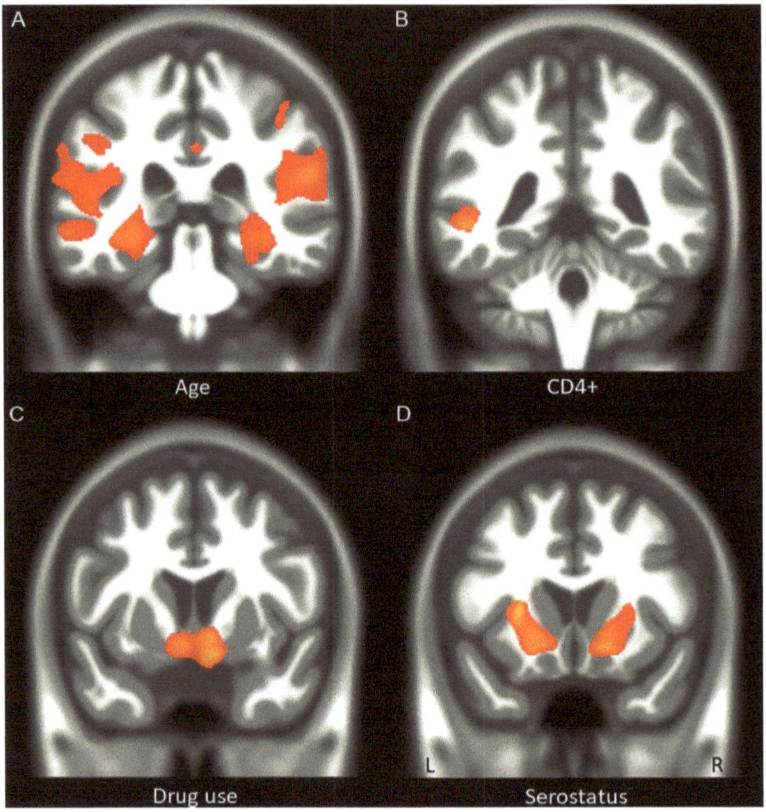

Fig. 3 Progressively removing effects of age, CD4+ count and illicit drug use reveals selective effects of HIV infection on striatal GMV. The sample consisted of 92 HIV+ and 84 HIV− controls. Voxel-wise maps show (A) HIV+ group age effects, (B) HIV+ group CD4+ effects, (C) HIV+ group substance use effects, and (D) between-group serostatus effects while controlling for age, CD4+ count, drug use, and MR system field strength. Details in O'Connor et al.[9]

with chronic or active substance abuse, found reduced TBV and reduced callosal anisotropy.[126] While those with drug use in the past 15 days were excluded, the characterization of drugs used was limited by the accuracy of self-report. Further, only a little over half of participants were treated with ART. Another study using this approach reported lower global and regional GMV and widespread lower anisotropy and higher diffusivity in a virally suppressed cohort with known duration of infection and ART treatment and no history of substance abuse. Altered diffusion characteristics in this study correlated with cognitive function. Data acquisition from two different scanners, if optimization of gray-white contrast on each scanner is not done, can introduce scanner specific biases in image measurements.[127] In addition, recreational drug use was not well characterized. The disadvantage to this study design is that the typical HIV population is not well represented, so associated ecological validity may be weak.

5. Measurement, design and analysis recommendations

Given the complexity of HIV neuroimaging studies related to both their observational nature and the numerous confounds likely to be encountered, developing a systematic approach to handling expected measurement, design and analysis issues can greatly benefit the study planning process. A flowchart for a hypothetical HIV neuroimaging study designed to detect serostatus effects against a background of physiological noise, varying participant clinical characteristics and other confounds is shown in Fig. 4.

Although simplified for visual presentation, the figure shows how three key areas of concern, measurement, design and analysis, are related to specific decisions about: (1) how the imaging protocol can be optimized for highest sensitivity to the tissue property of interest; (2) how participants will be selected, screened and characterized, and how the experiment will be structured; and (3) how the image processing and statistical modeling workflow will be constructed, respectively. The measurement and design branches of the workflow converge in the "collection" step, where information about image and participant character characteristics are aggregated with experimental design information to inform the subsequent construction of statistical models.

Statistical modeling in neuroimaging commonly takes a summary statistic approach in which a preliminary analysis carried out at the individual level results in estimates of voxel-wise or regional measures of the structural or functional variable of interest. For functional measures derived from time series data, time varying covariates, such as head motion, are generally incorporated in these models. Next, the regional parameter estimates from the individual participant analysis are used as inputs to group-level models that can incorporate grouping variables (e.g., serostatus), trait variables (e.g., age, sex, or intelligence), clinical variables (e.g., CD4+ count), or comorbidities (e.g., HCV co-infection or substance use). More recently, hierarchical linear modeling techniques have been used in neuroimaging research to better capture clustering effects in groups or within subjects over time.

This general approach reflects use of both experimental and statistical control principles and can be profitably employed to increase sensitivity

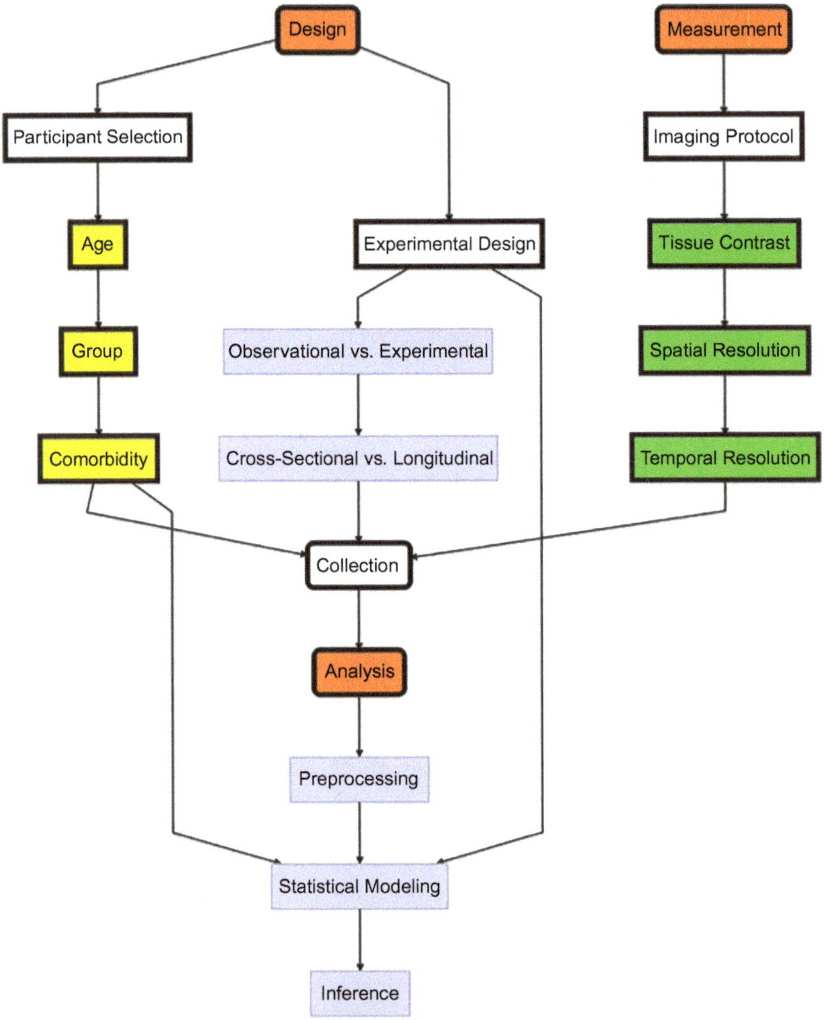

Fig. 4 A measurement, design and analysis flowchart for a hypothetical HIV neuroimaging study. Both experimental and statistical control techniques can be used in observational studies of HIV infection effects on the brain.

to serostatus effects, while optimally controlling for the various physiological, demographic and medical confounds encountered in particular datasets. In these sorts of experimental frameworks, the challenge is always to strike a reasonable balance among competing goals, including maintaining the largest possible sample for analysis, excluding or mitigating the effects of confounds, and achieving the highest sensitivity to the main effect of interest, usually serostatus.

Since the most recently published reviews, the HIV neuroimaging literature has moved towards greater emphasis on longitudinal studies. Although allowing examination of structural and functional change in individuals over time, most of these studies report data from relatively short observation periods (2–5 years), revealing baseline structural/microstructural differences between HIV infected cohorts and demographically matched controls, but no differences in rates of change in the cohorts.[18,25,28] As some of these samples are drawn from cohorts still being studied, we can look forward with anticipation to future reports derived from measurements over time scales perhaps more likely to show longitudinal change.

6. Conclusions

With implementation of ART therapy, HIV has become a treatable chronic disease, allowing more consideration of the multiple confounds that can affect quantitative neuroimaging measures. These confounds include demographic and clinical variables as well as image analysis methods and add layers of complexity to approaching HIV neuroimaging study design. The effects of HIV infection on brain structure and function appear to be diminishing with time, perhaps reflecting earlier, more widespread and more effective treatment. The persisting brain effects associated with HIV infection may be partly explained by co-morbid medical conditions. Continued use of rigorous experimental and statistical control procedures may facilitate efforts to uncover serostatus interactions and better isolate reproducible effects of HIV infection on the brain, making the added complexity well worth the associated effort.

References

1. Levy RM, Bredesen DE. Central nervous system dysfunction in acquired immunodeficiency syndrome. *J Acquir Immune Defic Syndr.* 1988;1(1):41–64.
2. Aylward EH, Henderer JD, McArthur JC, et al. Reduced basal ganglia volume in HIV-1-associated dementia: results from quantitative neuroimaging. *Neurology.* 1993;43(10):2099–2104.
3. d'Arminio Monforte A, Cinque P, Mocroft A, et al. Changing incidence of central nervous system diseases in the EuroSIDA cohort. *Ann Neurol.* 2004;55(3):320–328.
4. Becker JT, Sanders J, Madsen SK, et al. Subcortical brain atrophy persists even in HAART-regulated HIV disease. *Brain Imaging Behav.* 2011;5(2):77–85.
5. Gelman BB, Lisinicchia JG, Morgello S, et al. Neurovirological correlation with HIV-associated neurocognitive disorders and encephalitis in a HAART-era cohort. *J Acquir Immune Defic Syndr.* 2013;62(5):487–495.

6. Clark US, Walker KA, Cohen RA, et al. Facial emotion recognition impairments are associated with brain volume abnormalities in individuals with HIV. *Neuropsychologia.* 2015;70:263–271.
7. Pfefferbaum A, Rogosa DA, Rosenbloom MJ, et al. Accelerated aging of selective brain structures in human immunodeficiency virus infection: a controlled, longitudinal magnetic resonance imaging study. *Neurobiol Aging.* 2014;35(7):1755–1768.
8. Bonnet F, Amieva H, Marquant F, et al. Cognitive disorders in HIV-infected patients: are they HIV-related? *AIDS.* 2013;27(3):391–400.
9. O'Connor EE, Zeffiro T, Lopez OL, Becker JT, Zeffiro T. HIV infection and age effects on striatal structure are additive. *J. Neurovirol.* 2019 Apr 26: PMID: 31028692. https://doi.org/10.1007/s13365-019-00747-w. [Epub ahead of print].
10. Ances BM, Hammoud DA. Neuroimaging of HIV-associated neurocognitive disorders (HAND). *Curr Opin HIV AIDS.* 2014;9(6):545–551.
11. Canizares S, Cherner M, Ellis RJ. HIV and aging: effects on the central nervous system. *Semin Neurol.* 2014;34(1):27–34.
12. Chang L, Shukla DK. Imaging studies of the HIV-infected brain. *Handb Clin Neurol.* 2018;152:229–264.
13. Masters MC, Ances BM. Role of neuroimaging in HIV-associated neurocognitive disorders. *Semin Neurol.* 2014;34(1):89–102.
14. Thompson PM, Jahanshad N. Novel neuroimaging methods to understand how HIV affects the brain. *Curr HIV/AIDS Rep.* 2015;12(2):289–298.
15. Clifford DB, Ances BM. HIV-associated neurocognitive disorder. *Lancet Infect Dis.* 2013;13(11):976–986.
16. Cysique LA, Brew BJ. The effects of HIV and aging on brain functions: proposing a research framework and update on last 3 years' findings. *Curr Opin HIV AIDS.* 2014;9(4):355–364.
17. Zahr NM. The aging brain with HIV infection: effects of alcoholism or hepatitis C comorbidity. *Front Aging Neurosci.* 2018;10:56.
18. Sanford R, Fellows LK, Ances BM, Collins DL. Association of brain structure changes and cognitive function with combination antiretroviral therapy in HIV-positive individuals. *JAMA Neurol.* 2018;75(1):72–79.
19. Kallianpur KJ, Shikuma C, Kirk GR, et al. Peripheral blood HIV DNA is associated with atrophy of cerebellar and subcortical gray matter. *Neurology.* 2013;80(19):1792–1799.
20. McCutchan JA, Marquie-Beck JA, Fitzsimons CA, et al. Role of obesity, metabolic variables, and diabetes in HIV-associated neurocognitive disorder. *Neurology.* 2012;78(7):485–492.
21. Thompson PM, Dutton RA, Hayashi KM, et al. 3D mapping of ventricular and corpus callosum abnormalities in HIV/AIDS. *Neuroimage.* 2006;31(1):12–23.
22. Thompson PM, Dutton RA, Hayashi KM, et al. Thinning of the cerebral cortex visualized in HIV/AIDS reflects CD4+ T lymphocyte decline. *Proc Natl Acad Sci U S A.* 2005;102(43):15647–15652.
23. Clifford KM, Samboju V, Cobigo Y, et al. Progressive brain atrophy despite persistent viral suppression in HIV patients older than 60 years. *J Acquir Immune Defic Syndr.* 2017;76(3):289–297.
24. Cohen RA, Harezlak J, Schifitto G, et al. Effects of nadir CD4 count and duration of human immunodeficiency virus infection on brain volumes in the highly active antiretroviral therapy era. *J Neurovirol.* 2010;16(1):25–32.
25. Sanford R, Ances BM, Meyerhoff DJ, et al. Longitudinal trajectories of brain volume and cortical thickness in treated and untreated primary HIV infection. *Clin Infect Dis.* 2018;67(11):1697–1704. https://doi.org/10.1093/cid/ciy362.

26. O'Connor EE, Zeffiro TA, Zeffiro TA. Brain structural changes following HIV infection: meta-analysis. *AJNR Am J Neuroradiol*. 2018;39(1):54–62.
27. Maki PM, Rubin LH, Valcour V, et al. Cognitive function in women with HIV: findings from the women's interagency HIV study. *Neurology*. 2015;84(3):231–240.
28. Cole JH, Caan MWA, Underwood J, et al. No evidence for accelerated ageing-related brain pathology in treated HIV: longitudinal neuroimaging results from the comorbidity in relation to AIDS (COBRA) project. *Clin Infect Dis*. 2018;66(12):1899–1909. https://doi.org/10.1093/cid/cix1124.
29. Seider TR, Gongvatana A, Woods AJ, et al. Age exacerbates HIV-associated white matter abnormalities. *J Neurovirol*. 2016;22(2):201–212.
30. Guha A, Brier MR, Ortega M, Westerhaus E, Nelson B, Ances BM. Topographies of cortical and subcortical volume loss in HIV and aging in the cART era. *J Acquir Immune Defic Syndr*. 2016;73(4):374–383.
31. Kuhn T, Schonfeld D, Sayegh P, et al. The effects of HIV and aging on subcortical shape alterations: a 3D morphometric study. *Hum Brain Mapp*. 2017;38(2):1025–1037.
32. Pfefferbaum A, Zahr NM, Sassoon SA, Kwon D, Pohl KM, Sullivan EV. Accelerated and premature aging characterizing regional cortical volume loss in human immunodeficiency virus infection: contributions from alcohol, substance use, and hepatitis C coinfection. *Biol Psychiatry Cogn Neurosci Neuroimaging*. 2018;3(10):844–859. https://doi.org/10.1016/j.bpsc.2018.06.006. [Epub 2018 Jul 4].
33. Towgood KJ, Pitkanen M, Kulasegaram R, et al. Mapping the brain in younger and older asymptomatic HIV-1 men: frontal volume changes in the absence of other cortical or diffusion tensor abnormalities. *Cortex*. 2012;48(2):230–241.
34. Nir TM, Jahanshad N, Busovaca E, et al. Mapping white matter integrity in elderly people with HIV. *Hum Brain Mapp*. 2014;35(3):975–992.
35. Correa DG, Zimmermann N, Tukamoto G, et al. Longitudinal assessment of subcortical gray matter volume, cortical thickness, and white matter integrity in HIV-positive patients. *J Magn Reson Imaging*. 2016;44(5):1262–1269.
36. Cole JH, Caan MWA, Underwood J, et al. No evidence for accelerated aging-related brain pathology in treated human immunodeficiency virus: longitudinal neuroimaging results from the comorbidity in relation to AIDS (COBRA) project. *Clin Infect Dis*. 2018;66(12):1899–1909.
37. Kuhn T, Kaufmann T, Doan NT, et al. An augmented aging process in brain white matter in HIV. *Hum Brain Mapp*. 2018;39(6):2532–2540.
38. Chang L, Wong V, Nakama H, et al. Greater than age-related changes in brain diffusion of HIV patients after 1 year. *J Neuroimmune Pharmacol*. 2008;3(4):265–274.
39. Ances BM, Vaida F, Yeh MJ, et al. HIV infection and aging independently affect brain function as measured by functional magnetic resonance imaging. *J Infect Dis*. 2010;201(3):336–340.
40. Thomas JB, Brier MR, Snyder AZ, Vaida FF, Ances BM. Pathways to neurodegeneration: effects of HIV and aging on resting-state functional connectivity. *Neurology*. 2013;80(13):1186–1193.
41. Young AC, Yiannoutsos CT, Hegde M, et al. Cerebral metabolite changes prior to and after antiretroviral therapy in primary HIV infection. *Neurology*. 2014;83(18):1592–1600.
42. Towgood KJ, Pitkanen M, Kulasegaram R, et al. Regional cerebral blood flow and FDG uptake in asymptomatic HIV-1 men. *Hum Brain Mapp*. 2013;34(10):2484–2493.
43. Tolomeo S, Gray S, Matthews K, Steele JD, Baldacchino A. Multifaceted impairments in impulsivity and brain structural abnormalities in opioid dependence and abstinence. *Psychol Med*. 2016;46(13):2841–2853.

44. Ersche KD, Williams GB, Robbins TW, Bullmore ET. Meta-analysis of structural brain abnormalities associated with stimulant drug dependence and neuroimaging of addiction vulnerability and resilience. *Curr Opin Neurobiol.* 2013;23(4):615–624.
45. Ashtari M, Avants B, Cyckowski L, et al. Medial temporal structures and memory functions in adolescents with heavy cannabis use. *J Psychiatr Res.* 2011;45(8):1055–1066.
46. Yücel M, Solowij N, Respondek C, et al. Regional brain abnormalities associated with long-term heavy cannabis use. *Arch Gen Psychiatry.* 2008;65(6):694–701.
47. Cousijn J, Wiers RW, Ridderinkhof KR, van den Brink W, Veltman DJ, Goudriaan AE. Grey matter alterations associated with cannabis use: results of a VBM study in heavy cannabis users and healthy controls. *Neuroimage.* 2012;59(4): 3845–3851.
48. Bouso JC, Palhano-Fontes F, Rodriguez-Fornells A, et al. Long-term use of psychedelic drugs is associated with differences in brain structure and personality in humans. *Eur Neuropsychopharmacol.* 2015;25(4):483–492.
49. Fein G, Di Sclafani V, Cardenas VA, Goldmann H, Tolou-Shams M, Meyerhoff DJ. Cortical gray matter loss in treatment-naive alcohol dependent individuals. *Alcohol Clin Exp Res.* 2002;26(4):558–564.
50. Thayer RE, Hagerty SL, Sabbineni A, Claus ED, Hutchison KE, Weiland BJ. Negative and interactive effects of sex, aging, and alcohol abuse on gray matter morphometry. *Hum Brain Mapp.* 2016;37(6):2276–2292.
51. Pfefferbaum A, Lim KO, Zipursky RB, et al. Brain gray and white matter volume loss accelerates with aging in chronic alcoholics: a quantitative MRI study. *Alcohol Clin Exp Res.* 1992;16(6):1078–1089.
52. Pfefferbaum A, Adalsteinsson E, Sullivan EV. Supratentorial profile of white matter microstructural integrity in recovering alcoholic men and women. *Biol Psychiatry.* 2006;59(4):364–372.
53. Pfefferbaum A, Sullivan EV. Microstructural but not macrostructural disruption of white matter in women with chronic alcoholism. *Neuroimage.* 2002;15(3):708–718.
54. O'Daly OG, Trick L, Scaife J, et al. Withdrawal-associated increases and decreases in functional neural connectivity associated with altered emotional regulation in alcoholism. *Neuropsychopharmacology.* 2012;37(10):2267–2276.
55. Upadhyay J, Maleki N, Potter J, et al. Alterations in brain structure and functional connectivity in prescription opioid-dependent patients. *Brain.* 2010;133(pt 7): 2098–2114.
56. Pujol J, Blanco-Hinojo L, Batalla A, et al. Functional connectivity alterations in brain networks relevant to self-awareness in chronic cannabis users. *J Psychiatr Res.* 2014;51:68–78.
57. Muller-Oehring EM, Jung YC, Pfefferbaum A, Sullivan EV, Schulte T. The resting brain of alcoholics. *Cereb Cortex.* 2015;25(11):4155–4168.
58. Shokoohi M, Bauer GR, Kaida A, et al. Substance use patterns among women living with HIV compared with the general female population of Canada. *Drug Alcohol Depend.* 2018;191:70–77.
59. Mimiaga MJ, Reisner SL, Grasso C, et al. Substance use among HIV-infected patients engaged in primary care in the United States: findings from the Centers for AIDS Research Network of integrated clinical systems cohort. *Am J Public Health.* 2013;103(8):1457–1467.
60. Cook RL, Zhu F, Belnap BH, et al. Longitudinal trends in hazardous alcohol consumption among women with human immunodeficiency virus infection, 1995–2006. *Am J Epidemiol.* 2009;169(8):1025–1032.
61. Chander G, Josephs J, Fleishman JA, et al. Alcohol use among HIV-infected persons in care: results of a multi-site survey. *HIV Med.* 2008;9(4):196–202.

62. Tang VM, Lang DJ, Giesbrecht CJ, et al. White matter deficits assessed by diffusion tensor imaging and cognitive dysfunction in psychostimulant users with comorbid human immunodeficiency virus infection. *BMC Res Notes*. 2015;8:515.
63. Pfefferbaum A, Rosenbloom MJ, Adalsteinsson E, Sullivan EV. Diffusion tensor imaging with quantitative fibre tracking in HIV infection and alcoholism comorbidity: synergistic white matter damage. *Brain*. 2007;130(pt 1):48–64.
64. Pfefferbaum A, Rosenbloom MJ, Sassoon SA, et al. Regional brain structural dysmorphology in human immunodeficiency virus infection: effects of acquired immune deficiency syndrome, alcoholism, and age. *Biol Psychiatry*. 2012;72(5):361–370.
65. Liang H, Chang L, Chen R, Oishi K, Ernst T. Independent and combined effects of chronic HIV-infection and tobacco smoking on brain microstructure. *J Neuroimmune Pharmacol*. 2018;13(4):509–522.
66. MacDuffie KE, Brown GG, McKenna BS, et al. Effects of HIV infection, methamphetamine dependence and age on cortical thickness, area and volume. *Neuroimage Clin*. 2018;20:1044–1052.
67. Meade CS, Hobkirk AL, Towe SL, Chen NK, Bell RP, Huettel SA. Cocaine dependence modulates the effect of HIV infection on brain activation during intertemporal decision making. *Drug Alcohol Depend*. 2017;178:443–451.
68. Pfefferbaum A, Rosenbloom MJ, Rohlfing T, Kemper CA, Deresinski S, Sullivan EV. Frontostriatal fiber bundle compromise in HIV infection without dementia. *AIDS*. 2009;23(15):1977–1985.
69. Paul RH, Ernst T, Brickman AM, et al. Relative sensitivity of magnetic resonance spectroscopy and quantitative magnetic resonance imaging to cognitive function among nondemented individuals infected with HIV. *J Int Neuropsychol Soc*. 2008;14(5):725–733.
70. Sailasuta N, Ross W, Ananworanich J, et al. Change in brain magnetic resonance spectroscopy after treatment during acute HIV infection. *PLoS One*. 2012;7(11): e49272.
71. O'Connor EE, Jaillard A, Renard F, Zeffiro TA. Reliability of white matter microstructural changes in HIV infection: meta-analysis and confirmation. *AJNR Am J Neuroradiol*. 2017;38(8):1510–1519.
72. Canestri A, Lescure FX, Jaureguiberry S, et al. Discordance between cerebral spinal fluid and plasma HIV replication in patients with neurological symptoms who are receiving suppressive antiretroviral therapy. *Clin Infect Dis*. 2010;50(5):773–778.
73. Hjerrild S, Renvillard SG, Leutscher P, et al. Reduced cerebral cortical thickness in non-cirrhotic patients with hepatitis C. *Metab Brain Dis*. 2016;31(2):311–319.
74. Bladowska J, Zimny A, Knysz B, et al. Evaluation of early cerebral metabolic, perfusion and microstructural changes in HCV-positive patients: a pilot study. *J Hepatol*. 2013;59(4):651–657.
75. Thames AD, Castellon SA, Singer EJ, et al. Neuroimaging abnormalities, neurocognitive function, and fatigue in patients with hepatitis C. *Neurol Neuroimmunol Neuroinflamm*. 2015;2(1):e59.
76. Department of Health and Human Services. https://aidsinfo.nih.gov/understanding-hiv-aids/fact-sheets/26/88/hiv-and-hepatitis-c.
77. Hinkin CH, Castellon SA, Levine AJ, Barclay TR, Singer EJ. Neurocognition in individuals co-infected with HIV and hepatitis C. *J Addict Dis*. 2008;27(2):11–17.
78. Stebbins GT, Smith CA, Bartt RE, et al. HIV-associated alterations in normal-appearing white matter: a voxel-wise diffusion tensor imaging study. *J Acquir Immune Defic Syndr*. 2007;46(5):564–573.
79. Heaps-Woodruff JM, Wright PW, Ances BM, Clifford D, Paul RH. The impact of human immune deficiency virus and hepatitis C coinfection on white matter microstructural integrity. *J Neurovirol*. 2016;22(3):389–399.

80. Gongvatana A, Cohen RA, Correia S, et al. Clinical contributors to cerebral white matter integrity in HIV-infected individuals. *J Neurovirol.* 2011;17(5):477–486.
81. Robinson-Papp J, Navis A, Dhamoon MS, Clark US, Gutierrez-Contreras J, Morgello S. The use of visual rating scales to quantify brain MRI lesions in patients with HIV infection. *J Neuroimaging.* 2018;28(2):217–224.
82. Post WS, Budoff M, Kingsley L, et al. Associations between HIV infection and subclinical coronary atherosclerosis. *Ann Intern Med.* 2014;160(7):458–467.
83. Lo J, Abbara S, Shturman L, et al. Increased prevalence of subclinical coronary atherosclerosis detected by coronary computed tomography angiography in HIV-infected men. *AIDS.* 2010;24(2):243–253.
84. d'Ettorre G, Francone M, Mancone M, et al. Significant coronary stenosis detected by coronary computed angiography in asymptomatic HIV infected subjects. *J Infect.* 2012;64(1):82–88.
85. Becker JT, Kingsley L, Mullen J, et al. Vascular risk factors, HIV serostatus, and cognitive dysfunction in gay and bisexual men. *Neurology.* 2009;73(16):1292–1299.
86. Wright EJ, Grund B, Robertson K, et al. Cardiovascular risk factors associated with lower baseline cognitive performance in HIV-positive persons. *Neurology.* 2010;75(10):864–873.
87. Nakamoto BK, Valcour VG, Kallianpur K, et al. Impact of cerebrovascular disease on cognitive function in HIV-infected patients. *J Acquir Immune Defic Syndr.* 2011;57(3): e66–e68.
88. Rasmussen LD, Engsig FN, Christensen H, et al. Risk of cerebrovascular events in persons with and without HIV: a Danish nationwide population-based cohort study. *AIDS.* 2011;25(13):1637–1646.
89. van Harten B, de Leeuw FE, Weinstein HC, Scheltens P, Biessels GJ. Brain imaging in patients with diabetes: a systematic review. *Diabetes Care.* 2006;29(11):2539–2548.
90. Beauchet O, Celle S, Roche F, et al. Blood pressure levels and brain volume reduction: a systematic review and meta-analysis. *J Hypertens.* 2013;31(8):1502–1516.
91. Brody AL, Mandelkern MA, Jarvik ME, et al. Differences between smokers and nonsmokers in regional gray matter volumes and densities. *Biol Psychiatry.* 2004;55(1):77–84.
92. Schaare HL, Kharabian Masouleh S, Beyer F, et al. Association of peripheral blood pressure with gray matter volume in 19- to 40-year-old adults. *Neurology.* 2019;92(8):e758–e773.
93. Poels MM, Steyerberg EW, Wieberdink RG, et al. Assessment of cerebral small vessel disease predicts individual stroke risk. *J Neurol Neurosurg Psychiatry.* 2012;83(12):1174–1179.
94. Dufouil C, Godin O, Chalmers J, et al. Severe cerebral white matter hyperintensities predict severe cognitive decline in patients with cerebrovascular disease history. *Stroke.* 2009;40(6):2219–2221.
95. Haddow LJ, Dudau C, Chandrashekar H, et al. Cross-sectional study of unexplained white matter lesions in HIV positive individuals undergoing brain magnetic resonance imaging. *AIDS Patient Care STDS.* 2014;28(7):341–349.
96. Wu M, Fatukasi O, Yang S, et al. HIV disease and diabetes interact to affect brain white matter hyperintensities and cognition. *AIDS.* 2018;32(13):1803–1810.
97. Su T, Wit FW, Caan MW, et al. White matter hyperintensities in relation to cognition in HIV-infected men with sustained suppressed viral load on combination antiretroviral therapy. *AIDS.* 2016;30(15):2329–2339.
98. McMurtray A, Nakamoto B, Shikuma C, Valcour V. Small-vessel vascular disease in human immunodeficiency virus infection: the Hawaii aging with HIV cohort study. *Cerebrovasc Dis.* 2007;24(2–3):236–241.

99. Nakamoto BK, Jahanshad N, McMurtray A, et al. Cerebrovascular risk factors and brain microstructural abnormalities on diffusion tensor images in HIV-infected individuals. *J Neurovirol.* 2012;18(4):303–312.
100. Hammoud DA, Sinharay S, Steinbach S, et al. Global and regional brain hypometabolism on FDG-PET in treated HIV-infected individuals. *Neurology.* 2018;91(17):e1591–e1601.
101. Schacker TW, Hughes JP, Shea T, Coombs RW, Corey L. Biological and virologic characteristics of primary HIV infection. *Ann Intern Med.* 1998;128(8):613–620.
102. Cooper DA, Gold J, Maclean P, et al. Acute AIDS retrovirus infection. Definition of a clinical illness associated with seroconversion. *Lancet.* 1985;1(8428):537–540.
103. Gardner EM, McLees MP, Steiner JF, Del Rio C, Burman WJ. The spectrum of engagement in HIV care and its relevance to test-and-treat strategies for prevention of HIV infection. *Clin Infect Dis.* 2011;52(6):793–800.
104. Schwarcz S, Hsu L, Dilley JW, Loeb L, Nelson K, Boyd S. Late diagnosis of HIV infection: trends, prevalence, and characteristics of persons whose HIV diagnosis occurred within 12 months of developing AIDS. *J Acquir Immune Defic Syndr.* 2006;43(4):491–494.
105. Wright PW, Pyakurel A, Vaida FF, et al. Putamen volume and its clinical and neurological correlates in primary HIV infection. *AIDS.* 2016;30(11):1789–1794.
106. Keating SM, Hanson D, Lebedeva M, et al. Lower-sensitivity and avidity modifications of the vitros anti-HIV 1+2 assay for detection of recent HIV infections and incidence estimation. *J Clin Microbiol.* 2012;50(12):3968–3976.
107. Hestad K, McArthur JH, Dal Pan GJ, et al. Regional brain atrophy in HIV-1 infection: association with specific neuropsychological test performance. *Acta Neurol Scand.* 1993;88(2):112–118.
108. Desai R, Liebenthal E, Possing ET, Waldron E, Binder JR. Volumetric vs. surface-based alignment for localization of auditory cortex activation. *Neuroimage.* 2005;26(4):1019–1029.
109. Yotter RA, Nenadic I, Ziegler G, Thompson PM, Gaser C. Local cortical surface complexity maps from spherical harmonic reconstructions. *Neuroimage.* 2011;56(3):961–973.
110. Gaser C, Dahnke R. A computational anatomy toolbox for the analysis of structural MRI data. In: *Paper Presented at: OHBM*; 2016.
111. O'Connor EE, Zeffiro TA. Why is clinical fMRI in a resting state? *Frontiers in Neurology.* 2019;10:420. https://doi.org/10.3389/fneur.2019.00420.
112. Parkes L, Fulcher B, Yucel M, Fornito A. An evaluation of the efficacy, reliability, and sensitivity of motion correction strategies for resting-state functional MRI. *Neuroimage.* 2018;171:415–436.
113. Jenkinson M, Bannister P, Brady M, Smith S. Improved optimization for the robust and accurate linear registration and motion correction of brain images. *Neuroimage.* 2002;17(2):825–841.
114. Dagli MS, Ingeholm JE, Haxby JV. Localization of cardiac-induced signal change in fMRI. *Neuroimage.* 1999;9(4):407–415.
115. Murphy K, Birn RM, Bandettini PA. Resting-state fMRI confounds and cleanup. *Neuroimage.* 2013;80:349–359.
116. Wongsripuemtet J, Tyan AE, Carass A, et al. Preoperative mapping of the supplementary motor area in patients with brain tumor using resting-state fMRI with seed-based analysis. *AJNR Am J Neuroradiol.* 2018;39(8):1493–1498. https://doi.org/10.3174/ajnr.A5709. [Epub 2018 Jul 12].
117. Power JD, Plitt M, Laumann TO, Martin A. Sources and implications of whole-brain fMRI signals in humans. *Neuroimage.* 2017;146:609–625.

118. Chen G, Chen G, Xie C, et al. A method to determine the necessity for global signal regression in resting-state fMRI studies. *Magn Reson Med.* 2012;68(6):1828–1835.
119. Whitfield-Gabrieli S, Nieto-Castanon A. Conn: a functional connectivity toolbox for correlated and anticorrelated brain networks. *Brain Connect.* 2012;2(3):125–141.
120. Golestani AM, Chang C, Kwinta JB, Khatamian YB, Jean Chen J. Mapping the end-tidal CO2 response function in the resting-state BOLD fMRI signal: spatial specificity, test-retest reliability and effect of fMRI sampling rate. *Neuroimage.* 2015;104:266–277.
121. Parrish TB, Gitelman DR, LaBar KS, Mesulam MM. Impact of signal-to-noise on functional MRI. *Magn Reson Med.* 2000;44(6):925–932.
122. Lv H, Wang Z, Tong E, et al. Resting-state functional MRI: everything that nonexperts have always wanted to know. *AJNR Am J Neuroradiol.* 2018;39(8):1390–1399.
123. Hinkin CH, van Gorp WG, Mandelkern MA, et al. Cerebral metabolic change in patients with AIDS: report of a six-month follow-up using positron-emission tomography. *J Neuropsychiatry Clin Neurosci.* 1995;7(2):180–187.
124. Rottenberg DA, Moeller JR, Strother SC, et al. The metabolic pathology of the AIDS dementia complex. *Ann Neurol.* 1987;22(6):700–706.
125. Thurnher MM, Castillo M, Stadler A, Rieger A, Schmid B, Sundgren PC. Diffusion-tensor MR imaging of the brain in human immunodeficiency virus-positive patients. *AJNR Am J Neuroradiol.* 2005;26(9):2275–2281.
126. Ragin AB, Wu Y, Gao Y, et al. Brain alterations within the first 100 days of HIV infection. *Ann Clin Transl Neurol.* 2015;2(1):12–21.
127. Underwood J, De Francesco D, Post FA, et al. Associations between cognitive impairment and patient-reported measures of physical/mental functioning in older people living with HIV. *HIV Med.* 2017;18(5):363–369.

5b

Aberrant brain dynamics in neuroHIV: Evidence from magnetoencephalographic (MEG) imaging

Tony W. Wilson[a,b,*], Brandon J. Lew[a,b], Rachel K. Spooner[a,b], Michael T. Rezich[a,b], Alex I. Wiesman[a,b]

[a]Department of Neurological Sciences, University of Nebraska Medical Center (UNMC), Omaha, NE, United States

[b]Center for Magnetoencephalography, UNMC, Omaha, NE, United States

*Corresponding author: e-mail address: tony.w.wilson@gmail.com

Contents

Abstract

Magnetoencephalography (MEG) is a noninvasive, silent, and totally passive neurophysiological imaging method with excellent temporal resolution (~1 ms) and good spatial precision (~3–5 mm). While MEG studies of neuroHIV remain relatively rare, the number of studies per year has sharply increased recently and this trend will likely continue into the foreseeable future. The current in-depth review focuses on the studies that have been conducted to date, which include investigations of somatosensory and visual modalities, resting-state, as well as motor control and higher-level functions such as working memory and visual attention. The review begins with an introduction to the principles and methods of MEG, and then transitions to a review of each of the empirical studies that have been conducted to date, separated by sensory modality for the basic studies and cognitive domain for the higher-level investigations. As such, this review attempts to be exhaustive in its coverage of empirical MEG studies of neuroHIV.

Progress in Molecular Biology and Translational Science, Volume 165
ISSN 1877-1173
https://doi.org/10.1016/bs.pmbts.2019.04.008

Across studies major themes emerge including aberrant neural oscillatory activity in HIV-infected adults, both in primary sensory regions of the brain and higher-order executive regions. Many studies have also connected the amplitude of neural oscillations to behavioral and/or neuropsychological function in the study population, making a vital connection to performance and improving the veracity of the findings. One conspicuous emerging area is the use of MEG to distinguish cognitively-impaired from unimpaired HIV-infected adults, with major success reported and future studies sure to come. The review concludes with a summary of findings and suggested focus areas for future studies.

1. Introduction

With the availability of potent combination antiretroviral therapy (cART), the nature of HIV-infection dramatically shifted from a terminal illness to a chronic manageable condition, with a life expectancy that approaches that of seronegative persons.[1,2] Despite these advances, HIV-infected patients remain at a significantly increased risk of developing HIV-associated neurocognitive disorders (HAND),[3–8] and recent studies have shown that 35–70% of all HIV-infected patients are at least subtly impaired on neuropsychological tests of cognitive function.[1–7] This persistence is at least partially driven by people simply living longer with milder neurocognitive disorders, although the trend also reflects that cART is far more effective in the periphery than in the brain. Overall, the prevalence of HIV-associated dementia (HAD) has sharply decreased in the cART era,[3,4,9–12] while that of milder forms of HAND has remained largely the same or slightly increased.[3–11] Patients with HAND are more likely to be unemployed or under employed, have greater problems with medication adherence, and have lower quality of life.[13–15] Thus, the economic and societal burdens of HIV/AIDS are accentuated in cases of HAND.

Unfortunately, despite major scientific efforts, the neurobiological underpinnings of HAND remain incompletely understood. Numerous studies using structural and functional neuroimaging in humans have successfully identified the brain regions and networks most affected by the disease, but we still have a limited understanding of the inherent physiology in these brain regions, and how it is different in patients with HAND. There are also no current biomarkers of HAND, and hence diagnoses are based on standard neuropsychological testing and the exclusion of other contributing factors (e.g., comorbid conditions). Thus, there is a clear need for additional research using state-of-the-art tools to advance the field. One such tool that has seen increasing use in neuroHIV is magnetoencephalography (MEG). Compared to the other neuroimaging methods discussed in the accompanying chapters, MEG is unique in that it provides precise spatiotemporal data on the neural dynamics that serve cognitive processing. In other words, MEG is capable of characterizing neuronal activity on a millisecond temporal scale, which is critical given the extremely rapid time scale of neuronal traffic and circuit-level neural activity in the human brain.

In the current review, we start by providing some background information on MEG, including its physical and physiological origins, basic experimental design, terminology, and output variables. We then transition to empirical studies that have used MEG in the context of neuroHIV, which we subdivide into sections based on the cognitive and/or sensory process of interest. We then close by highlighting general conclusions and areas of consensus, pointing out areas in need of further study, and recommending future directions.

2. Magnetoencephalographic (MEG) imaging: A primer

MEG is a noninvasive human neurophysiological recording method that is primarily utilized in studies of systems-level brain function. The physical signal digitized in MEG is comprised of summated magnetic fields produced by ionic currents within active neural populations. Just as electric current within power lines produces magnetic fields, ionic currents within active neurons also produce magnetic fields according to the same physical principles (e.g., "the right-hand rule" of magnetism).[16] These so-called neuromagnetic fields are ultra-minute (10^{-15} T) relative to other magnetic fields in the environment (e.g., the earth's static field is about 10^{-5} T), and thus MEG signals are normally recorded in magnetically-shielded rooms using special sensors and/or noise-cancellation software.[16–21] In regard to the physiological basis of these ionic currents, MEG is most sensitive to dendritic currents in large populations of cortical pyramidal cells, as opposed to their action potentials.[16,22] However, other cortical neurons also contribute to the signal, as do cells in subcortical and cerebellar regions to a lesser extent.[22] The depth (i.e., distance from the MEG sensors), orientation, and synchronicity of the neural populations also affect the strength of MEG signals, although modern MEG systems are generally sensitive to responses across the entire brain.[23] Importantly, signal measurement in MEG occurs at the sensor-level and state-of-the-art systems are equipped with a large number

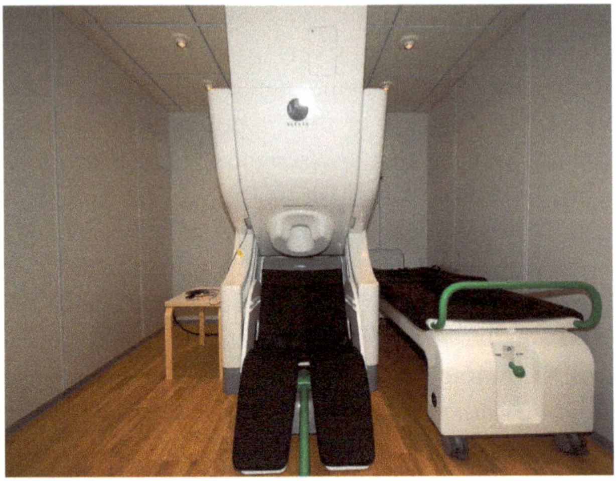

Fig. 1 Modern magnetoencephalography (MEG) system. A whole-head MEG system equipped with 306 cortical sensors is shown. MEG instruments are generally housed in magnetically-shielded rooms to reduce interference from magnetic fields in the local environment.

of sensors (e.g., ~300) arranged in a helmet-like array that surrounds the participants head (Fig. 1). Consequently, it is necessary to project these sensor measurements into source space (i.e., anatomical coordinates), which is generally accomplished using volume conductor models and optimization algorithms such as minimum-norm estimation, beamforming, and/or other related source estimation techniques.[16–21] This process of projecting the MEG signals into the brain is often termed MEG source reconstruction.

Although the first MEG measurements occurred over 50 years ago,[24] the analytical sophistication and instrumentation of the field have dramatically increased over the past decade.[17–21] MEG is currently the only noninvasive high-resolution imaging technique that measures the neurophysiology directly (i.e., does not rely on vascular responses) and the only functional brain imaging method to possess both high spatial (2–5 mm) and temporal resolution (i.e., millisecond level).[20,21] Another unique advantage of MEG over other electrophysiological and noninvasive imaging methods is that it does not require a reference electrode, which allows local spontaneous activity to be accurately quantified, greatly simplifies the interpretation of responses, and facilitates network modeling and connectivity analyses.

As mentioned above, this review focuses on the use of MEG in people living with HIV-infection. The main applications in this population include the identification of potential biomarkers of cognitive impairment secondary to HIV-infection (i.e., HAND), and the generation of basic knowledge concerning the neural origin and mechanisms of HAND. Of note, this review attempts to be exhaustive in its coverage of MEG in neuroHIV, as the first papers on this topic appeared in 2012 and to date there are only 13 published articles listed in PubMed (according to a search of "HIV and magnetoencephalography" and "HIV and MEG"). Of these 13 papers, 2 are review articles that include a MEG/neuroHIV component, 1 is a case study, and the other 10 are empirical studies. Interestingly, of the 10 empirical papers, 3 were published in 2018 (30%) so the frequency of study appears to be increasing.

Before beginning, a few words about the interpretation and nature of MEG results are in order. First, with the exception of resting-state studies, MEG experiments almost always use event-related designs and are laid out very similarly to invasive electrophysiological investigations in monkeys and rats. Briefly, there is a baseline period with no stimulation, followed by the occurrence of an event or stimulus at time 0.0 s, which leads to an "active" period when the brain's response to the stimulus occurs. This series of events constitutes a single trial, and the typical MEG experiment consists of 80–100 trials per condition, per participant.[17] The MEG signals are recorded throughout the experiment and the individual trials are aggregated (offline) in each participant to increase the signal-to-noise ratio of the primary brain responses. In interpreting the signals, it is generally the relative response (active compared with baseline) that is of paramount interest, as neural cells spontaneously fire and cortical regions are constantly undergoing local electrical fluctuations that are seemingly unrelated to the sensory and cognitive processes of interest. Second, as mentioned above, there is no reference electrode in MEG. Thus, unlike other electrophysiological methods, the signal amplitude is not a difference measure between two electrodes or the average across multiple electrodes. The amplitude is specific to the location and not relative to electrical activity at any other location. Third, over the past decade, neurophysiological studies have increasingly focused on population-level neural oscillatory activity.[25,26] In studies focusing on oscillatory analyses, rhythmic neural activity in discrete frequency bands, including delta (1–4 Hz), theta (4–7 Hz), alpha (8–14 Hz), beta (14–26 Hz), gamma (30–50 Hz), and high gamma (>50 Hz) are examined separately to quantify rhythmic changes in the underlying neural population during cognitive processing.[25] Note that MEG studies have generally focused on one or two frequency bands, based on their hypotheses, and that the precise definition of a band (e.g., alpha) is frequently determined statistically for a given study; for example, the definition of alpha may vary from 8–12 to 8–14 Hz in different studies and this range is commonly determined using a data-driven approach. Probing such frequency bands separately enhances the signal-to-noise ratio of each, and enables identification of high-frequency (e.g., gamma range) neural responses that are often of lower amplitude. These high-frequency responses are widely thought to be critical for regional interactions across active brain networks,[26] and thus characterizing them is of major importance for understanding overall brain function. Of note, oscillatory analyses largely rely on the same event-related experimental designs, except that each trial is generally longer to provide an adequately "silent" baseline, with the relative responses again being of primary interest (i.e., active period compared with pre-stimulus baseline). As will become apparent from the studies discussed in this review, the application of oscillatory analysis methods to MEG signals and subsequent source reconstruction is a powerful approach for identifying the dynamics of brain function, and has facilitated major discoveries in several distinct areas.

Finally, a succinct overview of MEG terminology and metrics will promote understanding of the critical aspects of the studies discussed below, as some terms in the MEG field are somewhat unintuitive. Critically, the most common metric is the amplitude or power of the neural signal, and this can expressed for a particular voxel in the brain, a whole brain region, or at the MEG sensor-level. Pseudo-t is another common MEG unit; in functional maps using this metric, each pseudo-t value reflects the noise-normalized differential power between an active and a baseline period for a specific voxel. Other common units include the phase of the signal, based on a sensor- or voxel-level time series, as well as an array of functional connectivity measures that are typically computed using the voxel time series data. Such functional connectivity measures are thought to reflect inter-regional interactions, and common metrics include coherence, phase-locking factor, phase-lag index, and many others. Lastly, for oscillatory approaches, two common terms are event-related desynchronization (ERD) and event-related synchronization (ERS). An ERD refers to a decrease in band-limited power or amplitude (relative to a baseline) in response to a specific stimulus event or movement, whereas an ERS refers to an increase in band-limited power or amplitude (relative to baseline) in response to a specific event. ERD and ERS can be computed at the sensor- or voxel-level and are typically expressed in percentage units, reflecting the percent increase or decrease relative to the baseline period.

3. NeuroHIV in the cART era

As mentioned above, the advent of combination antiretroviral therapy shifted the nature of HIV-infection from a terminal illness to a chronic manageable condition, but HIV-infected patients remain at a significantly elevated risk of developing cognitive impairments, with 35–70% of all patients exhibiting at least subtle deficits on tests of neuropsychological function.[1–7] Numerous neuropsychological studies have documented aberrations across a broad range of functional domains (e.g., attention, fine motor), but the basic pathophysiology remains largely unresolved. Below, we review the MEG studies of neuroHIV that have been conducted to date. Each subsection focuses on studies that targeted a distinct cognitive construct and/or sensory modality, which we hope improves clarity and helps to illuminate the specific areas most in need of further research.

3.1 Resting state and reliability studies

Consistent with the functional MRI (fMRI) literature, the most active area of MEG research in neuroHIV has been resting-state studies. Of course, resting-state studies have the advantage that they are easier to implement and that there is no chance of the imaging results being biased by differences in behavioral performance. Given these and other benefits, resting-state investigations have dominated the search for biomarkers across many different disease states, and this is certainly true for HAND and other neural alterations associated with neuroHIV.

One of the earliest studies in this field focused on the test-retest reliability of resting-state MEG signals in HIV-infected adults and matched healthy controls. All participants completed two 5 min sessions (i.e., eyes-open and eyes-closed) of resting-state MEG, as well as neuropsychological testing, and then returned 24 weeks later for a follow-up MEG session that was identical to the first.[27] For analysis, the authors divided the sensor-level MEG signals into eight sub-regions and five canonical frequency bands (e.g., left occipital alpha) and then focused on the power in each region. The study's primary findings were that the MEG measurements were highly stable in both groups, as determined by the intra-class correlation (ICC) coefficient. In fact, the median stability was $r=0.80$ and $r=0.72$ during the eyes-closed and eyes-open conditions, respectively, with the highest values in the temporal lobes and the lowest in the frontal lobes.[27] Spectrally, the gamma frequency range was the least reliable across both conditions, while the other canonical bands were far more reliable, although there was some variation by region.[27] Regardless, median ICC values of 0.80 and 0.72 indicate that the measurements are highly reliable and bode well for the promise of MEG as a marker of disease. Beyond the reliability analysis, this study also showed that resting gamma power in the right occipital and right frontal region was stronger in uninfected controls relative to HIV-infected patients, and that individuals with high theta-to-gamma ratios tended to have lower cognitive test performance, regardless of serostatus.[27] Overall, this study provided clear evidence that MEG is reliable, as well as the first indication that MEG signals may be capable of distinguishing adults with and without HIV-infection.

A later study by the same group went a step further and looked at functional connectivity during the resting state.[28] Specifically, the authors collected 5 min of MEG data during eyes-closed rest and computed the sensor-by-sensor mutual information, which is a metric thought to reflect functional connectivity. Critically, the authors identified a group of MEG sensor pairs where the mutual information metric distinguished HIV-infected and uninfected participants.[28] These sensor pairs were generally between left occipital and right frontal areas of the MEG helmet, and across these groupings the mutual information value was higher in the uninfected controls compared to the HIV-infected patients. Furthermore, the mean mutual information value averaged across all sensor pairs successfully discriminated the HIV-infected patients from the uninfected controls with only one exception, which suggests that measures of functional connectivity may have high sensitivity and specificity for identifying HIV-related neural alterations.[28]

Finally, a more recent MEG study focused on older HIV-infected participants and included a relatively larger sample than the previous resting-state investigations. In this study,[29] older participants (half infected with HIV) underwent 6-min of eyes-open rest during MEG, and the resulting data were projected into the brain and regional time series were extracted for 29 distinct brain regions.[29] These time series were then subject to spectral analyses to quantify power in five canonical frequency bands (i.e., delta, theta, alpha, beta, and gamma). The key findings were that resting-state beta activity was significantly reduced in bilateral regions of the supplementary motor area, paracentral lobule, posterior cingulate, and the superior parietal lobule of older HIV-infected participants compared with demographically-matched uninfected controls. Further, the authors found that beta activity in the posterior cingulate, a key region in the default-mode network, was significantly correlated with scores on the Hopkins Verbal Learning Test-Revised, such that higher (better) scores were associated with stronger beta activity.[29] Unlike previous studies,[27] the authors did not find group differences in gamma activity or any other band besides beta. However, this disparity could reflect that the more recent study[29] focused on older participants, and/or differences in the methodological approach adopted by the authors (e.g., sensor- vs. source-level analyses). Regardless, these studies have advanced understanding of resting-state neurophysiological alterations in neuroHIV, and illuminated key questions that remain unanswered; thus, future work is certainly needed.

3.2 Motor control and processing

While neuropsychological testing has documented abnormalities across a broad range of functional domains, deficits in fine motor control and reduced psychomotor speed have been some of the most common findings in HAND.[9,30–33] In fact, this was partially reflected in the original nomenclature, which included diagnostic categories for HAD (three subtypes) and a less severe condition termed minor cognitive motor disorder (MCMD).[34] The MCMD category was phased out in 2007 with the adoption of the current classification system, which includes HAD, mild neurocognitive impairment, and asymptomatic cognitive impairment.[10] Nonetheless, motor-related deficits remain a critical component of neuroHIV and are an essential feature of HAND assessment.[35] Furthermore, despite the frequency of motor control and psychomotor speed impairments in neuroHIV, very few studies have attempted to identify the neural bases of such deficits, and in fact only one such MEG study of motor control has been published to date. This is problematic as non-motor issues like abnormal executive functioning can sharply decrease performance on measures of motor coordination and psychomotor speed, and ultimately masquerade as pure motor deficits in the absence of extensive neuropsychological testing.

The one existing neuroHIV motor study focused on the oscillatory dynamics of motor planning operations using a finger-tapping task during MEG. Briefly, recent neurophysiological studies of motor planning in humans have largely focused on an event-related desynchronization (ERD) in the 14–28 Hz beta range, which has been termed the peri-movement beta ERD.[36–44] This response has been by far the most widely studied of the three main motor oscillatory responses (see below), and typically begins about 500 ms before movement onset and persists through the duration of the movement, with the strongest response peak in the motor hand knob region of the precentral gyrus contralateral to movement.[42–44] Other regions that generate a peri-movement beta ERD, especially during complex movements, include the supplementary motor area (SMA), premotor cortices, parietal cortices, and cerebellum.[36,38,39,42–45] Many independent studies have probed the functional role of the peri-movement beta ERD, and these studies have overwhelmingly supported the notion that this response is crucial to motor planning and movement selection. For example, the response is known to be modulated by the complexity of the movement to be executed,[38] the certainty of the pending movement direction,[46] the similarity between potential movement plans,[47] and other cue-related factors.[48] Beyond the beta ERD, there is a resynchronization in the beta band that follows movement termination, which has been termed the post-movement beta rebound (PMBR),[36,37,41–44,49] as well as a high-frequency gamma response that coincides with movement onset and has been linked to the motor execution signal.[43,50–55] However, neither of these responses have been examined in neuroHIV using MEG or any other modality, and thus we focus our background section on the beta ERD.

To investigate motor planning in neuroHIV, one study examined 24 participants (half with HIV—all virally suppressed) using MEG.[56] All participants completed an externally-paced finger tapping task using their right index finger during MEG, and the resulting magnetic signals were decomposed into the time-frequency domain and imaged using a beamformer. The key results indicated a robust peri-movement beta ERD from −350 to 250 ms (movement onset = 0 ms) in the bilateral precentral gyri of each group, with the strongest response being centered on the motor hand knob feature of the contralateral (left) precentral gyrus (Fig. 2).[56] As shown in Fig. 2, a strong beta ERD was also detected in the SMA of uninfected participants, but not seropositive participants. In regard to group differences, beta ERD responses were significantly stronger in controls relative to HIV-infected participants in the bilateral precentral gyri and the SMA (Fig. 3). Conversely, HIV-infected participants had stronger responses in the right dorsolateral prefrontal cortex, medial prefrontal cortices, and a small area of the left inferior frontal gyrus. Interestingly, the beta response amplitude in the left precentral gyrus and SMA was significantly correlated with performance on the Grooved Pegboard test (dominant hand), while performance on the Trailmaking Test A and B was correlated with response amplitude in multiple prefrontal regions of HIV-infected participants.[56]

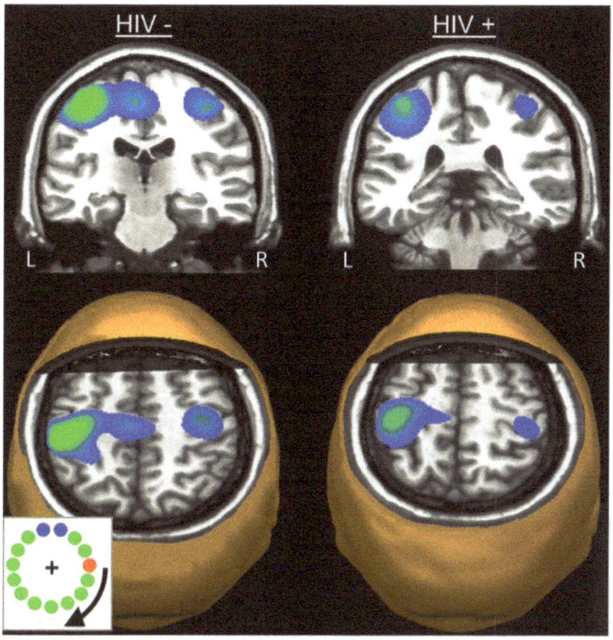

Fig. 2 Beta event-related desynchronizations (ERD) in each group. Strong beta ERD responses in the left precentral gyrus, extending posterior to include areas of the post-central gyrus were discerned in the statistical parametric maps (SPMs) of each group. Strong ERD responses in the supplementary motor area (SMA) and the right precentral gyrus were also apparent in the uninfected controls. Images are shown in neurological convention (Left = Left) and have been thresholded at $P < 0.001$, cluster-corrected. An image from the motor task is shown in the bottom left corner. Individuals made one movement when the red dot intersected area comprised of blue dots.

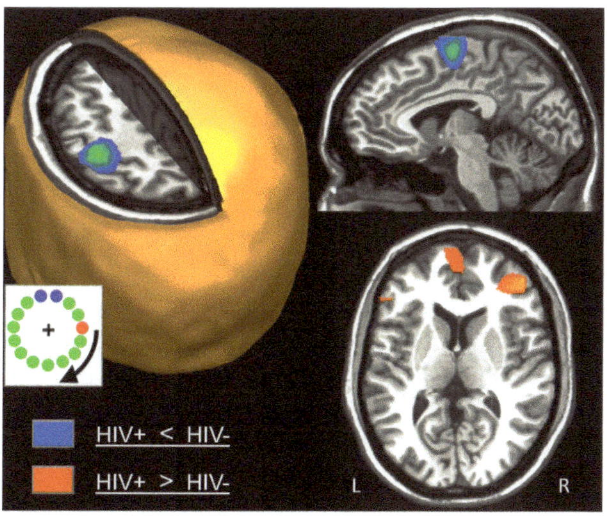

Fig. 3 Group differences in beta ERD activity. Uninfected adults exhibited significantly stronger beta ERD activity compared with HIV-infected adults in the left precentral gyrus, SMA, and a small area of the right precentral gyrus (blue areas; $P < 0.01$, cluster-corrected). In contrast, HIV-infected adults had significantly stronger beta ERD responses in the right (orange areas; $P < 0.01$, cluster-corrected). Again, an image of the task paradigm is shown in the lower left.

In summary, the one MEG study to date that has examined motor function in HIV-infection found robust neural differences during planning operations across a distributed motor network. These differences included reduced beta responses in the primary motor cortices (i.e., hypoactivity) of HIV-infected participants, as well as stronger responses in prefrontal regions involved in executive function. These responses correlated with neuropsychological tests of motor function in the infected group, which supports the veracity of these neurophysiological findings. Clearly, many questions remain concerning neuroHIV and cortical motor function, and studies that consider the full gamut of oscillatory motor responses (e.g., post-movement beta rebound and motor gamma responses) are direly needed. Studies that use motor tasks capable of eliciting behavioral differences would be especially informative.

3.3 Somatosensory processing

Deficits in motor control and processing generally need to be understood in the context of sensorimotor interactions and the degree of functioning within the somatosensory system. In other words, impaired somatosensory processing can have major negative consequences for motor control, and it is important to understand both systems to have a full grasp of motor function in any particular population. Thus far, there have been two MEG studies of somatosensory processing in neuroHIV. These studies were especially groundbreaking as there were no published fMRI or other functional imaging studies of somatosensation in HIV, although previous structural imaging studies had suggested widespread abnormalities in HIV-infected patients across the somatosensory strip and motor cortices.[57] Thus, there was clearly circumstantial evidence for aberrations in the somatosensory system in neuroHIV.

In the first study, 17 virally-suppressed HIV-infected adults and 17 demographically-matched uninfected controls completed a mechanoreceptor stimulation paradigm using an air-puff paradigm during MEG, and underwent high-resolution structural MRI.[58] The resulting MEG signals were transformed into the time-frequency domain, and time-frequency windows with significant responses (across all participants) relative to baseline were imaged using a beamformer. The whole-brain maps were then examined statistically for group differences and combined with structural gray matter maps derived from voxel-based morphometry using VBM8 and DARTEL implemented in the Statistical Parametric Mapping Software

(SPM8; Wellcome Trust Centre for Neuroimaging, http://www.fil.ion.ucl.ac.uk).[59] The key MEG findings were that all participants exhibited an increase (i.e., synchronization) in 4–8 Hz activity from 10 to 260 ms following stimulation onset, which mapped to the left postcentral gyrus directly posterior to the motor hand-knob feature. Interestingly, uninfected participants also exhibited a strong decrease in 8–14 Hz alpha activity during the same time period (Fig. 4),[58,60] which mapped to the left prefrontal cortex. In regard to group differences, the 4–8 Hz theta response was stronger in uninfected participants, which gave rise to a significant group difference in the left postcentral gyrus directly posterior to the hand-knob feature of the precentral gyrus. Likewise, the prefrontal alpha decrease was only observed in uninfected participants, which resulted in a relatively large group difference cluster centered on the left inferior frontal gyrus (Fig. 4).[58] The authors also extracted the peak group difference voxel from each of the significant clusters, and conducted a sensitivity/specificity analysis for both HIV-infection and cognitive impairment (9 of the 17 HIV-infected adults had HAND). This analysis revealed very high (>85%) sensitivity and specificity for identifying both HIV-infection and cognitive impairment based solely on these MEG metrics.[58] Finally, the VBM results indicated reduced gray matter in HIV-infected adults across several brain regions, including the left parahippocampal and lingual gyri, left postcentral gyrus, right cerebellum, and several other regions. Of particular interest, the cluster of reduced gray matter in the left postcentral gyrus was anatomically coincident with the reduced MEG theta responses observed in the HIV-infected adults, which suggests that at least some functional differences in this population may be attributable to structural aberrations, although others (e.g., prefrontal alpha) appear to be more purely functional in nature.[58]

A larger and more recent study has also examined somatosensory processing in adults with HIV, although this study used paired-pulse electrical stimulation to elicit somatosensory cortical responses, which targets decidedly different experimental questions.[61] Briefly, 71 participants (43 with HIV) completed a paired-pulse median nerve stimulation paradigm during MEG. The pulses were separated by 500 ms, which allowed the authors to look at sensory gating responses. Such gating is a neurophysiological phenomenon whereby the response to the second stimulus in a pair of identical stimuli is attenuated. This attenuation is thought to reflect the capacity of the CNS to filter less relevant information and preserve resources for behaviorally-relevant stimuli.[62] Substantial evidence suggests that patients with cerebral palsy, schizophrenia, and even healthy older adults exhibit

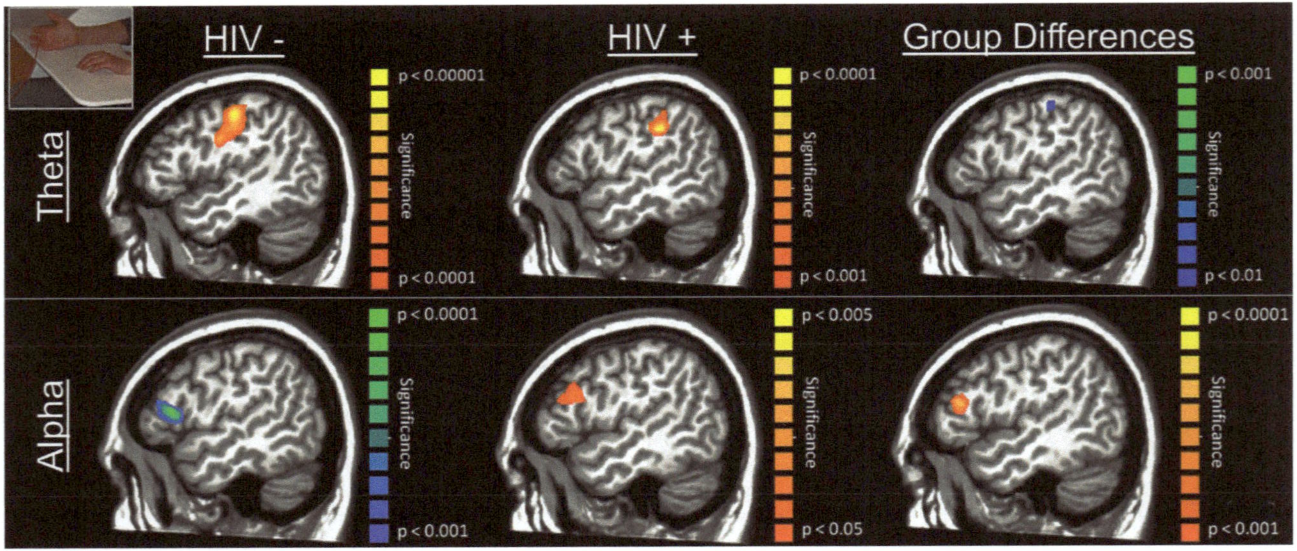

Fig. 4 Significant theta (4–8 Hz) and alpha (8–14 Hz) neuronal activity following tactile stimulation to the right hand. (Top): Uninfected controls (left) and HIV-infected adults (middle) exhibited theta responses in the left postcentral gyrus following mechanoreceptor stimulation, with the neural activity being stronger and more consistent in uninfected controls (note different *P*-value scales). The group effect (right; Controls > HIV-infected adults) was restricted to a small area of the left somatosensory cortex, and no brain areas had stronger 4–8 Hz activity in HIV-infected participants. Note the task design image in the top left corner. (Bottom): Uninfected controls exhibited a strong decrease in local alpha activity in the left prefrontal cortex (left) during the 0.01–0.26 s time window, whereas HIV-infected adults had a small increase in alpha activity in this same brain region (middle) and time window. Thus, the responses were in the opposite direction, which gave rise to a group effect of higher alpha activity in the HIV-infected participants (right panel). Again, note the different scales for *P*-values across images.

impaired gating of auditory and/or somatosensory stimuli, which is thought to reflect an impairment in bottom-up, pre-attentive inhibitory processing.[62–65] Since attentive and/or top-down inhibitory processing is thought to be a critical deficit in neuroHIV (see later sections), such gating studies should provide key information in this population. In the study by Spooner and colleagues,[61] the MEG signals were transformed into the time-frequency domain, revealing a robust response from 20 to 75 Hz in all participants during the 50 ms window following each stimulation (i.e., 0–50 and 500–550 ms). These responses were separately imaged in each participant and then averaged group-wise to reveal that the same region of primary somatosensory cortices generated the neural response in each group and to each stimulation. The authors then extracted the time series of the peak voxel in the grand averaged map for all subsequent analyses. The key findings indicated significantly reduced neural responses to both the first and second stimulation in HIV-infected participants relative to uninfected controls (Fig. 5, top panel). In regard to gating, both groups exhibited significant sensory gating responses (i.e., response to the second stimulus was significantly weaker than the response to the first), but there was no group difference in the magnitude of the gating effect. The latter finding was surprising, as numerous studies[62,63,65] have shown reduced gating in individuals with neurological and psychiatric disorders and such a reduction was hypothesized by the authors. Interestingly, follow-up analyses indicated that the peak gamma frequency of the second response relative to the first response was significantly higher in HIV-infected adults compared to controls (Fig. 5, bottom left panel), and the authors argued that this alteration could be compensatory and underlie the preserved sensory gating function seen in these patients.[61] Finally, since several recent studies have connected increased spontaneous activity during the baseline to aging,[39,64] and given evidence of accelerated aging in HIV,[66] the authors evaluated the power of spontaneous activity in the 20–75 Hz range in the same peak voxel as was used in their main analyses. This revealed that HIV-infected adults indeed had significantly elevated spontaneous activity relative to uninfected controls. Although future studies are needed to elucidate this effect, such elevated spontaneous gamma activity in HIV may underlie the reduced response amplitude observed in patients in the current study.[67]

In summary, only two studies have examined somatosensory function in adults with HIV and both of these investigations focused on virally-suppressed populations and found widespread evidence for dysfunction in key somatosensory circuitry. These studies used different stimulation

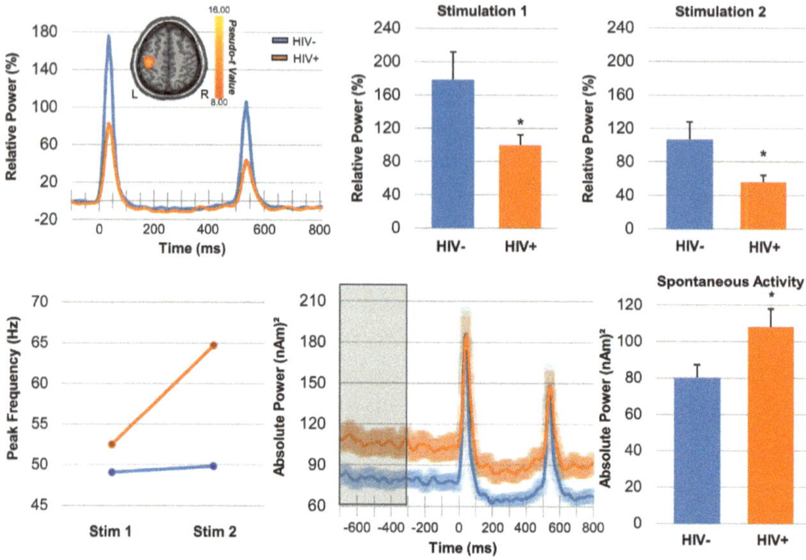

Fig. 5 Somatosensory gamma (20–75 Hz) activity following electrical stimulation. (Top): HIV-infected adults exhibited reduced response power to somatosensory stimulation. Disease-related power differences in response to each electrical stimulation were seen in the relative power envelope for the 20–75 Hz band (far left). Two-sample *t*-tests revealed reduced response power to stimulation 1 (middle) and 2 (far right) in HIV-infected adults compared to controls (* $P < 0.05$). (Bottom): HIV-infection modulated the peak gamma frequency in the somatosensory cortex (far left); a mixed-model ANOVA revealed a significant stimulation-by-group interaction, which indicated that HIV-infected adults had a higher peak gamma frequency in response to the second stimulation relative to controls ($P = 0.045$). The time series extracted from the peak voxel and averaged across each group revealed elevated spontaneous neural activity immediately preceding stimulus onset in HIV-infected adults (middle). Two-sample *t*-tests revealed significantly elevated gamma activity during the baseline period (−700 to −300 ms) in adults with HIV relative to controls (* $P < 0.05$).

approaches and focused on different spectral windows (i.e., theta and gamma), but ultimately reached many of the same conclusions. The recent work showing preserved gating along with increased spontaneous activity and higher peak frequencies is especially worthy of follow-up, as these parameters could directly contribute to aberrant primary sensory responses. Finally, these findings have direct implications for understanding motor dysfunction in neuroHIV and future studies should evaluate these together within the same study, as abnormalities within the somatosensory system could impact motor function directly and lead to key impairments in daily activities.

3.4 Working memory

Clinically-significant impairments in working memory are common in adults infected with HIV and are widely believed to contribute to other higher-level deficits in reasoning and decision making, which in-turn have a major impact on daily functioning and societal independence.[9,10,68] These deficits are known to extend across both verbal and spatial domains, but their precise cognitive and neurological origins remain largely unknown. Several neuropsychological studies have suggested that impairments in working memory arise from encoding deficiencies, whereas others have indicated that central executive processes are likely the origin.[33] Most functional neuroimaging studies have used fMRI and tended to support both views, but their conclusions remain at least partially tentative as aberrations in encoding stimuli would affect the maintenance of such memory representations, and these studies have not been able to temporally distinguish encoding and maintenance periods.[33,68–70] More broadly, previous functional imaging studies in HIV-infected adults have often reported abnormalities in both primary sensory and higher-order frontal and prefrontal cortices, which again conflates the encoding and maintenance operations serving working memory function.[69–74] Given these data, the precise timing of MEG could be especially valuable in identifying the cognitive (i.e., encoding vs. maintenance) and neural origins of working memory deficits in HIV, and to date only one such study has directly addressed these questions in the verbal domain of working memory.[75]

In this study, 36 older adults (half infected with HIV, all virally-suppressed) completed a verbal working memory task during MEG. Both groups were demographically-matched in regard to age, sex, ethnicity, and handedness, with the mean age across both groups being about 58 years at the time of scan.[75] The task was specifically designed to distinguish encoding and maintenance operations. Briefly, participants were instructed to memorize six letters that appeared in a grid (encoding). After 2.0 s these letters disappeared from the grid (maintenance period) and 3.0 s later a probe letter appeared in the grid (retrieval period), with the participants being instructed to respond "yes" or "no" via button press as to whether the probe letter was included in the original six-letter encoding grid (Fig. 6). This task had been validated by the investigators in many previous studies focuses on normative populations and those with other neurological and psychiatric diseases.[76–83] The key findings were that the uninfected adults performed significantly better on the task (~82.5% accurate) than did the adults with HIV (~76% accurate). In regard to the MEG data, all participants exhibited

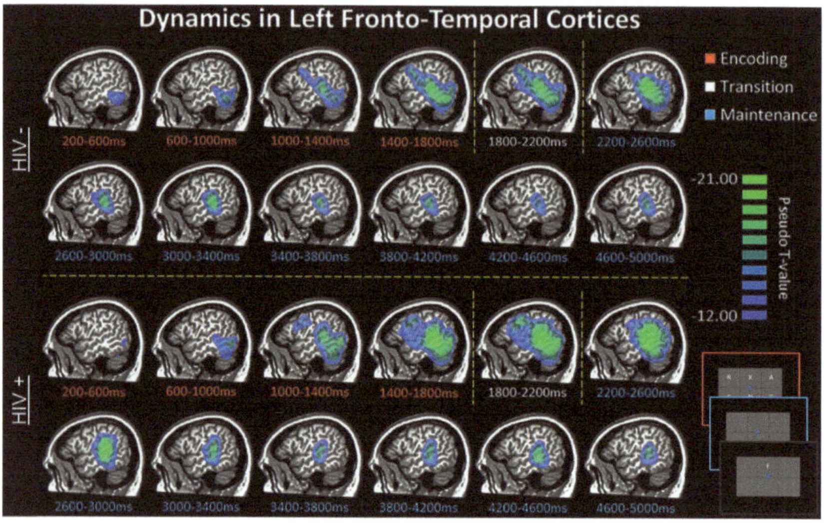

Fig. 6 Oscillatory activity during verbal working memory. Group mean beamformer images for uninfected older adults (top) and demographically-matched HIV-infected adults (bottom) are displayed across all time bins spanning the encoding (red labels), transition (white labels), and maintenance (blue labels) phases. An image of the task design is shown in the bottom right corner. Across both groups, there was a sustained decrease in alpha/beta oscillatory activity in left fronto-temporal cortices throughout most of the encoding and maintenance periods. This decrease began in posterior regions early in the encoding phase, and then spread anterior and superior to include language-related cortices in the left supramarginal and prefrontal areas during the latter half of encoding and early maintenance period. These responses slowly dissipated throughout the remainder of the maintenance phase.

a robust decrease in 9–16 Hz alpha/low-beta activity during the encoding period in left occipito-temporal sensors, as well as a robust increase in 9–12 Hz alpha activity in occipital cortices during maintenance.[75] To identify the time course of these responses, the spectral windows were imaged using a beamformer in 400 ms windows starting 200 ms after the onset of the encoding grid and extending until the onset of the probe letter (5000 ms). The results indicated a strong decrease in alpha/low-beta activity that began in the posterior occipital cortices and spread anteriorly to include the left inferior and superior temporal cortices, left supramarginal gyrus, and left inferior frontal cortices (Fig. 6). These dynamics appeared to be similar in HIV-infected and uninfected participants, especially during the encoding period as there were no group differences observed throughout this period.[75] In contrast, during memory maintenance, HIV-infected adults exhibited significantly stronger responses (i.e., alpha decreases) in the left supramarginal gyrus during the earliest time period, and this was sustained through the majority of the maintenance period (Fig. 7). Similar sustained group differences were also observed in the left inferior frontal gyrus and left cerebellum. In contrast, the opposite pattern was observed in the homologue right inferior frontal gyrus, as uninfected participants exhibited stronger alpha decreases relative to HIV-infected adults (Fig. 7). Controls also exhibited a significantly stronger alpha synchronization (i.e., increase) in the occipital cortices throughout the majority of the maintenance period

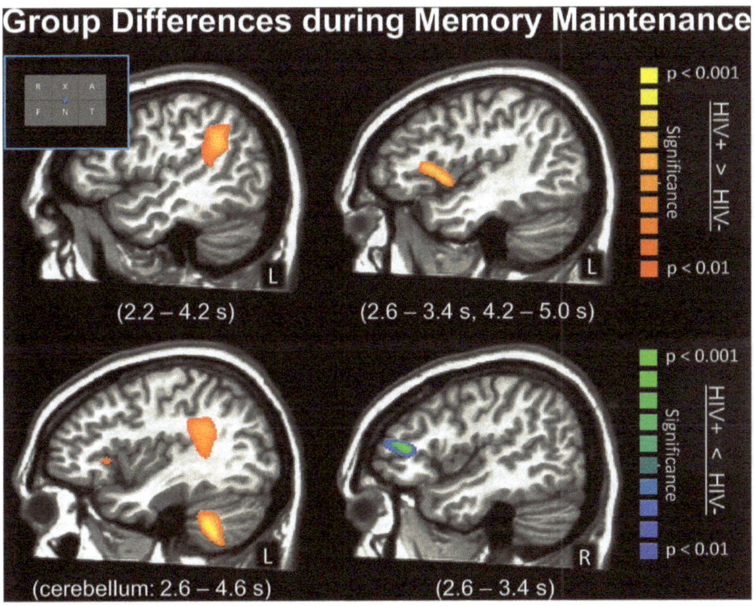

Fig. 7 Group oscillatory differences during working memory maintenance. Statistical parametric maps showing group differences during the maintenance period, with the time windows containing significant effects listed in the parentheses under each image. (Top Left) HIV-infected adults had significantly stronger alpha decreases (hyper-activation) in the left supramarginal cortices from 2.2 to 4.2 s after the onset of the encoding grid, which corresponded to the majority of the maintenance period. (Top Right) Stronger alpha decreases (hyper-activation) in HIV-infected adults were also observed in the left inferior frontal gyrus from 2.6 to 3.4 s and from 4.2 to 5.0 s. (Bottom Left) Stronger alpha decreases in HIV-infected adults were also observed in the left cerebellum from 2.6 to 4.6 s; the group differences in the supramarginal and left inferior frontal gyrus can also be seen in this image. (Bottom Right) In contrast, uninfected adults exhibited stronger alpha decreases in the right IFG compared to HIV-infected adults, which is consistent with healthy aging studies showing recruitment of homologue cortices in the right hemisphere to aid in ongoing processing. All maps have been thresholded at ($P < 0.01$, corrected).

and the amplitude of this response was significantly correlated with accuracy on the working memory task in controls, but not HIV-infected adults. Lastly, this study also found that the strength of the alpha decreases in the left supramarginal gyrus during maintenance were significantly correlated with performance on the working memory task, as well as global neuropsychological function and the attention and memory composite scores in HIV-infected patients.[75]

Given the widespread consequences of working memory dysfunction for other aspects of cognitive processing, as well as personal independence and quality of life, it is surprising that so few studies have examined this domain in HIV-infected adults. Beyond the limited number of fMRI studies, only one MEG study has been conducted to date. As mentioned above, this study clearly indicated that the impairments primarily reside in memory maintenance and not stimulus encoding, as there were no group differences during the 2.0 s encoding time window. Furthermore, during maintenance HIV-infected adults exhibited hyper-activations in the well-established left hemispheric verbal working memory networks, and did not widely utilize right hemispheric homologue regions to the same degree as age matched controls. The latter finding is particularly important, as studies have shown that healthy older adults more broadly utilize homologue regions during cognitively-demanding working memory paradigms relative to their younger peers.[81] In addition, this study showed that the well-known occipital alpha synchronization during maintenance is abnormal in HIV-infected adults, and this almost certainly contributed to their poor performance on the working memory task.[76–83] Obviously, much remains to be discovered in this area and future studies should focus on both spatial and verbal working memory to determine if both are affected equally.[84,85] Likewise, therapeutic studies targeting memory maintenance operations could be helpful in reducing the burden of working memory deficits in this population.

3.5 Visual-spatial processing and visual attention

Visual processing and visual attention has been one of the most frequent areas of investigation in functional neuroimaging studies of neuroHIV. Early fMRI studies reported aberrations in visual-perceptual and visual attention networks in HIV-infected adults, and these findings have generally been confirmed and expanded in more recent work. For example, Chang and colleagues found that HIV-infected patients exhibit markedly increased fMRI activation in lateral occipital, posterior parietal, and prefrontal regions during attention tasks.[86,87] More recent studies have focused more exclusively on visual cortex and found that HIV-infected adults have reduced activation in primary visual cortex during a visual stimulation task relative to healthy controls,[74] as well as reduced cerebral-blood-flow to the same visual regions during the resting-state.[73] Corruption of parietal- and fronto-visual functional connectivity in the resting state has also been reported in HIV-infected adults, and these fronto-visual attention networks align particularly closely with those that are compromised in advanced aging.[88] In regard to MEG studies, there have been a total of three to date, with the earliest focused on basic visual processing,[89] a second focused on visuo-spatial processing and attention,[90] and a third that examined visual selective attention.[91] We will review each of these studies in turn.

The first study included 23 participants (12 HIV-infected) who were group matched on major demographic factors. All participants completed a basic visual processing task during MEG, which involved fixating on a crosshair while a small checkerboard box appeared in the top right visual field quadrant.[89] All MEG data were transformed into the time-frequency domain, which revealed strong oscillations from 6 to 12 Hz during early stimulus processing (i.e., 50–225 ms, with 0 ms being stimulus onset). Source imaging of this time window revealed activity in the right dorsolateral prefrontal cortex, the posterior calcarine cortex, the posterior cingulate, and the frontal eye fields. In regard to group differences, the key findings were that controls showed greater 6–12 Hz power in the right frontal eye fields, right dorsolateral prefrontal cortex, and the posterior cingulate compared with HIV-infected adults.[89] Interestingly, the authors also found that the amplitude of responses in both the right dorsolateral prefrontal cortices and right frontal eye fields of HIV-infected patients was positively correlated with the retention index of the Hopkins Verbal Learning Test Revised (HVLT-R), and this effect survived stringent Bonferroni correction. In other words, the more "control-like" the responses in these regions, the higher the retention index score was in the HIV-infected group, which supports the notion that these oscillatory differences are integral to cognitive dysfunction.[89]

While this early study provided significant insight, later MEG studies of visual-spatial processing and selective attention have added a critical new dimension. These studies have focused on the key question of whether MEG measures of visual attention/processing can be used to distinguish HIV-infected adults with HAND from those who are cognitively unimpaired. One study focused on neural oscillatory activity in visual occipital cortices and examined 67 participants (43 HIV-infected) who were divided into three demographically-matched groups, including uninfected controls, unimpaired HIV-infected adults, and adults with HAND.[90] All participants underwent MEG during a visual-spatial processing task that required participants to respond (i.e., button press) regarding the spatial location of the visual stimulus. This task is known to elicit robust, multispectral oscillatory responses.[92–94] The resulting MEG data were probed for significant oscillatory visual responses and, in agreement with previous studies using this task,[92–94] the authors observed strong theta, alpha, and gamma responses in MEG sensors near occipital cortex during early stimulus processing. These three visual responses were imaged in each participant using a beamformer and the resulting images were averaged across groups to identify the peak voxels for each response. The time series for each of these peaks was then extracted for each person and used in the group-wise statistical analyses. The behavioral results indicated that uninfected controls responded significantly faster than both the unimpaired HIV-infected and HAND groups, and that the group with HAND also responded significantly slower than the unimpaired HIV-infected group (Fig. 8, top panel).[90] Adults with HAND were also significantly less accurate on the task compared with the other two groups. As per the MEG data, the key findings were that visual theta responses were significantly stronger in the uninfected group relative to both unimpaired HIV-infected and HAND groups, and that the relationship between accuracy on the task and the strength of occipital theta activity significantly differed between the groups (Fig. 9). In regard to the alpha and gamma responses, no group differences were observed in regard to the strength of each oscillatory response. However, given recent findings in the aging and HIV literatures,[39,61,64] the authors also examined the power of spontaneous neural activity in each of these spectral windows and their results were groundbreaking. Essentially, spontaneous alpha activity in visual occipital cortices was sharply elevated in the group with HAND, but not the control or the unimpaired HIV-infected group (Fig. 10, top left panel). In contrast, spontaneous occipital gamma activity was significantly elevated in both unimpaired HIV-infected adults and those with HAND relative to uninfected controls (Fig. 10, bottom left panel).[90] In other words, based on the power of spontaneous activity during the pre-stimulus baseline, one could distinguish each of the three groups from each other. These findings were of major importance as they suggested that specific MEG indices may have the capacity to serve as diagnostic markers of cognitive impairment in neuroHIV, which would have a major impact as HAND diagnoses are currently based on extensive neuropsychological testing and the exclusion of

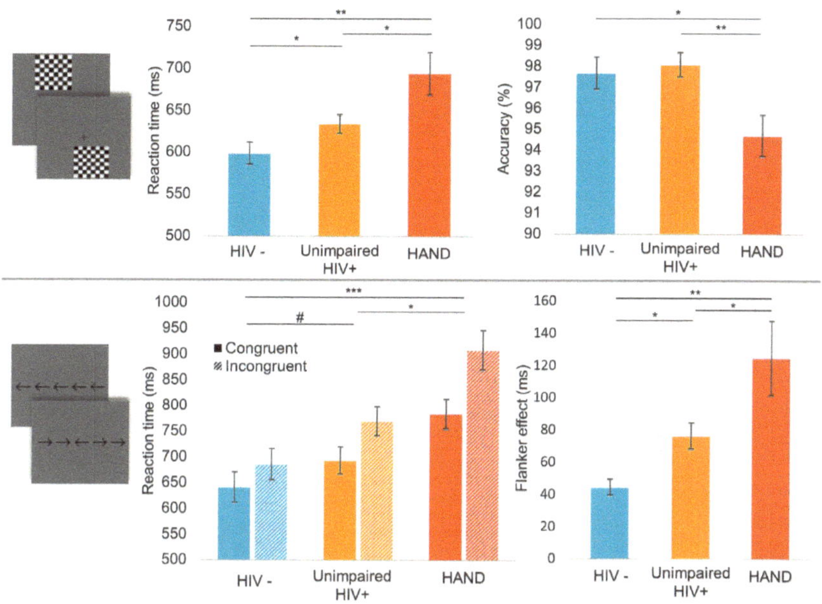

Fig. 8 Behavioral results from visuo-spatial processing and selective attention tasks. (Top): Images depicting the task design are shown to the far left, followed by reaction time (middle) and accuracy (right) results from the visuo-spatial task. In the bar graphs, participant group denoted on the *x*-axes. Reaction time (in milliseconds) is displayed on the *y*-axis of the graph in the middle, and accuracy (in % correct) is displayed on the *y*-axis of the graph to the right. Results indicated a clear stair-step effect whereby controls were significantly faster than unimpaired HIV-infected adults who were significantly faster than adults with HAND. As per accuracy, both controls and unimpaired HIV-infected adults were significantly more accurate than adults with HAND. (Bottom): As with the top, images depicting the task design are shown to the far left, with reaction time (middle) on the selective attention task in the middle. Data showing the flanker effect (Reaction time to Incongruent–Congruent trials) is shown to the far right. The results indicated that adults with HAND responded slower on both types of trials relative to both unimpaired HIV-infected adults and uninfected controls (middle). As per the flanker effect, a stair-step pattern was observed whereby controls were faster than both groups and unimpaired HIV-infected were faster than those with HAND ($^{\#}P<0.10$, $^{*}P<0.05$, $^{**}P<0.01$, $^{***}P<0.001$).

opportunistic infections and other confounding factors. Further, the overall pattern of increased spontaneous activity is consistent with the notion of accelerated aging in HIV-infected adults,[66] as multiple studies in healthy controls have supported the view that spontaneous cortical activity increases with age.[39,64] In sum, this MEG study provided critical new data and was the first to suggest that MEG may have the capacity to discern those HIV-infected adults with HAND from those who are cognitively unimpaired.

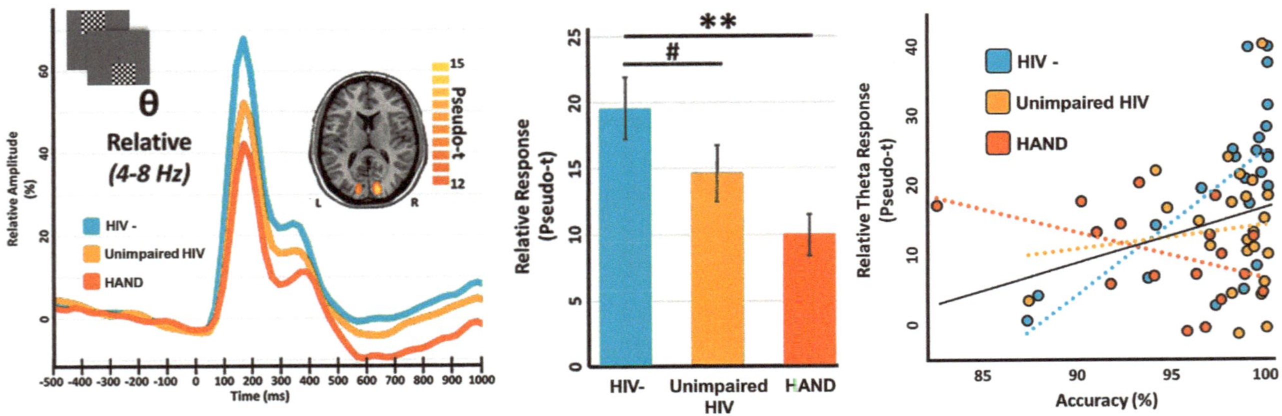

Fig. 9 Visual theta responses are reduced in HIV-infection and HAND. (Left): Peak voxel time series for each group; time (in millisecond) is denoted on the *x*-axis, with 0 ms defined as the onset of the visual stimulus, and relative amplitude (in % of baseline) denoted on the *y*-axis. Example task images are shown in the top left corner and the inlaid source map shows the peak-voxel locations used for pseudo-t extraction, along with its respective color scale legend. (Middle): Bar graph reveals extracted pseudo-t values averaged across participants per group, with significance levels of post-hoc statistical analyses indicated by asterisks above the bars ($^{\#}P < 0.10$, $^{**}P < 0.01$). (Right): Group-wise relationships between oscillatory theta responses and task accuracy. Theta oscillatory activity (in pseudo-t) is shown on the *y*-axis and task accuracy (in % correct) on the *x*-axis. The three dashed lines of best-fit represent the group-wise relationships between these variables, with the color legend inlaid above. The solid black overlaid line of best-fit represents this relationship across all participants.

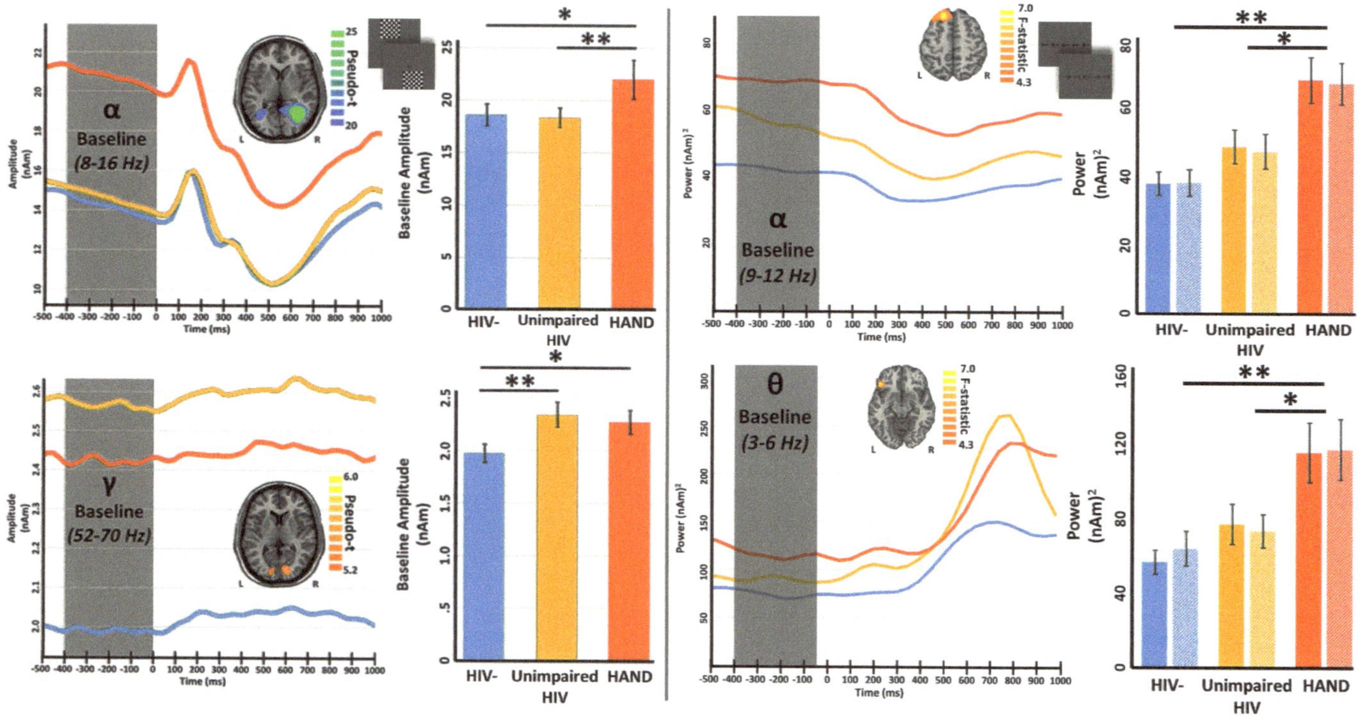

Fig. 10 Spontaneous neural activity is increased in HIV-infection and HAND. (Left): Peak voxel time series during a visuo-spatial processing task is separated by group for alpha (top) and gamma (bottom) on the far left. Time (in milliseconds) is denoted on the *x*-axes, with 0 ms defined as the onset of the visual stimulus, and absolute amplitude (in nAm) denoted on the *y*-axes. The inlaid maps show the peak-voxel location where data were extracted, along with its respective color scale legend. The bar graphs to the right represent pre-stimulus amplitude values averaged across participants per group, with significance levels of post-hoc statistical analyses indicated by asterisks above the bars (*$P < 0.05$, **$P < 0.01$). Spontaneous alpha distinguished the HAND group from the other two groups, while spontaneous gamma during the baseline distinguished controls from the other two groups. (Right): Peak voxel time series during a visual selective attention task is separated by group for alpha (top) and theta (bottom) activity. All axes are the same as above, except that the *y*-axes show power (nAm^2) instead of amplitude (nAm). As shown, alpha activity in the left superior frontal region differed between the HAND group and the other two groups, and the same was observed for theta activity in the left inferior frontal. Of note, spontaneous alpha in multiple regions distinguished all three groups (not shown).

The third MEG investigation in this area followed a similar model to the previous study in that there were three groups of participants (HAND, unimpaired HIV-infected, and uninfected controls), but this study focused on visual selective attention function and the impact of cognitive interference.[91] The study included 20 participants per group and the three groups were closely matched on major demographic factors, including age, sex, ethnicity, education, and handedness. Again, all HIV-infected participants were virally-suppressed. Participants completed a visual flanker task during MEG, which required them to respond as to the direction (left or right) of a middle arrow that was presented alongside flanking arrows that were either pointing in the same direction (Congruent) or the opposite direction (Incongruent; see bottom panel of Fig. 8). The classic finding in this task is that participants respond slower in the incongruent relative to the congruent condition due to the interference caused by the flanking arrows, and this difference in reaction time has been termed the flanker effect. Thus, the authors examined this behavioral effect as well as the associated neural oscillations using MEG. Their main behavioral findings were that adults with HAND were significantly slower than both unimpaired HIV-infected and control adults across both congruent and incongruent conditions, while the latter two groups marginally differed with controls being the fastest responders (Fig. 8, bottom panel).[91] In addition, all participants exhibited a significant flanker effect (i.e., responded slower in the incongruent condition) and the magnitude of this effect differed in a stepwise fashion across the three groups, being smallest in the uninfected controls and largest in the adults with HAND. As per the MEG signals, these data were decomposed into the time-frequency domain, which revealed strong responses in both the theta (3–6 Hz, 50–400 ms) and alpha (9–12 Hz, 250–650 ms) bands, in agreement with previous MEG studies using the flanker task.[95,96] The authors then imaged these responses separately for the congruent and incongruent condition, and subtracted the resulting congruent maps from the incongruent maps to derive a whole brain map of so-called flanker interference activity. An ANOVA of these maps revealed group differences in theta inference activity in the left inferior frontal cortices, left dorsolateral prefrontal cortices, left temporoparietal junction, and left inferior parietal regions (Fig. 11).[91] Likewise, group differences in alpha interference activity were found in the left superior frontal gyrus, middle frontal gyrus, supramarginal gyrus, and left angular gyrus, as well as the right dorsolateral prefrontal cortices (Fig. 11). Importantly, the authors found that left inferior frontal theta activity was a significant positive predictor of executive composite scores,

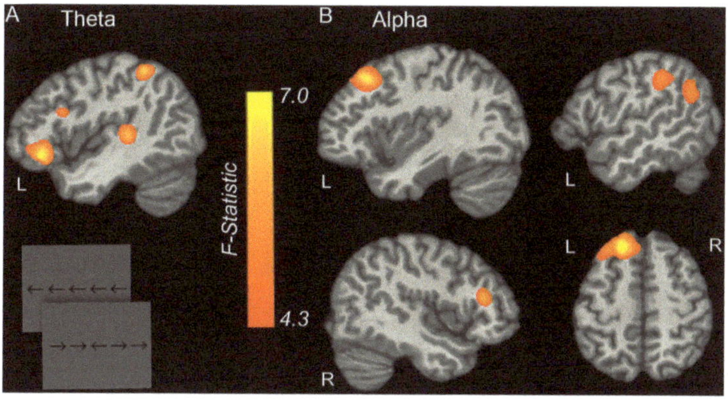

Fig. 11 Group differences in theta and alpha activity related to visual selective attention. (A) Whole-brain ANOVA of flanker interference activity (maps of Incongruent–Congruent trials) in the theta range showed significant group differences in the left inferior frontal gyrus, left dorsolateral prefrontal cortex, left temporoparietal junction, and the left inferior parietal cortex among uninfected controls, unimpaired HIV-infected adults, and adults with HAND. (B) Whole-brain ANOVA of alpha flanker interference activity across the same three groups revealed significant differences in the left superior frontal gyrus, left middle frontal gyrus, left supramarginal gyrus, left angular gyrus, and right dorsolateral prefrontal cortex. Both maps are shown at an alpha level of 0.05 ($F > 4.3$).

while the left dorsolateral prefrontal cortex theta response was a significant positive predictor of attention composite scores (Fig. 12). Further, groupwise post-hoc difference testing of these regions found that theta interference activity in both areas was significantly reduced in the group with HAND relative to the other two groups, and did not differ between controls and unimpaired HIV-infected adults.[91] Finally, the authors also examined whether there were group differences in spontaneous activity in each of the brain regions identified in the whole-brain theta and alpha ANOVAs. These analyses revealed that adults with HAND had increased spontaneous theta activity relative to infected controls in most of the regions identified in the whole-brain theta ANOVA, as well as increased spontaneous activity relative to unimpaired HIV-infected participants in the left inferior frontal and left temporoparietal junction regions (Fig. 10, right panel). In contrast, alpha activity was significantly elevated in the HAND group relative to both the unimpaired HIV-infected and uninfected controls groups in almost all regions identified in the whole-brain alpha ANOVA (Fig. 10, right panel).[91] Thus, in this study, the spontaneous alpha results alone (in several distinct brain regions) were enough to distinguish the three groups, and importantly made a clear distinction between the unimpaired HIV-infected adults and

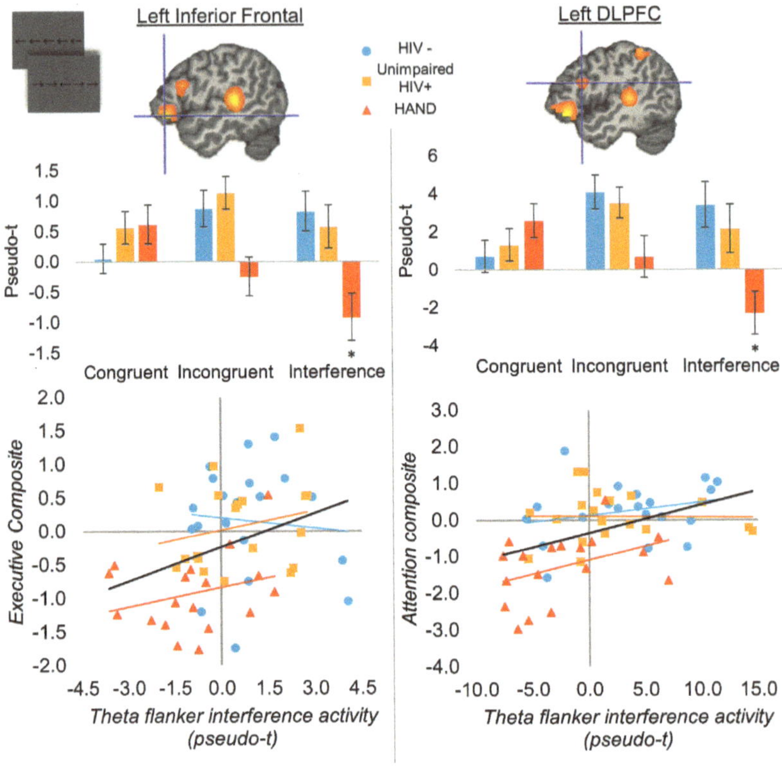

Fig. 12 Frontal theta activity predicts neuropsychological performance. (Left) Flanker interference activity (Incongruent–Congruent activity) in the left inferior frontal gyrus predicted executive function composite scores (*$P<0.05$). (Right) Flanker interference activity in the left dorsolateral prefrontal cortex predicted attention composite scores (*$P<0.05$).

those with HAND. Furthermore, the behavioral task was very sensitive to HIV-related changes in cognitive performance, as the flanker effect revealed a unique stepwise pattern of decline in visual selective attention function among the HIV-infected patients with and without cognitive impairment.

In conclusion, there have been three MEG studies of visual processing and/or visual attention function involving HIV-infected patients, and these studies have identified key aberrations across multiple nodes of the networks serving basic visual function, visuo-spatial processing, and executive function. Some of the most exciting findings have emerged in the past year with evidence that MEG markers of visual processing and visual selective attention may hold diagnostic value in distinguishing HIV-infected adults with and without cognitive impairment. The fact that these findings were observed in closely-matched, virally-suppressed individuals makes these data all the more exciting. Of course, further studies involving larger samples are necessary to confirm these findings, and to begin to move these data into prediction models for classifying groups. Regardless, visual processing studies in neuroHIV have provided critical new data in the past year, which have helped propel the functional neuroimaging of HIV community to new heights.

4. Conclusions and future directions

MEG is an ideal tool for examining the mechanisms of brain dysfunction in neuroHIV, as its spatiotemporal precision allows for accurate characterization of spectral alterations, as well as delays and slowing in specific neural systems. While most early studies focused on the resting-state, MEG studies of sensory processing in almost all modalities are now available and in the past 2 years multiple investigations of higher-order processes have now been published. Nonetheless, MEG studies of the neuronal aberrations associated with HIV-infection remain rare, with substantial work ongoing and many questions in need of further investigation.

Despite the sparsity of MEG investigations, several key conclusions can be drawn. First, aberrations in neural oscillatory activity stretch across all investigated sensory modalities, and in the same study are generally observed in both the primary sensory region (e.g., visual cortex) and higher-order brain areas such as the dorsolateral prefrontal cortex. These aberrations do not appear to be related to active viremia, as almost all MEG studies since 2013 have reported that their study populations were virally suppressed. Further, the notion that these abnormalities could be related to potential neurotoxic effects of certain antiretrovirals has not been supported in prior functional neuroimaging work.[97] Demographic factors also appear unlikely as virtually all studies to date closely matched their patient and control groups on age, sex, ethnicity, and other important factors. Alterations in neurovascular coupling in HIV are also unlikely,[98] as MEG is a direct measure of the physiology and should not be affected by such differences. This leads to the probable conclusion that such neural oscillatory aberrations are a key feature of the HIV-infected brain and not an artifact of inadequacies in study design or implementation.

Interestingly, the specific nature of these neural oscillatory abnormalities appears to differ by brain region and by cognitive/perceptual task. Some of these inter-regional disparities, especially the spectral findings, likely reflect low-level healthy differences in the underlying circuitry (e.g., healthy occipital neurons synchronize at the alpha rhythm when at rest), and thus the aberrations observed in HIV-infected adults likely reflect core dysfunction in these neural populations in regard to their specific abilities to compute, synchronize, and interact with neighboring and distant populations. Along these lines, it is interesting to note that HIV-related aberrations were often observed in the dominant spectral rhythm of the involved brain region. For example, widespread differences in beta activity were noted in the primary motor cortex of the only study to examine motor control in neuroHIV,[56] and beta is widely acknowledged to be the dominant spectral rhythm of the motor system. Likewise, tasks with major visual components all reported differences in alpha activity,[75,90,91] and differences in theta activity were often noted in the prefrontal cortex. Beyond these generalities, early MEG studies of the resting-state have highlighted deficits in fronto-visual circuits,[28] which were later replicated in studies directly examining visual selective attention.[91] Studies of somatosensory processing have universally shown deficits in the postcentral gyrus (primary somatosensory cortex) of HIV-infected adults.[58,61] The one investigation of working memory processing focused on the verbal domain and showed behavioral deficits, as well as aberrations in memory maintenance operations across a broad network of left hemispheric language regions and the right inferior frontal gyrus (i.e., homologue of Broca's area).[75] Several studies of visual processing and visual attention have also been conducted and these have shown multispectral deficits in the occipital visual cortices, as well as parietal and prefrontal regions.[89–91] Looking across studies it is interesting to note that studies of motor control,[56] somatosensory processing,[58] resting-state,[28] working memory,[75] visual processing,[89] and visual selective attention[91] all highlighted deficits in the prefrontal cortices of HIV-infected adults. Thus, conjunctive abnormalities in both modality-specific regions and prefrontal cortices appears to be a common thread across the majority of neuroHIV MEG studies to date. Finally, studies have shown that MEG measurements have high test-retest reliability in HIV-infected adults,[27] which is clearly important for the development of MEG-based diagnostic markers of HAND and other diseases.

In regard to future studies, one of the most exciting areas has been the recent discovery that specific MEG metrics may be able to distinguish unimpaired HIV-infected individuals from those with HAND.[90,91] This is a major breakthrough and future studies are needed to further refine the approach and test the predictive capacity of these markers in different HIV-infected populations (i.e., those not used in the development of the markers). The same and/or related MEG metrics may also be useful in monitoring HAND progression and establishing prognoses for HIV-infected individuals more broadly, but of course more work is needed. Regarding the latter, long-term longitudinal MEG studies would be the most informative route, so that the specific metric and clinical outcome could be continually monitored and compared. Concerning specific perceptual and cognitive domains of study, additional work is needed across the board. With no more than three studies in any specific area, there are numerous findings that need to be further confirmed and many lingering questions. Further, there are many domains that have yet to be investigated (e.g., auditory circuitry), and no studies to date have utilized executive function tasks, spatial working memory, nor multiple different forms of attention. Thus, many studies across many domains are required. The increasing availability of MEG should help facilitate this work over the next few years, as many new MEG sites are currently coming online. Finally, future studies should examine the impact of disease duration (time since diagnosis), CD4 nadir, substance abuse, and aging on HIV-related neural alterations, as these factors may be particularly important in understanding the pathophysiology and developing the most precise markers of HAND and its precursors.

Acknowledgments

This work was supported by NIH grants R01 MH103220 (T.W.W.), R01 MH116782 (T.W.W.), R01 MH118013 (T.W.W.), R01 DA047828 (T.W.W.), R03 DA041917 (T.W.W.), P30 MH062261 (T.W.W.), F31 AG055332 (A.I.W.), and NSF grant #1539067 (T.W.W.).

References

1. Nakagawa F, May M, Phillips A. Life expectancy living with HIV: recent estimates and future implications. *Curr Opin Infect Dis*. 2013;26(1):17–25. 23221765.
2. *Antiretroviral Therapy Cohort Collaboration*. Life expectancy of individuals on combination antiretroviral therapy in high-income countries: a collaborative analysis of 14 cohort studies. *Lancet*. 2008;372:293–299. 18657708. PMC3130543.
3. Heaton RK, Clifford DB, Franklin Jr DR, et al. HIV-associated neurocognitive disorders persist in the era of potent antiretroviral therapy: CHARTER Study. *Neurology*. 2010;75:2087–2096. 21135382. PMC2995535.
4. Heaton RK, Franklin DR, Ellis RJ, et al. HIV-associated neurocognitive disorders before and during the era of combination antiretroviral therapy: differences in rates, nature, and predictors. *J Neurovirol*. 2011;17:3–16. 21174240. PMC3032197.
5. Gannon P, Khan MZ, Kolson DL. Current understanding of HIV-associated neurocognitive disorders pathogenesis. *Curr Opin Neurol*. 2011;24:275–283. Review. 21467932.
6. Simioni S, Cavassini M, Annoni JM, et al. Cognitive dysfunction in HIV patients despite long-standing suppression of viremia. *AIDS*. 2010;24:1243–1250.

7. Grant I, Franklin Jr DR, Deutsch R, et al. Asymptomatic HIV-associated neurocognitive impairment increases risk for symptomatic decline. *Neurology*. 2014;82:2055–2062.
8. Sacktor N, McDermott MP, Marder K, et al. HIV-associated cognitive impairment before and after the advent of combination therapy. *J Neurovirol*. 2002;8:136–142. 11935465.
9. Cysique LA, Brew BJ. Neuropsychological functioning and antiretroviral treatment in HIV/AIDS: a review. *Neuropsychol Rev*. 2009;19:169–185. Review, 19424802.
10. Antinori A, Arendt G, Becker JT, et al. Updated research nosology for HIV-associated neurocognitive disorders. *Neurology*. 2007;69:1789–1799. 17914061.
11. Robertson KR, Smurzynski M, Parsons TD, et al. The prevalence and incidence of neurocognitive impairment in the HAART era. *AIDS*. 2007;21:1915–1921.
12. Dore GJ, McDonald A, Li Y, et al. Marked improvement in survival following AIDS dementia complex in the era of highly active antiretroviral therapy. *AIDS*. 2003;17: 1539–1545.
13. Heaton RK, Marcotte TD, Mindt MR, et al. The impact of HIV-associated neuropsychological impairment on everyday functioning. *J Int Neuropsychol Soc*. 2004;10: 317–331. 15147590.
14. Albert SM, Marder K, Dooneief G, et al. Neuropsychologic impairment in early HIV infection. A risk factor for work disability. *Arch Neurol*. 1995;52:525–530. 7733849.
15. Marcotte TD, Wolfson T, Rosenthal TJ, et al. A multimodal assessment of driving performance in HIV infection. *Neurology*. 2004;63:1417–1422. 15505158.
16. Hämäläinen M, Hari R, Ilmoniemi RJ, Knuutila J, Lounasmaa OV. Magnetoencephalography—theory, instrumentation, and applications to noninvasive studies of the working human brain. *Rev Mod Phys*. 1993;65(2):413–495.
17. Wilson TW. Noninvasive neurophysiological imaging with magnetoencephalography. In: Xiong H, Gendelman HE, eds. *Current Laboratory Methods in Neuroscience Research*. New York: Springer Press; 2014:293–311.
18. Supek S, Aine CJ, eds. *Magnetoencephalography: From Signals to Dynamic Cortical Networks*. Verlag: Springer Press; 2014.
19. Hansen PC, Kringelbach ML, Salmelin R, eds. *MEG: An Introduction to the Methods*. New York: Oxford University Press; 2010.
20. Hari R, Salmelin R. Magnetoencephalography: from SQUIDs to neuroscience. Neuroimage 20th anniversary special edition. *Neuroimage*. 2012;61(2):386–396.
21. Wilson TW, Heinrichs-Graham E, Proskovec AL, McDermott TJ. Neuroimaging with magnetoencephalography: a dynamic view of brain pathophysiology. *Transl Res*. 2016;175:17–36. 26874219. PMC4959997.
22. Murakami S, Okada Y. Contributions of principal neocortical neurons to magnetoencephalography and electroencephalography signals. *J Physiol*. 2006;575:925–936.
23. Hillebrand A, Barnes GR. A quantitative assessment of the sensitivity of whole-head MEG to activity in the adult human cortex. *Neuroimage*. 2002;16:638–650.
24. Cohen DS. Magnetoencephalography: evidence of magnetic fields produced by alpha rhythm currents. *Science*. 1968;161:784–786.
25. Jensen O, Spaak E, Zumer JM. Human brain oscillations: from physiological mechanisms to analysis and cognition. In: Supek S, Aine CJ, eds. *Magnetoencephalography: From Signals to Dynamic Cortical Networks*. Verlag: Springer Press; 2014:359–403.
26. Uhlhaas PJ, Pipa G, Lima B, et al. Neural synchrony in cortical networks: history, concept and current status. *Front Integr Neurosci*. 2009;3:17.
27. Becker JT, Fabrizio M, Sudre G, et al. Potential utility of resting-state magnetoencephalography as a biomarker of CNS abnormality in HIV disease. *J Neurosci Methods*. 2012;206(2):176–182.
28. Becker JT, Bajo R, Fabrizio M, et al. Functional connectivity measured with magnetoencephalography identifies persons with HIV disease. *Brain Imaging Behav*. 2012;6(3): 366–373.

29. Becker KM, Heinrichs-Graham E, Fox HS, et al. Decreased MEG beta oscillations in HIV-infected older adults during the resting-state. *J Neurovirol.* 2013;19:586–594. PMC3913174.
30. Hardy DJ, Vance DE. The neuropsychology of HIV/AIDS in older adults. *Neuropsychol Rev.* 2009;19:263–272.
31. Joska JA, Gouse H, Paul RH, Stein DJ, Flisher AJ. Does highly active antiretroviral therapy improve neurocognitive function? A systematic review. *J Neurovirol.* 2010;6: 101–114. Review.
32. Robertson K, Liner J, Heaton R. Neuropsychological assessment of HIV-infected populations in international settings. *Neuropsychol Rev.* 2009;19:232–249.
33. Woods SP, Moore DJ, Weber E, Grant I. Cognitive neuropsychology of HIV-associated neurocognitive disorders. *Neuropsychol Rev.* 2009;19:152–168. Review.
34. Janssen RS, Cornblath DR, Epstein LG, McArthur J, Price RW. Nomenclature and research case definitions for neurological manifestations of human immunodeficiency virus type-1 (HIV-1) infection. Report of a Working Group of the American Academy of Neurology AIDS Task Force. *Neurology.* 1991;41:778–785.
35. The Mind Exchange Working Group. Assessment, diagnosis, and treatment of HIV-associated neurocognitive disorder: a consensus report of the mind exchange program. *Clin Infect Dis.* 2013;56:1004–1017. 23175555.
36. Cheyne D, Bakhtazad L, Gaetz W. Spatiotemporal mapping of cortical activity accompanying voluntary movements using an event-related beamforming approach. *Hum Brain Mapp.* 2006;27:213–229.
37. Gaetz W, Macdonald M, Cheyne D, Snead OC. Neuromagnetic imaging of movement-related cortical oscillations in children and adults: age predicts post-movement beta rebound. *Neuroimage.* 2010;51:792–807.
38. Heinrichs-Graham E, Wilson TW. Coding complexity in the human motor circuit. *Hum Brain Mapp.* 2015;36:5155–5167.
39. Heinrichs-Graham E, Wilson TW. Is an absolute level of cortical beta suppression required for proper movement? Magnetoencephalographic evidence from healthy aging. *Neuroimage.* 2016;134:514–521.
40. Heinrichs-Graham E, Wilson TW, Santamaria PM, et al. Neuromagnetic evidence of abnormal movement-related beta desynchronization in Parkinson's disease. *Cereb Cortex.* 2014;24:2669–2678.
41. Jurkiewicz MT, Gaetz WC, Bostan AC, Cheyne D. Post-movement beta rebound is generated in motor cortex: evidence from neuromagnetic recordings. *Neuroimage.* 2006;32:1281–1289.
42. Wilson TW, Heinrichs-Graham E, Becker KM. Circadian modulation of motor-related beta oscillatory responses. *Neuroimage.* 2014;102(pt 2):531–539.
43. Wilson TW, Slason E, Asherin R, et al. An extended motor network generates beta and gamma oscillatory perturbations during development. *Brain Cogn.* 2010;73:75–84.
44. Wilson TW, Slason E, Asherin R, et al. Abnormal gamma and beta MEG activity during finger movements in early-onset psychosis. *Dev Neuropsychol.* 2011;36:596–613.
45. Kurz MJ, Becker KM, Heinrichs-Graham E, Wilson TW. Neurophysiological abnormalities in the sensorimotor cortices during the motor planning and movement execution stages of children with cerebral palsy. *Dev Med Child Neurol.* 2014;56:1072–1077.
46. Tzagarakis C, Ince NF, Leuthold AC, Pellizzer G. Beta-band activity during motor planning reflects response uncertainty. *J Neurosci.* 2010;30:11270–11277.
47. Grent-'t-Jong T, Oostenveld R, Jensen O, Medendorp WP, Praamstra P. Competitive interactions in sensorimotor cortex: oscillations express separation between alternative movement targets. *J Neurophysiol.* 2014;112:224–232.
48. Heinrichs-Graham E, Arpin DJ, Wilson TW. Cue-related temporal factors modulate movement-related beta oscillatory activity in the human motor circuit. *J Cogn Neurosci.* 2016;28:1039–1051.

49. Heinrichs-Graham E, Kurz MJ, Gehringer JL, Wilson TW. The functional role of post-movement beta oscillations in movement termination. *Brain Struct Funct*. 2017;222: 3075–3086.
50. Muthukumaraswamy SD. Functional properties of human primary motor cortex gamma oscillations. *J Neurophysiol*. 2010;104:2873–2885.
51. Grent-'t-Jong T, Oostenveld R, Jensen O, Medendorp WP, Praamstra P. Oscillatory dynamics of response competition in human sensorimotor cortex. *Neuroimage*. 2013;83:27–34.
52. Cheyne D, Bells S, Ferrari P, Gaetz W, Bostan AC. Self-paced movements induce high-frequency gamma oscillations in primary motor cortex. *Neuroimage*. 2008;42:332–342.
53. Gaetz W, Edgar JC, Wang DJ, Roberts TP. Relating MEG measured motor cortical oscillations to resting gamma-aminobutyric acid (GABA) concentration. *Neuroimage*. 2011;55:616–621.
54. Gaetz W, Liu C, Zhu H, Bloy L, Roberts TP. Evidence for a motor gamma-band network governing response interference. *Neuroimage*. 2013;74:245–253.
55. Heinrichs-Graham E, Hoburg JM, Wilson TW. The peak frequency of motor-related gamma oscillations is modulated by response competition. *Neuroimage*. 2018;165:27–34. https://doi.org/10.1016/j.neuroimage.2017.09.059. 28966082. PMC5826720.
56. Wilson TW, Heinrichs-Graham E, Robertson KR, et al. Functional brain abnormalities during finger-tapping in HIV-infected older adults: a magnetoencephalography study. *J Neuroimmune Pharmacol*. 2013;8(4):965–974. PMC3809128.
57. Thompson PM, Dutton RA, Hayashi KM, et al. Thinning of the cerebral cortex visualized in HIV/AIDS reflects CD4+ T lymphocyte decline. *Proc Natl Acad Sci USA*. 2005;102:15647–15652. PMC1266080.
58. Wilson TW, Heinrichs-Graham E, Becker KM, et al. Multimodal neuroimaging evidence of alterations in cortical structure and function in HIV-infected older adults. *Hum Brain Mapp*. 2015;36:897–910. PMC4491915.
59. Ashburner J. A fast diffeomorphic image registration algorithm. *Neuroimage*. 2007;38:95–113. 17761438.
60. Yousry TA, Schmid UD, Alkadhi H, et al. Localization of the motor hand area to a knob on the precentral gyrus: a new landmark. *Brain*. 1997;120:141–157.
61. Spooner RK, Wiesman AI, Mills MS, et al. Aberrant oscillatory dynamics during somatosensory processing in HIV-infected adults. *Neuroimage Clin*. 2018;20:85–91. PMC6070689.
62. Cromwell HC, Mears RP, Wan L, Boutros NN. Sensory gating: a translational effort from basic to clinical science. *Clin EEG Neurosci*. 2008;39:69–72.
63. Kurz MJ, Wiesman AI, Coolidge NM, Wilson TW. Children with cerebral palsy hyper-gate somatosensory stimulations of the foot. *Cereb Cortex*. 2017;28:1–8.
64. Spooner RK, Wiesman AI, Proskovec AL, Heinrichs-Graham E, Wilson TW. Rhythmic spontaneous activity mediates the age-related decline in somatosensory function. *Cereb Cortex*. 2019;29:680–688.
65. Thoma RJ, Meier A, Houck J, et al. Diminished auditory sensory gating during active auditory verbal hallucinations. *Schizophr Res*. 2017;188:125–131.
66. Gross AM, Jaeger PA, Kreisberg JF, et al. Methylome-wide analysis of chronic HIV infection reveals five-year increase in biological age and epigenetic targeting of HLA. *Mol Cell*. 2016;62(2):157–168.
67. Heinrichs-Graham E, McDermott TJ, Mills MS, et al. The lifespan trajectory of neural oscillatory activity in the motor system. *Dev Cogn Neurosci*. 2018;30:159–168. PMC5949086.
68. Sundermann EE, et al. Genetic predictor of working memory and prefrontal function in women with HIV. *J Neurovirol*. 2015;21:81–91.

69. Plessis SD, et al. HIV infection and the fronto-striatal system: a systematic review and meta-analysis of fMRI studies. *AIDS*. 2014;28:803–811. Review.
70. Ernst T, Chang L, Jovicich J, Ames N, Arnold S. Abnormal brain activation on functional MRI in cognitively asymptomatic HIV patients. *Neurology*. 2002;59:1343–1349.
71. Ernst T, et al. Declined neural efficiency in cognitively stable human immunodeficiency virus patients. *Ann Neurol*. 2009;65:316–325.
72. Chang L, Yakupov R, Nakama H, Stokes B, Ernst T. Antiretroviral treatment is associated with increased attentional load-dependent brain activation in HIV patients. *J Neuroimmune Pharmacol*. 2008;3:95–104.
73. Ances BM, et al. Resting cerebral blood flow: a potential biomarker of the effects of HIV in the brain. *Neurology*. 2009;73:702–708.
74. Ances BM, et al. HIV infection and aging independently affect brain function as measured by functional magnetic resonance imaging. *J Infect Dis*. 2010;201:336–340.
75. Wilson TW, Proskovec AL, Heinrichs-Graham E, et al. Aberrant neuronal dynamics during working memory operations in the aging HIV-infected brain. *Sci Rep*. 2017;7: 41568. PMC5290733.
76. Embury CM, Wiesman AI, Proskovec AL, et al. Altered brain dynamics in patients with type 1 diabetes during working memory processing. *Diabetes*. 2018;67:1140–1148.
77. Embury CM, Wiesman AI, Proskovec AL, et al. Neural dynamics of verbal working memory processing in children and adolescents. *Neuroimage*. 2018;185:191–197.
78. Heinrichs-Graham E, Wilson TW. Spatiotemporal oscillatory dynamics during the encoding and maintenance phases of a visual working memory task. *Cortex*. 2015;69: 121–130.
79. McDermott TJ, Badura-Brack AS, Becker KM, et al. Attention training improves aberrant neural dynamics during working memory processing in veterans with PTSD. *Cogn Affect Behav Neurosci*. 2016;16:1140–1149.
80. McDermott TJ, Badura-Brack AS, Becker KM, et al. Male veterans with PTSD exhibit aberrant neural dynamics during working memory processing: an MEG study. *J Psychiatry Neurosci*. 2016;41:251–260.
81. Proskovec AL, Heinrichs-Graham E, Wilson TW. Aging modulates the oscillatory dynamics underlying successful working memory encoding and maintenance. *Hum Brain Mapp*. 2016;37:2348–2361.
82. Proskovec AL, Heinrichs-Graham E, Wilson TW. Load modulates the alpha and beta oscillatory dynamics serving verbal working memory. *Neuroimage*. 2019;184:256–265.
83. Wiesman AI, Heinrichs-Graham E, McDermott TJ, Santamaria PM, Gendelman HE, Wilson TW. Quiet connections: reduced fronto-temporal connectivity in nondemented Parkinson's disease during working memory encoding. *Hum Brain Mapp*. 2016;37: 3224–3235.
84. Proskovec AL, Wiesman AI, Heinrichs-Graham E, Wilson TW. Beta oscillatory dynamics in the prefrontal and superior temporal cortices predict spatial working memory performance. *Sci Rep*. 2018;8:8488.
85. Proskovec AL, Wiesman AI, Heinrichs-Graham E, Wilson TW. Load effects on spatial working memory performance are linked to distributed alpha and beta oscillations. *Hum Brain Mapp* 2019. in press.
86. Chang L, Speck O, Miller EN, et al. Neural correlates of attention and working memory deficits in HIV patients. *Neurology*. 2001;57(6):1001–1007.
87. Chang L, Tomasi D, Yakupov R, et al. Adaptation of the attention network in human immunodeficiency virus brain injury. *Ann Neurol*. 2004;56(2):259–272.
88. Thomas JB, Brier MR, Snyder AZ, Vaida FF, Ances BM. Pathways to neurodegeneration: effects of HIV and aging on resting-state functional connectivity. *Neurology*. 2013;80(13):1186–1193.

89. Wilson TW, Fox HS, Robertson KR, et al. Abnormal MEG oscillatory activity during visual processing in the prefrontal cortices and frontal eye-fields of the aging HIV brain. *PLoS One*. 2013;8(6):e66241. PMC3688779.
90. Wiesman AI, O'Neill J, Mills MS, et al. Aberrant occipital dynamics differentiate HIV-infected patients with and without cognitive impairment. *Brain*. 2018;141(6):1678–1690. PMC5972635.
91. Lew BJ, McDermott TJ, Wiesman AI, et al. Neural dynamics of selective attention deficits in HIV-associated neurocognitive disorder. *Neurology*. 2018;91(20):e1860–e1869. PMC6260195.
92. Wiesman AI, Wilson TW. The impact of age and sex on the oscillatory dynamics of visuospatial processing. *Neuroimage*. 2019;185:513–520. PMC6290353.
93. Wiesman AI, Mills MS, McDermott TJ, Spooner RK, Coolidge NM, Wilson TW. Polarity-dependent modulation of multi-spectral neuronal activity by transcranial direct current stimulation. *Cortex*. 2018;108:222–233. PMC6234070.
94. Wiesman AI, Heinrichs-Graham E, Proskovec AL, McDermott TJ, Wilson TW. Oscillations during observations: dynamic oscillatory networks serving visuospatial attention. *Hum Brain Mapp*. 2017;38(10):5128–5140. PMC5593769.
95. McDermott TJ, Wiesman AI, Proskovec AL, Heinrichs-Graham E, Wilson TW. Spatiotemporal oscillatory dynamics of visual selective attention during a flanker task. *Neuroimage*. 2017;156:277–285. PMC5548621.
96. McDermott TJ, Wiesman AI, Mills MS, et al. tDCS modulates behavioral performance and the neural oscillatory dynamics serving visual selective attention. *Hum Brain Mapp*. 2019;40(3):729–740. PMC6328324.
97. Brier MR, Wu Q, Tanenbaum AB, Westerhaus ET, Kharasch ED, Ances BM. Effect of HAART on brain organization and function in HIV-negative subjects. *J Neuroimmune Pharmacol*. 2015;10(4):517–521. https://doi.org/10.1007/s11481-015-9634-9. 26446778. PMC4662631.
98. Ances B, Vaida F, Ellis R, Buxton R. Test-retest stability of calibrated BOLD-fMRI in HIV- and HIV+ subjects. *Neuroimage*. 2011;54(3):2156–2162. PMC3229916.

Chapter 6

Huntington's disease: Brain imaging in Huntington's disease

Eileanoir B. Johnson, Sarah Gregory*
Huntington's Disease Centre, UCL Queen Square Institute of Neurology, University College London, London, United Kingdom
*Corresponding author: e-mail address: s.gregory@ucl.ac.uk

Contents

Abstract

Huntington's disease (HD) gene-carriers show prominent neuronal loss by end-stage disease, and the use of magnetic resonance imaging (MRI) has been increasingly used to quantify brain changes during earlier stages of the disease. MRI offers an in vivo method of measuring structural and functional brain change. The images collected via MRI are processed to measure different anatomical features, such as brain volume, macro- and microstructural changes within white matter and functional brain activity. Structural imaging has demonstrated significant volume loss across multiple white and gray matter regions in HD, particularly within subcortical structures. There also appears to be increasing disorganization of white matter tracts and between-region connectivity with increasing disease progression. Finally, functional changes are thought to represent changes in brain activity underlying compensatory mechanisms in HD. This chapter will provide an overview of the principles of MRI and practicalities associated with using MRI in HD studies, and summarize findings from MRI studies investigating brain structure and function in HD.

Progress in Molecular Biology and Translational Science, Volume 165
ISSN 1877-1173
https://doi.org/10.1016/bs.pmbts.2019.04.004

The progression of Huntington's disease (HD) has long been associated with severe pathological changes reported in post-mortem brains.[1] However, to understand the progression of neural changes and mechanisms underlying these changes, it is essential that the brain is studied during earlier stages of the disease. Magnetic Resonance Imaging (MRI) offers an in vivo method of quantifying structural and functional brain changes during HD progression. It is a safe and well-tolerated technique that uses magnetic fields to collect three dimensional images of the brain. These images can then be processed via different tools to measure various anatomical features, such as brain volume, macro- and microstructural changes within white matter and functional brain activity. Extracted measures can then be used to develop our understanding of the underlying pathological changes occurring in HD, and linked to clinical, cognitive and psychiatric features of the disease. This chapter will cover a brief overview of the principles of MRI, provide a background of MRI research in HD, and then discuss practicalities and methods for the collection and analysis of MRI data in HD cohorts.

1. MRI theory

MRI uses magnetic fields, magnetic gradients and the application of radiofrequency pulses to change the state of hydrogen atoms in the brain; the energy created by these changes is subsequently measured and ultimately exported in the form of an MR image.[2]

Within each hydrogen nucleus there is a single proton with a positive electrical charge. Each proton is constantly spinning, and because a spinning electrical charge creates a current, each proton has a magnetic field called a magnetic moment.[2] Every magnetic moment is normally in a random orientation, but when a uniform magnetic field (such as that within an MRI scanner) is applied to these magnetic moments they align to the uniform field in either a parallel or an antiparallel fashion. More protons will align parallel to the magnetic field as this requires less energy. The difference between the number of protons in parallel vs antiparallel alignment is dependent on several factors, including the strength of the magnetic field.

The Maxwell equation stipulates that an electric current running through a wire results in a magnetic field that surrounds the wire.[2] An MRI scanner makes use of this principle, using a current that runs through a collection of superconducting metal coils to generate powerful magnetic fields (called B_0). When the external magnetic field is applied to the hydrogen protons they spin in a particular motion, called "precession." The frequency at which the protons precess is determined by the strength of B_0 multiplied by γ ($\omega_0 = \gamma B_0$), a constant value relative to the properties of each type of proton called the Larmor equation.[2] Protons in parallel and antiparallel directions cancel each other out in all directions except for that aligned with the z-axis of the B_0 magnetic field. When a person is placed in the B_0 magnetic field their hydrogen protons align with B_0. This magnetic field is called longitudinal magnetization.

To create a measurable signal from which to construct an image, a brief radiofrequency (RF) pulse is applied to create a transfer of energy from the RF pulse to the protons, which disturbs the protons causing them to fall out of line with B_0.[2] As the protons absorb RF energy, longitudinal magnetization is reduced and net magnetization is rotated into the x-y plane; this is called excitation. Excitation causes the protons to precess in phase, i.e., in the same direction, at the same time. This process is demonstrated in Fig. 1. Once the RF pulse has been switched off, the protons begin to realign with B_0 via two types of relaxation.[2]

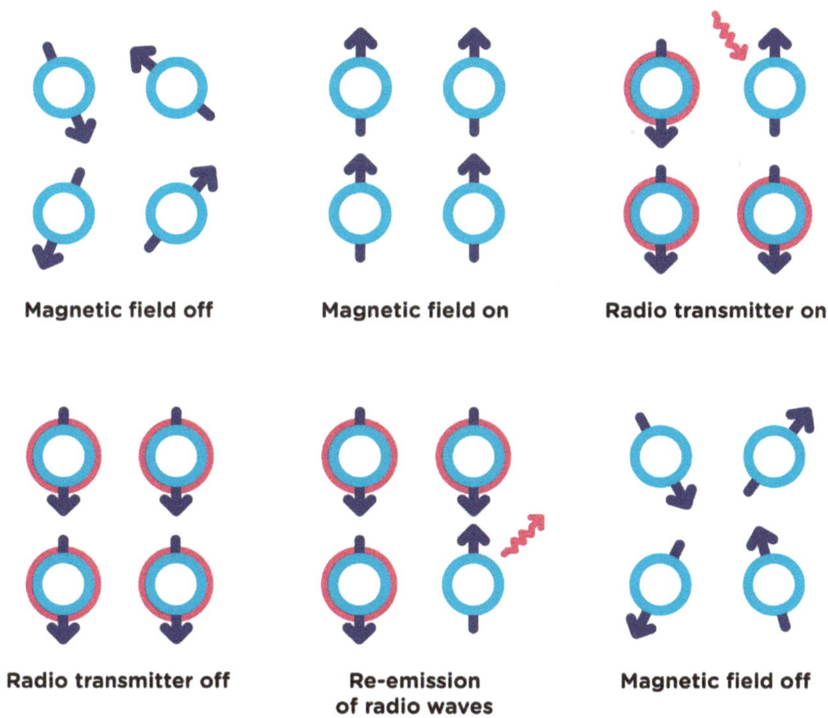

Fig. 1 A schematic illustrating the spin of hydrogen atoms before a magnetic field is applied, while in a magnetic field, and during and after an RF pulse is applied.

As the protons move back into the z plane in line with B_0, energy absorbed from the RF pulse is released; this is called T1 relaxation.[2] Every tissue within the body has a different T1 relaxation time that is dependent on how tightly bound the protons are in their environment. Within the cerebrospinal fluid (CSF), for example, hydrogen atoms and protons move freely resulting in a longer T1 relaxation time. The differences in relaxation time result in the ability to distinguish between different tissue types. In addition, once the RF pulse is switched off the magnetic fields of neighboring protons influence each other and push each other out of phase.[2] This is called T2 relaxation and when combined with small field inhomogeneities across the B_0 magnetic field results in differing degrees of T2 relaxation across the image. T1 and T2 relaxation occur at the same time but are independent to one another, with T1 relaxation happening more slowly than T2 relaxation. As the protons undergo T1 and T2 relaxation, they re-emit the energy absorbed from the RF pulse and this energy is measured by the MRI scanner to construct an image.

There are two main components required to collect MRI data, the MRI scanner and the control console. The control console performs the programming of scans, and is in a different room to the MRI scanner. The MRI scanner contains the magnet and technology to collect the images, with two main components required to create the images: the magnetic field and the gradient and radiofrequency coils. Gradient coils located within the scanner distort the main magnetic field in a predictable pattern that allows spatial encoding of the measured signal mapped in the x, y and z directions to create 3D images. The raw data are captured from the scanner and stored in a data matrix called k-space, which is then converted into an image via an inverse Fourier transform.

An MRI scanner also has RF coils that transmit an RF signal to the brain, and then in turn receive the induced signal from the tissue. Finally, shim coils are used to increase consistency within the magnetic field. This translates to increased homogeneity within the MRI images when processed and improves image quality. The strength of the magnetic field is described in units of Tesla (T), with higher Tesla scanners resulting in higher resolution images. Most scanners are either 1.5 T or 3 T in strength, with 7 T scanners becoming more common in research environments.

1.1 Image weighting

Different tissue types have different quantities of hydrogen (e.g., CSF vs white matter (WM)), resulting in variable properties of the MRI image, and furthermore different MRI acquisitions can be used to capture

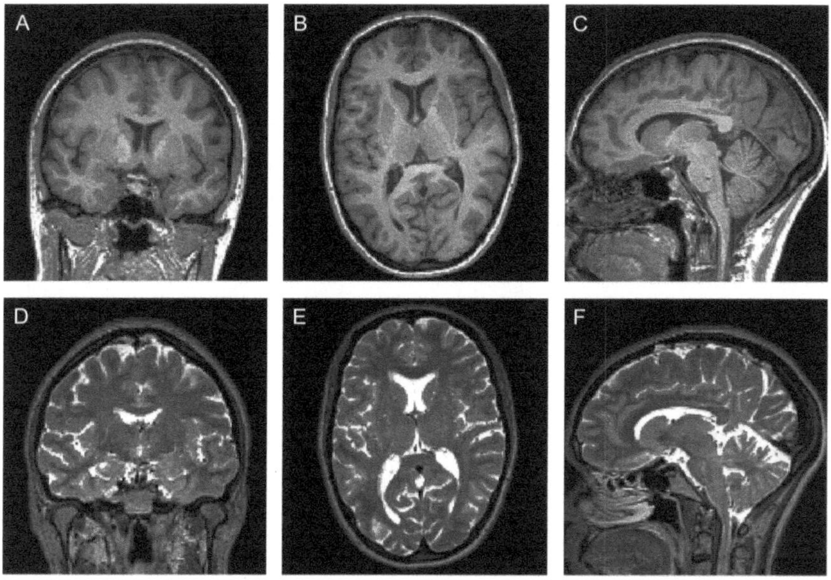

Fig. 2 An example of 3D T1 and T2 weighted images showing (A) A T1 scan in coronal view, (B) A T1 scan in axial view, (C) A T1 scan in sagittal view, (D) A T2 scan in coronal view, (E) A T2 scan in axial view and (F) A T2 scan in sagittal view.

macro- and micro-characteristics and brain function based on the properties exhibited by these atoms, such as how restricted they are within regions of the brain. T1 and T2 relaxation can be quantified independently to create the most commonly used types of structural images, T1 and T2 weighted images. T1 images measure the T1 relaxation and T2 images measure the T2 relaxation, with these two types of image exhibiting different contrast.[2] Fig. 2 shows an example of a T1 image and a T2 image. The weighting of a scan is determined by the acquisition specified on the control console, and is particularly influenced by two key parameters, the repetition time (TR), which is the length of time between RF pulses, and the echo time (TE), the duration between the RF pulse and the time of signal measurement. T1-weighted images tend to have clear boundary definitions and are thus often used for anatomical measures, whereas T2 weighting is often sensitive to pathological abnormalities and is frequently used for diagnostic scans.

2. Data acquisition and study design

When designing an MRI study for use in an HD cohort, the protocol should be developed with the study aims and cohort in mind. Firstly, the types of modality included in the study should be selected based on the overall purpose of the study and whether the researcher is interested in answering questions about brain volume, white matter, brain connectivity or task performance. Typically, a combination of scans is acquired to address various questions within one study. When considering different MRI modalities for HD research, there are various practical details that should be considered during the study setup phase. Furthermore, where possible, analysis pipelines should be pre-specified and tested before the main analysis is conducted. The process of data quality control is highly important and should be rigorous and performed within a short time-window after collection of each dataset. Here, a number of factors that bare consideration and can help to inform study design are discussed.

The collection of MRI data for HD studies can be complicated by disease related factors that result in issues associated with MRI scanning. These practical issues are mostly associated with the scanning of symptomatic participants and can mean that the design of an acquisition protocol may need to be specifically tailored to reduce the impact that these factors have on data quality.

One of the key difficulties with scanning HD cohorts is participant motion during data capture. Since manifest HD gene-carriers suffer from choreic movements, the data quality can be compromised. Movement within the magnitude of millimeters in any plane during acquisition can result in poor data. To reduce the effects of motion, it is helpful for participants to feel comfortable within the scanning environment and in the scanner. Explaining the process to participants and ensuring that all questions have been answered, can help to reduce anxiety. In addition, providing a blanket for warmth and padding underneath the participants' knees to reduce load on the lower back can help to increase comfort during the scan. Padding the head within the head coil to ensure a close fit in the coil can have a significant effect on reducing motion artifacts in the scanner. Some participants may also struggle to be correctly aligned within the head coil due to increased rigidity associated with HD,[3] resulting in poor alignment within the central bore of the scanner, and this should be handled with care. Finally, due to the difficulty associated with being restricted within the scanner bore, participants may not be able to tolerate long periods within the scanner. Acquisitions used to collect data for manifest HD participants should be restricted to essential sequences only where possible.

2.1 Acquisition parameters/pulse sequences

In addition to image weighting described earlier, different radiofrequency pulses and magnetic field gradients can be specified to create different sequences. Two commonly used sequences are spin echo (SE) and gradient echo (GE) sequences. An SE sequence applies two radiofrequency pulses, whereas a GE sequence applies radiofrequency pulse and then a gradient pulse. These two sequences can both be used to acquire T1- and T2-weighted scans. SE sequences take longer to collect and create higher quality scans, whereas GE sequences are fast but generally produce lower quality sequences. Echo planar imaging (EPI) sequences are also frequently used to collect data where many scans need to be acquired in rapid succession, for example, during functional MRI scans. EPI sequences use rapidly reversing gradients collect to multiple echoes via either GE or GE and SE sequences. When using any sequence, acquisition parameters should be optimized to ensure good signal within regions associated with HD. For example, for volumetric T1 scans, typically researchers use protocols that have good gray/white matter (GM/WM) contrast to allow accurate measurement of the subcortical regions.

In order to perform scans at multiple sites, further optimization of sequences is typically required. It is preferable that the sites included in a multi-site study have scanners which are the same field strength, make and model, however, this is not always possible. Acquisitions should be developed with the use of test-retest human data whereby one or more volunteers travel to all study scanners and data reliability and reproducibility tested.

2.2 Data quality control

Data quality control (QC) is an essential step in MRI pre-processing. This process involves the checking of data after it is acquired to determine whether the quality is sufficient to use in analysis, and also to ensure that there are no ongoing issues with data collection. If QC is postponed until the end of data collection there are risks that data quality can be compromised by ongoing hardware or technical errors that have not been detected, as well as by consistent mistakes in participant positioning. There are general qualities to look for during quality control, as well as issues specific to each modality.

Prior to any analysis, including QC, the data should be anonymized based on the study protocol. This helps to ensure that participants cannot be identified and maintains researcher status blinded to group designation. Anonymization can occur either on the MRI scanner, or on receipt of the data.

Typically, the first step in QC involves conversion of data from DICOM format, which is the format that is exported from the MRI scanner. There are a number of freely available tools that can be used to convert the data, and it is usually converted into either ANALYZE (.img and .hdr) or NIFTI (.nii) formats which can be easily visualized and analyzed. Following conversion, data should be visually examined for common MRI artifacts. There are a wide range of artifacts, such as motion artifacts, signal-dropout, intensity inhomogeneity, flow artifacts or wrap-around.

Motion artifacts are the most common artifacts seen in MRI images and can vary from mild to severe. Motion artifacts affect all modalities and in the worst cases the data will be unusable. To check for motion artifacts, the images should be examined for the appearance of "ringing" as shown in Fig. 3. Movement artifacts can also cause missing slices in raw diffusion data, discussed further below. Physiological noise such as cardiac rhythm and respiratory artifacts can result in ghosting, whereby there is the appearance of multiple images superimposed on top of each other. Wrap around, where the back of the image appears to wrap into the front of the image, can occur when the scanned area extends beyond the field of view (FOV) and thus the FOV should be positioned carefully. Susceptibility artifacts are caused by a metal object within the FOV and can cause signal drop-out. These artifacts result from large changes in magnetic susceptibility within that region, and the image typically has a dark patch with the metal object appearing bright white. They can cause significant drop-out and distortions within neighboring regions, but via careful screening they can mostly be reduced prior to scanning. Susceptibility artifacts are often be caused by dental work or

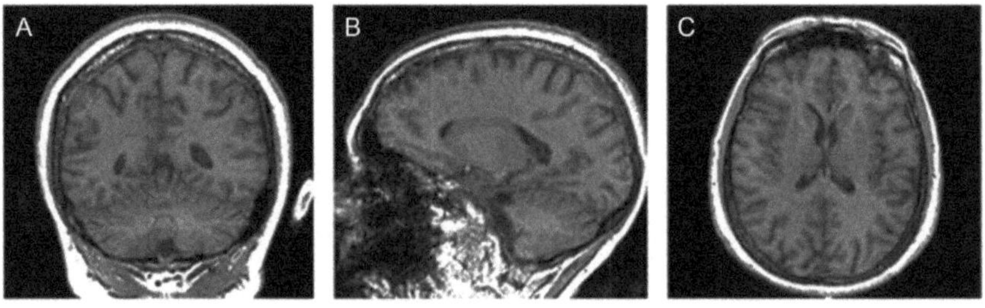

Fig. 3 An example of motion artifacts seen on a T1-weighted scan shown in (A) coronal, (B) sagittal, and (C) axial views. The motion causes the scan to appear blurred with rings around the edge of the cortex, making estimates of brain features unreliable.

implants but more recently, artifacts due to cosmetic procedures and products have been reported, with examples including artifacts due to dry shampoo and mascara.[4,5] While in most cases these are not severe enough to impact the brain region, participants should be asked to remove make-up and reduce the use of cosmetics before the scan. Tattoos have also been reported to cause signal drop-out as well as potentially damaging skin burns due to metallic ink,[6] although the likelihood of burns has been reported to be very low.[7] Participants should be carefully screened for metal implants prior to the scanning session with physicist or radiographers providing input on the decision to scan metallic implants.

One final artifact that is common in MRI scans is intensity inhomogeneity. This results from the varying magnetic field in the MRI bore and can vary between different scanners and acquisition types. It causes varying intensity across a smooth spatial gradient within an MRI scan and can be corrected after data collection but before analysis, described later. The tissue non-uniformity is caused by factors including local flip angle variations, participant anatomy as well as radiofrequency coil homogeneity[8,8a] (Boyes et al., 2008). While this artifact does not impact the ability to visually inspect an MRI scan, it can affect how well software can delineate different tissue types. This is especially true for automated processing tools. This inhomogeneity can be corrected for via a process called bias correction, explained below.

Finally, if using a multi-site setup or a scanner that has multiple different radiographers, it may be important to examine the header information for each scan and compare it against the expected values. This information contains metadata such as the TR and TE values and should be compared to a reference set of values extracted from the scanner on development of the final MRI protocol. This process acts as a check-point to ensure that the data are consistently collected using the correct protocol and that none of the parameters have been changed over time.

3. Modalities

There are many types of MRI data that can be collected from MRI scanners based on the principles described above, with each type used to quantify different neural characteristics. Three types of data frequently commonly collected for use in HD research are structural, diffusion and functional MRI data.

3.1 Structural MRI

Structural MRI (sMRI) scans can be used to measure static anatomical features of the brain, such as volume and cortical thickness (CT), and are either T1 or T2 weighted. Different tissue classes (such as GM, WM) have different intensities voxel (whereby voxels are the 3D array of elements that make up an MRI image), which are used to estimate qualities of the tissue classes, such as the volume of WM or the thickness of the cortical GM. Thus, it is essential that images have high between-tissue contrast and high spatial resolution. During the development of sequences, it is important to establish the best compromise between acquisition length, spatial resolution and tissue contrast.

3.1.1 Data processing

The first steps for T1 processing are typically conversion from DICOM format and bias correction. As discussed, bias correction reduces inhomogeneity artifacts within the data. The most common means of performing bias correction is via the N3 algorithm,[8] which results in more consistent signal intensity across the image. For standard T1 and T2 scans, the quality of tissue contrast should be checked, particularly for regions of interest, for example, within HD it is important that there is good contrast between WM and subcortical structures. The way in which these steps are performed is dependent on the type of analysis (whole-brain vs region of interest), with three of the most common tools used to process data being Statistical Parametric Mapping, SPM;[9] The FMRIB Software Library, FSL;[10] and FreeSurfer.[11,12] These three tools can all be used to perform whole-brain analyses on a voxel-wise level or to extract regional values (either volume or CT) from T1 scans. Common processing steps for sMRI data include segmentation, registration and smoothing, followed by statistical analysis. These steps will be discussed in more detail below, however, two techniques common to many tools and used not only in the processing of structural, but also DTI and functional MRI data, are image registration and data smoothing.

3.1.2 Registration

MRI scans can be processed in a range of "spaces," whereby image space refers to the location of the head within the FOV. The space in which the image was acquired is referred to as native space, and here the brain is in the same position within the FOV as on the scanner. Native space is different for each scan collected within a study. Even for the same participant scanned twice with the same sequence, it is likely that there will be slight differences in positioning due to participant movement or FOV shifts.

Standard space is a general term that refers to a common location to which to align different scans. The Montreal Neurological Institute space, MNI[13,14] is the most commonly used standard space. It is also possible to create a study-specific group template by concatenating all scans collected as part of a study into a single "customized" template. When performing cross-sectional group comparisons, scans are often registered to a common space to allow the direct comparison between multiple scans. Registration is a process of alignment whereby scans are moved from one image space to another. Registrations can be performed between individuals, or within individuals across multiple timepoints, for example, within-participant registrations can be used to quantify longitudinal change for imaging measures. There are different types of registration and different tools to perform registrations, however, all registrations consist of two steps: (a) calculation of registration parameters and (b) application of these parameters to an image.

Linear registrations aim to achieve a rough spatial alignment between scans, but not an exact correspondence. They are restricted in the number of deformations that are performed, and the same actions are applied to each voxel within the brain so that the content of the voxels are not changed. During a linear registration, a scan can be transformed via rotations, translations, shears and zooms. A rigid-registration is the most basic linear registration that can be performed, using 6 degrees of freedom (3 rotations, 3 translations) to align the moving image to the target image. An affine registration applies up to 12 degrees of freedom (3 translations, 3 rotations, 3 shears and 3 zooms) to provide more advanced co-registration of the scans.

To achieve a precise correspondence between two images, a non-linear registration should be performed. This type of registration warps the scan in many dimensions. First, the images are typically aligned roughly via a non-linear step, followed by a more comprehensive registration using a large number of degrees of freedom, warping the scan in any direction. There are different types of non-linear registrations available, including fluid registration, Diffeomorphic Anatomical Registration Through Exponential Lie Algebra (DARTEL) and longitudinal-serial registration.[15–18]

3.1.3 Smoothing

Smoothing is performed applying an isotropic Gaussian kernel to an image that averages the signal from neighboring voxels. This process not only improves signal-to-noise ratio but also normally distributes the data, which is essential for parametric statistical testing.[19] The kernel size is determined according to the size of the brain area of interest. The Kernel is Full Width Half Maximum (FWHM), with typical sizes between 4 and 8 mm. When deciding on a kernel size, it is important to recognize that there is a compromise between smoothing the data appropriately to establish normally distributed data and reducing noise within the data, but also losing spatial information and thus the ability to localize the signal. The kernel size is typically decided based on the resolution of the data along with the estimated regions of effect.

3.1.4 Data analysis

Region of interest (ROI) analyses require an a priori hypothesis to segment a structure of interest from which a volume measure is extracted for each participant (e.g., caudate volume). The extracted volumes can then be analyzed in an external statistical package. Conversely, whole-brain analysis involves registration and segmentation of tissue classes and then statistical comparison of every voxel of the brain (or a pre-specified tissue, such as GM) to test for group differences or associations with a variable of interest. These two methods provide complementary information, and within each method there are a multitude of different measurements that can be performed.

3.1.5 Region of interest (ROI) analysis

ROI analyses involve the delineation of structural regions, for example, the caudate, to provide precise measurements of the region. Measures such as volume, cortical thickness or gyrification can all be examined within ROIs. ROI studies often show high sensitivity for detecting group differences or demonstrating longitudinal change in HD, and thus can be very useful for examining specific hypotheses, especially when quantifying subcortical structures such as the caudate and putamen.

The delineation of ROIs can be performed manually or using automated software, and after a region is delineated a value for this region can be extracted. To obtain accurate data the ROIs need to be precisely defined. Manual measurement of ROIs can be a highly accurate method for measuring ROIs, yet can take a long time to develop, validate and perform, particularly when measuring multiple anatomically complex regions. In contrast, automated ROI measures can be performed more quickly and easily, however, they often bring an increased level of error whereby the segmentations spill into surrounding tissues. This is a particular problem in brains showing moderate to severe pathology.

Typically, to perform ROI analyses on sMRI data, after image conversion and bias correction, a scan is either manually segmented using tissue intensities and anatomical landmarks to establish boundaries between different regions[20,21] or is entered into an automated software for segmentation. There are numerous tools available for automated segmentation of ROIs, with some of the most widely used being FSL (FAST for subcortical, FIRST for WM/cortical) and FreeSurfer. There are also some more recently developed tools available, for example, ANTs, MALP-EM,[22,23] and it is important to recognize that different tools show variable performance on different datasets.[24,25] The selection of an appropriate segmentation tool should be based on which software provides the most visually accurate segmentations for that dataset. Following segmentation, regions can be extracted and the volumes from these regions used within statistical analyses.

3.1.6 Whole-brain analysis

Whole-brain methods enable exploratory analyses used to quantify differences across all regions of the brain. However, since they involve the statistical comparison of every voxel across the brain region via mass-univariate statistical testing, strict correction for multiple-comparisons is required. Whole-brain methods have been widely used in HD studies to measure brain volume and CT, with the results of these studies providing an understanding of brain-wide regions that are affected by atrophy in HD. Whole-brain analyses generally offer limited flexibility in terms of statistical design, because the analysis is conducted within an imaging software, and thus the options for using these techniques can be limited when using data sets with multiple time points or missing data.

Voxel-Based Morphometry (VBM) is the most commonly used whole-brain approach used in structural imaging research,[26] with SPM the most commonly used software, although VBM can also be performed using the FSL software.[27] VBM measures volumetric differences in the GM, WM or CSF between different groups (such as controls or HD gene carriers), or associations between volume and a covariate, such as cognitive scores of fluid biomarkers.[28–30] Within SPM, VBM processing involves a series of steps, while also segmenting images into three tissue types (GM, WM and CSF). Tissue segmentation models intensity distributions within an image using Gaussian models and tissue probability maps (tissue probability maps, TPMs) that weight tissue classification of each voxel.[9] TPMs use priors that provide the probability of any voxel being one of a series of tissue classes, most commonly GM, WM, or CSF. Following segmentation, each tissue class for every voxel is given a value ranging from 0 to 1 to represent the probability of that voxel belonging to that tissue class. At this point, QC should be performed to ensure that segmentation has been successful on all scans. Images are then aligned to the template created in the first step and smoothed to reduce errors associated with registration. An additional step that is recommended and should be performed prior to smoothing is to modulate all voxels within the scans. As the voxels within each scan are warped during registration to fit the template space, they are either contracted or expanded. Modulation accounts for this by adjusting the volume of the tissue type within each voxel using the warping parameters calculated in step one and aims to preserve the original volume of each voxel by accounting for volumetric changes that may have occurred during registration.[19] Following these processing steps, a general linear model (GLM) is specified to compare voxel-wise volumetric differences between participants, or to examine associations between volume and performance on a cognitive task, discussed below. The results are corrected for multiple comparisons, also described below.

Cortical thickness is another widely used whole-brain approach. This technique aims to measure the distance between the outer pial-CSF brain surface and the GM-WM surface, thus quantifying the thickness of the cortical layer. The most commonly used software to measure cortical thickness is FreeSurfer,[11,12] it has a volume-based stream to calculate volumetric measures and classify subcortical regions, as well as a surface-based stream which calculates cortical thickness.[31,32] In the volume-based stream the brain is affine registered to Talairach space (a frequently used standard space), and then an initial volumetric labeling step is performed followed by intensity correction. Following this, a high dimensional nonlinear volumetric alignment to a Talairach atlas is performed. In the final phase the volume is labeled. The volume-based stream has evolved separately to the surface-based stream and various stages are different (i.e., the registration and bias field correction). For the surface-based stream, after correction for field inhomogeneity, affine-registration and skull stripping, the WM voxels are classified based on their intensity, their neighbors' intensity, and their position. A triangular mesh is built around the WM, which is then smoothed and topologically corrected. The external cortical surface is created via expansion of the WM surface until it reaches a point of maximal contrast between the GM and CSF.

3.2 Structural MRI findings in HD

3.2.1 Region of interest studies

A large body of research has been focused on characterizing atrophy within basal ganglia regions. This is due to early pathology studies that highlighted the basal ganglia as regions showing particularly pronounced atrophy in postmortem HD brains.[1] Numerous findings from MRI studies indicate that the striatum is the earliest region to show volume loss.[28,33–36]

Cross-sectional studies comparing controls and HD gene carriers have shown significant group differences in caudate volume up to 15 years prior to disease onset,[37] and work is currently underway to examine participants who are even further from onset to determine when differences in the striatum can be detected.[38] In addition, significantly lower putamen volumes have been shown in pre-HD participants compared to control participants >15 years prior to estimated symptom onset.[37,39] There is currently a lack of clarity as to whether the caudate and putamen begin to undergo atrophy at the same stage and follow the same rate of change, or whether they undergo different atrophy trajectories. There have previously been divergent findings, with some studies showing greater atrophy in the caudate and others in the putamen. It is likely that the discrepancy between results is due to differing cohorts and measurement techniques,[40] with no study yet to directly compare data from different studies to reconcile these findings.

Longitudinal studies provide further information on subcortical change in HD, with significant striatal volume change detected HD over very short periods of time. While annual rates of change vary between studies (dependent on the cohort and analysis method), rates of around 3–4% reduction per year in the caudate and putamen have been reported.[40] Furthermore, significant volume reduction in both the caudate and putamen can be detected over 12 months in both pre-HD and manifest HD compared to controls,[36,41] and over 6 and 9 months in the caudate in manifest HD participants.[41]

Beyond the striatum, whole-brain volume (measured via an ROI approach) is significantly different between controls and pre-HD participants who are <10 years from disease onset.[28] Additionally, significant whole-brain volume change can be detected over 1 year in these participants, and over 6 months in early HD patients.[36,41,42] In addition, longitudinal change in total WM volume is detectable over 1 year in participants >10 years from disease onset, with the rate of atrophy appearing to increase closer to and after disease onset.[36] In manifest HD, WM volume shows significantly greater decline over 6, 9 and 15 months than in control participants.[41] Total GM has also been quantified (with total GM including both subcortical and cortical GM), and small yet significant group differences have been found between controls and pre-HD participants who are over 10 years from disease onset, however, the greatest GM changes occur in participants close to disease onset and those with manifest HD,[33,43] with total GM again showing significantly greater change in as little as 6 months in manifest HD compared to controls.[41] Atrophy of the frontal lobe was reported in an early ROI study, with a volume reduction of 17% in early manifest HD participants compared to controls.[44]

In addition to quantifying change over time, the relationship between atrophy within ROIs and disease progression has been studied. In the TRACK-HD cohort, striatal and GM volume measures were able to predict diagnosis in pre-HD individuals, and whole-brain, caudate, putamen, ventricular and GM volumes significantly predicted decline in total functional capacity in early HD participants, supporting a strong relationship between neural atrophy and clinical change.[45] Furthermore, in the PREDICT-HD cohort, both cross-sectional and longitudinal volume of the putamen, hippocampus, CSF, accumbens, globus pallidus and caudate could predict motor onset.[46]

3.2.2 Whole-brain studies

VBM is the most commonly used approach to measure whole-brain atrophy in HD. Within the TRACK-HD study, cross-sectional baseline results showed reduced volume of the subcortical regions, as well as the cingulate, precentral and prefrontal regions, occipital, parietal, and small regions of the temporal lobes in manifest HD compared to controls, as shown in Fig. 4.[28] In addition to the findings shown in the TRACK-HD cohort, group differences between pre-HD and control participants have been described in other cohorts within the motor regions, as well as within occipital and frontal regions.[47–49] Between controls and manifest HD participants, there have also been significant differences reported in the precentral and postcentral regions,[50–52] as well as in frontal and parietal regions.[52,53]

Whole-brain VBM studies have also established significant longitudinal change, with both pre-HD and manifest participants from the TRACK-HD cohort and manifest HD participants from the PADDINGTON cohort undergoing significantly greater subcortical and cortical change than control participants over 24 months, with cortical changes primarily in the occipital cortex in pre-HD and manifest groups, and manifest participants also showing atrophy in parietal and frontal regions.[43,54]

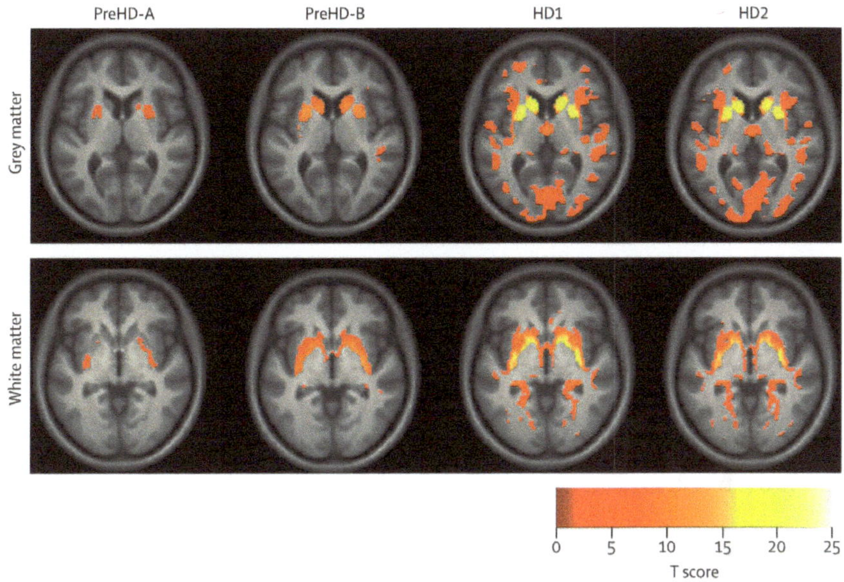

Fig. 4 An example of a voxel-based morphometry study demonstrating reduced gray and white matter volume across different stages of HD when compared to control participants, as shown in Tabrizi et al.[28]. Results are corrected for family-wise error at $P < 0.05$, and adjusted for age, sex, site and total intracranial volume.

A meta-analysis of 17 VBM studies attempted to collate and summarize the results of these analyses. After the results from all studies were combined, pre-HD participants showed significantly reduced volume in the basal ganglia, amygdala, thalamus, insula and right occipital lobe compared to controls.[55] In manifest HD, there were significant differences with controls found in subcortical regions, as well as the frontal cortex, primary motor, premotor and somatosensory regions, and in the intraparietal sulcus, the mid cingulate cortex and the secondary somatosensory cortex.[55] The results from this meta-analysis suggest a pattern of atrophy that begins prior to onset in the occipital lobe, and extends to the frontal lobe, key motor regions and further into the occipital lobe after onset of HD.

As well as volumetric change, a number of studies have measured cortical thickness in pre- and manifest HD compared to controls. Significant differences in the thickness of GM have been reported between controls and HD gene carriers in both pre-HD and manifest stages of the disease,[56–58] and results suggest that the earliest change in cortical thickness is in the occipital and posterior regions of the brain.[57,58] However, cortical thickness studies are almost exclusively cross-sectional. One longitudinal study of cortical thickness found significant change in early HD participants over 1 year, with the main changes seen in sensorimotor and fronto-parietal motor regions. A more recent study, using a large cohort to study the effect sizes of different imaging biomarkers, reports poor longitudinal sensitivity for cortical thickness in HD, especially when compared to volumetric measures, indicating that perhaps cortical thickness is not a sensitive measure of neural change in HD.[41]

Recently, more complex methods of characterizing GM volume change have been applied to HD cohorts with the aim of better understanding the progression of atrophy in HD. Two studies have attempted to map the co-variance of GM changes in HD. Co-variation analyses are designed to estimate the dependencies between regions of the brain showing volumetric reduction in HD, which are then thought to represent networks that undergo atrophy simultaneously. To measure these networks, VBM is performed followed by an independent component analysis (ICA) to identify regions that undergo atrophy at similar stages, called components.[59] One study conducted a large-scale analysis of pre-HD participants (N=831) compared to controls (N=219; Ciarochi et al. [60]) and found that a network of frontal and some motor regions (precentral, supplementary motor) was the earliest component showing atrophy in the far from onset group, with co-occurring parietal and occipital atrophy found in those participants closer to onset. A second, smaller study with 30 pre-HD participants, 30 HD participants and 30 controls identified 10 structural GM networks, with 2 of these networks showing significantly reduced network integrity in both pre-HD and HD compared to controls.[61] One network incorporated the caudate, the nucleus accumbens, putamen, pallidum and precuneus. The other included the parahippocampul gyrus, the cerebellum, the pallidum and the planum polare. Furthermore, one network made up of the precuneus, cuneus, lateral occipital and lingual regions showed significantly lower network integrity in HD compared to controls. The reduced structural integrity seen in these three networks was interpreted by the authors as a demonstration of the general breakdown of associated GM structures in HD. Finally, multivariate linear regression models have been applied to HD data to detect inflections points (change-points) in different GM and DTI measures in relation to predicted years to onset. The change-points represent changes in the linear trajectory of a measure, reflecting the disease stage at which each measure begins to show significant change in pre-HD. The data were a cross-sectional selection of 85 controls and 212 pre-HD participants for the sMRI analysis, and 79 controls and 178 pre-HD participants for the DTI measures. Significant volumetric loss and significant change points were seen in both subcortical and WM regions, suggesting that subcortical and WM regions undergo significant volumetric decline in different stages of pre-HD. Finally, one study has examined the longitudinal trajectory of cortical changes occurring during motor onset in 49 HD gene carriers.[62] There were distinct patterns of atrophy in frontal and motor regions in the decade surrounding HD motor diagnosis, with frontal regions undergoing linear rates of volumetric loss and, in contrast, motor regions showing accelerations of atrophy shortly after clinical motor diagnosis. These results suggest a variable pattern of cortical loss, with distinct regions linked to clinical progression showing change at different stages of the disease.

3.3 Diffusion weighted imaging

Diffusion Weighted Imaging (DWI) measures the coherence of water flow in biological tissue, reflecting the interactions of water movement with numerous molecular obstacles, such as cell membranes and macromolecules, in turn revealing information about underlying tissue microstructure.[63] Within brain imaging, DWI has focused on the investigation of water movement along white matter pathways and the nature of pathology-related WM microstructural change.

In an unrestricted space, water molecules can move in any number of directions, called isotropic movement. However, when the path of these molecules is restricted, such as along a white matter tract, water diffuses in an anisotropic way. As water molecules move more freely, such as in CSF, the MR signal is reduced. DWI capitalizes on this physical property measuring the signal variability that occurs due to differences in WM microstructure.[64] DWI can be used to measure both WM microstructure within regions of a pathway or along the tracts themselves, indexing connectivity strength between different brain regions, and providing an indication of the association (or breakdown of association) between brain regions.

Diffusion data are collected by acquiring a time series of images using different gradient directions that measure information about the coherence of water flow in different directions/planes of the 3D image; gradient strength is indexed using a b-value.[64] To quantify diffusion efficiently data should be collected in a minimum of 31 gradient-directions.[65] More recently, with the advent of advanced diffusion analysis techniques, data from 64 gradient-directions are generally acquired. Additionally, a B_0 image with no gradient is also acquired to capture physical anatomy and tissue contrasts without diffusion gradients. To speed up data collection due to the large number of images being acquired, EPI sequences are generally used in diffusion imaging.

3.3.1 Data processing

Diffusion processing requires different processing steps compared to sMRI data, with different software options also used to perform these processes. Firstly, each gradient volume should be inspected for artifacts; however, there is no consensus on the best technique to deal with a volume that has a significant artifact. For example, removal of affected volumes will improve data quality, but there is a risk of uneven data sampling since each volume collects data from a different direction. Thus, it is recommended that if data contains considerable artifacts the whole dataset should be excluded from analysis.

3.3.2 Distortion correction

Due to the varying susceptibility of different tissue types combined with the rapidly changing gradient echos, diffusion data are particularly susceptible to signal dropout and distortion. The susceptibility induced field can be estimated and corrected for, either using a field map or by collecting an additional diffusion scan in the reverse polarity that can be used to correct for the distortion. Diffusion data also require correction for the eddy currents that are introduced to the data by the changing magnetic field required to collect different gradients. These artifacts are worsened by motion, and thus can be particularly apparent in the data of HD gene carriers. The eddy currents result in a distorted image and vary by gradients.[66] Failure to correct for eddy currents can result in mis-aligned data as well as poorly estimated fractional anisotropy values.[66] There are different tools that can be used to correct for eddy currents, with FSL's "eddy" commonly used.[67]

3.3.3 Tensor estimation

Diffusion Tensor Imaging (DTI) is most commonly used to quantify the properties of diffusion. Following distortion and eddy correction, a tensor model is fitted at each voxel. Diffusivity is then characterized by three eigenvalues (λ_1, λ_2, and λ_3, Fig. 5), which represent diffusivity in three principal directions (axes) as modeled by the tensor, and their corresponding orientation eigenvector: ε_1, ε_2, and ε_3.[64] To estimate the tensor at each voxel, multivariate linear regression of the signal intensities is used. Following tensor estimation, the diffusion tensor maps should be visually checked.

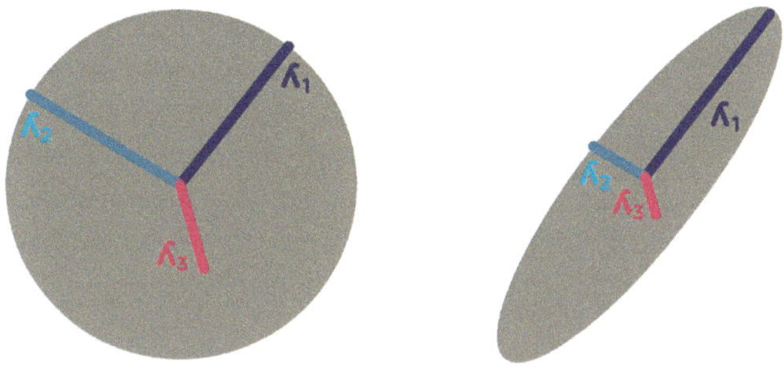

Fig. 5 An example of an isotropic and anisotropic tensor, with the three eigenvalues displayed.

When performing visual checking, it is important to determine that the fibers are delineated appropriately in the three image planes. Finally, diffusion metrics can then be extracted. Fractional anisotropy (FA) represents overall coherence or the degree of anisotropy, whereas mean diffusivity (MD) represents mean diffusion along the principal tracts in all three directions.[64] Parallel (axial) diffusivity (AD) and perpendicular (radial) diffusivity (RD) can also be measured. AD is a measure of diffusion parallel to the WM fibers, RD, in contrast, is a measure of perpendicular diffusion. Together, these measures provide information about the varying properties of WM tracts.

Standard DWI measures suffer from a lack of specificity, and it is unclear how biologically meaningful the metrics are. Recently, new methods have been developed that examine more specifically the biological properties underlying diffusion-based measures of microstructure. Neurite orientation dispersion and density imaging (NODDI), for example, applies a three-compartment tissue model to characterize both intra- and extracellular WM properties[68] and thus distinguish between the density of axons in the WM (neurite density index, NDI) and the spatial organization of axons (orientation dispersion index, ODI). In addition, this technique characterizes free water within the voxels.

3.3.4 Data analysis

Similarly to structural MRI, diffusion weighted images can be analyzed using both whole-brain and/or regional/tract-based methods. This is generally performed on diffusion metric maps following tensor fit, as described above.

3.3.5 Region of interest analysis

ROI metrics can be extracted from DTI data. ROIs are defined by either applying a standard atlas containing the regions of interest to diffusion space, or by performing atlas-based segmentation on a structural scan and then warping this to diffusion space. From this region, diffusion metrics can be extracted (see below). This method reduces the number of comparisons being performed and can result in more sensitive measures of diffusion.

3.3.6 Whole-brain analysis

Whole-brain comparisons of WM metrics can be performed, for example, using Tract-Based Spatial Statistics, TBSS.[69] TBSS is performed in FSL and is similar to VBM, in that it is used to compare voxel-wise values of diffusion metrics across whole-brain WM.

TBSS analysis is comprised of four steps and is performed on the FA maps that result from tensor fitting. Initially, the FA maps for each participant in the analysis are registered to a common space. This can either be a study-specific template (which is in standard-space) or, a standard space template that has already been created. Following registration, the standard space image for each participant is concatenated into a single 4D image and from this a mean FA image and a mean FA "skeleton" across all participants' FA values are created. The mean FA skeleton is then thresholded and binarized, thus defining the voxels to be used in the analysis. Finally, the 4D image containing each participant map is projected onto the mean FA skeleton and the resulting projected skeletonized data are used for statistical analysis. The statistical analysis makes use of a GLM framework to perform nonparametric permutation tests to compare groups for statistical differences across voxels within the mean FA skeleton, and a clustering tool is used to look for statistical differences within clusters of voxels and to correct for multiple comparisons. While TBSS is performed by default on FA maps, because tensor fitting also outputs mean diffusivity, axial diffusivity, and radial diffusivity maps, TBSS can be performed for these measures also.

3.3.7 Tract based analysis

In addition to ROI and whole-brain analyses, tract-based analyses can also be performed on diffusion-weighted data. Tractography is used to measure diffusivity between two or more regions of interest. There are two types of tractography, deterministic and probabilistic.[70] Deterministic fiber tracking is the earliest method of tract-based analysis, whereby a single seed region is used as a starting point for a WM tract that leads from that region.

Tractography is performed based on the assumption that the main eigenvector within each voxel represents the underlying fiber orientation of that tract. Tracts are reconstructed voxel-by-voxel, with the orientation of the main eigenvector within each voxel determining the orientation of the fiber toward the next voxel. This method only enables one tract to be traced from each seed region, however, and there is some uncertainty regarding the estimation of fiber orientation at each voxel and thus with the overall tract, particularly in areas with multiple overlapping tracts. In contrast, probabilistic tractography enables the propagation of multiple tracts from a seed region. While this method still assumes that within each voxel there is a dominant fiber orientation, at the boundary of each voxel there are different possibilities for the direction of the tract to continue. The streamlines which travel through each voxel are then given percentage values. Both methods require the use of an anisotropy threshold in order to stop the tracts when they reach GM. Generally when performing tractography, the seed ROI(s) are decided prior to analysis. This is usually based on empirical evidence, for example, within HD research motor or cognitive tracts are often examined. However, whole-brain tractography, which is purely data-driven, can be used to investigate the structural connectome, i.e., mapping of connections throughout the brain.[71] This method involves parcellating the brain into subregions, this can range from tens to thousands of regions, and then examining various quantitative qualities of the connections between these regions.

Within tractography, the detection of crossing fibers is a challenge. Since each voxel contains a number of WM fibers with differing orientations, this can result in the incorrect specification of a WM tract during analysis (false positive), or the lack of detection of a WM tract (false negative). There have been a number of solutions proposed to resolve this issue and these should be considered during analysis.[72] In addition, exclusion masks should be applied to ensure that reconstruction of WM tracts does not extend into regions of the GM or dura.

3.4 Diffusion weighted imaging findings in HD

A growing number of studies have used diffusion measures to quantify changes in WM microstructure over the course of HD progression. The results of early DWI studies were mixed, but a more coherent message has begun to appear within both ROI and whole-brain DWI studies. Overall, it seems that FA is reduced in widespread regions of the brain for HD gene carriers compared to controls, and this difference can be detected in pre-HD participants as well as manifest HD participants. Furthermore, radial diffusivity generally increases in pre-HD, and axial diffusivity measures tend to increase with increasing disease progression. The studies, summarized below, indicate progressive WM disorganization with increasing disease stage.

3.4.1 Region of interest studies

Early work looking at cross-sectional diffusion metrics within ROI's indicated altered DTI measures in striatal regions, periventricular WM, the corpus callosum, and internal capsule.[73–76] While the results of these early studies showed a conflicting pattern of increases and decreases in DWI metrics, increasingly there have been reductions in FA reported across the white matter, with the most commonly reported regions in the corpus callosum and external capsule in both pre-HD and manifest HD participants.[74,77,78] Reduced FA is often associated with either increases in AD or RD[78,79] indicating increased disorganization of WM tracts. There have been reported changes within the WM in additional areas, including the inferior and lateral prefrontal cortices, which correlate with cognitive scores in a number of regions,[80] as well as the sensorimotor cortex.[76]

Longitudinal ROI studies similarly show mixed results, with some finding no longitudinal change in DWI metrics in manifest HD participants.[75] In one study looking at diffusion metrics in controls, pre-HD and manifest HD participants, longitudinal changes were significant in early HD compared to controls in nine regions; the genu and body of the corpus callosum (reduced FA and increased RD and increased AD in the genu only), increased RD and AD in the left anterior corona radiata, and increases in AD also in the right superior corona radiata and anterior internal capsule.[78] Only RD within the corpus callosum showed any association with disease burden score.

3.4.2 Whole-brain studies

Many whole-brain analyses have been performed with the aim of detecting differences in diffusion in a data-driven using TBSS. One study showed decreased FA in the WM of superior frontal, middle frontal, postcentral and precentral regions for seven pre-HD participants compared to seven controls,[81] and widespread differences such as these have also been reported in a number of larger studies, with the corpus callosum, internal and external capsule often showing significant differences between both pre-HD and

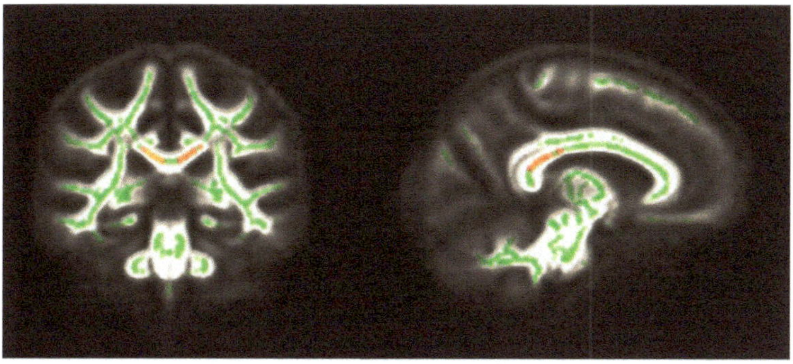

Fig. 6 An example of TBSS results showing tracts where reduced FA was associated with increased levels of depression in HD gene carriers, corrected for family-wise error at $P < 0.05$. *Adapted from Gregory S, Scahill RI, Seunarine KK, et al. Neuropsychiatry and white matter microstructure in Huntington's disease.* J Huntingtons Dis. *2015;4(3):239–249. doi:10.3233/JHD-150160.*

manifest HD and control participants.[82–85] These differences are also associated with predicted years to onset in pre-HD participants,[82] and reduced FA calculated via TBSS has also been associated with increased depression scores and irritability in HD (see Fig. 6 [86]). One paper using the NODDI method to measure whole-brain changes in NODDI metrics found that pre-HD participants had widespread reductions in NDI (neurite density), including within the corpus callosum and internal capsule among other regions, they also had reduced ODI (orientation dispersion) in regions surrounding the basal ganglia. The study also calculated standard diffusion metrics, demonstrating that pre-HD participants also had increased MD, AD and RD, and decreased FA compared with control participants in similar areas, and increased AD correlated with disease progression in the corpus callosum.[87]

A small number of longitudinal studies using TBSS to characterize change in diffusion measures over different stages of the disease have also been performed. The first longitudinal DTI study in HD showed change in FA and AD over 1 year in seven gene carriers compared to seven controls.[88] In addition, over 18 months in a larger cohort of manifest participants, there were significant reductions in FA within the corpus callosum and cingulum, as well as an increase in RD within the corpus callosum and projection pathways from the striatum Poudel et al. [89] The FA change within the corpus callosum was associated with baseline motor score in both pre-manifest and manifest participants and longitudinal change in FA could be predicted by TMS motor score.

3.4.3 Tractography studies

Finally, there have been a series of tractography studies that have investigated differences in the composition of WM tracts between HD and control participants. One study performed a detail analysis of the corpus callosum, dividing it into seven regions via manual segmentation and exploring diffusion metrics within these regions.[90] The results showed decreased FA and increased RD in pre-HD compared to controls across many regions of the corpus callosum, with even greater differences in similar regions in manifest HD compared to control participants. The FA and RD values were associated with both motor and cognitive scores in pre-HD and HD participants.

In addition, Poudel et al.[91] performed tractography between 40 brain regions within bilateral frontal and parietal regions, the striatum, and thalamus. They found that there were a significantly reduced number of streamlines between the putamen and prefrontal and motor cortices for pre-HD compared to controls. For manifest HD participants compared to both control and pre-HD participants, FA was reduced and RD was increased in the fronto-parietal and caudate-paracentral tracts, and RD was also increased in the putamen-prefrontal and putamen-motor tracts. In addition, for the pre-HD participants the increased RD in the bilateral putamen-prefrontal tracts was associated with worsening performance on the SDMT and Stroop cognitive tasks. In the symptomatic HD group, increased RD within the fronto-parietal tract was associated with worsening UHDRS motor score as well as poorer performance on the SDMT and Stroop tasks.

Two studies examined the links between natural variation of WM tracts and pathological change in HD gene carriers.[92,93] Initially, 243 participants were included in a multi-modal analysis demonstrated disturbed diffusion in sensory-motor tracts, associated with CAG length. While the characteristics of these tracts were similar between controls and HD participants, the HD participants displayed "enhanced" natural variation.[92] Furthermore, when this investigation was extended to motor, cognitive, and limbic tracts, there were three patterns of diffusivity that were seen in both controls and HD gene-carriers that could predict HD status.[93] One involved almost all tracts studied, the next pattern was limited to sensory-motor tracts and the final composed of cognitive network tracts. These studies suggest that, within HD-gene carriers, naturally occurring diffusivity patterns can become accentuated and impact on disease progression.

Finally, connectomics studies have examined at the whole-brain connectivity in HD. Connectivity between "rich club" regions, brain regions which show high connections to many other regions, and include the superior frontal, superior parietal, precuneus regions, as well as subcortical regions including the caudate, putamen and thalamus[94] was in both pre-HD and HD participants, with the networks appearing to have lower integration across rich club regions and demonstrated in Fig. 7.[95] In addition, the effects of these network differences were related to cognitive and motor deficits in HD participants, suggesting that the reduced connectivity between rich club regions is related to clinical components of the disease.

McColgan developed this further, by using longitudinal tractography to investigate global connectivity changes. The work indicated that longer between-region WM connections underwent greater deterioration over 24 months, with the cortico-striatal connections being the most vulnerable to early damage, followed by interhemispheric connections and finally intrahemispheric connections.[96] Furthermore, McColgan[97] used both cross-sectional and longitudinal tractography to investigate links between WM changes in HD and genetic transcription. The study demonstrated a dissociative pattern of gene transcription profiles associated with WM connectivity

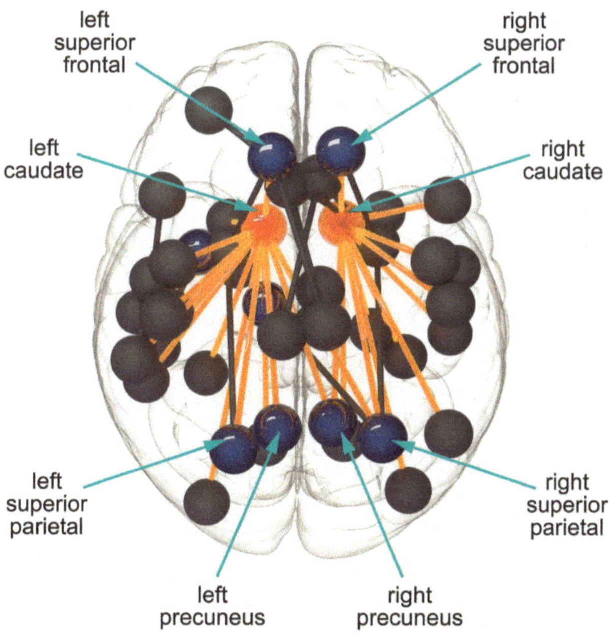

Fig. 7 An example of a connectivity analysis showing reduced connectivity between rich club regions, from McColgan et al.[95] Network-based statistics showed significantly reduced connectivity between pre-HD vs controls in cortico-caudate connections. Red, caudate; blue, cortical rich club regions; yellow, cortico-caudate connections.

loss between corticostriatal regions, interhemispheric regions and intrahemispheric regions at both baseline and occurring over 24 months was associated with synaptic and metabolic genes. The results suggest that the WM vulnerability seen in HD is related to the abnormal transcription of synaptic and metabolic genes.

3.5 Functional MRI

Functional MRI (fMRI) uses the bold oxygen level dependent (BOLD) signal to measure brain activity. As blood flow to a region increases, hemoglobin within the blood becomes deoxygenated and paramagnetic in comparison to oxygenated hemoglobin. The differences between oxyhemoglobin vs deoxyhemoglobin results from local changes in neural activity and give rise to the BOLD signal. The vascular changes follow a time-course (hemodynamic response function, HRF), which is very slow in comparison to the underlying neural activity occurring on a cellular level. However, the neurovascular coupling between blood flow and neural activity can be used for the spatial localization of the underlying neuronal activity.

fMRI data are collected via the rapid acquisition of a large number of images, ensuring that the BOLD signal can be accurately tracked over time. This is done via the use of EPI sequences. fMRI sequences use the T2* decay to quantify the BOLD signal, which is determined by the field inhomogeneity and proton-proton interaction, and as spin-echo sequences result in effectively no T2*, gradient echo sequences are typically used for fMRI data. fMRI can be acquired during rest or the performance of different tasks,[98] for example, increases or decreases in signal could be measured during a finger tapping task. Resting state fMRI (rs-fMRI), in contrast, measures BOLD signal during rest in the scanner. Processing of fMRI data is similar for both rs-fMRI data and task-based fMRI data.

3.5.1 Data processing

It should be noted that for fMRI, there is an accentuated risk of signal dropout, most often in frontal and orbital areas due to the different magnetic susceptibility of air within the facial cavities. Data should be checked for particularly bad signal dropout, but little can be done to improve or correct for this artifact other than positioning the FOV as previously described. In addition, if field maps have been collected during acquisition (an additional scan or set of scans that map the magnetic field intensity across the images) they can be used to correct for geometric distortion in functional MRI data.[99]

3.5.2 Realignment

During data collection for fMRI data, multiple volumes are collected. However, due to participant motion the volumes may undergo minor spatial shifts during data collection. The first step of fMRI analysis is thus realignment of these volumes. Typically, this is done in reference to the first volume and ensures that the voxels are aligned across the whole period of data collection. A linear, rigid-body registration is performed since the size/shape of the brain remains consistent during data collection.

3.5.3 Coregistration

Following realignment of the volumes, functional images for each participant are coregistered to a structural scan collected during the same scanning session. This is to improve the anatomical specificity of regions of interest within the functional data, since the functional data has poor contrast and low spatial resolution. Again, a rigid-body registration with six degrees of freedom can be used for this registration, but often another three degrees of freedom is also used for this registration to account for geometric distortions that occur within EPI sequences.

3.5.4 Normalization

To ensure that data from all participants included in the analysis are in the same space and can be statistically compared, each functional dataset is normalized. Normalization is a registration process that is used with functional data to align the individual datasets. First, the structural data are registered to a standard space image and deformation parameters (the adjustments made to the structural data to warp it to standard space) are saved. These parameters are calculated on the structural data to provide a more accurate registration, since the spatial resolution of functional data is poor. These deformation parameters are then applied to the functional data, thus applying the same warp to the functional data as the structural data and aligning it to standard space.

3.5.5 Smoothing

Finally, the data are smoothed using a Gaussian kernel, similarly to structural data. As in sMRI, the Gaussian smoothing kernel is selected based on the expected signal of interest.

3.5.6 Data analysis

fMRI data are analyzed using various approaches, dependent in part on whether the data are task-based or rs-fMRI. For all fMRI data, the BOLD signal is convolved with the HRF, meaning that the BOLD signal is modeled with the HRF. After the collection of fMRI data during completion of a task, data consist of a measured hemodynamic response that shifts in relation to the performance of a task in the scanner. Analysis aims to couple the hemodynamic signal with the changes in task condition to interpret which regions of the brain are experiencing increases or decreases in signal during different phases of the task.

3.5.7 Task-based fMRI analysis

During task design, the different task conditions should be contrasted repeatedly over time in order to establish brain activity related to each condition. Two types of paradigms are frequently used in fMRI studies: block design and event-based tasks.

Block designs alternate between periods of stimuli presentation and rest periods. The periods during which stimuli are presented are the "on" blocks and should last between 20 and 30 s in length in order to elicit a detectable hemodynamic signal. The periods of rest, "off" periods, act as the baseline by which to compare the on periods to. Using this design, multiple "on" conditions can be compared, as well as comparing the conditions to the baseline activity. The hemodynamic signal does not occur in response to a specific task during block designs, but it represents ongoing neural activity that is occurring within each block.

Event-related (ER) designs aim to measure changes across time in activity based on the response to a specific task or event. The events represent discrete intervals, and short trials are performed repeatedly in a variable order. Each trial is separated by an interstimulus event, during which the hemodynamic response returns to baseline. ER designs allow more flexibility than block designs, with increased possibility for the manipulation of the stimuli and presentation of these stimuli. However, the HRF tends to be weaker for ER design. Thus, ER sessions need to be longer than block design sessions to capture enough data in order to make sound inferences.

For both types of design, the data are analyzed using the GLM first at an individual level to define each condition, and then at a group level. With the output of analysis showing regions whereby activity is greater in one group compared to another group, or where activity is associated with a predictor variable.

3.5.8 Resting state fMRI analysis

Since in rs-fMRI data there is no task and therefore, no conditions to compare, connectivity within brain networks is investigated instead. It is important to consider issues that are common to rs-fMRI data. Since there is no task, the resting state signal tends to include more noise than other fMRI data. This noise includes physiological noise, and thus it is recommended that analyses consider as much of this non-neuronal noise as possible. There are two methods to do this: the first involves sampling time-series from WM or CSF during the first-level analysis, which is used to model cardiac or respiratory noise. Alternatively, respiratory and cardiac measures collected during the scan can also be used to correct for these artifacts. In addition, it is recommended that global signal across the brain is not regressed. While this recommendation has been controversial, it results in a trend toward negative correlations.[100,101]

There are two key approaches to perform connectivity analysis. Functional connectivity measures temporal correlations of activity between every voxel, e.g., seed connectivity or independent components analysis (ICA), but is non-directional and thus uninformative in terms of causality. Alternatively, effective connectivity can be used to examine the causal effect of activity within one region to activity in other regions via a number of techniques including structural equation modeling and dynamic causal modeling.

3.5.9 Functional connectivity

Functional connectivity examines non-directed relationships between regions during rest. The data can be analyzed via either hypothesis-driven seed connectivity methods or using data-driven independent components analysis. To perform seed-based connectivity, an ROI region is selected, defined using a mask or coordinates in standard space. For each participant, the time-series from this region is extracted and then a whole-brain voxel-wise analysis is performed which identifies voxels whereby the temporal pattern of activity is the same as the region of interest, that is, voxels which are temporally correlated. This analysis is thought to reveal networks that involve both the seed regions and the temporally correlated regions. The individual-level maps can then be analyzed together in a group-wise method. Similarly, ICA is a data-driven approach whereby across the brain, time-series from the data are temporally correlated. This results in functional networks maps (or noise maps). ICA can either be performed on individual data, to remove structured noise, or on a group-wise level to determine functional networks. Tools such as the SPM GIFT[102] toolbox associated and MELODIC in FSL[103] can be used to perform ICA.

3.5.10 Effective connectivity

Effective connectivity examines directed effects of activity between regions. It enables the understanding of causal relationships between regions that are temporally correlated. There are different techniques that can be used to measure effective connectivity, including structural equation modeling,[104] Granger causality[105] and dynamic causal modeling.[106] Dynamic causal modeling can be performed to test hypothesis-driven network models, but requires an a priori selection of biologically plausible regions to include in the model. Different directions of causality can be tested in the model, for example, top-down or bottom-up models, using the extracted time series from each region. The model with the best fit to the data is designated as the winning, and most biologically accurate, model. Based on this result, connectivity values can be extracted and compared across groups or used against other regressors.

3.6 Functional MRI studies findings in HD

Both resting state fMRI and task-based fMRI studies have aimed to elucidate changes in functional network performance during HD progression, with either group differences or longitudinal changes in functional networks thought to represent pathological changes due to the HD phenotype. While the results are often contradictory with both increases and decreases in BOLD signal having been described for similar studies, the different tasks, study design and analysis methods can all contribute to the varying results.

3.6.1 Resting state fMRI

There has been an increasing focus on rs-fMRI analyses within HD research due to the association between changes in rs-fMRI features and other neurodegenerative conditions.[107] Most studies have performed seed-based analyses or ICA to examine network-based brain activity, but more recent work has begun to focus more on analyses comparing the connectome between HD and non-HD participants.

Seed-based studies have frequently been applied within HD to examine motor network function, with an initial seed-based analyses demonstrated that HD participants had reduced connectivity between the premotor cortex and the caudate, providing early evidence of a disruption to the resting state network in HD.[108] Further support for motor network changes came from a study in which the primary motor cortex (M1) was used as a seed region, with decreasing connectivity between M1 and the ipsilateral M1 as well as other somatosensory regions which was associated with increasing disease burden score.[109] A number of studies have investigated the default mode network (DMN) in HD. The DMN refers to a network of midline brain regions that show increased temporally—correlated activity during periods of rest, i.e., when the brain is not performing a task or function[107] and which become less active when the brain is engaged in a task. The default mode network has been demonstrated as a network showing increasing disorganization with Alzheimer's disease progression, and thus has become a network of interest in HD. The posterior cingulate cortex, a DMN region, has been found to show decreased associations with the ventral medial prefrontal cortex, right dorso-medial prefrontal cortex and right inferior parietal cortex,[110] but the posterior cingulate cortex has also been reported to have higher connectivity to DMN regions in HD participants compared to controls.[111] While the concluding message is unclear, these regions are all part of the DMN and indicate disrupted DMN connectivity within HD gene carriers.

Data-driven ICA analyses have also been used to attempt to develop unbiased conclusions on resting network connectivity in HD. These studies have, however, also resulted in mixed results. With both increased and decreased connectivity in many regions in HD compared to controls. One study reporting increased connectivity within the thalamus, striatum, prefrontal and premotor cortices as well as within parietal regions,[112] but both increases and decreases in connectivity have also been shown within a number of these regions.[113] Another study using an ICA approach showed that pre-HD participants had reduced synchrony within sensorimotor network (associated with worsening motor score) and both pre-HD and manifest HD had reduced synchrony in the dorsal attention networks, which was related to poorer cognitive performance in the manifest group.[114] They also found that the putamen and superior parietal cortex had reduced connectivity from the frontal executive network within the manifest group compared to both control and pre-HD participants, and interestingly the fronto-parietal network had regions of increased connectivity within the manifest HD participants. Similarly, another ICA study reported both increases (bilateral inferior temporal gyrus (ITG) and left superior frontal gyrus) and decreases (right precuneus and angular gyrus) in the amplitude of low frequency fluctuations in HD gene carriers compared to controls.[115]

The increased fluctuations in the left inferior temporal gyrus were associated with worsening performance on the symbol digit modalities task (SDMT), and the decreased activity in the precuneus was associated with worse performance on the Stroop test and the SDMT. A recent study used an ICA and demonstrated a negative association between a putamen resting state network and CAG repeat length, as well as increased functional connectivity between the putamen and insular with increasing CAG length, which was also associated with motor and cognitive measures.[116] In contrast, the authors reported reduced functional connectivity within visual networks with increasing CAG repeat length. Finally, an early study which used an ICA analysis found that while the spatial distribution of the DMN was similar between control and pre-HD participants, the pre-HD participants had reduced connectivity in medial prefrontal cortex, inferior parietal and posterior cingulate when compared to the DMN of control participants.[107]

More recently, there has been a move toward understanding connectivity via connectomics. In one of the first papers using this method, HD participants closer to disease onset showed reduced connectivity between rich club regions of the brain.[117] In contrast, the results of this study indicated that local networks were maintained. Another study designed to investigate the connectome in HD looked at connectivity between 182 brain regions using graph theory measures.[118] They found that there were only significant differences in manifest HD rather than pre-HD individuals. The differences indicated that networks were less efficient (a reduction in path length), as well as reduction in the cluster strength for the somatosensory cortex which can indicated reduced network strength, and finally there were more connections (higher degree) across all networks studied.[118] There were no longitudinal changes detected in this study, although this could be due to a lack of sensitivity of the methodology to detect long-term change rather than an absence of change. Finally, within pre-manifest participants higher structural connectivity has been found to predict reduced functional connectivity.[119] This work also suggested that network-based pathology follows a gradient of change, with anterior brain regions having higher connectivity compared to posterior regions. Further work by the same group also indicated that increased connectivity within the DMN and basal ganglia were associated with higher levels of depression.[120]

3.6.2 Task-based fMRI studies

As well as resting-state fMRI studies, task-based fMRI studies have been used to measure the BOLD signal during task performance in a number of different domains. Motor tasks, working memory tasks and executive function have all been measured across different cohorts and studies. These tasks have been selected as they are designed to target the HD phenotype, which shows progressive deterioration of cognitive and motor skills.

Neural activity during a finger tapping task has been measured, with one study using a small cohort of pre-HD participants and control participants to participate in a finger tapping task involving varying speeds and complexity.[121] While there were no differences in task performance between the two groups, the pre-HD participants showed both increases and decreases in activity within the supplementary motor area and the parietal lobe for different levels of task complexity. However, these results were not replicated in a large cohort of 111 controls and 106 pre-HD participants, with no significant group effects in the finger tapping task.[122]

Working memory has been tested in various longitudinal functional MRI studies in HD, both measuring working memory and the other looking at verbal working memory. The TrackOn-HD study tested participants on a two-condition N-back test, whereby participants were instructed to respond to a letter on screen, deciding whether the letter was the same as the letter presented one previously (1-back), or two previously (2-back). The only significant difference between controls and HD participants in BOLD signal was in the cerebellum during the 2-back condition, results for both groups are shown in Fig. 8. The IMAGE-HD study also used the N-back (with 0-BACK, 1-BACK and 2-BACK conditions) to test working memory in 35 pre-HD, 23 manifest HD and 32 control participants at baseline.[123] There were no differences in performance between the groups, but there were significant differences in BOLD signal. In both 1-BACK and 2-BACK conditions, there were significant differences between both HD groups and controls in the working memory network, including the inferior frontal gyrus, anterior insula, caudate, putamen, and cerebellum, as well as the anterior cingulate gyrus, parietal lobe and thalamus in the 2-BACK. The dorsolateral prefrontal cortex showed increased activity in the 1-BACK, and decreased activity in the 2-BACK. An 18 month follow-up of this cohort indicated that the pre-HD participants had significantly increased activation longitudinally during 1-BACK vs 0-BACK in the anterior cingulate, lateral and medial prefrontal, primary motor, and temporal areas as well as in the caudate and putamen.[124] There were no significant differences between manifest HD and controls, and due to low power in the 2-BACK task this measure was not included in the analysis. Finally, in the 30 month follow-up, during the 1-BACK the pre-HD

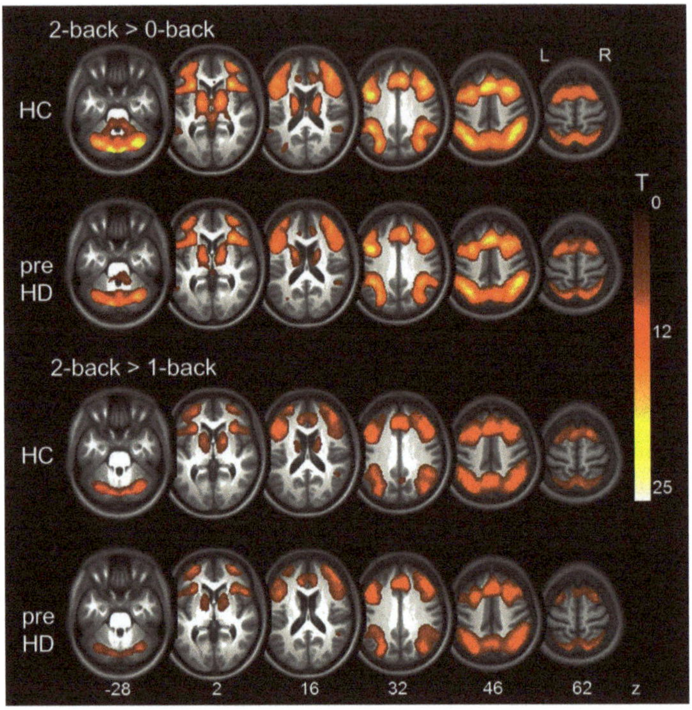

Fig. 8 An example of fMRI activity from a block-design task for both healthy controls and pre-HD participants. Participants were instructed to respond to a letter on screen, deciding whether the letter was the same as the letter presented one previously (1-back), or two previously (2-back).[122]

participants had significantly increased activation longitudinally in the left DLPFC and medial frontal cortex compared to controls, with no difference seen in the manifest HD-participants.[125] For the 2-BACK, the pre-HD group had increased activation in the bilateral caudate, putamen, and temporal cortex, with no difference seen in the manifest group. Furthermore, longitudinal changes in pre-HD were associated with disease burden and years to onset. In this cohort, deactivation was also associated with obsessive compulsive and depressive symptoms.[126] This series of studies demonstrates that functional signal during working memory is changing during disease progression in pre-HD participants, although it is unclear whether the lack of change in manifest participants is due to reduced power in this group or an actual lack of change.

The verbal working memory paradigm has again been used with both cross-sectional and longitudinal follow-up time points. 16 pre-HD participants and 16 control participants were recruited, using a verbal working memory task with 3 levels of difficulty. The initial analysis suggested that the pre-HD participants had significantly lower activation of the left dorsolateral prefrontal cortex in the medium and difficult conditions, and participants estimated to be close to motor onset had increased activity within the left inferior parietal lobule and the right superior frontal gyrus when compared to the pre-HD subjects far from symptom onset and controls.[127] In this group, there was also abnormal functional connectivity at the high working memory load for the left DLPFC, as well as lower connectivity in the left putamen, the right anterior cingulate and the left medial prefrontal cortex for pre-HD compared to controls.[128] Furthermore, an independent components analysis showed discrete bilateral frontostriatal and frontoparietal networks in the two groups, and the pre-HD group had lower functional connectivity in the left parietal and lateral prefrontal regions and bilateral caudate.[129] In a 24 month follow-up of these participants there were no further changes over time.[130] The authors later re-analyzed the data to study task negative networks, that is, networks more active during rest, such as the DMN. At baseline, the pre-HD participants had reduced functional connectivity in the posterior cingulate and higher connectivity in the left anterior prefrontal cortex compared to controls. Over 24 months, the pre-HD participants showed higher connectivity in the dorsal cingulate.[131] The same task was also used in a group of 12 manifest HD participants and 16 controls, and during the difficult conditions manifest participants had lower activation in the left dorso- and ventrolateral prefrontal cortex, the left inferior parietal cortex, the left putamen, and the right cerebellum compared to controls conditions.[52]

In addition to tapping and working memory, executive functioning tasks have been applied during fMRI. During completion of the Stroop naming task, there were no differences between 32 HD gene carriers in the "prodromal" phase of HD (i.e., pre-HD participants with some cognitive and psychiatric deficits) vs 52 control participants. However, there were correlations between activity in the ventromedial prefrontal cortex was more highly correlated with depressive symptoms in the HD participants than in controls.[132] The Tower of London test is another executive functioning task, and in 53 HD gene carriers (41 pre-HD and 12 manifest HD) compared to 52 control participants there was reduced functional connectivity between the medial prefrontal cortex and the left premotor area in HD gene carriers, with an effect of task complexity on this connectivity.[133] Set shifting tasks which measure executive functioning were included in the IMAGE-HD study. The task resulted in activation within the dorsolateral prefrontal cortex and the dorsal anterior cingulate cortex, as well as the premotor, parietal, and basal ganglia regions and deactivated the anterior cingulate cortex. The manifest participants had greater prefrontal activity compared to both controls and pre-HD.[134] There were no longitudinal differences in the BOLD signal between the pre-HD group and control participants over 30 months, however, there was a reduction in signal over this time in the manifest HD participants compared to the control group in both regions related to task performance, such as the anterior cingulate cortex and the caudate, as well as DMN regions including the medial prefrontal cortex and posterior cingulate. When the pre-HD and manifest HD participants were combined, longitudinal reductions in signal within the right dorsolateral prefrontal cortex and putamen were associated with executive dysfunction, and in fronto-striatal and default mode networks with disease severity.[135]

A series of other domains have also been measured, including emotional processing and facial recognition,[136–138] reward processing[139] and alertness.[107,140] With varying patterns of activation and deactivation demonstrated in HD gene carriers across studies, it is difficult to form a coherent understanding of functional brain changes in HD. The wide array of tasks, cohorts and group sizes further add to the confusion. However, there is evidence to suggest that functional changes reflect compensatory mechanisms occurring over the course of disease progression that cause increases in activity in pre-HD participants prior to decreases in activity with increasing disease progression in the manifest phase of HD.[122,141,142]

3.7 General analysis considerations

3.7.1 The general linear model

Multiple regression is one of the most commonly used statistical analyses techniques applied to analyze imaging data. Multiple regression uses a general linear model (GLM) to estimate the relationship between predictor/dependent variables (e.g., Huntington's disease stage) and measured (criterion/independent) variables (e.g., GM volume). The equation used to perform this analysis is $Y = X\beta + e$, where Y represents the dependent variable; X the regressors or independent variables; β is the estimated parameter values of these regressors (that is, their weighted contribution to the dependent variable) and e is the error or residual values (how much the individual data point varies from the mean value). The GLM is performed at every voxel within the brain to look for brain-wide patterns of interest.

Multiple regression enables the specification of multiple criterion variables so the unique contribution of variables to a predictor can be understood.

3.7.2 Covariates

Covariates are confounding variables that may be related to a variable of interest but are not of interest in themselves. They can be statistically controlled for during analysis, which results in a more direct measurement of the relationship between the variables of interest. Typically, in neuroimaging studies age and sex are controlled for due to the known effects of these two variables on brain size[143] (Barnes et al., 2010). In addition, in multi-site studies, site should be used as a covariate to control for effects of different MRI scanners on data. However, it should be noted that this does not fully account for effects of site, as sites are differentially affected by artifacts such as scanner drift, software upgrades and field inhomogeneity. Furthermore, often education is included as a covariate when examining relationships between neuroimaging measures and cognitive variables due to the relationship between education level and cognitive performance[144] (Beeri et al., 2006). Other co-variates should be included in the analyses where appropriate.

One measure that is frequently discussed as a potential covariate in structural neuroimaging literature is total intracranial volume (TIV;[145,143] Malone et al., 2015; Barnes et al., 2010). TIV acts as a proxy to "healthy" brain volume, since TIV measures the total intracranial space and has been demonstrated to remain constant despite increasing neural atrophy[146,147] (Whitwell et al., 2001; Matsumae et al., 1996). By controlling for TIV, variables of interest can be compared while accounting for differences in brain morphometry due to head size. TIV can be measured via manual delineation of the region or automated measures, such as SPM, which have also been validated [145].

3.7.3 Correction for multiple comparisons

The results of statistical analyses are susceptible to both Type I and Type II errors. A Type I error refers to the incorrect rejection of a true null hypothesis (a false positive). A Type II error is the acceptance of the null hypothesis when a true effect is present (a false negative). The more statistical comparisons performed in a given analysis, the more likely a Type I or Type II error is to occur. While an understanding of these two scenarios is necessary for all researchers undertaking statistical analysis, the nature of neuroimaging analyses and the volume of statistical comparisons conducted during many types of neuroimaging analyses mean that these errors are more likely to occur than within other fields[148,149] (Lindquist and Mejia, 2015; Hupé, 2015). To account for this, statistical adjustments can be made to correct for multiple comparisons. This is done by adjusting a statistical threshold dependent on the number of comparisons being performed, while the Bonferroni correction is the most widely known[148] (Lindquist and Mejia, 2015), within neuroimaging correction for family-wise error rate (FWE) and correction for false discovery rate (FDR) are the two most widely used methods.

FWE refers to the likelihood of making one or more Type I errors when conducting multiple tests (a family of tests), it includes the Bonferroni method. Correcting for FWE aims to limit the risk of a Type I error occurring in any of the statistical tests being performed, but is often too stringent, thus increasing the risk of Type II errors. In comparison, FDR controls for the number of Type I errors across all significant results, aiming to reduce the number of false positives only within the subset of voxels found to be significant. The choice between FWE and FDR is often dependent on the software used, since many software tools include one or the other as a default option to control for multiple comparisons.

4. Summary

MRI offers a non-invasive and safe method of measuring brain changes during disease progression in HD. While there are a number of practical considerations that may impact on the data, with careful study design many of these can be overcome.

Overall, the growing body of work using MRI techniques to measure brain changes in vivo in HD gene carriers indicates a progressively worsening neurodegenerative phenotype. Structural changes are the most well-established and robust results, with subcortical, WM and cortical atrophy occurring from at least a decade before motor diagnosis. Diffusion measures have begun to tell a consistent story, with reduced FA and increased RD apparent across pre-HD and manifest participants, particularly within the corpus callosum and internal capsule. Finally, while fMRI findings vary considerably and can be difficult to interpret due to the wide range of studies and findings, there is growing evidence to suggest that compensatory mechanisms may explain the apparent variable trajectory in fMRI changes across disease progression.

References

1. Vonsattel JP, Myers RH, Stevens TJ, Ferrante RJ, Bird ED, Richardson EP. Neuropathological classification of Huntington's disease. *J Neuropathol Exp Neurol.* 1985;44(6):559–577. Available at: http://www.ncbi.nlm.nih.gov/pubmed/2932539. Accessed 21.11.17.
2. Currie S, Hoggard N, Craven IJ, Hadjivassiliou M, Wilkinson ID. Understanding MRI: basic MR physics for physicians. *Postgrad Med J.* 2013;89(1050):209–223. https://doi.org/10.1136/postgradmedj-2012-131342.
3. Ross CA, Tabrizi SJ. Huntington's disease: from molecular pathogenesis to clinical treatment. *Lancet Neurol.* 2011;10:83–98. https://doi.org/10.1016/S1474-4422(10)70245-3.
4. Escher K, Shellock FG. Evaluation of MRI artifacts at 3 Tesla for 38 commonly used cosmetics. *Magn Reson Imaging.* 2013;31:778–782. https://doi.org/10.1016/j.mri.2012.11.002.
5. Chenji S, Wilman AH, Mah D, Seres P, Genge A, Kalra S. Hair product artifact in magnetic resonance imaging. *Magn Reson Imaging.* 2017;35:1–3. https://doi.org/10.1016/j.mri.2016.08.023.
6. Krupa K, Bekiesińska-Figatowska M. Artifacts in magnetic resonance imaging. *Pol J Radiol.* 2015;80:93–106. https://doi.org/10.12659/PJR.892628.
7. Callaghan MF, Negus C, Leff AP, et al. Safety of tattoos in persons undergoing MRI. *N Engl J Med.* 2019;380(5):495–496. https://doi.org/10.1056/NEJMc1811197.
8. Sled JG, Zijdenbos AP, Evans AC. A nonparametric method for automatic correction of intensity nonuniformity in MRI data. *IEEE Trans Med Imaging.* 1998;17:87–97. https://doi.org/10.1109/42.668698.

8a. Boyes RG, Gunter JL, Frost C, et al. Intensity non-uniformity correction using N3 on 3-T scanners with multichannel phased array coils. *Neuroimage.* 2008;39(4):1752–1762. https://doi.org/10.1016/j.neuroimage.2007.10.026.

9. Ashburner J, Friston KJ. Unified segmentation. *Neuroimage.* 2005;26(3):839–851. https://doi.org/10.1016/j.neuroimage.2005.02.018.
10. Jenkinson M, Beckmann C, Behrens TE, Woolrich MW, Smith SM. FSL. *Neuroimage.* 2012;62:782–790.
11. Dale AM, Fischl B, Sereno MI. Cortical surface-based analysis. I. Segmentation and surface reconstruction. *Neuroimage.* 1999;9:179–194.
12. Fischl B, Sereno MI, Dale AM. Cortical surface-based analysis. II. Inflation, flattening, and a surface-based coordinate system. *Neuroimage.* 1999;9:195–207.
13. Mazziotta J, Toga A, Evans A, et al. A probabilistic atlas and reference system for the human brain: International Consortium for Brain Mapping (ICBM). *Philos Trans R Soc Lond Ser B Biol Sci* 2001;356(1412):1293–1322. https://doi.org/10.1098/rstb.2001.0915.
14. Mazziotta J, Toga A, Evans A, et al. A four-dimensional probabilistic atlas of the human brain. *J Am Med Inform Assoc.* 2001;8(5):401–430. Available at: http://www.ncbi.nlm.nih.gov/pubmed/11522763. Accessed 21.2.19.
15. Ashburner J. A fast diffeomorphic image registration algorithm. *Neuroimage.* 2007;38(1):95–113. https://doi.org/10.1016/j.neuroimage.2007.07.007.
16. Ashburner J, Ridgway GR. Symmetric diffeomorphic modeling of longitudinal structural MRI. *Front Neurosci.* 2012;6:197. https://doi.org/10.3389/fnins.2012.00197.
17. Freeborough PA, Fox NC. Modeling brain deformations in Alzheimer disease by fluid registration of serial 3D MR images. *J Comput Assist Tomogr.* 1998;22(5):838–843. Available at: http://www.ncbi.nlm.nih.gov/pubmed/9754126. Accessed 3.8.17.
18. Crum WR, Scahill RI, Fox NC. Automated hippocampal segmentation by regional fluid registration of serial MRI: validation and application in Alzheimer's disease. *Neuroimage.* 2001;13(5):847–855. https://doi.org/10.1006/NIMG.2001.0744.

19. Henley SM, Ridgway GR, Scahill RI, et al. Pitfalls in the use of voxel-based morphometry as a biomarker: examples from Huntington disease. *AJNR Am J Neuroradiol*. 2010;31:711–719.
20. Freeborough PA, Fox NC, Kitney RI. Interactive algorithms for the segmentation and quantitation of 3-D MRI brain scans. *Comput Programs Biomed*. 1997;53:15–25. Available at: file://m/ION/TrackHD/Imaging/Eli/database/Interactive_alogorithms_for_the_segmentation.pdf.
21. Hobbs NZ, Henley SMD, Wild EJ, et al. Automated quantification of caudate atrophy by local registration of serial MRI: evaluation and application in Huntington's disease. *Neuroimage*. 2009;47(4):1659–1665. https://doi.org/10.1016/j.neuroimage.2009.06.003.
22. Avants BB, Tustison NJ, Wu J, Cook PA, Gee JC. An open source multivariate framework for n-tissue segmentation with evaluation on public data. *Neuroinformatics*. 2011;9(4):381–400. https://doi.org/10.1007/s12021-011-9109-y.
23. Ledig C, Heckemann RA, Hammers A, et al. Robust whole-brain segmentation: application to traumatic brain injury. *Med Image Anal*. 2015;21(1):40–58. https://doi.org/10.1016/j.media.2014.12.003.
24. Johnson EB, Gregory S, Johnson HJ, et al. Recommendations for the use of automated Gray matter segmentation tools: evidence from Huntington's disease. *Front Neurol*. 2017;8:519. https://doi.org/10.3389/fneur.2017.00519.
25. Katuwal GJ, Baum SA, Cahill ND, et al. Inter-method discrepancies in brain volume estimation may drive inconsistent findings in autism. *Front Neurosci*. 2016;10:439. https://doi.org/10.3389/fnins.2016.00439.
26. Ashburner J, Friston KJ. Voxel-based morphometry—the methods. *Neuroimage*. 2000;11(6):805–821. https://doi.org/10.1006/nimg.2000.0582.
27. Douaud G, Smith S, Jenkinson M, et al. Anatomically related grey and white matter abnormalities in adolescent-onset schizophrenia. *Brain*. 2007;130:2375–2386. https://doi.org/10.1093/brain/awm184.
28. Tabrizi SJ, Langbehn DR, Leavitt BR, et al. Biological and clinical manifestations of Huntington's disease in the longitudinal TRACK-HD study: cross-sectional analysis of baseline data. *Lancet Neurol*. 2009;8(9):791–801. https://doi.org/10.1016/S1474-4422(09)70170-X.
29. Scahill RI, Hobbs NZ, Say MJ, et al. Clinical impairment in premanifest and early Huntington's disease is associated with regionally specific atrophy. *Hum Brain Mapp*. 2013;34(3):519–529. https://doi.org/10.1002/hbm.21449.
30. Johnson EB, Byrne LM, Gregory S, et al. Neurofilament light protein in blood predicts regional atrophy in Huntington disease. *Neurology*. 2018;90(8):e717–e723. https://doi.org/10.1212/WNL.0000000000005005.
31. Fischl B, Salat DH, Busa E, et al. Whole brain segmentation: automated labeling of neuroanatomical structures in the human brain. *Neuron*. 2002;33(3):341–355. https://doi.org/10.1016/S0896-6273(02)00569-X.
32. Fischl B, van der Kouwe A, Destrieux C, et al. Automatically parcellating the human cerebral cortex. *Cereb Cortex*. 2004;14:11–22.
33. Nopoulos PC, Aylward EH, Ross CA, et al. Cerebral cortex structure in prodromal Huntington disease. *Neurobiol Dis*. 2010;40:544–554.
34. Aylward EH, Li Q, Stine OC, et al. Longitudinal change in basal ganglia volume in patients with Huntington's disease. *Neurology*. 1997;48(2):394–399.
35. Aylward EH, Sparks BF, Field KM, et al. Onset and rate of striatal atrophy in preclinical Huntington disease. *Neurology*. 2004;63:66–72.
36. Tabrizi SJ, Scahill RI, Durr A, et al. Biological and clinical changes in premanifest and early stage Huntington's disease in the TRACK-HD study: the 12-month longitudinal analysis. *Lancet Neurol*. 2011;10:31–42. https://doi.org/10.1016/S1474-4422(10)70276-3.

37. Paulsen JS, Nopoulos PC, Aylward EH, et al. Striatal and white matter predictors of estimated diagnosis for Huntington disease. *Brain Res Bull*. 2010;82(3–4):201–207. https://doi.org/10.1016/j.brainresbull.2010.04.003.
38. Zeun P, Lowe J, Osborne-Crowley K, et al. Huntington's disease young adult study (HD-YAS). *J Neurol Neurosurg Psychiatry*. 2018;89:A60–A61. https://doi.org/10.1136/jnnp-2018-ehdn.160. BMJ Publishing Group Ltd.
39. Majid DSA, Aron AR, Thompson W, et al. Basal ganglia atrophy in prodromal Huntington's disease is detectable over one year using automated segmentation. *Mov Disord*. 2011;26(14):2544–2551. https://doi.org/10.1002/mds.23912.
40. Georgiou-Karistianis N, Scahill R, Tabrizi SJ, Squitieri F, Aylward EH. Structural MRI in Huntington's disease and recommendations for its potential use in clinical trials. *Neurosci Biobehav Rev*. 2013;37(3):480–490. https://doi.org/10.1016/j.neubiorev.2013.01.022.
41. Hobbs NZ, Farmer RE, Rees EM, et al. Short-interval observational data to inform clinical trial design in Huntington's disease. *J Neurol Neurosurg Psychiatry*. 2015;86(12):1291–1298. https://doi.org/10.1136/jnnp-2014-309768.
42. Henley SM, Frost C, MacManus DG, Warner TT, Fox NC, Tabrizi SJ. Increased rate of whole-brain atrophy over 6 months in early Huntington disease. *Neurology*. 2006;67:694–696.
43. Tabrizi SJ, Reilmann R, Roos R, et al. Potential endpoints for clinical trials in premanifest and early Huntington's disease in the TRACK-HD study: analysis of 24 month observational data. *Lancet Neurol*. 2012;11(1):42–53. https://doi.org/10.1016/S1474-4422(11)70263-0.
44. Aylward EH, Anderson NB, Bylsma FW, et al. Frontal lobe volume in patients with Huntington's disease. *Neurology*. 1998;50(1):252–258. https://doi.org/10.1212/WNL.50.1.252.
45. Tabrizi SJ, Scahill RI, Owen G, et al. Predictors of phenotypic progression and disease onset in premanifest and early-stage Huntington's disease in the TRACK-HD study: analysis of 36-month observational data. *Lancet Neurol*. 2013;12:637–649. https://doi.org/10.1016/S1474-4422(13)70088-7.
46. Paulsen JS, Long JD, Ross CA, et al. Prediction of manifest Huntington's disease with clinical and imaging measures: a prospective observational study. *Lancet Neurol*. 2014;13(12):1193–1201. https://doi.org/10.1016/S1474-4422(14)70238-8.
47. Thieben MJ, Duggins AJ, Good CD, et al. The distribution of structural neuropathology in pre-clinical Huntington's disease. *Brain*. 2002;125(8):1815–1828. https://doi.org/10.1093/brain/awf179.
48. Gómez-Ansón B, Alegret M, Muñoz E, et al. Prefrontal cortex volume reduction on MRI in preclinical Huntington's disease relates to visuomotor performance and CAG number. *Parkinsonism Relat Disord*. 2009;15(3):213–219. https://doi.org/10.1016/j.parkreldis.2008.05.010.
49. Sormani MP, Rovaris M, Valsasina P, Wolinsky JS, Comi G, Filippi M. Measurement error of two different techniques for brain atrophy assessment in multiple sclerosis. *Neurology*. 2004;62(8):1432–1434. https://doi.org/10.1212/wnl.0b013e3181d8c20a.
50. Douaud G, Gaura V, Ribeiro M-J, et al. Distribution of grey matter atrophy in Huntington's disease patients: a combined ROI-based and voxel-based morphometric study. *Neuroimage*. 2006;32(4):1562–1575. Available at: http://www.sciencedirect.com/science/article/pii/S1053811906006422. Accessed 29.11.17.
51. Kassubek J, Juengling FD, Kioschies T, et al. Topography of cerebral atrophy in early Huntington's disease: a voxel based morphometric MRI study. *J Neurol Neurosurg Psychiatry*. 2004;75(2):213–220. https://doi.org/10.1136/JNNP.2002.009019.
52. Wolf RC, Vasic N, Schönfeldt-Lecuona C, Ecker D, Landwehrmeyer GB. Cortical dysfunction in patients with Huntington's disease during working memory performance. *Hum Brain Mapp*. 2009;30(1):327–339. https://doi.org/10.1002/hbm.20502.

53. Gavazzi C, Della NR, Petralli R, et al. Combining functional and structural brain magnetic resonance imaging in Huntington disease. *J Comput Assist Tomogr.* 2007;31(4):574–580. https://doi.org/10.1097/01.rct.0000284390.53202.2e.
54. Hobbs NZ, Henley SMD, Ridgway GR, et al. The progression of regional atrophy in premanifest and early Huntington's disease: a longitudinal voxel-based morphometry study. *J Neurol Neurosurg Psychiatry.* 2010;81(7):756–763. https://doi.org/10.1136/jnnp.2009.190702.
55. Dogan I, Eickhoff SB, Schulz JB, et al. Consistent neurodegeneration and its association with clinical progression in Huntington's disease: a coordinate-based meta-analysis. *Neurodegener Dis.* 2013;12(1):23–35. https://doi.org/10.1159/000339528.
56. Rosas HD, Salat DH, Lee SY, et al. Cerebral cortex and the clinical expression of Huntington's disease: complexity and heterogeneity. *Brain.* 2008;131:1057–1068.
57. Rosas HD, Liu AK, Hersch S, et al. Regional and progressive thinning of the cortical ribbon in Huntington's disease. *Neurology.* 2002;58:695–701.
58. Rosas HD, Reuter M, Doros G, et al. A tale of two factors: what determines the rate of progression in Huntington's disease? A longitudinal MRI study. *Mov Disord.* 2011;26:1691–1697.
59. Xu L, Groth KM, Pearlson G, Schretlen DJ, Calhoun VD. Source-based morphometry: the use of independent component analysis to identify gray matter differences with application to schizophrenia. *Hum Brain Mapp.* 2009;30(3):711–724. https://doi.org/10.1002/hbm.20540.
60. Ciarochi JA, Calhoun VD, Lourens S, et al. Patterns of co-occurring gray matter concentration loss across the Huntington disease prodrome. *Front Neurol.* 2016;7:147. https://doi.org/10.3389/fneur.2016.00147.
61. Coppen EM, van der Grond J, Hafkemeijer A, Rombouts SARB, Roos RAC. Early grey matter changes in structural covariance networks in Huntington's disease. *NeuroImage Clin.* 2016;12:806–814. https://doi.org/10.1016/j.nicl.2016.10.009.
62. Johnson EB, Ziegler G, Penny W, et al. Dynamics of cortical degeneration over a decade in Huntington's disease. *bioRxiv.* 2019:537977. https://doi.org/10.1101/537977.
63. Le Bihan D, Mangin JF, Poupon C, et al. Diffusion tensor imaging: concepts and applications. *J Magn Reson Imaging.* 2001;13(4):534–546. https://doi.org/10.1002/jmri.1076.
64. O'Donnell LJ, Westin C-F. An introduction to diffusion tensor image analysis. *Neurosurg Clin N Am.* 2011;22(2):185–196, viii. https://doi.org/10.1016/j.nec.2010.12.004.
65. Ni H, Kavcic V, Zhu T, Ekholm S, Zhong J. Effects of number of diffusion gradient directions on derived diffusion tensor imaging indices in human brain. *Am J Neuroradiol.* 2006;27:1776–1781.
66. Chan RW, von Deuster C, Giese D, et al. Characterization and correction of eddy-current artifacts in unipolar and bipolar diffusion sequences using magnetic field monitoring. *J Magn Reson.* 2014;244:74–84. https://doi.org/10.1016/J.JMR.2014.04.018.
67. Andersson JLR, Sotiropoulos SN. An integrated approach to correction for off-resonance effects and subject movement in diffusion MR imaging. *Neuroimage.* 2016;125:1063–1078. https://doi.org/10.1016/j.neuroimage.2015.10.019.
68. Zhang H, Schneider T, Wheeler-Kingshott CA, Alexander DC. NODDI: practical in vivo neurite orientation dispersion and density imaging of the human brain. *Neuroimage.* 2012;61(4):1000–1016. https://doi.org/10.1016/j.neuroimage.2012.03.072.
69. Smith SM, Jenkinson M, Johansen-Berg H, et al. Tract-based spatial statistics: voxelwise analysis of multi-subject diffusion data. *Neuroimage.* 2006;31:1487–1505. https://doi.org/10.1016/j.neuroimage.2006.02.024.

70. Khalsa S, Mayhew SD, Chechlacz M, Bagary M, Bagshaw AP. The structural and functional connectivity of the posterior cingulate cortex: comparison between deterministic and probabilistic tractography for the investigation of structure-function relationships. *Neuroimage*. 2014;102:118–127. https://doi.org/10.1016/j.neuroimage.2013.12.022.
71. Shi Y, Toga AW. Connectome imaging for mapping human brain pathways. *Mol Psychiatry*. 2017;22:1230–1240. https://doi.org/10.1038/mp.2017.92.
72. Tournier JD, Calamante F, Connelly A. MRtrix: diffusion tractography in crossing fiber regions. *Int J Imaging Syst Technol*. 2012;22:53–66. https://doi.org/10.1002/ima.22005.
73. Rosas HD, Tuch DS, Hevelone ND, et al. Diffusion tensor imaging in presymptomatic and early Huntington's disease: selective white matter pathology and its relationship to clinical measures. *Mov Disord*. 2006;21(9):1317–1325. https://doi.org/10.1002/mds.20979.
74. Mascalchi M, Lolli F, Della NR, et al. Huntington disease: volumetric, diffusion-weighted, and magnetization transfer MR imaging of brain. *Radiology*. 2004;232(3):867–873. https://doi.org/10.1148/radiol.2322030820.
75. Sritharan A, Egan GF, Johnston L, et al. A longitudinal diffusion tensor imaging study in symptomatic Huntington's disease. *J Neurol Neurosurg Psychiatry*. 2010;81(3):257–262. https://doi.org/10.1136/jnnp.2007.142786.
76. Dumas EM, van den Bogaard SJA, Ruber ME, et al. Early changes in white matter pathways of the sensorimotor cortex in premanifest Huntington's disease. *Hum Brain Mapp*. 2012;33(1):203–212. https://doi.org/10.1002/hbm.21205.
77. Rosas HD, Lee SY, Bender AC, et al. Altered white matter microstructure in the corpus callosum in Huntington's disease: implications for cortical "disconnection." *Neuroimage*. 2010;49(4):2995–3004. https://doi.org/10.1016/J.NEUROIMAGE.2009.10.015.
78. Gregory S, Cole JH, Farmer RE, et al. Longitudinal diffusion tensor imaging shows progressive changes in white matter in Huntington's disease. *J Huntingtons Dis*. 2015;4(4):333–346. https://doi.org/10.3233/JHD-150173.
79. Di Paola M, Luders E, Cherubini A, et al. Multimodal MRI analysis of the Corpus callosum reveals white matter differences in presymptomatic and early Huntington's disease. *Cereb Cortex*. 2012;22(12):2858–2866. https://doi.org/10.1093/cercor/bhr360.
80. Matsui JT, Vaidya JG, Johnson HJ, et al. Diffusion weighted imaging of prefrontal cortex in prodromal Huntington's disease. *Hum Brain Mapp*. 2014;35(4):1562–1573. https://doi.org/10.1002/hbm.22273.
81. Reading SAJ, Yassa MA, Bakker A, et al. Regional white matter change in pre-symptomatic Huntington's disease: a diffusion tensor imaging study. *Psychiatry Res Neuroimaging*. 2005;140(1):55–62. https://doi.org/10.1016/j.pscychresns.2005.05.011.
82. Stoffers D, Sheldon S, Kuperman JM, Goldstein J, Corey-Bloom J, Aron AR. Contrasting gray and white matter changes in preclinical Huntington disease: an MRI study. *Neurology*. 2010;74(15):1208–1216. https://doi.org/10.1212/WNL.0b013e3181d8c20a.
83. Bohanna I, Georgiou-Karistianis N, Sritharan A, et al. Diffusion tensor imaging in Huntington's disease reveals distinct patterns of white matter degeneration associated with motor and cognitive deficits. *Brain Imaging Behav*. 2011;5(3):171–180. https://doi.org/10.1007/s11682-011-9121-8.
84. Kincses ZT, Szabó N, Tóth E, et al. Diffusion MRI measured white matter microstructure as a biomarker of neurodegeneration in preclinical Huntington's disease. *Ideggyogy Sz*. 2013;66(11 – 12):399–405. Available at: http://www.ncbi.nlm.nih.gov/pubmed/24555239. Accessed 07.02.19.
85. Saba RA, Yared JH, Doring TM, Phys M, Borges V, Ferraz HB. Diffusion tensor imaging of brain white matter in Huntington gene mutation individuals. *Arq Neuropsiquiatr*. 2017;75(8):503–508. https://doi.org/10.1590/0004-282X20170085.

86. Gregory S, Scahill RI, Seunarine KK, et al. Neuropsychiatry and white matter microstructure in Huntington's disease. *J Huntingtons Dis*. 2015;4(3):239–249. https://doi.org/10.3233/JHD-150160.
87. Zhang J, Gregory S, Scahill RI, et al. In vivo characterization of white matter pathology in premanifest Huntington's disease. *Ann Neurol*. 2018;84(4):497–504. https://doi.org/10.1002/ana.25309.
88. Weaver KE, Richards TL, Liang O, Laurino MY, Samii A, Aylward EH. Longitudinal diffusion tensor imaging in Huntington's disease. *Exp Neurol*. 2009;216:525–529.
89. Poudel GR, Stout JC, Domínguez DJF, et al. Longitudinal change in white matter microstructure in Huntington's disease: the IMAGE-HD study. *Neurobiol Dis*. 2015;74:406–412. https://doi.org/10.1016/j.nbd.2014.12.009.
90. Phillips O, Sanchez-Castaneda C, Elifani F, et al. Tractography of the Corpus callosum in Huntington's disease. Tang Y-P, ed. *PLoS One*. 2013;8(9):e73280. https://doi.org/10.1371/journal.pone.0073280.
91. Poudel GR, Stout JC, Domínguez DJF, et al. White matter connectivity reflects clinical and cognitive status in Huntington's disease. *Neurobiol Dis*. 2014;65:180–187. https://doi.org/10.1016/j.nbd.2014.01.013.
92. Orth M, Gregory S, Scahill RI, et al. Natural variation in sensory-motor white matter organization influences manifestations of Huntington's disease. *Hum Brain Mapp*. 2016;37:4615–4628. https://doi.org/10.1002/hbm.23332.
93. Gregory S, Crawford H, Seunarine K, et al. Natural biological variation of white matter microstructure is accentuated in Huntington's disease. *Hum Brain Mapp*. 2018;39:3516–3527. https://doi.org/10.1002/hbm.24191.
94. van den Heuvel MP, Sporns O. Rich-club organization of the human connectome. *J Neurosci*. 2011;31(44):15775–15786. https://doi.org/10.1523/JNEUROSCI.3539-11.2011.
95. McColgan P, Seunarine KK, Razi A, et al. Selective vulnerability of Rich Club brain regions is an organizational principle of structural connectivity loss in Huntington's disease. *Brain*. 2015;138(pt 11):3327–3344. https://doi.org/10.1093/brain/awv259.
96. McColgan P, Seunarine KK, Gregory S, et al. Topological length of white matter connections predicts their rate of atrophy in premanifest Huntington's disease. *JCI Insight*. 2017;2(8). https://doi.org/10.1172/jci.insight.92641.
97. McColgan P, Gregory S, Seunarine KK, et al. Brain regions showing white matter loss in Huntington's disease are enriched for synaptic and metabolic genes. *Biol Psychiatry*. 2017;83:456–465. https://doi.org/10.1016/j.biopsych.2017.10.019.
98. Ogawa S, Lee TM, Kay AR, Tank DW. Brain magnetic resonance imaging with contrast dependent on blood oxygenation. *Proc Natl Acad Sci U S A*. 1990;87(24):9868–9872. https://doi.org/10.1073/pnas.87.24.9868.
99. Togo H, Rokicki J, Yoshinaga K, et al. Effects of field-map distortion correction on resting state functional connectivity MRI. *Front Neurosci*. 2017;11:656. https://doi.org/10.3389/fnins.2017.00656.
100. Jenkinson M, Smith S. A global optimisation method for robust affine registration of brain images. *Med Image Anal*. 2001;5(2):143–156. Available at: http://www.ncbi.nlm.nih.gov/pubmed/11516708. Accessed 11.02.15.
101. Gotts SJ, Saad ZS, Jo HJ, Wallace GL, Cox RW, Martin A. The perils of global signal regression for group comparisons: a case study of autism spectrum disorders. *Front Hum Neurosci*. 2013;7. https://doi.org/10.3389/fnhum.2013.00356.
102. Calhoun VD, Adali T, Pearlson GD, Pekar JJ. A method for making group inferences from functional MRI data using independent component analysis. *Hum Brain Mapp*. 2001;14:140–151. https://doi.org/10.1002/hbm.1048.
103. Beckmann CF, Smith SM. Probabilistic independent component analysis for functional magnetic resonance imaging. *IEEE Trans Med Imaging*. 2004;23:137–152. https://doi.org/10.1109/TMI.2003.822821.

104. Kim J, Zhu W, Chang L, Bentler PM, Ernst T. Unified structural equation modeling approach for the analysis of multisubject, multivariate functional MRI data. *Hum Brain Mapp*. 2007;28:85–93. https://doi.org/10.1002/hbm.20259.
105. Seth AK, Barrett AB, Barnett L. Granger causality analysis in neuroscience and neuroimaging. *J Neurosci*. 2015;35:3293–3297. https://doi.org/10.1523/JNEUROSCI.4399-14.2015.
106. Friston KJ, Harrison L, Penny W. Dynamic causal modelling. *Neuroimage*. 2003;19(4):1273–1302. https://doi.org/10.1016/S1053-8119(03)00202-7.
107. Wolf RC, Sambataro F, Vasic N, et al. Default-mode network changes in preclinical Huntington's disease. *Exp Neurol*. 2012;237(1):191–198. https://doi.org/10.1016/j.expneurol.2012.06.014.
108. Unschuld PG, Joel SE, Liu X, et al. Impaired cortico-striatal functional connectivity in prodromal Huntington's disease. *Neurosci Lett*. 2012;514(2):204–209. https://doi.org/10.1016/J.NEULET.2012.02.095.
109. Koenig KA, Lowe MJ, Harrington DL, et al. Functional connectivity of primary motor cortex is dependent on genetic burden in prodromal Huntington disease. *Brain Connect*. 2014;4(7):535–546. https://doi.org/10.1089/brain.2014.0271.
110. Quarantelli M, Salvatore E, Giorgio SMDA, et al. Default-mode network changes in Huntington's disease: an integrated MRI study of functional connectivity and morphometry. Zang Y-F, ed. *PLoS One*. 2013;8(8):e72159. https://doi.org/10.1371/journal.pone.0072159.
111. Sánchez-Castañeda C, de Pasquale F, Caravasso CF, et al. Resting-state connectivity and modulated somatomotor and default-mode networks in Huntington disease. *CNS Neurosci Ther*. 2017;23(6):488–497. https://doi.org/10.1111/cns.12701.
112. Werner CJ, Dogan I, Saß C, et al. Altered resting-state connectivity in Huntington's disease. *Hum Brain Mapp*. 2014;35(6):2582–2593. https://doi.org/10.1002/hbm.22351.
113. Wolf RC, Sambataro F, Vasic N, et al. Abnormal resting-state connectivity of motor and cognitive networks in early manifest Huntington's disease. *Psychol Med*. 2014;44(15):3341–3356. https://doi.org/10.1017/S0033291714000579.
114. Poudel GR, Egan GF, Churchyard A, Chua P, Stout JC, Georgiou-Karistianis N. Abnormal synchrony of resting state networks in premanifest and symptomatic Huntington disease: the IMAGE-HD study. *J Psychiatry Neurosci*. 2014;39(2):87–96. https://doi.org/10.1503/JPN.120226.
115. Liu W, Yang J, Chen K, et al. Resting-state fMRI reveals potential neural correlates of impaired cognition in Huntington's disease. *Parkinsonism Relat Disord*. 2016;27:41–46. https://doi.org/10.1016/J.PARKRELDIS.2016.04.017.
116. Espinoza FA, Turner JA, Vergara VM, et al. Whole-brain connectivity in a large study of Huntington's disease gene mutation carriers and healthy controls. *Brain Connect*. 2018;8(3):166–178. https://doi.org/10.1089/brain.2017.0538.
117. Harrington DL, Rubinov M, Durgerian S, et al. Network topology and functional connectivity disturbances precede the onset of Huntington's disease. *Brain*. 2015;138(8):2332–2346. https://doi.org/10.1093/brain/awv145.
118. Gargouri F, Messé A, Perlbarg V, et al. Longitudinal changes in functional connectivity of cortico-basal ganglia networks in manifests and premanifest Huntington's disease. *Hum Brain Mapp*. 2016;37(11):4112–4128. https://doi.org/10.1002/hbm.23299.
119. McColgan P, Gregory S, Razi A, et al. White matter predicts functional connectivity in premanifest Huntington's disease. *Ann Clin Transl Neurol*. 2017;4(2):106–118. https://doi.org/10.1002/acn3.384.
120. McColgan P, Razi A, Gregory S, et al. Structural and functional brain network correlates of depressive symptoms in premanifest Huntington's disease. *Hum Brain Mapp*. 2017;38(6):2819–2829. https://doi.org/10.1002/hbm.23527.

121. Klöppel S, Draganski B, Siebner HR, Tabrizi SJ, Weiller C, Frackowiak RSJ. Functional compensation of motor function in pre-symptomatic Huntington's disease. *Brain*. 2009;132(6):1624–1632. https://doi.org/10.1093/brain/awp081.
122. Klöppel S, Gregory S, Scheller E, et al. Compensation in preclinical Huntington's disease: evidence from the Track-on HD study. *EBioMedicine*. 2015;2(10): 1420–1429. https://doi.org/10.1016/j.ebiom.2015.08.002.
123. Georgiou-Karistianis N, Stout JC, Domínguez DJF, et al. Functional magnetic resonance imaging of working memory in Huntington's disease: cross-sectional data from the IMAGE-HD study. *Hum Brain Mapp*. 2014;35(5):1847–1864. https://doi.org/10.1002/hbm.22296.
124. Georgiou-Karistianis N, Poudel GR, Domínguez DJF, et al. Functional and connectivity changes during working memory in Huntington's disease: 18 month longitudinal data from the IMAGE-HD study. *Brain Cogn*. 2013;83(1):80–91. https://doi.org/10.1016/j.bandc.2013.07.004.
125. Poudel GR, Stout JC, Domínguez DJF, et al. Functional changes during working memory in Huntington's disease: 30-month longitudinal data from the IMAGE-HD study. *Brain Struct Funct*. 2015;220(1):501–512. https://doi.org/10.1007/s00429-013-0670-z.
126. Poudel GR, Driscoll S, Domínguez DJF, et al. Functional brain correlates of neuropsychiatric symptoms in presymptomatic Huntington's disease: The IMAGE-HD study. *J Huntingtons Dis*. 2015;4(4):325–332. https://doi.org/10.3233/JHD-150154.
127. Wolf RC, Vasic N, Schonfeldt-Lecuona C, Landwehrmeyer GB, Ecker D. Dorsolateral prefrontal cortex dysfunction in presymptomatic Huntington's disease: evidence from event-related fMRI. *Brain*. 2007;130(11):2845–2857. https://doi.org/10.1093/brain/awm210.
128. Wolf RC, Sambataro F, Vasic N, Schönfeldt-Lecuona C, Ecker D, Landwehrmeyer B. Altered frontostriatal coupling in pre-manifest Huntington's disease: effects of increasing cognitive load. *Eur J Neurol*. 2008;15(11):1180–1190. https://doi.org/10.1111/j.1468-1331.2008.02253.x.
129. Wolf RC, Sambataro F, Vasic N, Schönfeldt-Lecuona C, Ecker D, Landwehrmeyer B. Aberrant connectivity of lateral prefrontal networks in presymptomatic Huntington's disease. *Exp Neurol*. 2008;213(1):137–144. https://doi.org/10.1016/j.expneurol.2008.05.017.
130. Wolf RC, Sambataro F, Vasic N, et al. Longitudinal functional magnetic resonance imaging of cognition in preclinical Huntington's disease. *Exp Neurol*. 2011;231(2):214–222. https://doi.org/10.1016/j.expneurol.2011.06.011.
131. Wolf RC, Sambataro F, Vasic N, et al. Longitudinal task-negative network analyses in preclinical Huntington's disease. *Eur Arch Psychiatry Clin Neurosci*. 2014;264(6): 493–505. https://doi.org/10.1007/s00406-013-0447-7.
132. Unschuld PG, Joel SE, Pekar JJ, et al. Depressive symptoms in prodromal Huntington's disease correlate with Stroop-interference related functional connectivity in the ventromedial prefrontal cortex. *Psychiatry Res Neuroimaging*. 2012;203(2–3):166–174. https://doi.org/10.1016/J.PSCYCHRESNS.2012.01.002.
133. Unschuld PG, Liu X, Shanahan M, et al. Prefrontal executive function associated coupling relates to Huntington's disease stage. *Cortex*. 2013;49(10):2661–2673. https://doi.org/10.1016/j.cortex.2013.05.015.
134. Gray MA, Egan GF, Ando A, et al. Prefrontal activity in Huntington's disease reflects cognitive and neuropsychiatric disturbances: the IMAGE-HD study. *Exp Neurol*. 2013;239(1):218–228. https://doi.org/10.1016/j.expneurol.2012.10.020.
135. Domínguez DJF, Poudel G, Stout JC, et al. Longitudinal changes in the fronto-striatal network are associated with executive dysfunction and behavioral dysregulation in Huntington's disease: 30 months IMAGE-HD data. *Cortex*. 2017;92:139–149. https://doi.org/10.1016/j.cortex.2017.04.001.

136. Novak MJ, Warren JD, Henley SM, Draganski B, Frackowiak RS, Tabrizi SJ. Altered brain mechanisms of emotion processing in pre-manifest Huntington's disease. *Brain*. 2012;135:1165–1179.
137. Hennenlotter A, Schroeder U, Erhard P, et al. Neural correlates associated with impaired disgust processing in pre-symptomatic Huntington's disease. *Brain*. 2004;127(6):1446–1453. https://doi.org/10.1093/brain/awh165.
138. Van den Stock J, De Winter FL, Ahmad R, et al. Functional brain changes underlying irritability in premanifest Huntington's disease. *Hum Brain Mapp*. 2015;36(7):2681–2690. https://doi.org/10.1002/hbm.22799.
139. Malejko K, Weydt P, Süßmuth SD, Grön G, Landwehrmeyer BG, Abler B. Prodromal Huntington disease as a model for functional compensation of early neurodegeneration. *PLoS One*. 2014;9(12). https://doi.org/10.1371/journal.pone.0114569.
140. Wolf RC, Grön G, Sambataro F, et al. Brain activation and functional connectivity in premanifest Huntington's disease during states of intrinsic and phasic alertness. *Hum Brain Mapp*. 2012;33(9):2161–2173. https://doi.org/10.1002/hbm.21348.
141. Gregory S, Long JD, Klöppel S, et al. Operationalizing compensation over time in neurodegenerative disease. *Brain*. 2017;140(4):1158–1165. https://doi.org/10.1093/brain/awx022.
142. Gregory S, Long JD, Klöppel S, et al. Testing a longitudinal compensation model in premanifest Huntington's disease. *Brain*. 2018;141(7):2156–2166. https://doi.org/10.1093/brain/awy122.
143. Barnes J, Ridgway GR, Bartlett J, et al. Head size, age and gender adjustment in MRI studies: a necessary nuisance? *Neuroimage*. 2010;53(4):1244–1255. https://doi.org/10.1016/j.neuroimage.2010.06.025.
144. Beeri MS, Schmeidler J, Sano M, et al. Age, gender, and education norms on the CERAD neuropsychological battery in the oldest old. *Neurology*. 2006;67(6): 1006–1010. https://doi.org/10.1212/01.wnl.0000237548.15734.cd.
145. Malone IB, Leung KK, Clegg S, et al. Accurate automatic estimation of total intracranial volume: a nuisance variable with less nuisance. *Neuroimage*. 2015;104:366–372. https://doi.org/10.1016/j.neuroimage.2014.09.034.
146. Whitwell JL, Crum WR, Watt HC, Fox NC. Normalization of cerebral volumes by use of intracranial volume: implications for longitudinal quantitative MR imaging. *AJNR Am J Neuroradiol* 2001;22:1483–1489.
147. Matsumae M, Kikinis R, Mórocz IA, et al. Age-related changes in intracranial compartment volumes in normal adults assessed by magnetic resonance imaging. *J Neurosurg*. 1996;84(6):982–991. https://doi.org/10.3171/jns.1996.84.6.0982.
148. Lindquist MA, Mejia A. Zen and the art of multiple comparisons. *Psychosom Med*. 2015;77(2):114–125. https://doi.org/10.1097/PSY.0000000000000148.
149. Hupé J-M. Statistical inferences under the Null hypothesis: common mistakes and pitfalls in neuroimaging studies. *Front Neurosci*. 2015;9:18. https://doi.org/10.3389/fnins.2015.00018.

Chapter 7

Development of brain PET imaging agents: Strategies for imaging neuroinflammation in Alzheimer's disease

Bieneke Janssen, Robert H. Mach*
Department of Radiology, Perelman School of Medicine, University of Pennsylvania, Philadelphia, PA, United States
*Corresponding author: e-mail address: rmach@pennmedicine.upenn.edu

Contents

Abstract

Neuroinflammation is thought to play a key role in the development and progression of neurodegenerative diseases such as Alzheimer's disease. Positron emission tomography (PET) is an in vivo imaging technique capable of studying the biochemical processes which provide the molecular basis of disease. PET imaging of neuroinflammation, which is characterized by the activation of glial cells in the central nervous system (CNS), has been traditionally studied using radiotracers targeting the translocator protein 18 kDa (TSPO). However, limitations in the properties of PET radiotracers for imaging the TSPO has led to the search for alternative strategies for imaging neuroinflammation. The current review presents the current status of PET radiotracers for imaging the TSPO, and describes alternative biological targets that have gained interest for imaging of

Progress in Molecular Biology and Translational Science, Volume 165
ISSN 1877-1173
https://doi.org/10.1016/bs.pmbts.2019.04.005

microglial activation over recent years. These include cannabinoid receptor type-2, cyclooxygenase-2, the $P2X_7$ receptor, CSF1R and reactive oxygen species and some promising radiotracers for these targets. Although a lot of advances have been made in the field of imaging neuroinflammation, current radiotracers all target the pro-inflammatory (M1) phenotype of activated microglia, since the number of known biological targets specific for the anti-inflammatory (M2) phenotype, is still limited. In spite of these limitations, PET imaging studies comparing the temporal relationship between Aβ plaque formation, microglial activation, and the formation of tau-based neurofibrillary tangles would provide valuable information on the "neuroinflammatory hypothesis" of neurodegeneration. The development of a suitable radiotracer for imaging the anti-inflammatory, M2 microglia would represent a great advance in imaging neuroinflammation since this would allow for imaging of the dynamics of microglial activation in different diseases.

1. Introduction

The accumulation of insoluble protein aggregates is a hallmark feature of neurodegenerative disorders. The most prominent example of this is Alzheimer's disease (AD), which is characterized by the formation of two different insoluble protein aggregates: (1) beta amyloid plaques, which consist of aggregated beta amyloid protein, $A\beta_{1-42}$; and (2) neurofibrillary tangles (NFTs), which consist of aggregates of hyperphosphorylated tau. The development of [^{11}C]PiB, a ^{11}C-labeled analog of thioflavin-T, the fluorescent dye used to visualize Aβ plaques in postmortem samples of AD brain, ushered in an entirely new era of PET in studying the molecular mechanisms underlying AD.[1] The development of ^{18}F-labeled beta amyloid imaging agents has expanded PET studies in AD patients at imaging centers without an on-site cyclotron facility. A more recent development has been identification of a number of ^{18}F- and ^{11}C-labeled tau specific agents for imaging NFTs.[2] The use of this panel of Aβ and tau imaging agents has provided valuable information on the temporal separation between the formation of Aβ plaques and NFT formation in AD.[3] For example, PET studies have clearly shown that [^{11}C]PiB(+) beta amyloid plaques occur early in the disease process, and amyloid plaques likely represent a preclinical or antecedent biomarker of AD whereas tau-based NFTs develop near the time of the neurodegeneration that results in a rapid decline of cognitive function.[3,4] The ability to measure the pathological time course of Aβ plaque and NFT formation in AD patients will be critical in the evaluation of disease-modifying therapeutics aimed at slowing the progression of this disease.[5]

Although much has been gained in the understanding of the temporal progression of Aβ plaque and NFT formation, there is still a significant gap in knowledge in the molecular mechanisms of neurodegeneration in AD. One process that is thought to play a prominent role in the "gap period" between Aβ plaque and NFT formation is neuroinflammation. Consequently, the development of PET probes for imaging neuroinflammation has been an active area of research in recent years.

2. Neuroinflammation and AD

Microglia are the major resident immune cells in the CNS. Their primary function is to survey the microenvironment and release factors that influence surrounding neurons and astrocytes, a second type of glial cell in the CNS with support functions. Microglia are usually in a "resting" or ramified state. However, under pathological conditions, they convert to an "amoeboid-like" or activated state.[6] Activated microglia are polarized into two forms, which have been termed M1 and M2 microglia (Fig. 1).[7] M1 microglia are proinflammatory and release reactive oxygen and reactive nitrogen species (ROS/RNS); the primary role of M1 microglia is to engage and remove the agent (i.e., bacterium or virus particle) responsible for triggering neuroinflammation. M2 microglia protect the neurons from ROS/RNS by releasing growth factors and other neuroprotective peptides. However, recent studies have suggested that this classification of microglia into M1 and M2 states is an oversimplification, and multiple states of "M2" microglia have been identified based on gene expression analysis.[8,9] Finally, the prolonged activation of microglia results in a "second wave" of inflammation via infiltration of monocyte-derived macrophages (mo-MΦ) from blood (Fig. 1).[7,10]

Neuroinflammation is currently believed to play a key role in AD.[6,10,11] The *neuroinflammatory hypothesis* proposes that misfolded aggregates of Aβ are recognized as a foreign substance" such as bacteria and responds by activating the neuroinflammatory mechanisms outlined above. The persistent elevation of Aβ levels in AD results in uncontrolled neuroinflammation and elevated ROS/RNS leading to neuronal cell death. The mechanism by which elevated levels of ROS/RNS triggers cell death is likely through noncaspase-mediated apoptosis, parthanatos or necrosis.[7] Furthermore, neuroinflammation induced by either viral or bacterial infection has been shown to exacerbate tau pathology in transgenic models of AD.[12–14] Since elderly individuals are more prone to viral and bacterial infections, the

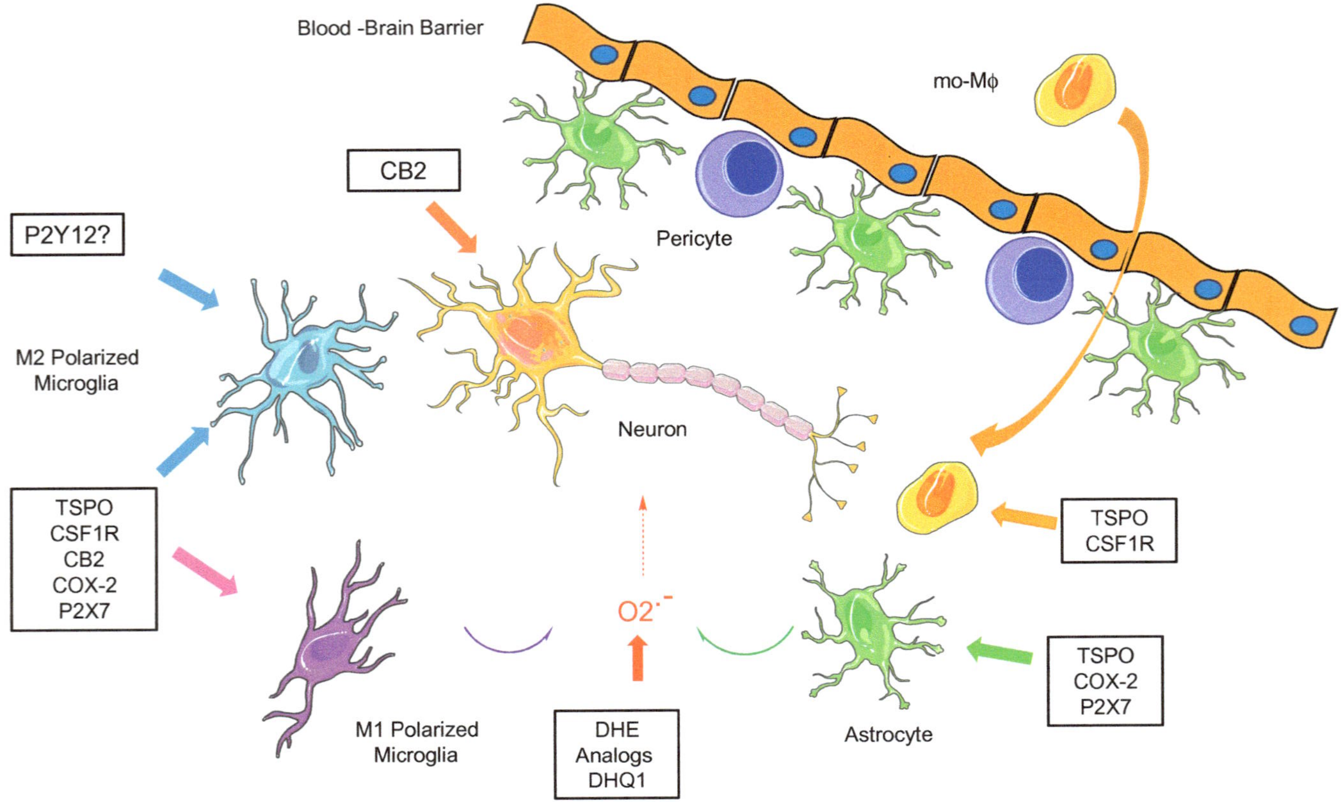

Fig. 1 Cellular components of neuroinflammation and molecular targets for PET radiotracer development.

combination of increased Aβ burden with a neuroinflammatory insult may accelerate the process of neurodegeneration in patients who are at risk for AD.[15]

From the above discussion, it should not be surprising to find that there has been an international effort to develop radiotracers for imaging neuroinflammation and AD with PET. The sections below provide a brief review of the different strategies these groups have taken in developing PET radiotracers targeting the different proteins and other molecular species that have been observed in the process of neuroinflammation.

3. Translocator protein (TSPO)

The translocator protein is an 18 kDa protein whose function is poorly understood. It was originally termed the peripheral benzodiazepine receptor or PBR based on the binding properties of the benzodiazepine, [^{3}H]diazepam.[16–19] The benzodiazepine clonazepam could displace [^{3}H]diazepam from binding sites associated with the GABAa receptor in brain tissue but not those found in peripheral tissues. On the other hand, Ro5-4864 could displace [^{3}H]diazepam binding from peripheral tissues but not those associated with the GABAa receptor. Furthermore, the GABAa-associated binding sites were located on the plasma membrane, whereas the binding sites in peripheral tissues were localized on the mitochondria. This led to the definition of the GABAa-associated benzodiazepine receptor as the "central benzodiazepine receptor" (CBR), and the mitochondrial-associated receptor as the "peripheral benzodiazepine receptor" (PBR). However, since the mitochondria of glial cells also express this protein, the term "peripheral benzodiazepine receptor" was a misnomer because the protein had high expression levels in CNS.

In 2006, this protein was renamed the translocator protein 18 kDa or TSPO because of its role in the translocation of cholesterol from the outer to inner membrane of the mitochondria, which is a key step in the biosynthesis of pregnenolone and other neuroactive steroids.[19,20] A second function of TSPO that was recently proposed is its association with NOX2, a protein that regulates reactive oxygen species (ROS) levels in mitochondria as a means of regulating redox homeostasis.[21] In vitro binding studies with (*R*)-[^{3}H]PK11195, an isoquinoline analog having a high affinity for the PBR and low affinity for the CBR, provided the first evidence that this protein could be used to study glial cell activation and neuroinflammation in animal models of brain injury.[17,22,23] Although the TSPO is often referred to as a marker for imaging microglial activation in neuroinflammation, it is important to note that this protein is also expressed in activated astrocytes and infiltrating monocyte-derived macrophages (mo-MΦ).[24,25] A comprehensive presentation of the role of the TSPO as a marker of neuroinflammation, and the caveats on the interpretation of its use in this capacity, can be found in an excellent review by Guilarte.[19]

4. PET imaging studies of the TSPO in AD

4.1 [^{11}C]PK11195

The use of (*R*)-[^{3}H]PK11195 in radioligand binding and autoradiography studies of tissue obtained from animal models of neuroinflammation, led to the development of its C-11 isotopologue, (*R*)-[^{11}C]PK11195, as a radiotracer for PET imaging studies of neuroinflammation in AD (Fig. 2).[26] Although initial imaging studies revealed an increase in uptake of (*R*)-[^{11}C]PK11195 in multiple brain regions in AD versus age-matched healthy controls, other studies did not observe any differences between the two groups.[26–32] In studies where both [^{11}C]PiB and (*R*)-[^{11}C]PK11195 imaging studies were performed, a positive correlation between amyloid load and (*R*)-[^{11}C]PK11195 uptake was observed.[31,33] These results are consistent with the neuroinflammatory hypothesis of AD described above.

One of the limitations of (*R*)-[^{11}C]PK11195 is its low signal-to-noise ratio, which limits the sensitivity in differentiating between AD and healthy control groups. In the cases where differences were observed, there was extensive overlap between the AD and healthy control subjects.[26] This low sensitivity of (*R*)-[^{11}C]PK11195 led to the investigation of a number of "second generation" radioligands possessing a higher affinity for TSPO and lower nonspecific binding.

4.2 Second generation TSPO ligands

The second generation TSPO radioligands can be separated into two different chemical structures: (1) phenoxyacetamides, which have been labeled with either C-11 ([^{11}C]PBR28, [^{11}C]DAA1106)[34,35] or F-18 ([^{18}F]FEDAA1106; [^{18}F]FEPPA, [^{18}F]FEMPA)[36–38]; and (2) imidazopyrimidines ([^{18}F]CLINME)[39] and pyrazolopyrimidines ([^{18}F]DPA-714).[40] Although these generally exhibit an improved signal-to-noise ratio relative to (*R*)-[^{11}C]PK11195, their relative uptake is determined by the presence or absence of a single nucleotide polymorphism (SNP) in the TSPO gene.[26]

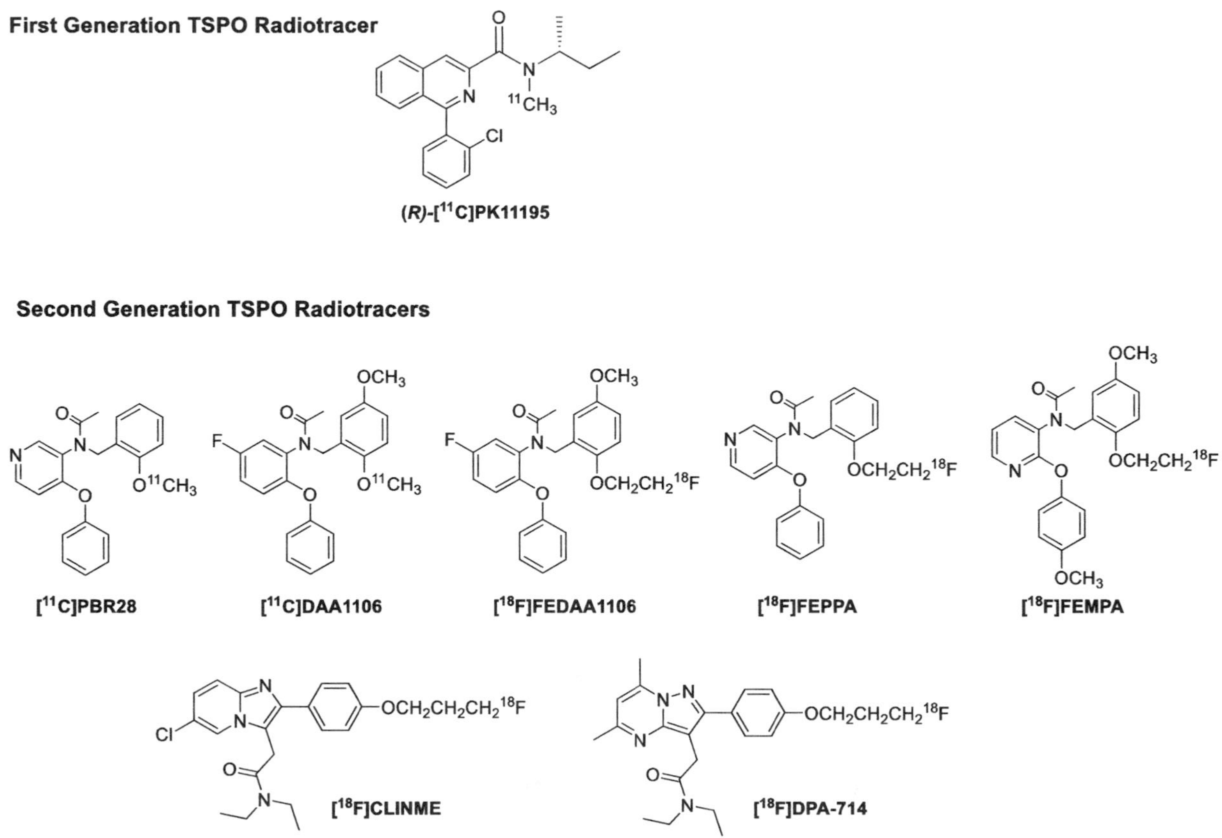

Fig. 2 Structures of (*R*)-[^{11}C]PK11195 and the second generation TSPO PET radiotracers.

This SNP, which involves a single amino acid substitution (A147T) in the protein, results in a complete loss of binding of these second generation radiotracers to the TSPO. That is, individuals homozygous for this SNP (low affinity binders or LABs) display only nonspecific binding of the radiotracer relative to subjects that do not possess the SNP (high affinity binders or HABs). Individuals that are heterozygous for the SNP (mixed affinity binders or MABs) have specific binding (i.e., binding potentials or BPs) intermediate to the two homozygous groups (SNP and non-SNP). However, one can control for this "low binder effect" by conducting a genetic screen of the SNP prior to conducting the PET studies, and stratifying the patient and control groups according to the presence or absence of the SNP. Of course, only those who are heterozygous for the SNP can be included in a PET study since the homozygous group will display only nonspecific binding of the radiotracer. It is interesting to note that the "low binder effect" has not been observed with (*R*)-[^{11}C]PK11195, and appears to be limited to the second generation TSPO ligands.

A number of PET imaging studies have been conducted in AD patients using the second generation TSPO radiotracers. In the majority of the studies, an increase in radiotracer uptake was observed in AD versus age-matched healthy controls.[38,41–47] Two studies have been reported in the literature where no increase was observed between the two groups.[48,49] However, the second generation analogs used in these studies had a relatively high level of nonspecific binding, which likely reduced the sensitivity in detecting differences between the two groups. This was used to explain the differences observed with [^{18}F]FEDAA1106 (no difference between AD and controls) and [^{18}F]FEMPA (higher uptake in AD versus controls) by Varrone et al.[38] Comparisons between the uptake of these second generation TSPO radiotracers and amyloid status will be useful in testing the neuroinflammatory hypothesis of AD.

4.3 Third generation TSPO ligands

The limitations of the second generation TSPO ligands led to the search for a class of TSPO ligands having both improved properties over (*R*)-[^{11}C]PK11195 (i.e., better signal-to-noise ratio) and avoided the low binder effect caused by the A147T SNP. Two recent compounds, which have been termed "third generation TSPO radiotracers" have been reported in the literature (Fig. 3). The first ligand, [^{11}C]ER176, is an aza-analogue of (*R*)-[^{11}C]PK11195.[50] As expected, this radiotracer displays a similar BP in HAB, MAB and LAB subjects, and an improved signal-to-noise ratio over

Third Generation TSPO Radiotracers

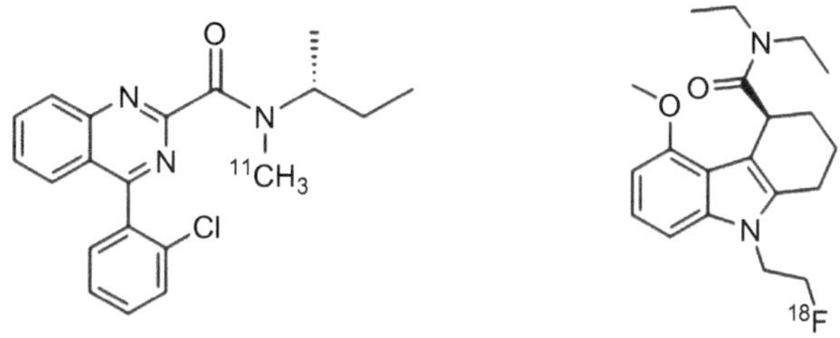

[^{11}C]ER176 [^{18}F]GE-180

Fig. 3 Structures of the third generation PET radiotracers for imaging TSPO.

(*R*)-[^{11}C]PK11195. An ^{18}F-labeled compound, [^{18}F]Flutriciclamide ([^{18}F] GE180), was found to have similar BPs in HAB and MAB subjects.[51] These third generation PET radiotracers represent an improvement over the second generation TSPO ligands since it is not necessary to perform the genetic screen for the A147T SNP. However, it is not clear if they have the sensitivity needed to differentiate between AD and age-matched healthy controls as described above for the second generation TSPO radiotracers.

5. Other targets for imaging neuroinflammation

The above noted limitations of the TSPO ligands (i.e., low binder effect and expression on multiple cell types) has led to the pursuit of other protein targets that are expressed in activated glial cells which could serve as a molecular marker of neuroinflammation. The sections below present an overview of the newer targets researchers have pursued in the development of PET radiotracers for imaging neuroinflammation.

5.1 Cannabinoid receptor type-2

Two subtypes of cannabinoid receptors are known, of which type 1 (CB1) is constitutively expressed in the CNS and type 2 (CB2) is mostly expressed in periphery.[52] CB2 expression is also found on microglia and neurons, and its expression increases significantly upon neuroinflammation, mainly in microglia.[52,53] Therefore, CB2 is a potential marker for the neuroinflammatory response, and hence a number of tracers have been developed and investigated to date. This review mainly focuses on the tracers that have been developed in a model of neuroinflammation, for a complete overview of all developed tracers targeting CB2 the reader is referred to recent reviews by Ametamey and coworkers.[54,55] Using [^{11}C]A-836339

(Fig. 4, K_i = 0.7 nM, 425-fold selective over CB1[56,57]), Horti et al. were able to show increased specific brain uptake in mice that were subjected to a systemic injection of lipopolysaccharide (LPS) 5 days after treatment compared with untreated mice.[57] When another group used the same tracer in LPS-treated rats, no marked increase of tracer uptake in brain was observed.[58] Moreover, two other rat models of inflammation (α-amino-3-hydroxy-5-methyl-4-isoxazolepropionic acid (AMPA) and experimental stroke) used by the same group did not show increased brain uptake of [^{11}C]A-836339 either.[58] Similar results were obtained using [^{11}C]NE40 (Fig. 4, K_i = 9.6 nM, 100-fold selective over CB1[59]) in a rat model of experimental stroke, and no significant increase in tracer uptake was observed compared with sham-operated rats.[60] In a later study, Hosoya et al. did observe an increase in rat brain uptake of [^{11}C]NE40 1 day after experimental stroke, but at later time-points the uptake decreased back to baseline levels.[61]

A fluorine-18 labeled analog of A-836339, [^{18}F]**29** (Fig. 4, K_i = 0.4 nM, 1000-fold selective over CB1[62]), displayed increased brain uptake in LPS-treated mice compared with control mice, but was found to be rapidly metabolized with only 7% of intact tracer left in plasma at 30 min post injection (p.i.).[62] In an attempt to enhance metabolic stability, Caillé et al. designed a small library of fluorinated analogs of A-836339, of which **2f** (Fig. 4, K_i = 0.1 nM, 300-fold selective over CB1[63]) was radiolabeled and further investigated based on its favorable pharmacological and physicochemical profile.[63] Specificity of [^{18}F]**2f** was shown by blocking the observed binding in autoradiography experiments on rat spleen (high CB2 expression) with non-labeled **2f** and selective CB2 agonist GW405833. Indeed, in vivo metabolic stability of [^{18}F]**2f** seems to be improved when compared with [^{18}F]**29**, with 25% of [^{18}F]**2f** in rat plasma still intact at 30 min p.i.,[63] although interspecies comparison (metabolite analysis of [^{18}F]**29** was performed in mice) makes it difficult to draw any definite conclusions.

Two carbon-11 labeled tracers developed by Slavik et al. were used in LPS-treated mice (systemic administration of LPS), and both [^{11}C]RS-016 (Fig. 4, K_i = 0.7 nM, >10,000-fold selective over CB1[64]) and [^{11}C]RSR-056 (Fig. 4, K_i = 2.5 nM, >1000-fold selective over CB1[65]) showed increased tracer uptake in mouse brain, but only little blocking of the uptake was achieved for the latter, suggesting high non-specific binding of [^{11}C]RSR-056.[64] In the same model, fluorine-18 labeled analog [^{18}F]RS-126 (Fig. 4, K_i = 1.2 nM, >10,000-fold selective over CB1[66]) did not show increased brain uptake, possibly due to a high level of non-specific binding,

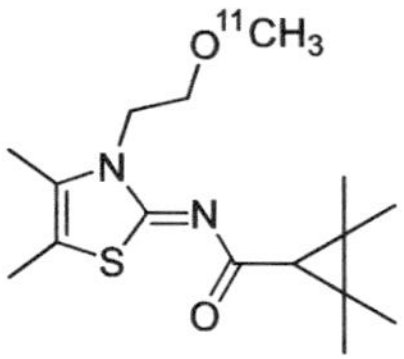

[¹¹C]A-836339

K_i = 0.7 nM
$K_{i,CB1}/K_{i,CB2}$ = 425

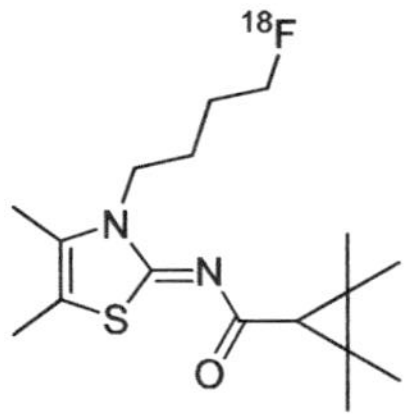

[¹⁸F]29

K_i = 0.4 nM
$K_{i,CB1}/K_{i,CB2}$ = 1,000

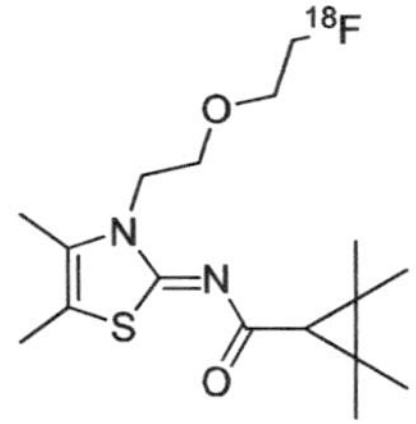

[¹⁸F]2f

K_i = 0.1 nM
$K_{i,CB1}/K_{i,CB2}$ = 300

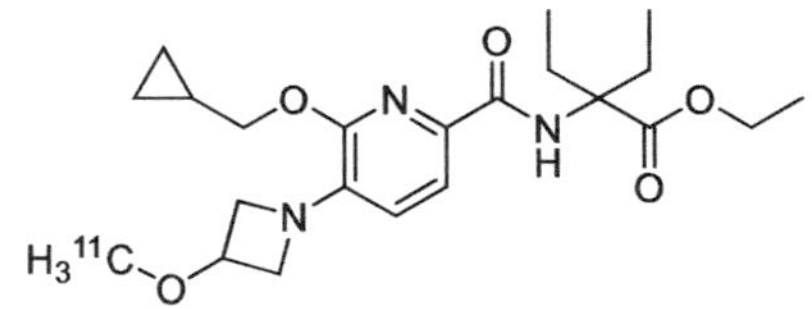

[¹¹C]RSR-056

K_i = 2.5 nM
$K_{i,CB1}/K_{i,CB2}$ > 1,000

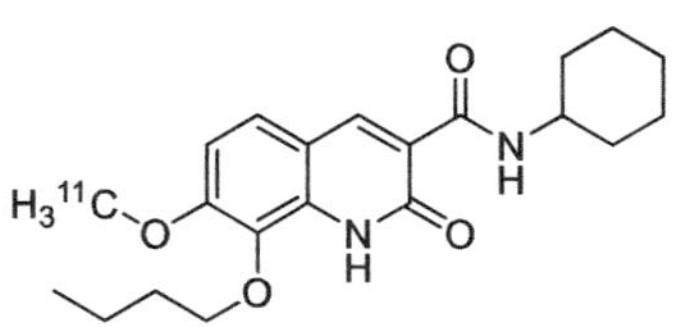

[¹¹C]NE40

K_i = 9.6 nM
$K_{i,CB1}/K_{i,CB2}$ = 100

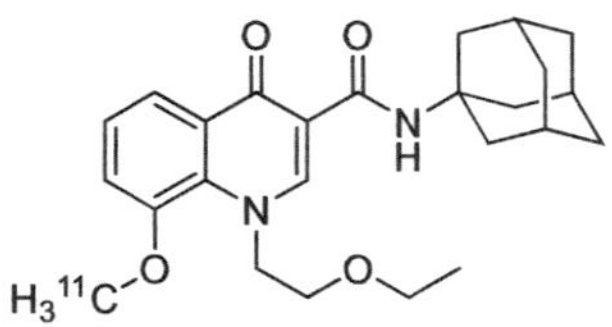

[¹¹C]RS-016

K_i = 0.7 nM
$K_{i,CB1}/K_{i,CB2}$ > 10,000

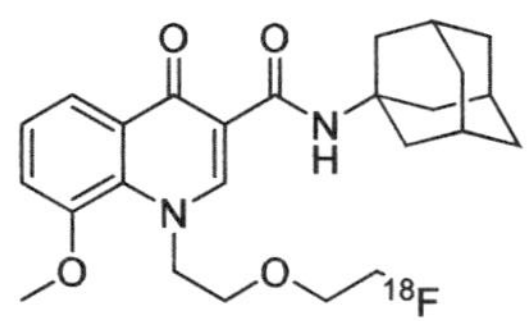

[¹⁸F]RS-126

K_i = 1.2 nM
$K_{i,CB1}/K_{i,CB2}$ > 10,000

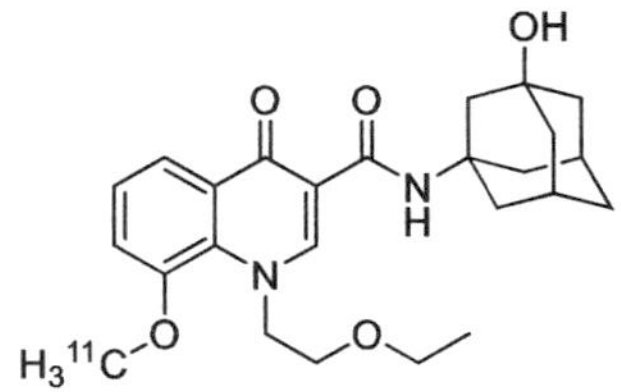

[¹¹C]RS-028

K_i = 0.8 nM
$K_{i,CB1}/K_{i,CB2}$ > 10,000

Fig. 4 Structures of radiotracers for imaging CB2.

that was later shown in post mortem tissue of amyotrophic lateral sclerosis (ALS) patients.[67] The same group reported another carbon-11 labeled analog in this compound class, [^{11}C]RS-028 (Fig. 4, *K*i = 0.8 nM, >10,000-fold selective over CB1[67]), which showed significantly higher binding to ALS patient spinal cord tissue compared with control tissue; however, in vivo in mice fast washout in spleen (CB2-rich organ) was observed, and so authors concluded [^{11}C]RS-028 is not optimal for use in vivo.[67]

[^{11}C]NE40 is thus far the only CB2 ligand evaluated in human, and showed fast brain uptake and washout in healthy subjects.[68] However, in a follow-up study brain uptake of [^{11}C]NE40 was lower in patients with Alzheimer's disease (AD) than brain uptake in healthy control subjects.[69] This is possibly due to neuronal loss in late-stage AD, leading to an overall decreased CB2 expression in whole brain,[69] since increased expression of CB2 in microglia has been demonstrated in post mortem AD tissue before.[70] A study in transgenic AD mice using [^{11}C]A-836339 supports this hypothesis, even though increased tracer uptake was observed in AD mice compared with control mice.[53] Using immunohistochemical staining, Savonenko et al. showed upregulated expression of CB2 in microglia in the transgenic AD mice, which accounts for the increased tracer uptake. Notably, however, the transgenic AD mouse model is not accompanied by neuronal loss, and there was no difference in expression of CB2 in neurons between AD and control mice.[53]

To summarize, given the high abundance of CB1 in CNS, a PET tracer targeting CB2 needs to display a high selectivity for CB2 over CB1, demonstrated by the studies described in this section. In addition, CB2 may only be a useful target for PET in the earliest stages of neurodegenerative diseases, as more advanced stages of the disease are accompanied by significant neuronal loss, thereby possibly decreasing the overall expression of CB2, even though in microglia CB2 expression is upregulated.

5.2 Cyclooxygenase-2 (COX-2)

The cyclooxygenases (COX) are enzymes involved in the arachidonic acid cascade, but also have a role in the activation of inflammatory pathways (e.g., Nuclear Factor (NF)-κB) that lead to the release of prostaglandins, chemokines, cytokines and ROS.[71] Known isoforms COX-1 and COX-2 are both expressed in brain, but are localized in different cell types, and have different regulatory functions. Although COX-1 has also been suggested as playing a role in neuroinflammation, of the two isoforms, main focus as a neuroinflammatory target lies on COX-2.[71] The interest in COX-2 is mainly due to its low expression in healthy conditions, and marked and rapid increase under neuroinflammatory conditions, even though some conflicting results about the latter have been reported.[71,72] Although radiotracers targeting either COX isoform should be highly selective for one isoform over the other to be able to accurately investigate the potential of both targets in neuroinflammation, only a limited amount of selective tracers have been reported. Most COX-2 targeting tracers also have relatively low affinities (>50 nM), and show high blood pool retention, limited uptake in target organs, high levels of non-specific binding and are rapidly metabolized.[72,73]

Fig. 5 depicts some novel promising COX-2 radiotracers, one of which is [^{11}C]MC1 (IC_{50} COX-2 = 3 nM; COX-1 > 1000 nM),[74] which has been evaluated in non-human primates together with COX-1 selective ligand [^{11}C]PS13 (IC_{50} COX-1 = 1 nM; COX-2 > 1000 nM).[75,76] Whereas [^{11}C]PS13 displays specific uptake in rhesus monkeys, the uptake of [^{11}C]

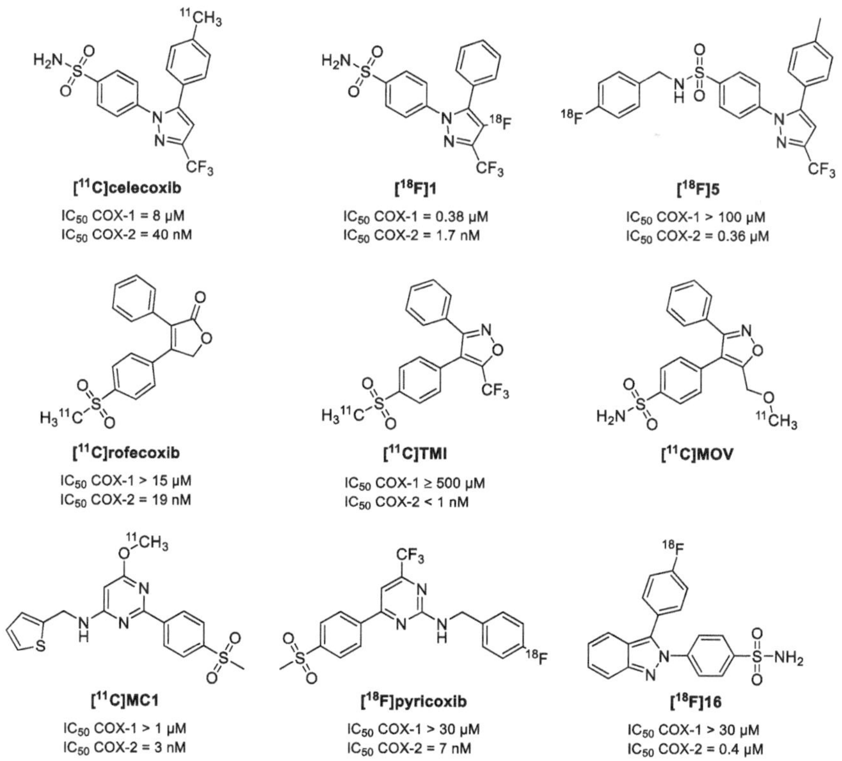

Fig. 5 Structures of radiotracers for imaging COX-2.

MC1 showed only minimal specific uptake, as the only significant blocking effect was observed in the ovaries.[77] In addition, brain uptake and retention of [^{11}C]MC1 was low, however, this is in agreement with COX-2 not being constitutively expressed in healthy brain.[77] Interestingly, in a study using these same two selective tracers, upregulation of COX-2, but not COX-1 was demonstrated in rhesus monkey brains after administration of LPS to induce neuroinflammation.[78,79]

Even though previous results with both [^{11}C]celecoxib and [^{18}F]celecoxib were not encouraging,[80,81] [^{11}C]celecoxib was recently evaluated in baboon, but was found to be rapidly metabolized with <25% left as soon as 10 min p.i.[82] The same group evaluated two carbon-11 isoxazole derivatives in baboon: [^{11}C]TMI (IC_{50} COX-2 < 1 nM; COX-1 ≥ 500 μM)[83] and [^{11}C]MOV[84] (no IC_{50} values reported, IC_{50} of desmethyl-MOV COX-2 = 0.18 μM; COX-1 = 1120 μM[85]). Both tracers entered baboon brain, but while [^{11}C]MOV was rapidly metabolized (27% of intact tracer at 30 min p.i.),[84] [^{11}C]TMI was metabolically stable, significant metabolites were only observed in blocking conditions using meloxicam (78% intact tracer at 90 min p.i.).[83]

Two fluorine-18 labeled analogues of celecoxib, [^{18}F]**1** (IC_{50} COX-2 = 1.7 nM; COX-1 = 0.38 μM)[86] and [^{18}F]**5** (IC_{50} COX-2 0.36 μM; COX-1 > 100 μM)[87] both showed excellent metabolic stability in mice. However, [^{18}F]**5** failed to show tumor uptake in COX-2 expressing tumor-bearing mice.[87] [^{18}F]**1** showed high uptake in mouse brain (2%ID/g at 60 min p.i.), but the tracer could only be obtained in low yields (2% RCY) and low molar activity (~110 MBq/μmol) due to the labeling method (electrochemical radiofluorination).[86] Once a more efficient radiolabeling method can be established, [^{18}F]**1** warrants further in vivo evaluation.

Another tracer that has been evaluated in COX-2 expressing tumor-bearing mice is [^{18}F]pyricoxib (IC_{50} COX-2 = 7 nM; COX-1 > 30 μM).[88,89] [^{18}F]pyricoxib is metabolically relatively stable with 60% of intact tracer left at 2 h p.i.,[88] and shows good uptake in COX-2 expressing tumors.[88,89] Interestingly, a minor modification of [^{18}F]pyricoxib (replacement of benzylamine by benzyloxy moiety), that only had a small effect on affinity toward COX-2 (20 nM versus 7 nM for pyricoxib), had a big impact in brain kinetics, making [^{18}F]**3a** the preferred ligand to study in neuro-inflammatory conditions.[89]

In an effort to develop a novel class of COX-2 inhibitors with enhanced brain penetrance, Elie et al. performed a SAR study around a series of (aza)indazoles.[90] The most potent COX-2 selective compound of the series (**16**, IC_{50} COX 2 = 409 nM; COX-1 > 30 μM) was labeled with fluorine-18 via the boronic acid precursor and evaluated in rats injected unilaterally in striatum with quinolinic acid to induce neuroinflammation. Unfortunately, brain uptake at 60 min p.i. was low (0.046–0.078%ID/g) and no significant difference in in brain uptake was observed between ipsi- and contralateral sides and uptake could not be blocked with unlabeled **16** or celecoxib.[90] A limitation is, however, that uptake of the tracer was only evaluated in an ex vivo biodistribution at a relatively late time point of 60 min p.i.

In the past, the inability to demonstrate COX-2 upregulation in neuro-inflammatory conditions may have been due to suboptimal tracer properties such as poor selectivity over COX-1 and high non-specific binding. Nevertheless, more recent results in animal models of neuroinflammation or COX-2 overexpression with more selective COX-2 ligands (e.g., [^{11}C] MC1, [^{18}F]**3a**) warrant evaluation of these novel, more metabolically stable tracers in human disease to evaluate the potential of COX-2 as a target in neuroinflammation.

5.3 P2X$_7$ receptor

Purinergic receptor P2X$_7$ (P2X$_7$R) is mainly expressed in microglia in the CNS, however, the receptor is expressed in multiple cell types of the myeloid lineage and has also been found in astrocytes and neurons. Its endogenous agonist adenosine triphosphate (ATP) needs to be present in high (mM) concentrations in order for P2X$_7$R to be functionally upregulated, and the receptor is therefore regarded as silent in normal physiology.[91] P2X$_7$R activation is associated with several signaling pathways with a link to neuroinflammation[92] and plays a pivotal role in activation of the inflammasome, which leads to the release of pro-inflammatory cytokines (e.g., IL-1β).[91] Following P2X$_7$R activation by (Bz)ATP, the generation of ROS has also been reported for multiple cell types, including microglia.[92,93] Hence, P2X$_7$R activation is associated with the pro-inflammatory phenotype of activated microglia, and is an interesting drug target due to its functional upregulation in CNS disease.[91]

Two cyanoguanidine containing PET tracers have been reported to date (Fig. 6), [^{11}C]A-740003[94] and [^{18}F]EFB,[95] but both of these tracers show very limited brain uptake in rodents. Nevertheless, tritiated analog [^{3}H] A-740003 could be used for in vitro autoradiography studies to demonstrate P2X$_7$R upregulation at the peak of the disease in a rat model of MS (experimental autoimmune encephalomyelitis; EAE) and in active MS lesions in post mortem human brain sections.[96]

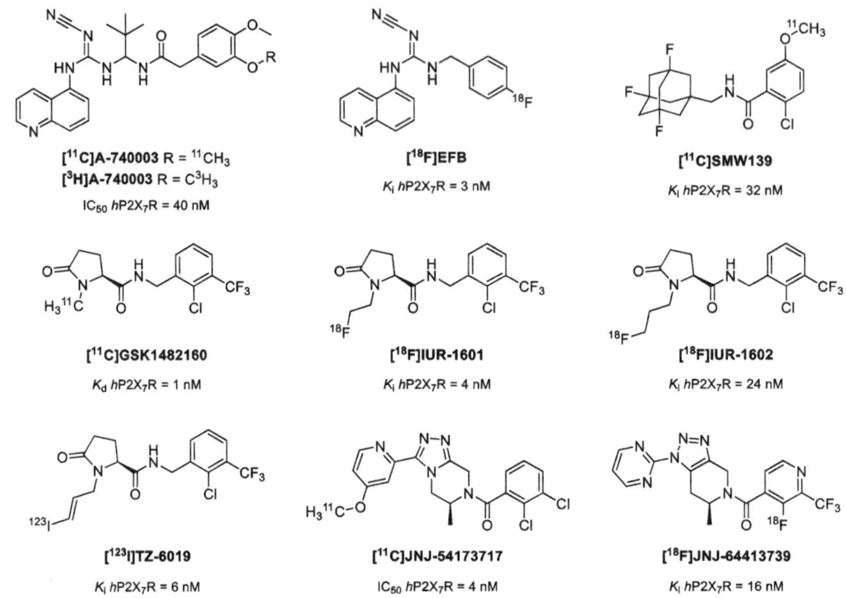

Fig. 6 Structures of radiotracers for imaging $P2X_7R$.

[^{11}C]GSK1482160 (Fig. 6) was also evaluated in the EAE rat model,[97] and additionally in a mouse model of LPS-induced neuroinflammation.[98] Increased tracer binding was observed in vitro in lumbar spinal cord in EAE rats at the peak of the disease (12–14 days after immunization) compared with healthy rats, but the increase could not be demonstrated in vivo because of the low affinity of [^{11}C]GSK1482160 for rat $P2X_7R$.[97] In the PET study using LPS-induced neuroinflammation in mice, tracer uptake in brain of LPS-injected mice was 3.6-fold higher compared with saline-injected animals, and this difference was reached within 10 min after tracer injection.[98] Of note, is that the mice were subjected to PET scanning 72 h after treatment with LPS, as this group found highest Iba1 expression at this time-point,[98] whereas other groups reported peak expression of Iba1 and $P2X_7R$ as early as 12 h after LPS injection,[95,99] while yet another study found the highest $P2X_7R$ expression at 24 h p.i.[100] These differences are likely due to varying doses of LPS used between the studies, route of administration of LPS, or strain of LPS used, as this particular model is known to be highly variable.[101] [^{11}C]GSK1482160 was also shown to enter rhesus macaque brain with maximum uptake (SUV 2.7) at around 70 min p.i.[97] Production of [^{11}C]GSK1482160 has been validated[102] and first-in-human results in 10 healthy volunteers showed low brain uptake of around 2% of the injected dose (SUV 1.0) at 20 min p.i. and high uptake in liver, but clearance appears suitable for further evaluation.[103] Two fluorine-18 labeled analogs of GSK1482160 have also been designed ([^{18}F]IUR-1601[104] and [^{18}F] IUR-1602[105]), but despite optimization of the reaction conditions, for both compounds the radiochemical yield was not sufficient to facilitate animal studies due to instability of the product in these conditions. Using an iodinated analog of GSK1482160, [^{123}I]TZ-6019, for in vitro autoradiography in the P301S mouse model of AD, significantly higher tracer binding was observed (1.3-fold higher in whole brain; 1.7-fold higher in hippocampus and 2.4-fold higher in cortex) compared with binding in brains of wild-type mice.[106] In contrast, another autoradiography study in post mortem brain sections of AD patients using [^{11}C]SMW139 (Fig. 6), a difference in tracer binding in patients compared with healthy control could not be demonstrated.[107] The same tracer however, did cross the BBB in rats and showed excellent uptake in a PET study in rats locally overexpressing human $P2X_7R$ in the treated striatum (SUV 2.1 versus 1.4 in contralateral striatum at 10 min p.i.). In the same rat model of local *h*$P2X_7R$ overexpression, [^{11}C]JNJ-54173717 (Fig. 6) also showed specific uptake in the *h*$P2X_7R$ overexpressing striatum (SUV 0.8 versus 0.6 in contralateral striatum at 10 min p.i.).[108] In addition, high initial brain uptake (SUV 3.3) was observed for [^{11}C]JNJ-54173717 in non-human primates, likely enabling successful translation to humans. Another fluorinated ligand developed by Janssen R&D, [^{18}F] JNJ-64413739 (Fig. 6) was used in an imaging study in LPS-induced rats.[100] First, the optimal time-point for imaging was determined to be at 2 days after unilateral stereotactic injection of LPS (20 μg). PET imaging, ex vivo autoradiography and immunohistochemical staining for $P2X_7R$ were all performed in the same animals, and all showed a significant increase in $P2X_7R$ levels in the LPS-injected hemisphere compared to the non-injected contralateral side (PET: 29% increase in SUV at 20 min p.i.).[100] Levels of microglial marker Iba1 were also found to correlate with tracer binding, and the signal in the PET scans could be decreased by pre-treatment or displacement with clinical candidate JNJ-54175446. Interestingly, although an increased expression of $P2X_7R$ in the LPS-injected hemisphere using immunohistochemical staining, in vitro autoradiography studies failed to show binding of [^{18}F]JNJ-64413739 in that same hemisphere, which is suggestive of a difference in binding mechanism in vitro an in vivo.[100] In rhesus macaques, peak brain uptake of [^{18}F]JNJ-64413739 was reached within 10 min, and the PET signal could be reduced by pre-treatment with JNJ-54175446.[109] From occupancy plots, non-displaceable radioligand binding (V_{ND}) was estimated (1.2), which suggested some level of non-specific binding. Metabolic rate was found to be relatively fast, with 40% of intact tracer left at 30 min p.i., but no defluorination was observed.[109] A high free fraction was observed in both non-human primate (72%) and human plasma (63%). A first-in-human study in healthy volunteers was performed with [^{18}F]JNJ-64413739, and found that a dynamic 90 min acquisition showed limited bias, good test-retest variability and excellent reliability.[110] Shortening acquisition time to 60 min provided too much bias for clinical applicability. In both non-human primate and human studies, no appropriate reference region in brain could be determined.[109,110] It should also be noted that a high inter-individual signal variability across human subjects was observed, which is possibly an effect of $P2X_7R$ polymorphism, but further studies are required to verify if that is indeed the case.[110]

In recent years, $P2X_7R$ has gained interest as a target for PET imaging of microglial activation. Some $P2X_7R$ targeting tracers developed to date have shown great promise in animal models of diseases associated with a neuroinflammatory component both in vitro and in vivo. Encouraging results have also been obtained in post mortem brain sections of human patients in vitro, but these results still need to be validated in a clinical situation. As currently a number of $P2X_7R$ tracers are being evaluated in the clinic, first results in human patients are expected to be reported shortly.

5.4 Colony stimulating factor 1 receptor (CSF1R)

Colony stimulating factor 1 receptor (CSF1R) is a class III receptor tyrosine kinase that is expressed on the cell surface of microglia and macrophages.[111] CSF1R is directly involved in controlling development, survival and maintenance of microglia and these effects are associated with neuroinflammation.[112] For instance in transgenic AD mice, high expression of CSF1R in microglia in close proximity to Aβ deposits has been demonstrated.[113,114] Therefore, CSF1R is an interesting emerging target for PET imaging of neuroinflammation, for which to date only a few radiotracers have been developed (Fig. 7). The first radioligand reported, [^{18}F]**10** is a dual inhibitor for both CSF1R and tropomyosin receptor kinase, but no in vitro or in vivo preclinical evaluation has been published to date.[115] CSF1R specific tracer [^{11}C]AZ683 was evaluated in rats and rhesus macaques, but showed poor brain influx and little retention in both animals.[111] Even though uptake in the non-human primates was slightly higher than in rodents, whether the

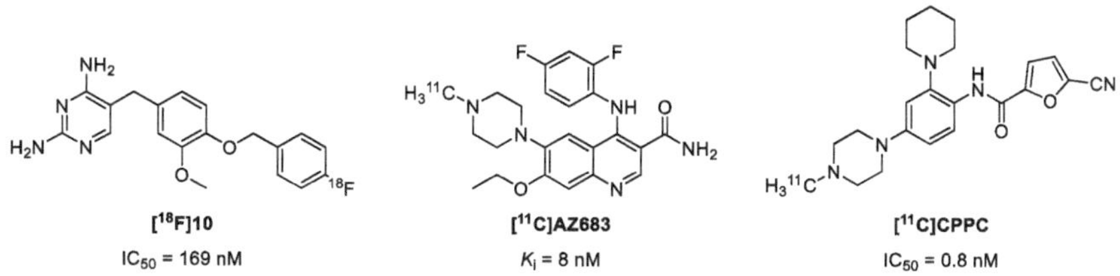

Fig. 7 Structure of radiotracers for imaging CSF1R.

influx is sufficient to get an accumulated signal in an inflammation model remains to be elucidated. The most extensively evaluated CSF1R tracer yet is [^{11}C]CPPC.[112] A neuroinflammatory response was induced by injection of LPS in mice and baboon, and in both animal models tracer uptake was increased in comparison to healthy animals (up to 59% and 120% in mice and baboon, respectively). Two more physiologically relevant mouse models were also studied, and in both transgenic AD and EAE mice an increased PET signal was observed (31% and up to 100% for AD and EAE, respectively). In addition, in the EAE model, tracer uptake in brain correlated with disease severity.[112] Lastly, in vitro autoradiography studies on post mortem brain sections of AD patients showed a 75–99% increase in tracer binding compared with healthy control brain sections, and binding could also be blocked using multiple CS1FR specific ligands. This latter study with [^{11}C]CPPC indicates that indeed, CS1FR may be an important target for PET imaging of neuroinflammation.

6. PET radiotracers for reactive oxidative species (ROS)

Oxidative stress refers to a condition characterized by the formation of partially reduced and highly reactive oxygen species termed ROS. The most prevalent form of ROS is superoxide, which is formed by the one electron reduction of molecular oxygen by the enzymes xanthine oxidase and NADPH oxidase (NOX). A second molecular mediator of oxidative stress is nitric oxide (NO), which is formed by the enzymatic activity inducible nitric oxide synthase (iNOS). The combination of NO with superoxide leads to the formation of a highly reactive species peroxynitrate (also termed reactive nitrogen species or RNS), which can damage DNA and lead to the nitration of tyrosine residues in proteins. In neuroinflammation, oxidative stress results from the formation of M1 polarized microglia and astrocyte activation, which are a source of NO and superoxide due to their increased expression of iNOS and high levels of NADPH oxidase activity.[7]

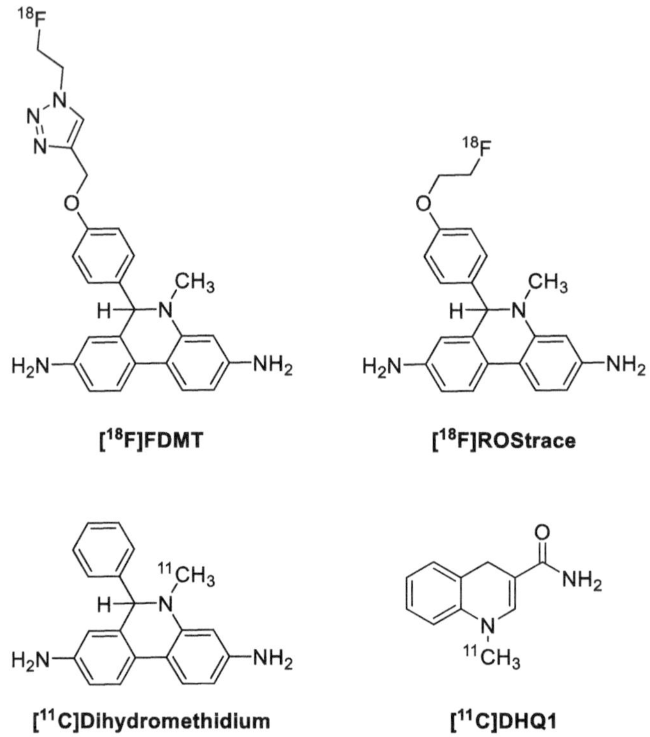

Fig. 8 Structures of the radiotracers for imaging ROS.

Therefore, the development of PET probes capable of imaging superoxide levels in the CNS is expected to provide a sensitive means for imaging M1-mediated neuroinflammation in neurodegenerative disorders such as AD.

The lead compound for PET radiotracer development for imaging superoxide in vivo is the Dihydroethidium or DHE (Fig. 8).[116] DHE is selectively oxidized by superoxide to form a charged species, ethidium, which is trapped within the cell. Ethidium is also capable of intercalating into DNA, which results in a strong fluorescent signal. Therefore, DHE has been largely used as probe for the optical imaging of superoxide formation in cells and in optical imaging studies. Two ^{18}F-labeled analogs of DHE have been reported in the literature, [^{18}F]FDMT and [^{18}F]ROStrace (Figs. 8 and 9).[116,117] Although [^{18}F]FDMT was able to image elevated levels of superoxide in an animal model of adriamycin-induced model of cardiotoxicity, this radiotracer does not cross the blood-brain barrier (BBB) and is not able to image elevated levels of superoxide that occur in neuroinflammation. On the other hand, [^{18}F]ROStrace readily crosses

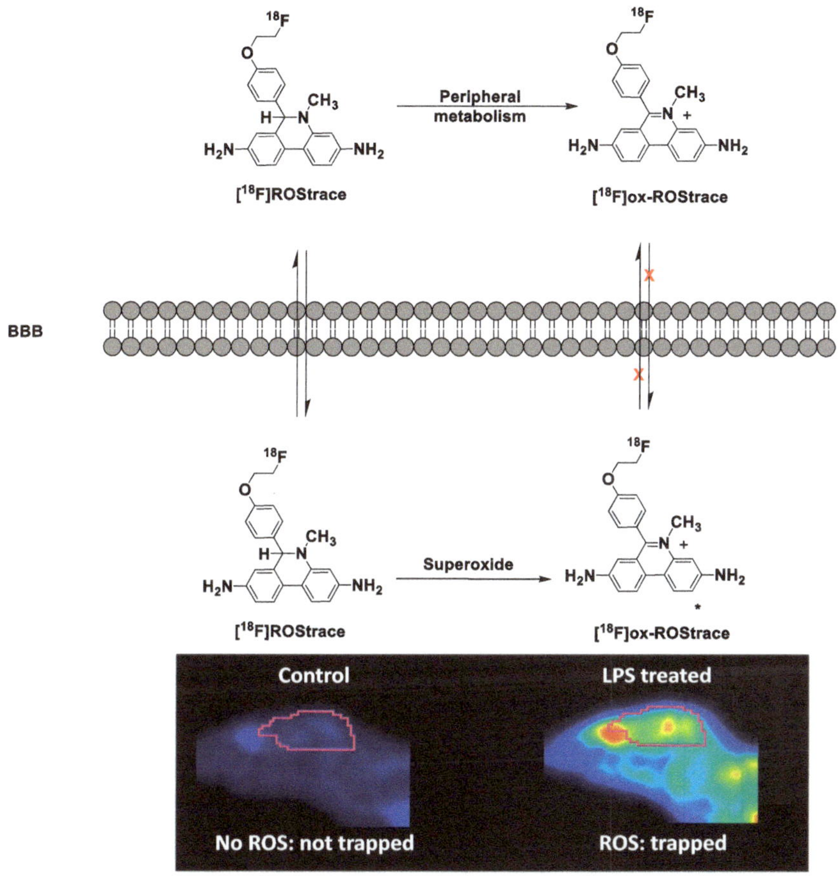

Fig. 9 Mechanism of trapping of [^{18}F]ROStrace. The * represents a site of hydroxylation of [^{18}F]*ox*-ROStrace which has also been observed in the CNS of LPS-treated mice.

the BBB and has been fully validated in the LPS-induced model of neuroinflammation.[117] Studies are currently ongoing evaluating this radiotracer in transgenic models of neurodegeneration. A C-11 labeled analog of DHE, termed [^{11}C]dihydromethidium has also been reported.[118] This compound shows high brain uptake in microPET imaging studies, but a detailed analysis of the mechanism of uptake and trapping of this radiotracer as it relates to neuroinflammation was not conducted.

Another compound that has shown promise in imaging oxidative stress is the dihydroisoquinoline analog, [^{11}C]DHQ1 (Fig. 8), which is an analog of NADH/NADPH.[119] This class of compounds has been used previously as a redox carrier in the delivery of drugs to the CNS. [^{11}C]DHQ1 crosses the

BBB and is trapped in the brain by oxidation to a charged species. Pretreatment with the NOX2 inhibitor apocyanin resulted in a reduction in uptake of [^{11}C]DHQ1, suggesting that its trapping mechanism can be partly attributed to the oxidation of the dihydro species to the corresponding *N*-methylquinolinium species. This compound is also thought to be a substrate for enzymes for the oxidation of NADH/NADPH in oxidative phosphorylation; consequently, its trapping mechanism may not be limited to the presence of elevated levels of superoxide. In addition, its relative reactivity to other oxidizing species such as hydrogen peroxide (H_2O_2) or hypochlorous acid (HOCl) is not known, and this tracer may likely provide a nonselective measure of oxidative stress.

7. Conclusion

Being the first target investigated for potential imaging of neuroinflammation with PET, over 80 ligands targeting TSPO have been developed to date. While first generation tracer (*R*)-[^{11}C]PK11195 suffers from low brain uptake and a poor signal-to-noise ratio, binding of second generation tracers is affected by the existence of a SNP in human subjects, requiring genetic stratification of patient populations. Third generation tracers have overcome this issue, but it is yet to be shown if their sensitivity is sufficient to show diffuse neuroinflammation in AD.

Despite these limitations of TSPO PET imaging, to date this is still the gold standard, as the development of tracers developed for other neuroinflammatory targets is still emerging, and thus far less clinical validation has been performed to date. However, it is to be expected that in the coming years, multiple tracers and targets will enter the clinical stage of research in human disease. Whereas COX-2 tracers do not show sufficient brain uptake, and preclinical evaluation is limited to in vitro results, ROS could successfully be detected in brain of rodents with LPS-induced neuroinflammation, and evaluation in transgenic models of human disease is currently ongoing. For CB2 and $P2X_7R$, first-in-human studies in healthy volunteers have already been reported, and while the newly developed tracers for $P2X_7R$ still need to be evaluated in human disease, CB2 targeting tracer [^{11}C]NE40 was unable to detect an increase in tracer uptake in AD patients. On the contrary, the PET signal in AD patients was decreased, likely due to neuronal loss occurring in advanced disease stages. This illustrates the fact that not every tracer or target may be suitable for every disease with a neuroinflammatory component, as none of the targets to date are specific for microglia only.

CSF1R may solve this problem for some diseases, but that remains to be elucidated, as the receptor is also expressed on macrophages, and in some neurological diseases the BBB is compromised, leading to immune cell infiltration from periphery. Nevertheless, first results in preclinical evaluation are promising. Another interesting target that in brain is exclusively expressed on microglia, and is even specific for the anti-inflammatory (M2) phenotype, is purinergic receptor $P2Y_{12}$.[96,120] Unfortunately, the only developed tracer for this target thus far does not cross the BBB, and was therefore not discussed in detail in the current review. Nevertheless, in vitro results in stroke and MS patient tissue seem encouraging.[96,120] The development of phenotype-specific, brain-penetrant ligands would mean a great advance in PET imaging of the neuroinflammatory response, since microglial activation is a dynamic process, and in vivo evidence of these dynamics in health, disease, and even treatment would be extremely valuable.

References

1. Klunk WE, Engler H, Nordberg A, et al. Imaging brain amyloid in Alzheimer's disease with Pittsburgh compound-B. *Ann Neurol.* 2004;55(3):306–319.
2. Saint-Aubert L, Lemoine L, Chiotis K, Leuzy A, Rodriguez-Vieitez E, Nordberg A. Tau PET imaging: present and future directions. *Mol Neurodegener.* 2017;12(1):19.
3. Bateman RJ, Xiong C, Benzinger TL, et al. Clinical and biomarker changes in dominantly inherited Alzheimer's disease. *N Engl J Med.* 2012;367(9):795–804.
4. Vlassenko AG, Benzinger TL, Morris JC. PET amyloid-beta imaging in preclinical Alzheimer's disease. *Biochim Biophys Acta.* 2012;1822(3):370–379.
5. Mach RH. New targets for the development of PET tracers for imaging neurodegeneration in Alzheimer disease. *J Nucl Med.* 2014;55(8):1221–1224.
6. Graeber MB. Changing face of microglia. *Science.* 2010;330(6005):783–788.
7. Glass CK, Saijo K, Winner B, Marchetto MC, Gage FH. Mechanisms underlying inflammation in neurodegeneration. *Cell.* 2010;140(6):918–934.
8. J-I S. Gene expression profiles of M1 and M2 microglia characterized by comparative analysis of public datasets. *Clin Exp Neurol.* 2018;9(2):124–138.
9. Franco R, Fernández-Suárez D. Alternatively activated microglia and macrophages in the central nervous system. *Prog Neurobiol.* 2015;131:65–86.
10. London A, Cohen M, Schwartz M. Microglia and monocyte-derived macrophages: functionally distinct populations that act in concert in CNS plasticity and repair. *Front Cell Neurosci.* 2013;7:34.
11. Wyss-Coray T, Rogers J. Inflammation in Alzheimer disease-a brief review of the basic science and clinical literature. *Cold Spring Harb Perspect Med.* 2012;2(1):a006346.
12. Sheng JG, Bora SH, Xu G, Borchelt DR, Price DL, Koliatsos VE. Lipopolysaccharide-induced-neuroinflammation increases intracellular accumulation of amyloid precursor protein and amyloid beta peptide in APPswe transgenic mice. *Neurobiol Dis.* 2003;14(1):133–145.
13. Lee JW, Lee YK, Yuk DY, et al. Neuro-inflammation induced by lipopolysaccharide causes cognitive impairment through enhancement of beta-amyloid generation. *J Neuroinflammation.* 2008;5:37.

14. Sy M, Kitazawa M, Medeiros R, et al. Inflammation induced by infection potentiates tau pathological features in transgenic mice. *Am J Pathol.* 2011;178(6):2811–2822.
15. Newcombe EA, Camats-Perna J, Silva ML, Valmas N, Huat TJ, Medeiros R. Inflammation: the link between comorbidities, genetics, and Alzheimer's disease. *J Neuroinflammation.* 2018;15(1):276.
16. Braestrup C, Squires RF. Specific benzodiazepine receptors in rat brain characterized by high-affinity (3H)diazepam binding. *Proc Natl Acad Sci.* 1977;74(9):3805–3809.
17. Benavides J, Quarteronet D, Imbault F, et al. Labelling of "peripheral-type" benzodiazepine binding sites in the rat brain by using [3H]PK 11195, an isoquinoline carboxamide derivative: kinetic studies and autoradiographic localization. *J Neurochem.* 1983;41(6):1744–1750.
18. Bonsack F, Sukumari-Ramesh S. TSPO: an evolutionarily conserved protein with elusive functions. *Int J Mol Sci.* 2018;19(6):1694.
19. Guilarte TR. TSPO in diverse CNS pathologies and psychiatric disease: a critical review and a way forward. *Pharmacol Ther.* 2019;194:44–58.
20. Papadopoulos V, Baraldi M, Guilarte TR, et al. Translocator protein (18 kDa): new nomenclature for the peripheral-type benzodiazepine receptor based on its structure and molecular function. *Trends Pharmacol Sci.* 2006;27(8):402–409.
21. Guilarte TR, Loth MK, Guariglia SR. TSPO finds NOX2 in microglia for redox homeostasis. *Trends Pharmacol Sci.* 2016;37(5):334–343.
22. Guilarte TR, Kuhlmann AC, O'Callaghan JP, Miceli RC. Enhanced expression of peripheral benzodiazepine receptors in trimethyltin-exposed rat brain: a biomarker of neurotoxicity. *Neurotoxicology.* 1995;16(3):441–450.
23. Miyazawa N, Diksic M, Yamamoto Y. Chronological study of peripheral benzodiazepine binding sites in the rat brain stab wounds using [3H] PK-11195 as a marker for gliosis. *Acta Neurochir.* 1995;137(3):207–216.
24. Lavisse S, Guillermier M, Herard AS, et al. Reactive astrocytes overexpress TSPO and are detected by TSPO positron emission tomography imaging. *J Neurosci.* 2012;32(32):10809–10818.
25. Venneti S, Lopresti BJ, Wiley CA. Molecular imaging of microglia/macrophages in the brain. *Glia.* 2013;61(1):10–23.
26. Dupont AC, Largeau B, Santiago Ribeiro MJ, Guilloteau D, Tronel C, Arlicot N. Translocator protein-18 kDa (TSPO) positron emission tomography (PET) imaging and its clinical impact in neurodegenerative diseases. *Int J Mol Sci.* 2017;18(4):785.
27. Cagnin A, Brooks DJ, Kennedy AM, et al. In-vivo measurement of activated microglia in dementia. *Lancet.* 2001;358(9280):461–467.
28. Edison P, Archer HA, Gerhard A, et al. Microglia, amyloid, and cognition in Alzheimer's disease: an [11C](R)PK11195-PET and [11C]PIB-PET study. *Neurobiol Dis.* 2008;32(3):412–419.
29. Wiley CA, Lopresti BJ, Venneti S, et al. Carbon 11-labeled pittsburgh compound B and carbon 11-labeled (R)-PK11195 positron emission tomographic imaging in Alzheimer disease. *Arch Neurol.* 2009;66(1):60–67.
30. Okello A, Edison P, Archer HA, et al. Microglial activation and amyloid deposition in mild cognitive impairment: a PET study. *Neurology.* 2009;72(1):56–62.
31. Yokokura M, Mori N, Yagi S, et al. In vivo changes in microglial activation and amyloid deposits in brain regions with hypometabolism in Alzheimer's disease. *Eur J Nucl Med Mol Imaging.* 2011;38(2):343–351.
32. Schuitemaker A, Kropholler MA, Boellaard R, et al. Microglial activation in Alzheimer's disease: an (R)-[(1)(1)C]PK11195 positron emission tomography study. *Neurobiol Aging.* 2013;34(1):128–136.

33. Fan Z, Aman Y, Ahmed I, et al. Influence of microglial activation on neuronal function in Alzheimer's and Parkinson's disease dementia. *Alzheimers Dement.* 2015;11(6): 608–621.e607.
34. Imaizumi M, Kim HJ, Zoghbi SS, et al. PET imaging with [11C]PBR28 can localize and quantify upregulated peripheral benzodiazepine receptors associated with cerebral ischemia in rat. *Neurosci Lett.* 2007;411(3):200–205.
35. Zhang MR, Kida T, Noguchi J, et al. [(11)C]DAA1106: radiosynthesis and in vivo binding to peripheral benzodiazepine receptors in mouse brain. *Nucl Med Biol.* 2003;30(5):513–519.
36. Wilson AA, Garcia A, Parkes J, et al. Radiosynthesis and initial evaluation of [18F]-FEPPA for PET imaging of peripheral benzodiazepine receptors. *Nucl Med Biol.* 2008;35(3):305–314.
37. Zhang MR, Maeda J, Furutsuka K, et al. [18F]FMDAA1106 and [18F]FEDAA1106: two positron-emitter labeled ligands for peripheral benzodiazepine receptor (PBR). *Bioorg Med Chem Lett.* 2003;13(2):201–204.
38. Varrone A, Oikonen V, Forsberg A, et al. Positron emission tomography imaging of the 18-kDa translocator protein (TSPO) with [18F]FEMPA in Alzheimer's disease patients and control subjects. *Eur J Nucl Med Mol Imaging.* 2015;42(3):438–446.
39. Fookes CJ, Pham TQ, Mattner F, et al. Synthesis and biological evaluation of substituted [18F]imidazo[1,2-a]pyridines and [18F]pyrazolo[1,5-a]pyrimidines for the study of the peripheral benzodiazepine receptor using positron emission tomography. *J Med Chem.* 2008;51(13):3700–3712.
40. James ML, Fulton RR, Vercoullie J, et al. DPA-714, a new translocator protein-specific ligand: synthesis, radiofluorination, and pharmacologic characterization. *J Nucl Med.* 2008;49(5):814–822.
41. Yasuno F, Ota M, Kosaka J, et al. Increased binding of peripheral benzodiazepine receptor in Alzheimer's disease measured by positron emission tomography with [11C]DAA1106. *Biol Psychiatry.* 2008;64(10):835–841.
42. Yasuno F, Kosaka J, Ota M, et al. Increased binding of peripheral benzodiazepine receptor in mild cognitive impairment-dementia converters measured by positron emission tomography with [(1)(1)C]DAA1106. *Psychiatry Res.* 2012;203(1):67–74.
43. Kreisl WC, Lyoo CH, McGwier M, et al. In vivo radioligand binding to translocator protein correlates with severity of Alzheimer's disease. *Brain J Neurol.* 2013; 136(Pt 7):2228–2238.
44. Lyoo CH, Ikawa M, Liow JS, et al. Cerebellum can serve as a pseudo-reference region in Alzheimer disease to detect neuroinflammation measured with PET radioligand binding to translocator protein. *J Nucl Med.* 2015;56(5):701–706.
45. Suridjan I, Pollock BG, Verhoeff NP, et al. In-vivo imaging of grey and white matter neuroinflammation in Alzheimer's disease: a positron emission tomography study with a novel radioligand, [18F]-FEPPA. *Mol Psychiatry.* 2015;20(12):1579–1587.
46. Kreisl WC, Lyoo CH, Liow JS, et al. (11)C-PBR28 binding to translocator protein increases with progression of Alzheimer's disease. *Neurobiol Aging.* 2016;44:53–61.
47. Hamelin L, Lagarde J, Dorothee G, et al. Early and protective microglial activation in Alzheimer's disease: a prospective study using 18F-DPA-714 PET imaging. *Brain J Neurol.* 2016;139(Pt 4):1252–1264.
48. Varrone A, Mattsson P, Forsberg A, et al. In vivo imaging of the 18-kDa translocator protein (TSPO) with [18F]FEDAA1106 and PET does not show increased binding in Alzheimer's disease patients. *Eur J Nucl Med Mol Imaging.* 2013;40(6):921–931.
49. Golla SS, Boellaard R, Oikonen V, et al. Quantification of [18F]DPA-714 binding in the human brain: initial studies in healthy controls and Alzheimer's disease patients. *J Cereb Blood Flow Metab.* 2015;35(5):766–772.

50. Ikawa M, Lohith TG, Shrestha S, et al. 11C-ER176, a radioligand for 18-kDa translocator protein, has adequate sensitivity to robustly image all three affinity genotypes in human brain. *J Nucl Med.* 2017;58(2):320–325.
51. Fan Z, Calsolaro V, Atkinson RA, et al. Flutriciclamide (18F-GE180) PET: first-in-human PET study of novel third-generation in vivo marker of human translocator protein. *J Nucl Med.* 2016;57(11):1753–1759.
52. Navarro G, Morales P, Rodríguez-Cueto C, Fernández-Ruiz J, Jagerovic N, Franco R. Targeting cannabinoid CB2 receptors in the central nervous system. Medicinal chemistry approaches with focus on neurodegenerative disorders. *Front Neurosci.* 2016;10:406.
53. Savonenko AV, Melnikova T, Wang Y, et al. Cannabinoid CB2 receptors in a mouse model of Aβ amyloidosis: immunohistochemical analysis and suitability as a PET biomarker of neuroinflammation. *PLoS One.* 2015;10(6):e0129618.
54. Spinelli F, Mu L, Ametamey SM. Radioligands for positron emission tomography imaging of cannabinoid type 2 receptor. *J Label Compd Radiopharm.* 2018;61(3): 299–308.
55. Ni R, Mu L, Ametamey S. Positron emission tomography of type 2 cannabinoid receptors for detecting inflammation in the central nervous system. *Acta Pharmacol Sin.* 2019;40(3):351–357.
56. Yao BB, Hsieh G, Daza AV, et al. Characterization of a cannabinoid CB2 receptor-selective agonist, A-836339 [2,2,3,3-tetramethyl-cyclopropanecarboxylic acid [3-(2-methoxy-ethyl)-4,5-dimethyl-3H-thiazol-(2Z)-ylidene]-amide], using in vitro pharmacological assays, in vivo pain models, and pharmacological magnetic resonance imaging. *J Pharmacol Exp Ther.* 2009;328(1):141–151.
57. Horti AG, Gao Y, Ravert HT, et al. Synthesis and biodistribution of [11C]A-836339, a new potential radioligand for PET imaging of cannabinoid type 2 receptors (CB2). *Bioorg Med Chem.* 2010;18(14):5202–5207.
58. Pottier G, Gomez-Vallejo V, Padro D, et al. PET imaging of cannabinoid type 2 receptors with [(11)C]A-836339 did not evidence changes following neuroinflammation in rats. *J Cereb Blood Flow Metab.* 2017;37(3):1163–1178.
59. Evens N, Vandeputte C, Coolen C, et al. Preclinical evaluation of [11C]NE40, a type 2 cannabinoid receptor PET tracer. *Nucl Med Biol.* 2012;39(3):389–399.
60. Vandeputte C, Casteels C, Struys T, et al. Small-animal PET imaging of the type 1 and type 2 cannabinoid receptors in a photothrombotic stroke model. *Eur J Nucl Med Mol Imaging.* 2012;39(11):1796–1806.
61. Hosoya T, Fukumoto D, Kakiuchi T, et al. In vivo TSPO and cannabinoid receptor type 2 availability early in post-stroke neuroinflammation in rats: a positron emission tomography study. *J Neuroinflammation.* 2017;14(1):69.
62. Moldovan RP, Teodoro R, Gao YJ, et al. Development of a high-affinity PET radioligand for imaging cannabinoid subtype 2 receptor. *J Med Chem.* 2016;59(17): 7840–7855.
63. Caillé F, Cacheux F, Peyronneau M-A, et al. From structure–activity relationships on thiazole derivatives to the in vivo evaluation of a new radiotracer for cannabinoid subtype 2 PET imaging. *Mol Pharm.* 2017;14(11):4064–4078.
64. Slavik R, Herde AM, Bieri D, et al. Synthesis, radiolabeling and evaluation of novel 4-oxo-quinoline derivatives as PET tracers for imaging cannabinoid type 2 receptor. *Eur J Med Chem.* 2015;92:554–564.
65. Slavik R, Grether U, Herde AM, et al. Discovery of a high affinity and selective pyridine analog as a potential positron emission tomography imaging agent for cannabinoid type 2 receptor. *J Med Chem.* 2015;58(10):4266–4277.
66. Slavik R, Muller Herde A, Haider A, et al. Discovery of a fluorinated 4-oxo-quinoline derivative as a potential positron emission tomography radiotracer for imaging cannabinoid receptor type 2. *J Neurochem.* 2016;138(6):874–886.

67. Haider A, Spinelli F, Herde AM, et al. Evaluation of 4-oxo-quinoline-based CB2 PET radioligands in R6/2 chorea huntington mouse model and human ALS spinal cord tissue. *Eur J Med Chem*. 2018;145:746–759.
68. Ahmad R, Koole M, Evens N, et al. Whole-body biodistribution and radiation dosimetry of the cannabinoid type 2 receptor ligand [11C]-NE40 in healthy subjects. *Mol Imaging Biol*. 2013;15(4):384–390.
69. Ahmad R, Postnov A, Bormans G, Versijpt J, Vandenbulcke M, Van Laere K. Decreased in vivo availability of the cannabinoid type 2 receptor in Alzheimer's disease. *Eur J Nucl Med Mol Imaging*. 2016;43(12):2219–2227.
70. Benito C, Nunez E, Tolon RM, et al. Cannabinoid CB2 receptors and fatty acid amide hydrolase are selectively overexpressed in neuritic plaque-associated glia in Alzheimer's disease brains. *J Neurosci*. 2003;23(35):11136–11141.
71. Aid S, Bosetti F. Targeting cyclooxygenases-1 and -2 in neuroinflammation: therapeutic implications. *Biochimie*. 2011;93(1):46–51.
72. Tietz O, Marshall A, Wuest M, Wang M, Wuest F. Radiotracers for molecular imaging of cyclooxygenase-2 (COX-2) enzyme. *Curr Med Chem*. 2013;20(35):4350–4369.
73. Pacelli A, Greenman J, Cawthorne C, Smith G. Imaging COX-2 expression in cancer using PET/SPECT radioligands: current status and future directions. *J Label Compd Radiopharm*. 2014;57(4):317–322.
74. Cortes-Salva MY, Shrestha S, Singh P, et al. 2-(4-Methylsulfonylphenyl)pyrimidines as prospective radioligands for imaging cyclooxygenase-2 with PET—synthesis, triage, and radiolabeling. *Molecules*. 2018;23(11):2850.
75. Singh P, Shrestha S, Cortes-Salva MY, et al. 3-substituted 1,5-diaryl-1H-1,2,4-triazoles as prospective PET radioligands for imaging brain COX-1 in monkey. Part 1: synthesis and pharmacology. *ACS Chem Neurosci*. 2018;9(11):2610–2619.
76. Shrestha S, Singh P, Cortes-Salva MY, et al. 3-substituted 1,5-diaryl-1H-1,2,4-triazoles as prospective PET radioligands for imaging brain COX-1 in monkey. Part 2: selection and evaluation of [11C]PS13 for quantitative imaging. *ACS Chem Neurosci*. 2018;9(11):2620–2627.
77. Kim M-J, Shrestha SS, Cortes M, et al. Evaluation of two potent and selective PET radioligands to image COX-1 and COX-2 in rhesus monkeys. *J Nucl Med*. 2018;.
78. Kim MJ, Shrestha S, Eldridge M, et al. Novel PET radioligands show that, in rhesus monkeys, COX-1 is constitutively expressed and COX-2 is induced by inflammation. *J Nucl Med*. 2017;58:2.
79. Gallagher E, Shrestha S, Eldridge M, et al. T80. novel PET radioligands show that COX-2, but not COX-1, is induced by neuroinflammation in rhesus macaque. *Biol Psychiatry*. 2018;83(9, Supplement):S160.
80. Prabhakaran J, Underwood MD, Parsey RV, et al. Synthesis and in vivo evaluation of [18F]-4-[5-(4-methylphenyl)-3-(trifluoromethyl)-1H-pyrazol-1-yl] benzenesulfonamide as a PET imaging probe for COX-2 expression. *Bioorg Med Chem*. 2007;15(4):1802–1807.
81. Ji B, Kumata K, Onoe H, et al. Assessment of radioligands for PET imaging of cyclooxygenase-2 in an ischemic neuronal injury model. *Brain Res*. 2013;1533: 152–162.
82. Kumar JSD, Bai B, Zanderigo F, et al. In vivo brain imaging, biodistribution, and radiation dosimetry estimation of [11C]celecoxib, a COX-2 PET ligand, in nonhuman primates. *Molecules*. 2018;23(8):1929.
83. Kumar JSD, Zanderigo F, Prabhakaran J, Rubin-Falcone H, Parsey RV, Mann JJ. In vivo evaluation of [11C]TMI, a COX-2 selective PET tracer, in baboons. *Bioorg Med Chem Lett*. 2018;28(23):3592–3595.
84. Prabhakaran J, Underwood M, Zanderigo F, et al. Radiosynthesis and in vivo evaluation of [11C]MOV as a PET imaging agent for COX-2. *Bioorg Med Chem Lett*. 2018; 28(14):2432–2435.

85. Talley JJ, Brown DL, Carter JS, et al. 4-[5-Methyl-3-phenylisoxazol-4-yl]-benzenesulfonamide, Valdecoxib: a potent and selective inhibitor of COX-2. *J Med Chem*. 2000;43(5):775–777.
86. Lebedev A, Jiao J, Lee J, et al. Radiochemistry on electrodes: synthesis of an 18F-labelled and in vivo stable COX-2 inhibitor. *PLoS One*. 2017;12(5):e0176606.
87. Kaur J, Tietz O, Bhardwaj A, et al. Design, synthesis, and evaluation of an (18)F-labeled radiotracer based on celecoxib-NBD for positron emission tomography (PET) imaging of cyclooxygenase-2 (COX-2). *ChemMedChem*. 2015;10(10):1635–1640.
88. Tietz O, Wuest M, Marshall A, et al. PET imaging of cyclooxygenase-2 (COX-2) in a pre-clinical colorectal cancer model. *EJNMMI Res*. 2016;6(1):37.
89. Tietz O, Marshall A, Bergman C, Wuest M, Wuest F. Impact of structural alterations on the radiopharmacological profile of 18F-labeled pyrimidines as cyclooxygenase-2 (COX-2) imaging agents. *Nucl Med Biol*. 2018;62–63:9–17.
90. Elie J, Vercouillie J, Arlicot N, et al. Design of selective COX-2 inhibitors in the (aza) indazole series. Chemistry, in vitro studies, radiochemistry and evaluations in rats of a [18F] PET tracer. *J Enzyme Inhib Med Chem*. 2019;34(1):1–7.
91. Bhattacharya A, Biber K. The microglial ATP-gated ion channel P2X7 as a CNS drug target. *Glia*. 2016;64(10):1772–1787.
92. Bartlett R, Stokes L, Sluyter R. The P2X7 receptor channel: recent developments and the use of P2X7 antagonists in models of disease. *Pharmacol Rev*. 2014;66(3):638–675.
93. Bartlett R, Yerbury JJ, Sluyter R. P2X7 receptor activation induces reactive oxygen species formation and cell death in murine EOC13 microglia. *Mediat Inflamm*. 2013;2013:271813.
94. Janssen B, Vugts DJ, Funke U, et al. Synthesis and initial preclinical evaluation of the P2X7 receptor antagonist [(1)(1)C]A-740003 as a novel tracer of neuroinflammation. *J Label Compd Radiopharm*. 2014;57(8):509–516.
95. Fantoni ER, Dal Ben D, Falzoni S, Di Virgilio F, Lovestone S, Gee A. Design, synthesis and evaluation in an LPS rodent model of neuroinflammation of a novel (18)F-labelled PET tracer targeting P2X7. *EJNMMI Res*. 2017;7(1):31.
96. Beaino W, Janssen B, Kooij G, et al. Purinergic receptors P2Y12R and P2X7R: potential targets for PET imaging of microglia phenotypes in multiple sclerosis. *J Neuroinflammation*. 2017;14(1):259.
97. Han J, Liu H, Liu C, et al. Pharmacologic characterizations of a P2X7 receptor-specific radioligand, [11C]GSK1482160 for neuroinflammatory response. *Nucl Med Commun*. 2017;38(5):372–382.
98. Territo PR, Meyer JA, Peters JS, et al. Characterization of (11)C-GSK1482160 for targeting the P2X7 receptor as a biomarker for neuroinflammation. *J Nucl Med*. 2017;58(3):458–465.
99. Choi HB, Ryu JK, Kim SU, McLarnon JG. Modulation of the purinergic P2X7 receptor attenuates lipopolysaccharide-mediated microglial activation and neuronal damage in inflamed brain. *J Neurosci*. 2007;27(18):4957–4968.
100. Berdyyeva T, Xia C, Taylor N, et al. PET imaging of the P2X7 ion channel with a novel tracer [18F]JNJ-64413739 in a rat model of neuroinflammation. *Mol Imaging Biol* 2019; Epub ahead of print. https://doi.org/10.1007/s11307-018-01313-2.
101. Catorce MN, Gevorkian G. LPS-induced murine neuroinflammation model: main features and suitability for pre-clinical assessment of nutraceuticals. *Curr Neuropharmacol*. 2016;14(2):155–164.
102. Wissmann CL, Wang M, Gao M, Zheng Q-H, Green MA. Development, validation and implementation of radio-HPLC methods for the P2X7-receptor-targeted [11C] GSK1482160 radiopharmaceutical. *Appl Radiat Isot*. 2018;142:8–11.
103. Green M, Hutchins G, Fletcher J, et al. Distribution of the P2X7-receptor-targeted [11C]GSK1482160 radiopharmaceutical in normal human subjects. *J Nucl Med*. 2018; 59(Suppl. 1):1009.

104. Gao M, Wang M, Glick-Wilson BE, et al. Synthesis and preliminary biological evaluation of a novel P2X7R radioligand [18F]IUR-1601. *Bioorg Med Chem Lett.* 2018;28(9):1603–1609.
105. Gao M, Wang M, Glick-Wilson BE, et al. Synthesis and initial in vitro characterization of a new P2X7R radioligand [18F]IUR-1602. *Appl Radiat Isot.* 2019;144:10–18.
106. Jin H, Han J, Resing D, et al. Synthesis and in vitro characterization of a P2X7 radioligand [123I]TZ6019 and its response to neuroinflammation in a mouse model of Alzheimer disease. *Eur J Pharmacol.* 2018;820:8–17.
107. Janssen B, Vugts DJ, Wilkinson SM, et al. Identification of the allosteric P2X7 receptor antagonist [11C]SMW139 as a PET tracer of microglial activation. *Sci Rep.* 2018;8(1):6580.
108. Ory D, Celen S, Gijsbers R, et al. Preclinical evaluation of a P2X7 receptor-selective radiotracer: PET studies in a rat model with local overexpression of the human P2X7 receptor and in nonhuman primates. *J Nucl Med.* 2016;57(9):1436–1441.
109. Kolb H, Barret O, Bhattacharya A, et al. Preclinical evaluation and non-human primate receptor occupancy study of (18)F-JNJ-64413739, a novel PET radioligand for P2X7 receptors. *J Nucl Med.* 2019. Epub ahead of print. https://doi.org/10.2967/jnumed.118.212696.
110. Koole M, Schmidt M, Hijzen A, et al. 18F-JNJ-64413739, a novel PET ligand for the P2X7 ion channel: radiation dosimetry, kinetic modeling, test-retest variability and occupancy of the P2X7 antagonist JNJ-54175446. *J Nucl Med.* 2018. Epub ahead of print. https://doi.org/10.2967/jnumed.118.216747.
111. Tanzey SS, Shao X, Stauff J, et al. Synthesis and initial in vivo evaluation of [11C]AZ683—a novel PET radiotracer for colony stimulating factor 1 receptor (CSF1R). *Pharmaceuticals.* 2018;11(4):136.
112. Horti AG, Naik R, Foss CA, et al. PET imaging of microglia by targeting macrophage colony-stimulating factor 1 receptor (CSF1R). *Proc Natl Acad Sci.* 2019;116(5):1686–1691.
113. Boissonneault V, Filali M, Lessard M, Relton J, Wong G, Rivest S. Powerful beneficial effects of macrophage colony-stimulating factor on beta-amyloid deposition and cognitive impairment in Alzheimer's disease. *Brain J Neurol.* 2009;132(Pt 4):1078–1092.
114. Murphy Jr GM, Zhao F, Yang L, Cordell B. Expression of macrophage colony-stimulating factor receptor is increased in the AbetaPP(V717F) transgenic mouse model of Alzheimer's disease. *Am J Pathol.* 2000;157(3):895–904.
115. Bernard-Gauthier V, Schirrmacher R. 5-(4-((4-[18F]fluorobenzyl)oxy)-3-methoxybenzyl)pyrimidine-2,4-diamine: a selective dual inhibitor for potential PET imaging of Trk/CSF-1R. *Bioorg Med Chem Lett.* 2014;24(20):4784–4790.
116. Chu W, Chepetan A, Zhou D, et al. Development of a PET radiotracer for non-invasive imaging of the reactive oxygen species, superoxide, in vivo. *Org Biomol Chem.* 2014;12(25):4421–4431.
117. Hou C, Hsieh CJ, Li S, et al. Development of a positron emission tomography radiotracer for imaging elevated levels of superoxide in neuroinflammation. *ACS Chem Neurosci.* 2017;9(3):578–586.
118. Wilson AA, Sadovski O, Nobrega JN, et al. Evaluation of a novel radiotracer for positron emission tomography imaging of reactive oxygen species in the central nervous system. *Nucl Med Biol.* 2017;53:14–20.
119. Okamura T, Okada M, Kikuchi T, Wakizaka H, Zhang MR. A (11)C-labeled 1,4-dihydroquinoline derivative as a potential PET tracer for imaging of redox status in mouse brain. *J Cereb Blood Flow Metab.* 2015;35(12):1930–1936.
120. Villa A, Klein B, Janssen B, et al. Identification of new molecular targets for PET imaging of the microglial anti-inflammatory activation state. *Theranostics.* 2018;8(19):5400–5418.

Chapter 8

Disclosure of amyloid PET scan results: A systematic review

Hyejin Kim[a,b], Jennifer H. Lingler[a,b,*]
[a]Department of Health and Community Systems, University of Pittsburgh School of Nursing, Pittsburgh, PA, United States
[b]Alzheimer Disease Research Center, University of Pittsburgh School of Medicine, Pittsburgh, PA, United States
*Corresponding author: e-mail address: linglerj@pitt.edu

Contents

Abstract

The increasing use of biomarker tests for Alzheimer's disease (AD) in research and, to a much lesser extent, specialty care settings has led to questions concerning how individuals may react to learning of their AD biomarker status in the absence of a cure or preventative treatment. The purpose of this chapter is to systematically review the published evidence regarding amyloid imaging results disclosure and to synthesize findings across studies with a focus on the psychological, social, and behavioral outcomes of such results disclosure. Following the Preferred Reporting Items for Systematic Review and Meta-Analyses (PRISMA) guidelines, we searched six electronic databases, screened 265 articles, and reviewed seven publications in depth. Most studies were descriptive in nature and lack control groups. However, as a group, these articles provide important early insights into the psychological safety of disclosing amyloid imaging results to cognitively normal persons, and highlight the need for rigorously designed studies that address social and behavioral outcomes and extend to symptomatic populations.

In the past two decades, neuroscientists have made remarkable progress in the development and testing of biological markers, or biomarkers, of dementia disorders, particularly Alzheimer's disease (AD). Biomarker evidence of the intracellular and extracellular pathological hallmarks of AD can be detected

Progress in Molecular Biology and Translational Science, Volume 165
ISSN 1877-1173
https://doi.org/10.1016/bs.pmbts.2019.05.002

in cerebrospinal fluid[1] and on various neuroimaging scans of persons with and without symptomatic manifestations of the disease.[2] Techniques for the in vivo visualization of abnormal neurofibrillary tau and amyloid-β (Aβ) accumulation are revolutionizing the approaches to the evaluation of treatment effects and have propelled the field toward a biologically based, rather than syndrome focused, definition of AD.[3] Efforts to identify robust blood-based biomarkers for AD are well underway, and if successful, would significantly minimize the logistical burden of biomarker testing for AD.[4] All of these advances are occurring against a backdrop of repeated failures in clinical drug studies targeting AD,[5] raising critical questions what it may mean for individuals and family members to contemplate testing for AD pathology in the absence of curative or preventative measures to alter the course of disease progression.[6]

Given the stigma that is associated with mental health conditions in general,[7,8] and the documented status of AD as among the most feared diseases of aging,[8] researchers and other commentators have raised concerns about potential unintended adverse outcomes of testing for AD biomarkers in the absence of a treatment. Such concerns have been primarily theoretical and have focused on the potential for negative psychological outcomes (like depression or suicidality) and discrimination in social, employment, insurance, and other settings in which an individual may be treated prejudicially if their AD biomarker status were known to third parties.

Informed largely by studies of the impact of genetic testing for AD risk, a small but growing body of literature has begun to examine the impact of disclosing AD biomarker test results, particularly amyloid PET, in subspecialty and research settings. The purpose of this chapter is to systematically review the evidence base on amyloid PET results disclosure and to synthesize findings across studies with a focus on the psychological, social, and behavioral outcomes of such results disclosure.

1. Methods

1.1 Search strategy

Literature searches were informed by consultation with a medical reference librarian and included the electronic databases: PubMed, PsycINFO®, Cumulative Index to Nursing and Allied Health Literature (CINAHL®), Cochrane Central Register of Controlled Trials, EMBASE, and Web of

Science. Using both subject headings and free-text terms, the keywords of "dementia" and "Alzheimer's disease" were paired with combinations of the terms "Positron Emission Tomography," "PET," "PET/CT," "Amyloid," "Beta Amyloid," "Amyloid beta-Peptides," "Amyloidosis," "Biomarkers," "Biological Marker," "Disclosure," "Truth Disclosure," "Patient Satisfaction," "Patient Preference," "Patient Education as Topic," "Patient Participation," "Disclosure," "Caregivers," and "Family." Truncation was used to capture both noun and verb forms of each term. Databases were searched for the 10-year period of January 1, 2018 to December 31, 2018 and hand searching was performed to identify the additional relevant articles that may have been missed during the indexing process.

1.2 Inclusion and exclusion criteria

The inclusion criteria for articles in this review were: (1) original research published in a peer-reviewed journal, (2) sample included participants with normal cognition or a cognitive impairment diagnosis (e.g., mild cognitive impairment (MCI), AD), (3) report described outcomes following the disclosure of amyloid PET scan results, (4) classified as human subject research, and (5) written in English. Articles were excluded if they were: (1) focused on other biomarkers, including blood tests and tests of cerebrospinal fluid, (2) a single case study, (3) a clinical guideline, (4) a conference abstract or proceeding, (5) a commentary, or (6) research involving animals or cells.

1.3 Study selection and data extraction for review

This systematic review was conducted in accordance with the Preferred Reporting Items for Systematic Review and Meta-Analyses (PRISMA) guidelines.[9] First, all of the literature search results were uploaded to the reference management software library, EndNote. As articles were uploaded, one reviewer screened each study title. Studies involving duplicated data were identified, and only the larger dataset was included. Next, two independent reviewers screened each abstract to determine inclusion in a full-text review. Full-text reviews were conducted to finalize the decision to include an article. Inclusion of an article required agreement by two reviewers that the article meets the inclusion criteria. Any discrepancies arising at this final stage of screening were resolved through a consensus process involving all of the review team members. Upon inclusion in the review, key information was extracted from each article, including the first author, year of the study, country, study methodology, study population, sample size, assessment time-points, key outcomes, measures, and key findings. In two instances, corresponding authors were contacted for additional information about the results reported in their articles. Data were extracted into a Microsoft Excel® spreadsheet (Microsoft, Redmond, WA).

2. Results

As shown in Fig. 1, our search strategy yielded 265 unique articles and seven met criteria for inclusion in this review. Sample sizes of the studies meeting our inclusion criteria from 11 to 97, with an average sample size of 41. Most of the reports ($k=4$) focused on cognitively normal older adults and one included caregiver perspectives. Five of the seven reports originated in the United States, one in Belgium, and one in Japan.

Procedures for disclosing amyloid PET results to participants and/or their caregivers varied across studies. For example, the first published report on this topic described sending the amyloid PET result to each participant's treating neurologist, with an educational session on the meaning of the result being performed by the research team at either a 9-month or 18-month follow-up visit.[10] Later reports of prospective studies detail attempts to standardize procedures for results disclosure. In particular, Mozersky and colleagues reference a multi-site protocol outlining the content and process of results disclosure, with education being offered concurrently with results disclosure.[11] The former approaches may more closely reflect clinical practice and therefore have greater ecological validity. However, protocolized approaches ease the interpretation of findings by ensuring that participants within a given study of results disclosure are reacting to the same stimulus.

In terms of methodological approaches, several of the study designs were qualitative, meaning that semi-structured interviews, rather than forced-choice surveys, were employed to gather participants' perspectives on amyloid PET disclosure. Qualitative methods are by definition, exploratory, and have been recommended for addressing research questions that seek to describe rather than to measure a phenomenon.[12] In addition to using qualitative interviews to explore participant perspectives, four of the studies included at least one survey instrument to quantify participant reactions to results disclosure. These surveys typically included measures of test related distress and mood, following in the tradition of studies of reactions to learning of one's genetic risk for Alzheimer's disease. Table 1 summarizes the key characteristics and findings of each article.

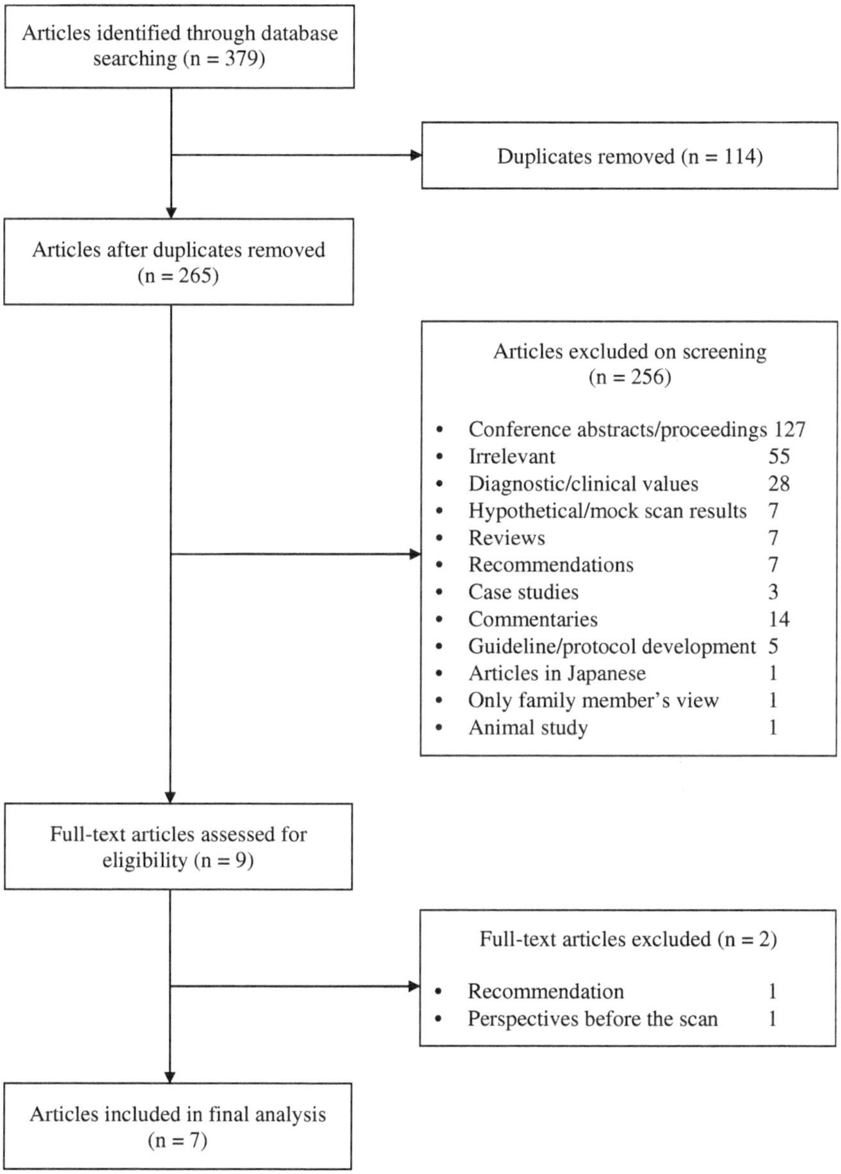

Fig. 1 Study selection.

2.1.1 Qualitative characterizations of responses to disclosure

Across qualitative studies, participants who received negative amyloid PET scan results consistently described feeling relieved that their scans did not show evidence of elevated amyloid levels. This finding was also present in the 2017 report by Grill and colleagues, which extended the finding to

Table 1 Overview of studies included in the systematic review.

First author (year) country	Methodology	Study population	Assessment time-points	Outcome measures	Key findings
Lim (2016)[10] United States	Semi-structured interview	11 Cognitively normal adults (Aβ− [n=8]; Aβ+ [n=3]) with two risk factors for AD (first-degree family history of AD and subjective cognitive impairment)	Baseline, and 9 or 18 months follow up visit	Subjective memory complaints (MAC-Q[13]), depression (GDS[14]; DASS-D[15]), anxiety (DASS-A[15]), stress symptoms (DASS-S[15]), perceived risk of AD, impact of events scale (IES-R[16]), emotional reactions, interpretation of the PET scan results, and consequences of learning about the PET scan results	• Aβ− participants felt relief • Aβ+ participants felt anxious but were not surprised • Disclosure of Aβ status did not significantly impact individuals' mood, subjective sense of memory impairment, or perceived risk of developing AD in both Aβ− and Aβ+participants • Aβ+ participant were more likely than Aβ− participants to make changes to their lifestyle
Grill (2018)[17] United States	Interview included forced choice and open-ended questions	33 Cognitively normal older adults who were made ineligible for the A4 study due to a "not elevated" brain amyloid imaging biomarker	3–30 Months after the scan result disclosure	Expectations for the scan, impression of the scan results, meaning of the scan result, test-related distress (IES[18]), role of study partner, sharing the scan results, changes in health and planning behaviors	• Most participants understood the scan's purpose and result • The most frequent emotional reaction to learning of the result was relief (n=17) • Most participants (n=31) had subclinical distress related to learning the scan results • All participants shared the PET result with others (e.g., spouses, adult children) • No changes in health and planning behaviors

Vanderschaeghe (2017)[19] Belgium	Semi-structured interview	38 Individuals with amnestic MCI (Aβ− [n=8]; Aβ+ [n=30])	2 Weeks and 6 months after the scan result disclosure	Recall of the scan result, emotional responses related to the scan result, sharing the result with others, and potential advantages and disadvantages of result disclosure	• Most participants recalled the gist of their scan result • Emotional responses were mixed, including relief, sadness, and worry, but not surprise • Most participants shared the result with family members • The experienced advantages and disadvantages of result disclosure depended on the result (Aβ+ or Aβ−) and the interval of the interview (2 weeks or 6 months after amyloid PET disclosure)
Wake (2018)[20] Japan	Survey	42 Individuals with subjective cognitive decline (Aβ− [n=32]; Aβ+ [n=10])	Baseline and 6 weeks after result disclosure	Anxiety (STAI[21]), depression (BDI-II[22]), and test-related distress (IES-R[16])	• Anxiety, depression, and, test-related distress did not change over time, and were not different between Aβ+ and Aβ− groups
Mozersky (2017)[11] United States	Semi-structured phone interview	50 Cognitively normal older adults of an elevated Aβ (part of the SOKRATES which is an A4 sub-study)	4–12 Weeks after result disclosure	Understanding of the scan results, implications for sense of self, memory, and future, sharing the results with others and subsequent effect on these relationships, and risk reduction behaviors	• Most participants (62%) understood the meaning of "elevated" amyloid in their brain • 54% of the participants expected their result • Some participants wanted clear information about their result using the numbers such as "percentages," or "scales" rather than the term "elevated"

Continued

Table 1 Overview of studies included in the systematic review.—cont'd

First author (year) country	Methodology	Study population	Assessment time-points	Outcome measures	Key findings
Burns (2017)[23] United States	Survey by e-mail or phone	97 Cognitively normal older adults (non-elevated Aβ [n=70]; elevated Aβ [n=27])	Pre-disclosure (baseline), at disclosure, 6 weeks, and 6 months post-disclosure	Depression (CES-D[24]), anxiety (BAI[25]), psychological impact of disclosure (modified version of the distress subscale from the IGT-AD[26]; 6 weeks and 6 months after result disclosure)	• No significant group differences in levels of depression or anxiety at any time point • Participants with elevated Aβ had higher levels of test-related distress compared to those with non-elevated amyloid at 6 weeks and 6 months after result disclosure • Baseline anxiety and depression were associated with total distress 6 weeks and 6 months after result disclosure in the overall group combined
Grill (2017)[27] United States	Telephone interview	26 Patient-caregiver dyads with whom a neurologist discussed amyloid imaging. Of those 26 patient participants,	• After having a discussion with a neurologist about the option of the amyloid imaging	• Rationale for the decision whether to have the scan (for both scan and non-scan groups) • Only scan group: expectations for the scan,	• Reasons to undergo the scan: to receive a definitive diagnosis, to learn the cause of cognitive impairment, to learn more information

	20 underwent the scan (Aβ− [n=2]; Aβ+ [*n*=18]). 10 patients and 23 caregivers completed the interview	• Mean time between the scan and the interview was 234 days	meaning of the scan results, how the neurologist communicate the scan results, impact of learning the scan results, and benefits and negatives of having the scan	• Meaning of the scan result: the participants felt that they needed to engage in future planning • Emotion response: most participants felt relieved after learning their scan result • Benefits of the scan: the most frequent response was the information gained • Most of participants in scan group reported that there were no negative aspects to getting the scan • Interactions with the neurologist: most participants reported that neurologists informed them the scan was supportive of or confirmed an AD diagnosis • Nearly all scan group participants reported that they would repeat the decision to have the scan

Note. A4, Anti-Amyloid treatment in Asymptomatic Alzheimer's disease; *Aβ*, Amyloid beta; *AD*, Alzheimer's disease; *BAI*, Beck Anxiety Inventory; *BDI II*, Beck Depression Inventory-II; *DASS-A*, depression, anxiety, and stress scale, anxiety subscore; *DASS-D*, depression, anxiety, and stress scale, depression subscore; *DASS-S*, depression, anxiety, and stress scale, stress subscore; *GDS*, Geriatric Depression Screening scale; *IES-R*, Impact of events scale, revised edition; *IGT-AD*, Impact of Genetic Testing for Alzheimer's Disease; *MAC-Q*, Memory Complaints Questionnaire; *MCI*, mild cognitive impairment; *SOKRATES*, Study of Knowledge and Reactions to Amyloid Testing; *STAI*, State-Trait Anxiety Inventory.

family caregivers of patients who were imaged.[27] Participants who received positive amyloid PET results were consistently described as reporting themselves to be unsurprised by their result. This finding was reported in both cognitively normal samples[10,11] and in the symptomatic population of persons with amnestic mild cognitive impairment (MCI).[19] Qualitative descriptions of distress in response to receiving positive amyloid PET scan results were reported in samples of symptomatic individuals and by caregivers of symptomatic individuals. For example, Vanderschaeghe and colleagues describe one woman's report of crying upon receipt of her results,[19] while multiple caregivers of amyloid positive patients in the Grill 2017 study characterized their emotional states as sad, depressing, or hopeless.[27] Both of those reports included rich descriptions of participants who endorsed mixed feelings about their scan results, citing the opportunity to plan ahead as a positive factor despite its association with concerns about future cognitive decline.

Qualitative approaches were also used to assess participants' comprehension and recollection of their amyloid scan results. Findings from these analyses were positive, with most amnestic MCI patients in the Vanderschaeghe sample[19] and nearly all of the amyloid negative respondents from the Anti-Amyloid Treatment in Asymptomatic Alzheimer's (A4) trial expressing adequate comprehension and/or recollection.[11] Of note, a substantial minority (38%) of the amyloid positive A4 participants demonstrated inadequate comprehension of the results disclosure session, with six individuals expressing an overt misunderstanding that they were either at imminent risk for, or now diagnosed with, AD.[11]

2.1.2 Quantitative findings of responses to disclosure

Test related distress was measured using three different instruments, including the original Impact of Events Scale (IES),[18] the Impact of Events Scale—Revised (IES-R),[16] and the Impact of Genetic Testing—AD scale (IGT-AD).[26] While each of these measures is designed to capture reactions to a potentially traumatic life event and was administered with the goal of assessing amyloid imaging test-related distress, the use of three measures across four studies precludes meta-analysis of this outcome. In addition to variation in the measure of test-related distress, these assessments were administered at a wide range of time points following disclosure, ranging from the day of disclosure to over 1 year later these. It also important to note that these four studies were conducted in three different populations.

For example, Grill and colleagues reported specifically on cognitively normal individuals who were amyloid PET negative upon eligibility screening for the A4 trial.[17] The Grill team's finding of an average IES score of 3.3 is consistent with reports from the REVEAL trial showing that asymptomatic persons who are APOE 4 negative score an average of 4.4 on the IES.[28] Both of these values are well below published threshold of an IES score of 9, which represents the lower limit for a mild level of clinically concerning event-related distress. For reference, scores of 44 or higher are considered to be indicative of a severely impactful stressful event.

Wake and colleagues' were the only group to quantify the impact of amyloid PET results disclosure in the subjective cognitive disorder (SCD) population.[20] At 6 weeks post-disclosure, the Wake team reported mean IES-R scores of 7.9 and 13.6 among amyloid negative and positive participants, respectively. These scores are below the cut point of 24 which has been associated with clinical concern in populations at risk for post-traumatic stress disorder (PTSD) following a traumatic event.[29] In Lim and colleagues' sample of cognitively normal individuals, IES-R scores ranged from 0 to 26, with higher scores noted among amyloid positive participants and one individual scoring above the cut point of 24.[10] Similarly, the Burns team found higher impact of event scores among amyloid positive participant as compared to amyloid negative participants, but even the highest mean total IGT-AD score in that sample (7.17 in amyloid positive participants at week-6), was far below the mean score among participants in the Risk Evaluation & Education for Alzheimer's Disease (REVEAL) II trial of genetic risk disclosure, which was 16.9 (S.D. 9.9).[26]

Findings regarding mood outcomes were relatively consistent across studies. None of the three studies that measured symptoms of depression or anxiety found statistically significant differences from baseline to follow-up or between amyloid positive and amyloid negative participants. When reported as means, no studies showed mood levels exceeding published cut points for clinical intervention. Notably, none of these studies included caregiver or cognitively impaired participants. Thus, Grill and colleagues' qualitative finding of lower mood among caregivers of amyloid positive individuals have not been cross-validated using quantitative methods.

Examination of other post-disclosure behaviors varied from one study to the next precluding integration of findings. Some studies reported that participants typically shared their results with family members. Yet scrutiny of those articles revealed that the involvement of a study partner (typically a relative) was often part of the trial design, raising the question of whether formal involvement of a study partner may have influenced this outcome. Other studies described self-reports of intention to initiate lifestyle changes or plan for the future based on positive scan results, yet the actual performance of such behaviors was not assessed. Post-disclosure discrimination in social, employment, or insurance settings—a concern frequently raised in the literature[30]—was not examined in these studies.

2.1.3 Limitations

Several limitations are shared across the articles included in this review. Although these studies, as a group, included patients from three continents whose diagnoses ranged from normal cognitive to mild dementia, none of the studies were designed to produce high quality evidence that can be generalized to clinical care.[31] Randomization and control groups were lacking across studies and several were retrospective in nature. Further, research settings were generally academic medical centers with limited racial and ethnic diversity among participants and most articles in this review described the exclusion of individuals with clinically significant mood symptoms. Taken together, these limitations suggest that it is inappropriate to assume that findings from this body of literature will translate to typical individuals who may present to clinical practice settings seeking testing to determine their risk of developing ADRD.

2.1.4 Implications for future research

Despite the above-mentioned methodological limitations, the studies included in this review provide important early insights into the patient's experience of receiving amyloid PET scan results. Studies of cognitively normal individuals, who are euthymic at baseline, clearly and consistently suggest that the disclosure of amyloid PET results has no discernable impact on psychological well-being. Whether the psychological safety of disclosing amyloid PET results extends to other populations is unknown and warrants investigation. Indeed, the two studies in this review that included participants with cognitive impairment provide a signal that emotional responses may be more variable within symptomatic populations. Given the striking lack of evidence about social (e.g., discriminatory) or behavioral (e.g., lifestyle changes) outcomes of amyloid imaging results disclosure, there is a need within both asymptomatic and cognitively impaired populations, to explore this in future studies. Finally, future research on this topic should implement more rigorous designs, including studies with control groups and randomization, as well as studies within more racially and ethnically diverse samples.

References

1. Huynh RA, Mohan C. Alzheimer's disease: biomarkers in the genome, blood, and cerebrospinal fluid. *Front Neurol.* 2017;8:102.
2. Villain N, Chételat G, Grassiot B, et al. Regional dynamics of amyloid-β deposition in healthy elderly, mild cognitive impairment and Alzheimer's disease: a voxelwise PiB–PET longitudinal study. *Brain.* 2012;135(7):2126–2139.
3. Jack Jr CR, Bennett DA, Blennow K, et al. NIA-AA research framework: toward a biological definition of Alzheimer's disease. *Alzheimers Dement.* 2018;14(4):535–562.
4. O'Bryant SE, Mielke MM, Rissman RA, et al. Blood-based biomarkers in Alzheimer disease: current state of the science and a novel collaborative paradigm for advancing from discovery to clinic. *Alzheimers Dement.* 2017;13(1):45–58.
5. Khachaturian AS, Hayden KM, Mielke MM, et al. Future prospects and challenges for Alzheimer's disease drug development in the era of the NIA-AA Research Framework. *Alzheimers Dement.* 2018;14(4):532–534.
6. Lingler JH, Roberts JS, Kim H, et al. Amyloid positron emission tomography candidates may focus more on benefits than risks of results disclosure. *Alzheimers Dement (Amst).* 2018;10:413–420.
7. Corrigan PW, Watson AC. Understanding the impact of stigma on people with mental illness. *World Psychiatry.* 2002;1(1):16–20.
8. Garand L, Lingler JH, Conner KO, Dew MA. Diagnostic labels, stigma, and participation in research related to dementia and mild cognitive impairment. *Res Gerontol Nurs.* 2009;2(2):112–121.
9. Moher D, Liberati A, Tetzlaff J, Altman DG, PRISMA Group. Preferred reporting items for systematic reviews and meta-analyses: the PRISMA statement. *Ann Intern Med.* 2009;151(4):264–269.
10. Lim YY, Maruff P, Getter C, Snyder PJ. Disclosure of positron emission tomography amyloid imaging results: A preliminary study of safety and tolerability. *Alzheimers Dement.* 2016;12(4):454–458.
11. Mozersky J, Sankar P, Harkins K, Hachey S, Karlawish J. Comprehension of an elevated amyloid positron emission tomography biomarker result by cognitively normal older adults. *JAMA Neurol.* 2018;75(1):44–50.
12. Qual text here. (n.d.).
13. Crook TH, Feher EP, Larrabee GJ. Assessment of memory complaint in age-associated memory impairment: the MAC-Q. *Int Psychogeriatr.* 1992;4(2):165–176.
14. Yesavage JA, Brink TL, Rose TL, et al. Development and validation of a geriatric depression screening scale: a preliminary report. *J Psychiatr Res.* 1982–1983;17(1):37–49.
15. Lovibond SH, Lovibond PF. *Manual for the Depression Anxiety Stress Scales (DASS).* Sydney, NSW: Psychology Foundation of Australia; 1993.
16. Weiss DS, Marmar CR. The impact of event scale-revised. In: Wilson JP, Keane TM, eds. *Assessing Psychological Trauma and PTSD.* New York, NY: The Guilford Press; 1997:399–411.
17. Grill JD, Cox CG, Harkins K, Karlawish J. Reactions to learning a "not elevated" amyloid PET result in a preclinical Alzheimer's disease trial. *Alzheimers Res Ther.* 2018;10(1):125.

18. Horowitz M, Wilner N, Alvarez W. Impact of event scale: a measure of subjective stress. *Psychosom Med.* 1979;41(3):209–218.
19. Vanderschaeghe G, Schaeverbeke J, Bruffaerts R, et al. Amnestic MCI patients' experiences after disclosure of their amyloid PET result in a research context. *Alzheimer Res Ther.* 2017;9(1):92.
20. Wake T, Tabuchi H, Funaki K, et al. The psychological impact of disclosing amyloid status to Japanese elderly: a preliminary study on asymptomatic patients with subjective cognitive decline. *Int Psychogeriatr.* 2018;30(5):635–639.
21. Spielberger CD, Gorsuch RL, Lushene R, Vagg PR, Jacobs GA. *Manual for the State-Trait Anxiety Inventory (Form Y).* Palo Alto, CA: Consulting Psychologists Press; 1983.
22. Beck AT, Steer RA, Brown GK. *Manual for the Beck Depression Inventory-II.* 2nd ed. San Antonio, TX: Psychological Corporation; 1996.
23. Burns JM, Johnson DK, Liebmann EP, Bothwell RJ, Morris JK, Vidoni ED. Safety of disclosing amyloid status in cognitively normal older adults. *Alzheimers Dement.* 2017;13(9):1024–1030.
24. Radloff LS. The CES-D scale: a self-report depression scale for research in the general population. *Appl Psychol Measur.* 1977;(3):385–401.
25. Beck AT, Epstein N, Brown G, Steer RA. An inventory for measuring clinical anxiety: psychometric properties. *J Consult Clin Psychol.* 1988;56(6):893–897.
26. Chung WW, Chen CA, Cupples LA, et al. A new scale measuring psychologic impact of genetic susceptibility testing for Alzheimer disease. *Alzheimer Dis Assoc Disord.* 2009;23(1):50–56.
27. Grill JD, Cox CG, Kremen S, et al. Patient and caregiver reactions to clinical amyloid imaging. *Alzheimers Dement.* 2017;13(8):924–932.
28. Cassidy MR, Roberts JS, Bird TD, et al. Comparing test-specific distress of susceptibility versus deterministic genetic testing for Alzheimer's disease. *Alzheimers Dement.* 2008; 4(6):406–413.
29. Asukai N, Kato H, Kawamura N, et al. Reliability and validity of the Japanese-language version of the impact of event scale-revised (IES-RJ): four studies of different traumatic events. *J Nerv Ment Dis.* 2002;190(3):175–182.
30. Arias JJ, Karlawish J. Confidentiality in preclinical Alzheimer disease studies: when research and medical records meet. *Neurology.* 2014;82(8):725–729.
31. West S, King V, Carey TS, et al. Systems to Rate the Strength of Scientific Evidence. *Evid Rep Technol Assess (Summ).* 2002;47:1–11 [AHRQ Publication No. 02-E016].

Further reading

32. Harris Interactive. What America Thinks: MetLife Foundation Alzheimer's Survey. https://www.metlife.com/content/dam/microsites/about/corporate-profile/alzheimers-2011.pdf. Accessed 29 April 2019.

Index

Note: Page numbers followed by "*f*" indicate figures, and "*t*" indicate tables.

A

B

C

D

E

F

G

O

P

S

T

U

V

W

FSC
www.fsc.org
MIX
Papier | Fördert gute Waldnutzung
FSC® C083411

Kolibri 360 GmbH
Ferdinand-Jühlke-Straße 7
99095 Erfurt, Deutschland
produktsicherheit@kolibri360.de